Lecture Notes in Computer Science 14628

The series Lecture Notes in Computer Science (LNCS), including its subseries Lecture Notes in Artificial Intelligence (LNAI) and Lecture Notes in Bioinformatics (LNBI), has established itself as a medium for the publication of new developments in computer science and information technology research, teaching, and education.

LNCS enjoys close cooperation with the computer science R & D community, the series counts many renowned academics among its volume editors and paper authors, and collaborates with prestigious societies. Its mission is to serve this international community by providing an invaluable service, mainly focused on the publication of conference and workshop proceedings and postproceedings. LNCS commenced publication in 1973.

Lars Michael Kristensen ·
Jan Martijn van der Werf
Editors

Application and Theory of Petri Nets and Concurrency

45th International Conference, PETRI NETS 2024 Geneva, Switzerland, June 26–28, 2024 Proceedings

Editors
Lars Michael Kristensen
Western Norway University of Applied Sciences
Bergen, Norway

Jan Martijn van der Werf
Utrecht University
Utrecht, The Netherlands

ISSN 0302-9743 ISSN 1611-3349 (electronic)
Lecture Notes in Computer Science
ISBN 978-3-031-61432-3 ISBN 978-3-031-61433-0 (eBook)
https://doi.org/10.1007/978-3-031-61433-0

This Springer imprint is published by the registered company Springer Nature Switzerland AG
The registered company address is: Gewerbestrasse 11, 6330 Cham, Switzerland

Preface

This volume contains the proceedings of the 45th International Conference on Application and Theory of Petri Nets and Concurrency (Petri Nets 2024). The aim of this series of conferences is to create an annual opportunity to discuss and disseminate the latest results in the field of Petri nets and related models of concurrency, including tools, applications and theoretical developments.

The 45th conference and affiliated events were organized by the SMV (Semantics, Modeling and Verification) team at the Computer Science Department of the Faculty of Sciences of the University of Geneva, Switzerland, jointly with members of the Centre Universitaire d'Informatique. The conference took place at the Campus Biotech Geneva.

This year, 42 papers were submitted. Each paper was reviewed by at least three reviewers. The discussion phase and final selection process by the Program Committee (PC) were supported by the EasyChair conference system. From 39 regular papers and 3 tool papers, the PC selected 19 papers for presentation: 17 regular papers and 2 tool papers. After the conference, some of these authors were invited to submit an extended version of their contribution for consideration in a special issue of a journal. We thank the PC members and other reviewers for their careful and timely evaluation of the submissions and the fruitful constructive discussions that resulted in the final selection of papers. The Springer LNCS team provided excellent support in the preparation of this volume.

The keynote presentations were given by

- Rob van Glabeek, University of Edinburgh: Just Distributability
- José Manuel Colom, University of Zaragoza: Harnessing Structure Theory of Petri Nets in Discrete Event System Simulation
- Gabriele Taentzer, Philipps-Universität Marburg: On the Application of Model-Driven Optimization to Business Processes

Alongside Petri Nets 2024, the following workshops took place:

- Model Checking Contest 2024
- Workshop on Petri Nets and Software Engineering (PNSE) 2024
- Workshop on Petri Net Games, Examples and Quizzes for Education, Contest and Fun (PENGE) 2024

coordinated by Susanna Donatelli and Karsten Wolf. Other co-located events included the Petri Net Course and Tutorials, coordinated by Jörg Desel as well as a Tool Exhibition.

We greatly appreciate the efforts of all members of the Local Organizing Committee chaired by Didier Buchs for the time spent in the organization of this event. We hope you enjoy reading the contributions in this LNCS volume.

June 2024

Lars Michael Kristensen
Jan Martijn van der Werf

Organization

Organising Chair

Didier Buchs	University of Geneva, Switzerland

Program Committee Chairs

Lars Michael Kristensen	Western Norway University of Applied Sciences, Norway
Jan Martijn van der Werf	Utrecht University, The Netherlands

Steering Committee

Will van der Aalst	RWTH Aachen University, Germany
Gianfranco Ciardo	Iowa State University, USA
Jörg Desel	Fern Universität in Hagen, Germany
Susanna Donatelli	University of Turin, Italy
Serge Haddad	ENS Paris-Saclay, France
Kunihiko Hiraishi	Japan Advanced Institute of Science and Technology
Jetty Kleijn	Leiden University, The Netherlands
Fabrice Kordon (Co-chair)	Sorbonne Université, France
Maciej Koutny	Newcastle University, UK
Lars Michael Kristensen	Western Norway University of Applied Sciences, Norway
Chuang Lin	Tsinghua University, China
Lukasz Mikulski	Nicolaus Copernicus University, Poland
Wojtek Penczek	Polish Academy of Sciences, Poland
Laure Petrucci (Co-chair)	Université Sorbonne Paris Nord, France
Lucia Pomello	University of Milano-Bicocca, Italy
Wolfgang Reisig	Humboldt-Universität zu Berlin, Germany
Grzegorz Rozenberg	Leiden University, The Netherlands
Antti Valmari	University of Jyväskylä, Finland
Karsten Wolf	University of Rostock, Germany
Alex Yakovlev	Newcastle University, UK

Program Committee

Didier Buchs	University of Geneva, Switcherland
Arnaud Sangnier	University Paris Diderot, France
Irina Lomazova	University Higher School of Economics, Russia
Ryszard Janicki	McMaster University, Canada
Remigiusz Wisniewski	University of Zielona Gora, Poland
Elvio Gilberto Amparore	University of Turin, Italy
Susanna Donatelli	University of Turin, Italy
Loic Helouet	Inria, France
Lars Michael Kristensen	Western Norway University of Applied Sciences, Norway
Robert Lorenz	University of Augsburg, Germany
Michael Köhler-Buß Ÿmeier	University of Applied Science Hamburg, Germany
Luis Gomes	Universidade NOVA de Lisboa, Portugal
Lukasz Mikulski	Nicolaus Copernicus University, Poland
Natalia Sidorova	Technische Universiteit Eindhoven, The Netherlands
David Frutos Escrig	Universidad Complutense de Madrid, Spain
Andrew Miner	Iowa State University, USA
Joao Paulo Barros	Instituto Politecnico de Beja, Portugal
Andrey Rivkin	Technical University of Denmark, Denmark
Xudong He	Florida International University, USA
Marco Montali	Free University of Bozen-Bolzano, Italy
João M. Fernandes	University of Minho, Portugal
Jan Martijn van der Werf	Utrecht University, The Netherlands
Boudewijn Van Dongen	Eindhoven University of Technology, The Netherlands
Guillermo Perez	University of Antwerp, The Netherlands
Artem Polyvyanyy	The University of Melbourne, Australia
Anna Kalenkova	The University of Adelaide, Australia
Benoît Delahaye	LS2N de Nantes, France
Abel Armas Cervantes	The University of Melbourne, Australia
Wojtek Jamroga	Polish Academy of Sciences, Poland

Additional Reviewers

Marcin Wojnakowski
Christoph Welzel
Eric Verbeek
Teofil Sidoruk
Patrizia Schalk
Fernando Rosa-Velardo
Michael Raskin
Shrisha Rao
Dimi Racordon
Wojciech Penczek
Roman Nesterov
Damien Morard
Didier Lime
Damian Kurpiewski
Stefan Klikovits
Andrei Karatkevich
Elena Gómez-Martínez
Roland Guttenberg
Anna Gogolinska
Catalin Dima
Aurélien Coet
Robin Bergenthum
Grzegorz Bazydło
Vladimir Bashkin
Kamila Barylska
Christian Attiogbé
Federica Adobbati

Just Distributability (Abstract of Invited Talk)

Rob van Glabbeck

In this talk I investigate which systems can be implemented in a distributed fashion, by sequential components that interact solely through asynchronous communication. Naturally, such an implementation should satisfy the same safety and liveness properties as the original system. Here I use Petri nets to model the systems under investigation, and find a way to implement any net in a distributed fashion, provided we allow read or listen arcs in our implementations.

Contents

Reachability

Verification and Model Checking

Applications and Tools

Invited Papers

Harnessing Structure Theory of Petri Nets in Discrete Event System Simulation

José-Manuel Colom(✉)

Aragón Institute of Engineering Research (I3A), University of Zaragoza, Zaragoza, Spain
jm@unizar.es

Abstract. Nowadays Discrete Event Systems (DESs) require complex and large models, for which distributed simulation engines become, in practice, the tools used to understand and analyze their behavior. The feasibility and efficiency of a distributed simulation of these large-scale models is strongly dependent of the information that can be obtained from the models, previously to the simulation process itself. This information can give assistance to the generation of an initial partition of the model, allowing a well balanced workload among the individual simulation engines deployed, or in the generation of the predicates to be evaluated in order to determine the enabling of transitions; or the computation of look-ahead information in conservative strategies of distributed simulation. Petri nets allow to obtain information from the structure that can be used to advance conclusions or properties about the course of a simulation. This information can be usefull either independently of the considered initial marking, or parameterised by its initial choice. This structural information can be obtained in modelling phase, completed in simulation time and re-elaborated from the simulation results, and therefore associated to the model or modules of the model in such a way that can be harnessed in further simulations where these nets will be used. Last but no least, the maintenance of the structure of the Petri net during the simulation (in an interpreted simulation instead of a compiled one) allows to make load balancing during the simulation or to federate with legacy simulators, in an easier way than using other kind of specification models or simulation schemes.

Keywords: Petri Nets · Structural Analysis · Distributed Simulation · Discrete Event Systems

1 Introduction

Discrete Event Systems (DES) are systems whose state is modified in a timely and distinguishable way in their evolution. DESs have acquired great relevance due to the multitude of them that are inserted into the daily life: from the Internet of Things (IoT) itself to the evacuation of a sports stadium for a serious incident, passing through the public health system, or highly automated

L. M. Kristensen and J. M. van der Werf (Eds.): PETRI NETS 2024, LNCS 14628, pp. 3–23, 2024.
https://doi.org/10.1007/978-3-031-61433-0_1

and geographically dispersed manufacturing systems. The main characteristic of these systems is their complexity and size, which can scale to large values. These properties, together with the fact that prototypes cannot be built to test solutions or introduce improvements, make model building an essential necessity. These models must abstract the behavior of the system as faithfully as possible to address the problem of its design, analysis or improvement.

These complex and scalable DES models are used to simulate the behavior of the system, under different scenarios, to investigate its evolution. The analysis of results will allow to better understand the system, help make decisions in real situations or even design optimization systems that improve its behavior. Petri nets (PNs) are models that have proven to be particularly suitable for describing the behavior of DESs. In addition to their conceptual simplicity and graphic representation, they add the characteristic of being executable models: the state is represented by variables called places, and the state change by transitions that modify the token contens in the places, in a local and atomic way.

In a general context, DESs Parallel Simulation (DESPS) slices a single execution of a discrete event simulation program across multiple processes on a high-performance computing system [12]. A discrete event simulation captures the behavior of a real or conceived system over time and evolves the state of the system to a new state at distinguishable, typically irregular, points in simulation time. A sequential discrete event simulation program includes two fundamental concepts: state variables that capture the state of the system being modeled and events that allow the transition of state variables from one state to the next. In a discrete event simulation, changes in the simulation state occur only through event processing. Each event contains a timestamp that represents a point in simulation time at which the state transition occurs.

A DESPS program can be viewed as a collection of sequential discrete event simulations (called Logical Processes - LP) that interact by exchanging time-stamped messages. Each message represents an event scheduled (sent) by one LP to another. Synchronization of DESPS programs must ensure that parallel execution of the LPs produces exactly the same results as a sequential execution of the same program where all events take effect in timestamp order. Synchronization schemes (see [11]) that guarantee this property are classified into conservative and optimistic. Some conservative schemes [7] use a mechanism of locking an LP until it can guarantee that no events with a lower timestamp will be received in the future. To do this, each LP must maintain a global lookahead value (LA) that guarantees that it will not receive messages with a timestamp lower than its current simulation time plus LA. Optimistic schemes allow events to be processed out of order, but use a rollback mechanism to recover from such errors.

Distributed simulation is concerned with running simulations on computing platforms of a wider geographical extent than parallel computers. Distributed simulation allows to enable the exploitation of geographically distributed resources in the simulation operation, or to run simulations close to source data streams. Moreover, using distributed simulation allows the integration of simulators operating on different computers into a single simulation environment. LP synchronization (time management) is also a problem in distributed simulation, but new and no less important problems appear: (1) distribution of information between simulators participating in distributed simulation in an efficient and timely manner; or (2) reduction in the amount of data and messages communicated. The search for interoperability between simulators developed separately has led to standards to interconnect simulations such as the High Level Architecture (HLA) [1–3].

The domain of distributed simulation has grown and evolved over the past years, but has new challenges arising from new applications and changes in the hardware and software systems on which they operate. R.Fujimoto pointed out six research challenges in [12] that still remain today:

C1. Scalable simulations based on large and complex real-world applications.
C2. Exploitation of heterogeneous machine architectures.
C3. Distributed simulation more accessible thanks to simpler models and cloud computing.
C4. Online decision making through real-time distributed simulation.
C5. Distributed and parallel simulation with energy efficiency.
C6. Fast composition of distributed simulations.

This article propose the use PNs in all phases of the DES Simulation Engineering process. The characteristic to exploit of the PNs is its structural information that the model provides. In the following sections some of the contributions of the Structure Theory of PNs to the distributed simulation of scalable DES models will be presented. Section 2 presents the justification of a modular methodology for the construction of scalable models and the most appropriate simulation schemes for the challenges in [12]. Section 3 presents Linear Enabling Functions as a mechanism for eliminating the need to maintain the global state of the system in a distributed simulation and facilitating federation with legacy simulators. Section 4 presents how to take advantage of Vector Simulators for calculating and managing the lookahead of LPs, and how to deal with scalability, dynamic evolution, and interoperability of models. Section 5 presents an overview of the approach and Sect. 6, highlights some conclusions.

2 Petri Nets for Distributed Simulation of Scalable DESs

Petri Nets (PN) are used as models to specify the behaviour of DESs. In this work flat Place/Transition Nets will be considered. Its definition is based on two main pillars. The first one is that they are models with an *explicit structure* expressed by static relationships between *state variables* (places) and *transformers of the values of the state variables* (transitions). The second pillar is the rule for the *occurrence of transitions* that transforms the net state. For the ocurrence of a transition a precondition named *enabling condition*, must be fullfilled: all input places to the transition have a token contents greater than a threshold specified by the *weight of the arc*. If this is the case, the classic token game happens by updating the token contents of the places connected to the transition in an atomic form. These characteristics make PNs a kind of *executable model*.

2.1 Construction of PNs for Simulation

Given a system modelled with a PN, the *system behaviour* is obtained from the PN through the occurrence of its transitions. An *execution* in the PN is a *sequence of transition occurrences*. It is possible that from an initial net state (initial marking) different sequences of transitions can be followed. It will be assumed a *weak fairness property* in the occurrence of transitions [13]: if a transition is persistently enabled, it will eventually progress.

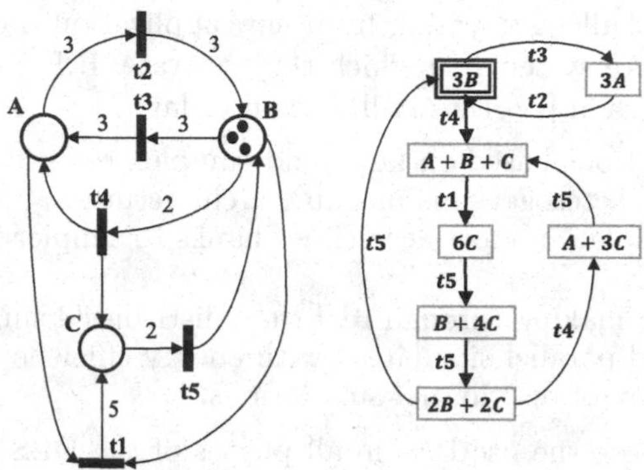

Fig. 1. A bounded Petri Net [20]. The net becomes unbounded with the initial marking $\mathbf{m_0}[A] = \mathbf{m_0}[C] = 0$ and $\mathbf{m_0}[B] = 4$.

The PN in Fig. 1 has the finite Reachability Graph (RG), depicted on the right in the Figure, with initial marking $\mathbf{m_0}[B] = 3$. The same net, but with $\mathbf{m_0}[B] = 4$, becomes an unbounded PN system that has an infinite RG. This example leads to drawing several considerations that affect the development of simulation techniques for this class of models. The first one is that, in the general case, **it is not feasible to simulate unbounded PNs**, whatever the approach adopted. Simulation of unbounded nets requires storing and maintaining the content of tokens of unbounded places (or lists of projected events), so problems of overflow the memory dedicated to storing them can arise. The second consideration arises from how to ensure that the PN to be simulated is bounded. If this certainty is wanted before starting to simulate the model, then some analysis will have to be made to determine if the boundedness property is verified by the PN. This can be done by the construction of the RG with an Algorithm such as that in [15], including the condition to detect unboundedness of a net system: the system $\mathcal{S} = \langle \mathcal{N}, \mathbf{m_0} \rangle$ is unbounded iff there exists $\mathbf{m}''$, reachable from $\mathbf{m_0}$ and $\sigma \in T^*$, such that $\mathbf{m}'' \xrightarrow{\sigma} \mathbf{m}'$ and $\mathbf{m}' \gneqq \mathbf{m}''$.

The PN in Fig. 1, with $\mathbf{m_0}[B] = 3$, is a bounded net system that would be simulatable, but in the case of $\mathbf{m_0}[B] = 4$, it is unbounded in which case it would not be simulatable. Therefore, before modifying the marking of any place to launch a new simulation (common action when exploring the behavior of a model with different initial markings that, for example, represent different resource contents) it is necessary to carry out an analysis of the boundedness property, which in the general case requires building the RG.

The integration of the Karp and Miller condition [15] to detect unboundedness of a net system, within the simulation algorithm it is also not feasible. That is, every time a transition t occurs, verify that there is no predecessor marking of the one reached by the occurrence of t, which token contents is less than and non-equal to the new marking obtained. This verification requires the storage of all the markings reached during the simulation, and the time costs grow as the simulation progresses (since the list of markings grows). Moreover, the consistency of the information of the states reached must be guaranteed, since the PN has been partitioned into several components that have been separated for its distributed simulation.

From all of the above, it can be concluded that the *PN models that are well-formed for simulation* are those that, for any initial marking, the resulting net system is bounded. That is, this property must be guaranteed by the net structure, and is known as **structural boundedness property** [21]. The property can be verified from the incidence matrix as the following results shows.

Proposition 1 ([23]). *The following three statements are equivalent:*

1. *$\mathcal{N} = (P, T, \mathbf{Pre}, \mathbf{Post})$ is structurally bounded.*
2. *There is no $\mathbf{x} \geq 0$ such that $\mathbf{C} \cdot \mathbf{x} \gneq 0$ (transition-based characterization)*
3. *There exists $\mathbf{y} > 0$ such that $\mathbf{y} \cdot \mathbf{C} \leq 0$. (place-based characterization)*

To verify the previous property, linear algebra algorithms are required that determine if a system of inequalities has at least one solution. There are Linear Algebra Software Packages such as [9], in which the matrices they can deal with are limited to a few tens of thousands of rows and columns. This number is small when it concerns scalable PNs modeling detailed logistics systems, for example, which may contain millions of places and transitions.

The practical impossibility of verifying the structural boundedness property for a complete large PN makes it necessary to define a strategy to guarantee that the simulation is possible. In [4,14,25] a strategy is proposed that aims to obtain structurally bounded models by construction, from small-sized components that each one has the property. The construction proceeds through component integration operations, which preserve the property (closed operators with respect to the class of structurally bounded PNs). For example, components can be Petri subnets that can be integrated through the merging of some interface transitions declared into each component. The following result shows that the transition merging operator between PN components is closed for the class of structurally bounded nets.

Proposition 2. *Let $\mathcal{F} = \{\mathcal{C}_i | \mathcal{C}_i = (\mathcal{S}_i, \Theta_i), i = 1, \ldots, k\}$ be a finite family of PN components, where $\mathcal{S}_i = (\mathcal{N}_i, \mathbf{m_0}_i)$, $\mathcal{N}_i = (P_i, T_i, \mathbf{C_i})$, and $\Theta_i \subseteq T_i$ the set of transitions that $\mathcal{N}_i$ offers to merge with other components. If $\mathcal{N}_i = (P_i, T_i, \mathbf{C_i})$, for all $i = 1, \ldots, k$, is structurally bounded, any PN obtained by composition of elements of $\mathcal{F}$ via transition merging operations is structurally bounded.*

In order to exploit the above result to obtain a structurally bounded net by construction, by merging transitions, some previous works with the components are needed: (1) Independent verification that each component is structurally bounded using standard software packages. It will be affordable because the size of a component is modest compared to the global model; (2) Associate the property to the component when adding it to the library (for a component only once must be verified the property).

The methodology for building models from components can be extended including new operators. Similarly to the merging operator, they must be closed with respect to the class of structurally bounded nets. For example, a transition replication operator in a PN component (a replicated transition is a distinguishable copy of a original transition, with a different name, and connected to the same places as the transition it replicates and with the same weights in arcs) is closed with respect to the class of structurally bounded nets.

2.2 Strategy for Distributed Simulation of PNs

The simulation of a DES is based on the construction of a PN model that will be executed following occurrence sequences of transitions of the PN. The execution of a PN requires an engineering task that obtains a computer program from the PN which execution is the simulation of the DES. This engineering task is called *PN implementation* and, in the case of parallel or distributed simulation, it will give rise to a set of Logical Processes (LP) that are synchronized interchanging messages. These programs must take into account, to generate them, the representation of the state of the PN, the rules of enabling of transitions and the rules for the occurrence of transitions playing the token game. Transitions can incorporate a temporal interpretation that can be deterministic or stochastic—timed PNs (TPN) or stochastic PNs (SPN).

Implementations of a PN are classified as *compiled* or *interpreted*. A compiled implementation of the PN generates programs (LPs) whose execution sequences mimic the sequences of transition occurrences in the PN. The PN is only a kind of blue-print that dissapears at the moment the control flow of the LPs is generated using the instructions of a programming language. This kind of implementation makes that an initial partition of the model must be done to assign each part to a LP in compilation-time, and this partition cannot be changed in execution-time. This means that the level of parallelism and distribution is locked at compilation-time [14], and dynamic load balancing or reconfiguration can be only done at the level of LP. This initial partition introduces rigidities in the load re-balancing during the simulation, which may be necessary to do for multiple factors [10]. This rigidity is due to the fact that the transitions encapsulated in an LP are not externally distinguishable and are treated as a unit. Compiled implementations have been the option for discrete event control systems [18,19].

An interpreted implementation of a PN is based on a generic program, named interpreter or simulation engine (simbot), that implements the basic algorithm for occurrence of transitions in a PN: (1) detection of enabled transitions; (2)

solving conflicts between transitions; (3) occurrence of transitions and updating of the net marking. The PN remains during execution under the form of data structures representing each one of the PN objects that must be used and updated by the interpretation cycle of the simbot for the occurrence of transitions. In this implementations, the model specification is separated from the program used to simulate the model, which is essential for scaling simulations and simplifying the dynamic deployment of simulation programs on distributed execution platforms [27]. When the model is not embedded in the program instructions of the simulator, this enables easier model portability to other simulators or hardware platforms.

The use of simulation engines specialised in the interpreted implementation of PNs (simbots) avoids the development of the systems entirely from scratch [17]. Simbots provide a core permanent portion of services that can be executed in heterogeneous infrastructures such as mini clusters, the cloud, and even allow to embed these simulation services into IoT devices.

A simbot follows an execution cycle, in three stages, for the occurence of a set of concurrent transitions from a state: (1) detection of enabled transitions; (2) construction of the list of transitions to occur in the cycle, solving previously the existing conflicts; (3) occurrence of the transitions with the updating of the net marking. The first stage is the most time-consuming operation, and the efficiency of a PN interpreter is mainly related to the transition enabling test implemented. There are two large families of enabling tests known [8,20].

1. *Place-driven approaches* [19,22,24,26]. In this type of tests an explicit representation of the marking of the places of the PN is required. The enabling test for a transition is an assertion that verifies that each input place to it has a token contents above the threshold that specifies the input arc. Normally, for each transition a representative input place is chosen, that in case of not having enough tokens avoids continuing to test input places. This choice reduces the number of marking tests, and therefore, a good representative place is one that is not frequently marked.
2. *Transition-driven approaches* [20,24]. In this case, a representation of the places and their marking is not required. Instead, an enabling function is created for each transition that discriminates whether the transition is enabled or not. The value of this function is updated when the occurrence of a neighboring transition changes the enabling state of the transition. In [5,6], a transition-driven method was presented for P/T nets characterising the transition enabling by means of a Linear Enabling Functions (LEF).

As a conclusion to this section, the adopted decisions for the definition of the simulation strategy of PNs are shown below. These decisions are supported by the advantages provided by PNs and its Structure Theory, offering answers to the research challenges posed by R. Fujimoto and listed in the introduction.

- Answers to challenge C1. Methodologies for building scalable models based on components that are structurally bounded (to ensure that they are simulatable for all initial marking) and composition operators closed with respect to structurally bounded nets.

- Answers to challenge C2. Interpreted implementations of PNs based on generic simulation engines executing the transition occurrence cycle, that can be provided for any type of computing platform and software architecture. These simulation engines are separated from the models to be simulated, which are data structures portable to different platforms.
- Answers to C3 challenge. Transition-driven approaches for enabling transition tests that do not require an explicit representation of the net marking, avoiding costly mechanisms to guaranteeing data consistency and availability in distributed environments. The grouping of each set of transitions in structural coupled conflict relation within a same simulation engine to avoid costly protocols for distributed decision making.
- Answers to challenges C4, C5 and C6. The points indicated here follow from the content of the following section. The intensive use of Linear Enabling Functions (LEFs) to characterize the transitions enabling. This reduces the number of messages (events) exchanged between simbots to the minimum necessary to collect the changes in the enabling of a transition due to the occurrence of neighboring transitions. This leads to a better energy efficiency of the computational platforms where the simulations are executed. Use of simulators with vector of LEFs to link model transitions, in a natural and simple way, with agents in the simulation environment for decision making or connect with legacy simulators that perform specialized functions.

3 Enabling Functions of Transitions and Event Distribution

Given a net system $(\mathcal{N}, \mathbf{m_0})$ and a transition t of $\mathcal{N}$, a *Linear Enabling Function (LEF)* of t, $f_t : \mathrm{RS}(\mathcal{N}, \mathbf{m_0}) \to \mathbb{Z}$, is a linear function depending on the marking $\mathbf{m}$, that characterizes the enabling of t at $\mathbf{m}$ according to the following rule: $f_t(\mathbf{m}) \leq 0$, $\mathbf{m} \in \mathrm{RS}(\mathcal{N}, \mathbf{m_0})$, if and only if for all $p \in {}^\bullet t$, $\mathbf{m}[p] \geq \mathbf{Pre}[p,t]$.

The use of LEFs simplifies the enabling test, because only a scanning for transitions with LEFs less than or equal to zero is needed, i.e. only one test per transition over an integer value without the maintenance of the net marking (the state). Moreover, the LEF of t only must be updated if some transition belonging to ${}^{\bullet\bullet}t$ or $({}^\bullet t)^\bullet$ has occurred. The updating only requires the addition of a predefined constant to the current function value. In the case of conflicts, preemption of some enabled transitions can appear.

For each $t \in T$ the following sets are defined: (1) $S_t^{\mathrm{eqPre}} = \{p \in {}^\bullet t \mid \mathbf{sb}(p) = \mathbf{Pre}[p,t]\}$; and (2) $S_t^{\mathrm{upPre}} = \{p \in {}^\bullet t \mid \mathbf{sb}(p) > \mathbf{Pre}[p,t]\}$. These sets are defined using $\mathbf{sb}(p)$ instead of $\mathbf{b}(p)$ because $\mathbf{b}(p) \leq \mathbf{sb}(p)$, and it can be computed in polynomial time (with respect to the net size) by solving a Linear Programming Problem [21]: $\mathbf{sb}(p) = \max\{\mathbf{m}[p] \mid \mathbf{m}[p] = \mathbf{m_0} + C \cdot \boldsymbol{\sigma}, \mathbf{m} \geq 0, \boldsymbol{\sigma} \geq 0\}$. The computation of all $\mathbf{sb}(p)$ requires an Algorithm that proceeds in a compositional way over the components used to construct the complete model. Therefore, a transition t, ${}^\bullet t = S_t^{\mathrm{eqPre}} \cup S_t^{\mathrm{upPre}}$, belongs to one of following Classes:

Class 1. $|{}^\bullet t| \geq 1$, and $|S_t^{\mathrm{upPre}}| \leq 1$.

Class 2. $|^\bullet t| \geq 1$, and $|S_t^{\text{upPre}}| > 1$.

3.1 Linear Enabling Functions (LEFs) for Class 1 Transitions

Let $(\mathcal{N}, \mathbf{m_0})$ be a net system and t a transition of Class 1. In the case that $S_t^{\text{upPre}} \neq \emptyset$, let p_π be the place belonging to this set. The function $f_t : \text{RS}(\mathcal{N}, \mathbf{m_0}) \rightarrow \mathbb{Z}$, is called *Linear Enabling Function (LEF)* of t, where $\forall \mathbf{m} \in \text{RS}(\mathcal{N}, \mathbf{m_0})$,

$$f_t(\mathbf{m}) = \mathbf{sb}(p_\pi) \cdot \textstyle\sum_{p \in S_t^{\text{eqPre}}} (\mathbf{Pre}[p,t] - \mathbf{m}[p]) + \mathbf{Pre}[p_\pi, t] - \mathbf{m}[p_\pi],$$
$$\text{with } \mathbf{sb}(p_\pi) \cdot \textstyle\sum_{p \in S_t^{\text{eqPre}}} \mathbf{Pre}[p,t] + \mathbf{Pre}[p_\pi, t] \geq f_t(\mathbf{m}) \geq \mathbf{Pre}[p_\pi, t] - \mathbf{sb}(p_\pi)$$

Previous definition assumes: $S_t^{\text{eqPre}} \neq \emptyset$ and $S_t^{\text{upPre}} \neq \emptyset$. Particular cases are,

$$S_t^{\text{eqPre}} = \emptyset\text{: } f_t(\mathbf{m}) = \mathbf{Pre}[p_\pi, t] - \mathbf{m}[p_\pi];\ \mathbf{Pre}[p_\pi, t] \geq f_t(\mathbf{m}) \geq \mathbf{Pre}[p_\pi, t] - \mathbf{sb}(p_\pi)$$
$$S_t^{\text{upPre}} = \emptyset\text{: } f_t(\mathbf{m}) = \textstyle\sum_{p \in S_t^{\text{eqPre}}} (\mathbf{Pre}[p,t] - \mathbf{m}[p]); \textstyle\sum_{p \in S_t^{\text{eqPre}}} \mathbf{Pre}[p,t] \geq f_t(\mathbf{m}) \geq 0$$

For a given marking $\mathbf{m} \in \text{RS}(\mathcal{N}, \mathbf{m_0})$, $f_t(\mathbf{m})$ can be used to discriminate when a transition t is enabled. The following result characterizes this situation.

Proposition 3. *A Class 1 transition, t, is enabled at marking* $\mathbf{m}$ *iff* $f_t(\mathbf{m}) \leq 0$.

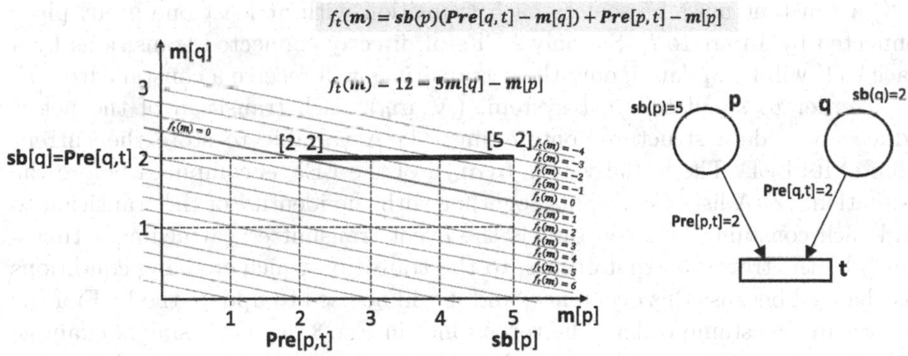

Fig. 2. Transition t of Class 1. Geometric representation of its LEF $f_t(\mathbf{m})$.

Figure 2 illustrates the geometric interpretation of LEFs for a transition t of Class 1 with two input places p and q. In the cartesian plane in Figure, taking into account the $\mathbf{sb}(p)$ and $\mathbf{sb}(q)$, the potential marking vectors are inside the rectangle with vertices $[0,0]$, $[0,2]$, $[5,0]$, and $[5,2]$. Nevertheless, the marking vectors enabling t are only those contained in the segment defined by the points $[2,2]$ and $[5,2]$. The LEF for this transition is $f_t(\mathbf{m}[p], \mathbf{m}[q]) = 12 - 5\mathbf{m}[q] - \mathbf{m}[p]$. The hyperplane $f_t(\mathbf{m}[p], \mathbf{m}[q]) = 0$ separates the markings enabling t from the rest of possible markings.

All LEFs need to be re-calculated if a new marking is reached. This re-evaluation has three drawbacks: (1) The marking of the net must explicitly represented; (2) The marking must be updated each time a transition occurs; (3) The value of the LEFs must be re-computed each time the marking changes. Thus, the use of LEFs increases the computational cost of the simulation cycle of a classical simulation based on the markings. Proposition 4 presents an updating method of the values of the LEFs to cope the previous drawbacks: it does not require to keep the marking of the net, the updating of a LEF is made only if the number of tokens of some input place of the transition has been modified, and the updating uses the current value of the LEF adding a constant supplied by the occurred transition. This constant summarizes the changes in the enabling of the transition by the occurrence of its neighbour transition.

Proposition 4. *Let t be a transition of Class 1, where $S_t^{eqPre} \neq \emptyset$ and $S_t^{upPre} \neq \emptyset$, and t' a transition occurring from $\mathbf{m} \in \mathrm{RS}(\mathcal{N}, \mathbf{m_0})$, $\mathbf{m}[t'\rangle\mathbf{m}'$,*

$$f_t(\mathbf{m}') = f_t(\mathbf{m}) - \mathbf{sb}(p_\pi) \cdot \textstyle\sum_{p \in S_t^{eqPre}} \mathbf{C}[p, t'] - \mathbf{C}[p_\pi, t']$$

In particular: 1) If $S_t^{eqPre} = \emptyset$, $f_t(\mathbf{m}') = f_t(\mathbf{m}) - \mathbf{C}[p_\pi, t']$; 2) If $S_t^{upPre} = \emptyset$, $f_t(\mathbf{m}') = f_t(\mathbf{m}) - \sum_{p \in S_t^{eqPre}} \mathbf{C}[p, t']$. □

The constants in Proposition 4 can be obtained before simulation from the net structure and the structural bounds of places, and they are associated to the transition that occurs, t'. From an operative point of view, after the occurrence of t', a constant must be sent to each transition with at least one input place connected by an arc to t'. So, only LEFs of directly connected transitions by a place to t' will be updated; only these transitions will receive a constant from t'.

In order to simulate a net system, $(\mathcal{N}, \mathbf{m_0})$, each transition of the net is reduced to a data structure containning: (1) A variable to store the current value of its LEF. The initial value, $f_t(\mathbf{m_0})$, of the LEF is computed before the simulation; (2) A list of constants together with the identity of the transition to send each constant. This constant is the event transmitted (attaching a timestamp) by the transition that occurs, to the transition which enabling conditions has changed because this occurence, and it will be used to update the LEF of the receiver in timestamp order. The top subnet in Fig. 3 depicts a simple example with the data of two transitions α and β. The occurrence of α requires the updating of the LEFs of α and β because α modifies the contents of tokens of places J and K, input places of α and β, respectively. So, two constants appear in the data structure of α indicating the destination transitions after its occurrence.

3.2 Vector Linear Enabling Functions for Class 2 Transitions

The enabling of a Class 2 transitions cannot be characterized by means of a unique LEF, as in the case of Class 1 transitions. Nevertheless, *Non-Linear Enabling Functions (NLEF)* for the characterization of its enabling can be defined. The non-linearity of NLEFs introduces computational costs and representation problems that make disappear all advantages obtained from the use

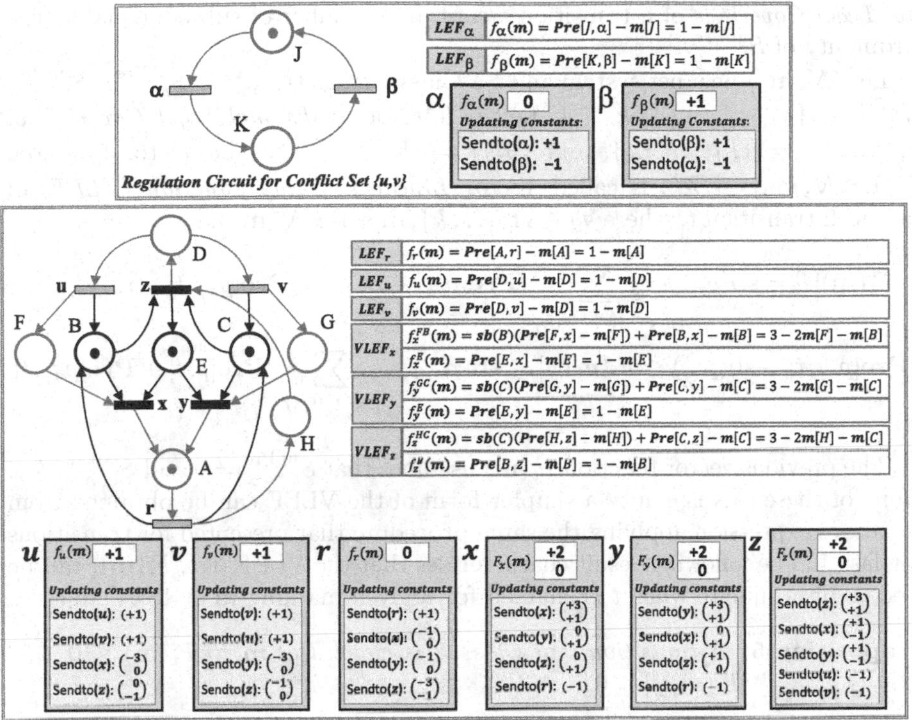

Fig. 3. Construction of a PN model by composition of two subnets via fusion of transition u with the transition α and the transition v with the transition β.

of Enabling Functions in distributed simulation, as in the case of Class 1 transitions.

The enabling of Class 2 transitions can be characterized by means of a family of LEFs. Each LEF determines the contribution of a subset of input places to the enabling of the transition. A subset of transition input places enables the transition if and only if the LEF constructed with their marking variables has a value less than or equal to zero. To build this LEF, the subset of places must satisfy the same conditions that in the case of Class 1 transitions. A transition is completely enabled, if all place subsets (used to construct the LEFs family) enable the transition individually. So, all LEFs must have a value less than or equal to zero for the marking evaluated.

A transition t of Class 2 satisfies: $|{}^\bullet t| \geq 1$, ${}^\bullet t = S_t^{\text{eqPre}} \cup S_t^{\text{upPre}}$, $S_t^{\text{eqPre}} \cap S_t^{\text{upPre}} = \emptyset$ and $|S_t^{\text{upPre}}| > 1$. A *Legal Covering* of ${}^\bullet t$, $LC({}^\bullet t)$, is a family of place subsets, $LC({}^\bullet t) = \{L_1, \ldots, L_k\}$, that verifies: (1) $L_i \subseteq {}^\bullet t$, $i = 1, \ldots, k$; (2) ${}^\bullet t = \bigcup_{i=1}^{k} L_i$; and (3) $|L_i \cap S_t^{\text{upPre}}| \leq 1$, $i = 1, \ldots, k$.

LCs with the minimum number of subsets are preferred, because give rise to a minimum number of LEFs and therefore optimizes the number of test to determine the enabling of a given transition. It is easy to see, that the *Mini-*

mal Legal Coverings of a transition t contain a number of subsets equal to the cardinality of S_t^{upPre}.

Let $(\mathcal{N}, \mathbf{m_0})$ be a net system and t a Class 2 transition, ${}^\bullet t = S_t^{\mathrm{eqPre}} \cup S_t^{\mathrm{upPre}}$, $S_t^{\mathrm{upPre}} = \{p_{\pi_1}, \ldots, p_{\pi_k}\}$, $k > 1$. Let $LC({}^\bullet t)$ be a *Minimal Legal Covering* of ${}^\bullet t$, $LC({}^\bullet t) = \{L_1, \ldots, L_k\}$, and $\{p_{\pi_j}\} = L_j \cap S_t^{\mathrm{upPre}}$. The vector function, $F_t : \mathrm{RS}(\mathcal{N}, \mathbf{m_0}) \to \mathbb{Z}^k$, is called *Vector Linear Enabling Function (VLEF)* of a Class 2 transition t, where $\forall j \in \{1, \ldots, k\}, \mathbf{m} \in \mathrm{RS}(\mathcal{N}, \mathbf{m_0})$,

$$F_t(\mathbf{m})[j] = \mathbf{sb}(p_{\pi_j}) \cdot \sum_{p \in S_t^{\mathrm{eqPre}} \cap L_j} (\mathbf{Pre}[p,t] - \mathbf{m}[p]) + \mathbf{Pre}[p_{\pi_j}, t] - \mathbf{m}[p_{\pi_j}],$$

$$\mathbf{Pre}[p_{\pi_j}, t] - \mathbf{sb}(p_{\pi_j}) \leq F_t(\mathbf{m})[j] \leq \mathbf{sb}(p_{\pi_j}) \cdot \sum_{p \in S_t^{\mathrm{eqPre}} \cap L_j} \mathbf{Pre}[p,t] + \mathbf{Pre}[p_{\pi_j}, t]$$

The previous vector function $F_t(\mathbf{m})$ assumes that $S_t^{\mathrm{upPre}} \neq \emptyset$ and $S_t^{\mathrm{eqPre}} \neq \emptyset$. If one of these sets is empty a simpler form of the VLEF can be obtained from the above expression applying the same procedure that presented for transitions of Class 1. The following result characterizes that the VLEF of t, $F_t(\mathbf{m})$, can be used to discriminate when t is enabled for a given marking $\mathbf{m} \in \mathrm{RS}(\mathcal{N}, \mathbf{m_0})$.

Proposition 5. *A transition t of Class 2 is enabled at $\mathbf{m}$ iff $F_t(\mathbf{m}) \leq \mathbf{0}$ (i.e. $\forall j \in \{1, \ldots, k\}$, $F_t(\mathbf{m})[j] \leq 0$).*

Proposition 6 presents an updating method, similar to Class 1 transitions, of the VLEFs with the same advantages than there, but in this case the constants are vectors (*Updating Factors, UF*) instead of scalars.

Proposition 6. *Let t be a transition of Class 2, with VLEF $F_t(\mathbf{m}) \in \mathbb{Z}^k$ derived from the Minimal Legal Covering $LC({}^\bullet t) = \{L_1, \ldots, L_k\}$, and let t' be a transition occurring from $\mathbf{m} \in \mathrm{RS}(\mathcal{N}, \mathbf{m_0})$, $\mathbf{m}[t'\rangle\mathbf{m}'$,*

$$F_t(\mathbf{m}') = F_t(\mathbf{m}) + UF(t' \to t), UF(t' \to t) \in \mathbb{Z}^k,$$
$$UF(t' \to t)[j] = -\mathbf{sb}(p_{\pi_j}) \cdot \sum_{p \in S_t^{eqPre} \cap L_j} \mathbf{C}[p,t'] - \mathbf{C}[p_{\pi_j}, t'], \forall j = 1 \ldots k,$$

The previous updating of the VLEF from $F_t(\mathbf{m})$ to $F_t(\mathbf{m}')$ assumes that $S_t^{\mathrm{upPre}} \neq \emptyset$ and $S_t^{\mathrm{eqPre}} \cap L_j \neq \emptyset$. Simplified versions of the updating constants can be obtained for the cases $S_t^{\mathrm{eqPre}} \cap L_j = \emptyset$ or $S_t^{\mathrm{upPre}} = \emptyset$, in a similar way to the case of transitions of Class 1.

The Updating Factors, $UF(t' \to t)$, can be obtained directly from the net structure and the structural bounds of places, and they are associated to the transition that occurs, t'. After the occurrence of t', an UF must be sent to each transition with at least one input place connected by an arc to t'. So, only the VLEFs of the directly connected transitions by a place to t' will be updated. In the bottom net in Fig. 3, the transitions r, u and v admit a single LEF since they are Class 1 transitions and therefore only require handling scalars. However, the transitions x, y, and z are Class 2 transitions whose minimal LCs require 2 LEFs, i.e. vectors of dimension 2 must be handled. In this example exist functions and constants of various dimensions, but it is possible to *normalize the representation* to the dimension imposed by the transition with the largest cardinality of its minimal LC.

4 Impact of Enabling Functions in Distributed Simulation of PNs

Distributed simulators specialized in Class 1 transitions will be called *Scalar Simulators*. PNs that contain Class 2 transitions, need *Vector Simulators* that are able of operating, representing and managing vectors.

The deployment of a PN for distributed simulation requires having a battery of simulation interpreters (symbots) generated to run on the different machines of the platform. Then we proceed to partition the PN into sets of transitions where each of them is represented by its LEF/VLEF and the updating constants that include the identity of the recipient transitions. When generating this partition, it will be necessary to include all transitions in conflict relation with a given one in the same set (to avoid complex distributed decision-making protocols). Each one of these sets of transitions will be loaded into a different simbot. Each simbot with its loaded transitions constitutes a Logical Process of the Distributed Simulator. Load balancing techniques in execution platforms that redistribute Logical Processes to improve efficiency can be used with the simbot and its loaded transitions. Additionally, load balancing can be performed between transitions hosted in different symbots. To do this, it is enough to move the data structures of the transitions from the source simbot to the destination simbot since this data are independent of the machine where the simbot that interprets them is executed.

In [4,6,14], several versions of the simbot interpretation cycle are presented. In this cycle, the standard techniques for calculating lookahead within conservative simulation techniques, can be improved by adding information obtained from the structure of the PN using the Synchrony Theory. For example, the computation of the synchronic distance [21] between the groups of transitions included in two different symbots, allows to generate an implicit place that informs when a group should not wait for messages from the other group, or to predict the updating constants yet to arrive. This place can be easily incorporated to the corresponding VLEFs.

The VLEF mechanism of Vector Simulators can see extended its use to support: (1) the deployment of large models to simulate (scalability); (2) the dynamic evolution of the model being simulated (dynamic models); (3) interaction with legacy simulators, external systems or environments not included in the PN model (interoperability). In essence, a VLEF represents a family of preconditions that must be satisfied simultaneously for the transition to occur. The preconditions described so far come from the transition input places, but with the same support is possible to include preconditions based on the result returned by a function, or a program or simply the fusion with a new transition belonging to a external model that synchronizes during simulation. These aspects are developed below.

Vector Simulators and the Deployment of Scalable and Large PNs. The simulation of PNs of scalable systems presents a first problem concerning the description of the model. This can be alleviated by using modular/hierarchical component-oriented description languages [4,16] that allows models to be built from modules of manageable dimensions, and later composed by means of transition synchronization mechanisms, message passing or competition for shared resources. The second problem concerns the compilation and elaboration of these models to generate the transitions data that will finally be loaded into the different simbots of the distributed simulator. This is a challenging problem that requires a large amount of computational resources (time and memory). This makes that the sizes of the nets that can be simulated are very restrictive in practical situations, frustrating one of the objectives pursued with distributed simulation and using cloud infrastructures, even before the simulation can be launched.

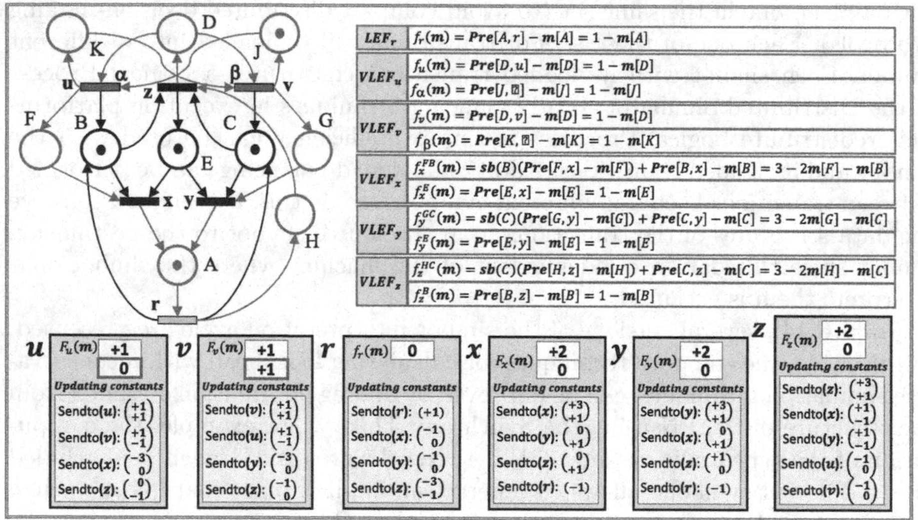

Fig. 4. PN model for simulation after composition of two subnets of Fig. 3.

Vector Simulators offer the adequate mechanisms to support the *s*eparated compilation of modules, of a manageable size, that can later be linked when loading the model in the simbots of the distributed simulator via merging of transitions over the corresponding VLEFs. The two nets presented in Fig. 3 are compiled separately and later composed via fusion of transition α with transition u, and transition β with transition v. The integration of the two compiled modules is an operation restricted to: (1) the transitions that are composed via fusion, by changing the definition of the VLEFs; (2) the transitions whose occurrence modifies the enabling state of the transitions used in the fusion, by adapting the constants to be sent to the merged transitions. The rest of the transitions are not affected. Figure 4 shows the result of the composition.

Vector Simulators and Models that can Change Dynamically During Simulation. In the previous paragraph, a method has been presented to link compiled models, prior to their simulation. Linking is done by merging transitions that belong to different modules. The presented method mainly sought to support the separate compilation of large PNs that would be impossible to directly compile the entire model, or that would require high resource consumption.

Nothing prevents this operation of linking the two nets from being carried out at simulation time, as long as this operation is carried out from a *safe state* so that the modification does not affect the simulation itself. If all transitions involved in the fusion of the two nets are assigned to a same simulation engine, then the data structures affected by the fusion reside in the same simulation engine and the fusion only requires the stacking of the data structures of the VLEFs and the constants. The rest of transitions whose occurrence modifies the enabling state of the fused transitions only require the modification of the constants to be sent. This modification can be done in any moment. This means that the linking can be done at the beginning of the simulation cycle in the simulation engine.

Vector Simulators and the Interaction with External Agents or Legacy Simulators in Order to Federate a Set of Simulators. In some cases, the modelling of a given system by means of PNs makes that some parts of the system cannot be represented in terms of nets. For example, predicates associated to the occurrence of a transition which truth value requires the consultation of some external variables corresponding to: inputs from an unspecified environment or simply inputs coming from human agents to be integrated in the behaviour of the system. A typical example of a predicate of this kind can be a function to solve a conflict between transitions.

Giving support to the simulation of the resulting models of this class of systems requires equipping the simulation engines with capacities to evaluate the predicates associated with the transitions, and whose truth value is a condition of enabling of the transition but not based on terms of Petri nets. Vector Simulators allow integrating these predicate-based enabling conditions in a natural way. To do this, simply introduce the following mechanisms in the simulation engines and in the data structures associated with the transitions:

- In the vectors that contain the compiled VLEFs, it is necessary to introduce an extra entry for each predicate associated with the occurrence of the transition. The value stored in this entry will be the call to an external function that will evaluate the associated predicate. The function will return 0 if the predicate is true (the transition is enabled by the predicate) and 1 if the predicate is false (the transition is not enabled because of the predicate), and so, the enabling of the transition follows the rules previously stated for VLEFs.

- In the constants that must be transmitted to the transitions with predicates, from transitions whose occurrence modifies their degree of enabling, the components related to the predicates are equal to zero, in general. However, in some cases, the transmission of some parameter to a data repository concerning simulation or a data collection for decision making in some part of the simulation can be done including in these entries invocation to transmission functions making this sending of data.

Integration of a PN simulator with external agents like legacy simulators or simulators to be federated, in general, can be done in the same way. To do this, it is enough, as in the case of the predicates associated with the transitions, to introduce an entry in the VLEF that allows invoking a function or procedure related to the simulator to be federated. Summarising, Vector Simulators of Petri Nets present well characteristics for a natural interoperability to work with other products or systems.

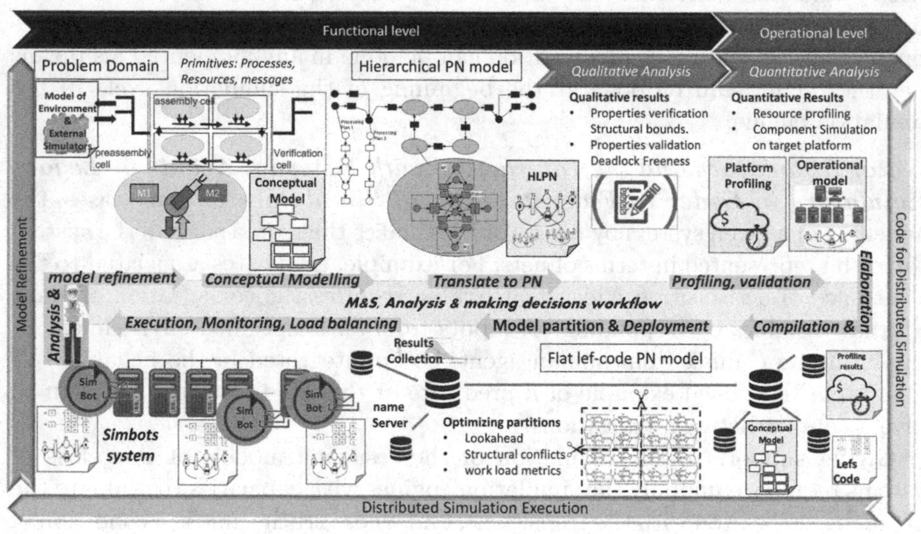

Fig. 5. Large scale M&S supporting decision-making workflow on a PN-based MDE approach.

5 A Model-Driven Approach for Scalable DESs

The analysis and design of real DESs today is inextricably associated with the property of scalability. The systems have an extraordinarily large size (both in size and in the specified details of the system) and therefore it is necessary to have models that allow this complexity to be managed, tools that support analysis and design tasks, and methods that organize and systematize the work. Figure 5 summarizes the approach proposed for this objective in [4,14], and is configured as a Model-Driven Engineering approach based on PNs as a model and Distributed Simulation as a tool. The most relevant parts of this proposal are briefly discussed below.

Conceptual PN-Modelling. Modelling with a Domain-Specific Language (DSL) that provides the basic conceptual entities of an application in a specific domain. Interactions with the environment or external simulators are also modelled.

Timing and Functional Model Annotations. The model can be annotated with labels that extend its interpretation. These annotations are required when the semantics of the model does not cover some characteristics or functionalities of the system to be modeled. For example, PN models can be annotated and configured with deterministic time, probability distributions, or declarative knowledge that mimics the system's behaviour.

Operational Model. In order to provide assistance in the next phase of deployment and configuration, it is essential to develop an operational model of the simulation platform. The *operational model* enriches the functional model with characteristics of the execution platform to develop a *quantitative analysis*.

Elaboration and Compilation. An *elaboration* process translates high level PN specifications into a *flat model* for an efficient interpretation. The conceptual model is *compiled* generating data for each transition (following the LEF/VLEF convention) that will be uploaded in the simbots devoted to execute the cycle of transition occurrences. This is the coding phase of the model and it can require a separated compilation of modules in large PNs.

Deployment and Distributed System Configuration. The set of transitions coded following the compilation process must be partitioned into groups for deployment among the simbots available on the execution platform. The criteria for constructing these groups take into account the characteristics of the platform (minimization of messages exchanged), the efficiency of the simulation (number of concurrent LPs), etc. The PN structure provides information to satisfy these criteria at a low cost. For example, locating embedded state machines in the PN or marking invariants. The only strict constraint to be fulfilled in the partition process is to keep transitions in structural conflicts in the same group.

Dynamic Configuration and Load-Balancing. In a large computing infrastructure, resources are shared by many agents, resulting in a much greater variability in the allocation of computing and communication resources. Load-balancing can be based on LPs (simbots as locked containers of transitions) as interacting entities that can move between parts of the platform. LEF/VLEF-coded model does not lock simbots, which enables the portability of transitions (data coding transitions) between them.

Execution, Data-Collection, Monitoring, Fault Detection. The architectural proposal for distributed simulation is based on simbots specialised in an interpreted simulation of the model. A *simbot* provides an efficient LEF-coded PN interpreter, simulation services such as conservative synchronisation protocols, load-balancing mechanisms, and monitoring and result collection [14].

Large-scale distributed simulation also requires a *Simulation Information System* to support an MDE approach. Conceptual models, compiled code, and collected information must be supported by an information system. Compilation and elaboration processes store the correspondence between events in the flat model and the conceptual model in the information system to translate the simulation results in terms of the model users. A distributed *Name Service* is also required to provide the location of any transition in order to send the events generated for it.

6 Conclusions

Today there are many Discrete Event Systems of interest both for analysis (because they are already created) and for their design from the beginning. All of them share the need to be scalable; and the characteristic of their large size. That is why a Model-Driven approach allows managing the complexity of the system; and a tool such as distributed simulation allows to efficiently address the tasks of analysis and design. This work proposes the use of PNs for DES modeling, with the necessary extensions to cover characteristics of the system to be modeled that are not directly supported by the PNs or that are not sufficiently specified to be modeled with a PN. However, being a model with a sound, well-developed and widely studied theory, when faced with large-dimensional systems it encounters practical limitations solving problems (e.g. verifying the structural boundedness of a net with millions of transitions) that, in many cases, makes unfeasible it use and therefore interest in a model like PNs can decline. To prevent this negative perspective towards the model, this work have tried to take advantage of the Structure Theory of PNs: a relevant characteristic of PNs, that marks the difference with other existing models. Throughout the article, it has been made evident, from Fujimoto's challenges, that the exploitation of the information extracted from the net structure allows to answer to these challenges in an efficient way. The first step was to adopt a model building methodology based on the composition of structurally bounded modules by merging transitions, which allows large models to be built from small components (helping efficient property verification). After that, solutions have been developed for a separate compilation strategy that makes feasible to generate the executable model and its deployment among the PN interpreters (LPs that constitute the distributed simulator). Central to all these methods of exploiting structural information for simulation have been techniques for characterizing transition enabling and that allow PN simulation to be adapted to distributed environments by facilitating portability, load balancing, or federation with legacy simulators. Among other features.

Acknowledgments. This work was co-financed by the Aragonese Government and the European Regional Development Fund "Construyendo Europa desde Aragón" (COSMOS research group); and by the Spanish program "Programa estatal del Generación de Conocimiento y Fortalecimiento Científico y Tecnológico del Sistema de I+D+i", project PGC2018-099815-B-100.

A Basic Concepts and Notations on Petri Nets

A *Place/Transition (P/T) net*, $\mathcal{N}$, is a 4-tuple $\mathcal{N} = (P, T, \mathbf{Pre}, \mathbf{Post})$, where P and T are the sets of places and transitions, and **Pre** and **Post** are the $|P| \times |T|$ sized, natural valued, *pre- and post-incidence matrices*. $\mathbf{Post}[p, t] = w$ means that there is an arc from t to p with weight w, and $\mathbf{Pre}[p, t] = 0$ indicates no arc from p to t.

A *marking* is a $|P|$ sized, natural valued, vector. A *P/T system* is a pair $\mathcal{S} = (\mathcal{N}, \mathbf{m_0})$, where $\mathbf{m_0}$ is the *initial marking*. A transition t is enabled at $\mathbf{m}$ iff $\mathbf{m} \geq \mathbf{Pre}[P, t]$; its *occurrence*, denoted by $\mathbf{m} \xrightarrow{t} \mathbf{m}'$ or $\mathbf{m}[t\rangle\mathbf{m}'$, transform the marking $\mathbf{m}$ into the marking $\mathbf{m}' = \mathbf{m} + \mathbf{C}[P, t]$, where $\mathbf{C} = \mathbf{Post} - \mathbf{Pre}$ is called the *token flow matrix* or *incidence matrix*.

An *occurrence sequence* from $\mathbf{m}$ is a sequence of transitions $\sigma = t_1 \cdots t_k \cdots$ such that $\mathbf{m} \xrightarrow{t_1} \mathbf{m}_1 \cdots \mathbf{m}_{k-1} \xrightarrow{t_k} \mathbf{m}_k \cdots$. The set of all the occurrence sequences, or *language of the net system*, from $\mathbf{m_0}$, is denoted by $\mathrm{L}(\mathcal{N}, \mathbf{m_0})$, and the set of all the reachable markings, or *reachability set*, from $\mathbf{m_0}$, is denoted by $\mathrm{RS}(\mathcal{N}, \mathbf{m_0})$. The reachability relation is represented by a *reachability graph* $\mathrm{RG}(\mathcal{N}, \mathbf{m_0})$ where the nodes are the reachability set, and the directed arcs connect nodes $\mathbf{m}$ and $\mathbf{m}'$, with a label t, iff $\mathbf{m} \xrightarrow{t} \mathbf{m}'$.

The conventional dot notation is used for pre- and postsets of a place or a transition, e.g., ${}^\bullet t = \{p \in P | \mathbf{Pre}[p, t] \neq 0\}$. Places with more than 1 output transition are required to model conflicts. The output transitions of a same place are said to be in *structural conflict relation*. The *coupled conflict relation* is defined as the transitive closure of the structural conflict relation.

References

1. IEEE Standard for Modeling and Simulation (M&S) High Level Architecture (HLA) – Federate Interface Specification. IEEE Std 1516.1-2010 (Revision of IEEE Std 1516.1-2000), pp. 1–378 (2010)
2. IEEE Standard for Modeling and Simulation (M&S) High Level Architecture (HLA) – Framework and Rules. IEEE Std 1516-2010 (Revision of IEEE Std 1516-2000), pp. 1–38 (2010)
3. IEEE Standard for Modeling and Simulation (M&S) High Level Architecture (HLA) – Object Model Template (OMT) Specification. IEEE Std 1516.2-2010 (Revision of IEEE Std 1516.2-2000), pp. 1–110 (2010)
4. Arronategui, U., Bañares, J.Á., Colom, J.M.: A MDE approach for modelling and distributed simulation of health systems. In: Djemame, K., Altmann, J., Bañares, J.Á., Agmon Ben-Yehuda, O., Stankovski, V., Tuffin, B. (eds.) GECON 2020. LNCS, vol. 12441, pp. 89–103. Springer, Cham (2020). https://doi.org/10.1007/978-3-030-63058-4_9

5. Briz, J., Colom, J., Silva, M.: Simulation of Petri nets and linear enabling functions. In: Proceedings of IEEE International Conference on Systems, Man and Cybernetics, vol. 2, pp. 1671–1676 (1994)
6. Briz, J.L., Colom, J.M.: Implementation of weighted place/transition nets based on linear enabling functions. In: Valette, R. (ed.) ICATPN 1994. LNCS, vol. 815, pp. 99–118. Springer, Heidelberg (1994). https://doi.org/10.1007/3-540-58152-9_7
7. Chandy, K., Misra, J.: Distributed simulation: a case study in design and verification of distributed programs. IEEE Trans. Softw. Eng. **SE-5**(5), 440–452 (1979)
8. Colom, J., Silva, M., Villarroel, J.: On software implementation of Petri Nets and colored Petri Nets using high level concurrent languages. In: Proceedings of 7th European Workshop on Application and Theory of Petri nets, Oxford, England, pp. 207–241 (1986)
9. Dongarra, J.: Basic linear algebra subprograms technical (BLAST) forum standard II. Int. J. High Perform. Comput. Appl. (IJHPCA) **16**, 1–111 (2002)
10. D'Angelo, G.: The simulation model partitioning problem: an adaptive solution based on self-Clustering. Simul. Model. Pract. Theory **70**, 1–20 (2017)
11. Fujimoto, R.: Parallel and Distributed Simulation Systems. Wiley Interscience (2000)
12. Fujimoto, R.: Research challenges in parallel and distributed simulation. ACM Trans. Model. Comput. Simul. **26**(4), 1–29 (2016). Article 22
13. Glabbeek, R.V., Höfner, P.: Progress, justness, and fairness. ACM Comput. Surv. **52**(4), 1–38 (2020). Article 69
14. Bañares, J.Á., Colom, J.M.: Model and simulation engines for distributed simulation of discrete event systems. In: Coppola, M., Carlini, E., D'Agostino, D., Altmann, J., Bañares, J.Á. (eds.) GECON 2018. LNCS, vol. 11113, pp. 77–91. Springer, Cham (2019). https://doi.org/10.1007/978-3-030-13342-9_7
15. Karp, R.M., Miller, R.E.: Parallel program schemata. J. Comput. Syst. Sci. **3**(2), 147–195 (1969)
16. Merino, A., Tolosana-Calasanz, R., Bañares, J.Á., Colom, J.-M.: A specification language for performance and economical analysis of short term data intensive energy management services. In: Altmann, J., Silaghi, G.C., Rana, O.F. (eds.) GECON 2015. LNCS, vol. 9512, pp. 147–163. Springer, Cham (2016). https://doi.org/10.1007/978-3-319-43177-2_10
17. Perumalla, K.S.: μsik - a micro-kernel for parallel/distributed simulation systems. In: Workshop on Principles of Advanced and Distributed Simulation (PADS 2005), pp. 59–68 (2005)
18. Piedrafita, R., Tardioli, D., Villarroel, J.L.: Distributed implementation of discrete event control systems based on Petri Nets. In: Proceedings of the IEEE International Symposium on Industrial Electronics, pp. 1738–1745 (2008)
19. Piedrafita, R., Villarroel, J.: Performance evaluation of Petri nets centralized implementation. The execution time controller. Discrete Event Dyn. Syst. **21**(2), 139–169 (2011)
20. Silva, M.: Las Redes de Petri en la Informática y en la Automática. AC, Madrid (1985)
21. Silva, M., Colom, J.M.: On the computation of structural synchronic invariants in P/T nets. In: Rozenberg, G. (ed.) APN 1987. LNCS, vol. 340, pp. 386–417. Springer, Heidelberg (1988). https://doi.org/10.1007/3-540-50580-6_39
22. Silva, M., David, R.: Synthèse programmée des automatismes logiques décrits par réseaux de Petri: une méthode de mise en oeuvre sur microcalculateurs. Rairo-Automatique **13**(4), 369–393 (1979)

23. Silva, M., Teruel, E., Colom, J.M.: Linear algebraic and linear programming techniques for the analysis of place/transition net systems. In: Reisig, W., Rozenberg, G. (eds.) ACPN 1996. LNCS, vol. 1491, pp. 309–373. Springer, Heidelberg (1998). https://doi.org/10.1007/3-540-65306-6_19
24. Silva, M., Velilla, S.: Programmable logic controllers and Petri Nets: a comparative study. In: Proceedings of the Third IFAC/IFIP Symposium. Software for Computer Control 1982, pp. 83–88. Pergamon Press (1982)
25. Tolosana, R., Bañares, J., Colom, J.: Model-driven development of data intensive applications over cloud resources. Future Gener. Comput. Syst. **87**, 888–909 (2018)
26. Valette, R.: Nets in production systems. In: Brauer, W., Reisig, W., Rozenberg, G. (eds.) ACPN 1986. LNCS, vol. 255, pp. 191–217. Springer, Heidelberg (1987). https://doi.org/10.1007/3-540-17906-2_26
27. Zeigler, B., Praehofer, H., Kim, T.: Theory of Modeling and Simulation: Integrating Discrete Event and Continuous Complex Dynamic Systems. Academic Press (2000)

On the Application of Model-Driven Optimization to Business Processes

Gabriele Taentzer[1(✉)], Jens Kosiol[1,2], and Leen Lambers[3]

[1] Philipps-Universität Marburg, Marburg, Germany
{taentzer,kosiolje}@mathematik.uni-marburg.de
[2] Universität Kassel, Kassel, Germany
[3] BTU Cottbus-Senftenberg, Cottbus, Germany
leen.lambers@b-tu.de

Abstract. The optimization of business processes is an important task to increase the efficiency of the described workflows. Metaheuristic optimization, such as evolutionary search, has been used to optimize business process models, but it requires a high level of expertise that not all process designers have. *Model-driven optimization* (MDO) promises to make the use of metaheuristic optimization accessible to domain experts without in-depth technical expertise by allowing them to specify the optimization algorithm directly at the model level. Because this approach is less technical, the process designers can focus on the business process models and their properties. Using concrete business process optimization problems as a starting point, we discuss how MDO can be applied to these problems, what MDO would offer for business process optimization, and how the application to business processes could stimulate research on MDO.

Keywords: Business process models · Optimization · Model-driven engineering

1 Introduction

The optimization of business processes is a strategic activity in organizations because it can increase the efficiency of work. A number of metrics have been developed to analyze the quality of business process models [47]. Granularity, which is reflected in the size of activities, is crucial for the design of balanced processes; it can be measured with coupling and cohesion metrics [36]. To speed up work, workflows can be further optimized by increasing parallelism within tasks [12]. However, optimization is a difficult task when performed manually, especially when multiple objectives must be considered. It requires a well-suited optimization algorithm; the development of such an algorithm requires a high level of expertise that not all process designers have. For example, the implementation of an evolutionary algorithm usually requires (aspects of) business process models to be encoded in integer representations in order to perform an

L. M. Kristensen and J. M. van der Werf (Eds.): PETRI NETS 2024, LNCS 14628, pp. 24–35, 2024.
https://doi.org/10.1007/978-3-031-61433-0_2

evolutionary search [46]. To make optimization accessible to domain experts, an approach is needed that allows optimization tasks to be specified and executed without deep technical expertise.

Various software engineering problems, such as software modularization [6], process optimization, and release planning [2], have already been considered as optimization problems. They have often been solved by using evolutionary algorithms [20] which mimic the evolution in nature to solve optimization problems. Model-driven engineering (MDE) [38] aims at representing domain knowledge in models and solving problems through model transformations. MDE can be used in the context of evolutionary optimization to minimize the expertise required by users of optimization techniques. This combination of optimization and MDE is referred to in the literature as model-based or model-driven optimization (MDO) [1,3,9,18,22,48]. It applies evolutionary optimization to models. MDO can simplify the application of evolutionary search to software engineering problems, because models are not encoded, but the search space consists directly of models that are evolved by model transformations. A conceptual framework that precisely defines all the main concepts of MDO based on graphs and graph transformation is presented in [23]. It is intended to assist the modeler in using MDO to solve such optimization problems.

Since we are focusing on evolutionary algorithms as the optimization technique in this paper, we will briefly recall them. With reference to, for example, [5,16,23,48], an *evolutionary algorithm* is used to solve an optimization problem. Usually, such a problem is formally defined by means of an *objective* (or *fitness*) *function* that expresses the objective that is to be optimized. In practical applications, multiple objective functions often must be addressed simultaneously, leading to the concepts of *multi-objective problems* and *many-objective problems* and *Pareto optimization* [41]. For the optimization process, one needs a representation of possible solutions to the problem at hand; the solutions constitute the *search space*. Practical optimization problems usually come with additional *constraints* that clarify which of the represented solutions are *feasible* (i.e., constitute valid solutions to the optimization problem). Given a *constrained optimization problem* and a representation for solutions, the key ingredients of an evolutionary algorithm are a generator for an initial population of solutions, a mechanism for generating new solutions from existing ones (e.g., by *mutation* or *crossover*), a selection mechanism that typically implements the evolutionary concept of survival of the fittest, and a condition for stopping evolutionary computations.

MDO applies ordinary evolutionary algorithms, such as the well-known NSGA-II algorithm [10], or other metaheuristic search techniques to (constrained) optimization problems but uses models, e.g., [8,23,48], or model transformation sequences, e.g., [1,3,18], as representation for solutions. With MDEOptimiser [8] and MOMoT [3], tool support for both approaches is available so that optimizations can be performed. In this paper, we propose the use of MDO for business process optimization. We focus on the model-based approach to MDO (as it tends to perform better [22]) and propose to optimize models of workflow processes directly, e.g., in the standardized workflow modeling language BPMN [34].

Business processes have been optimized in various ways [47]. In Sect. 2, we present a selection of typical optimization problems that have been considered for them. For each problem, we discuss the type of model being optimized, the objectives for which the models are being optimized, and the constraints that a feasible solution must satisfy. In Sect. 3, we describe for each of the selected problems how MDO can be applied to it and what the benefits and challenges are. Section 4 concludes the paper.

2 Optimization Problems on Business Processes

The selected optimization problems we consider are the clustering of information elements (Sect. 2.1), the multi-objective optimization of non-functional properties of a business process (Sect. 2.2), and the parallelization of tasks of a business process (Sect. 2.3).

2.1 Clustering of Information Elements in Workflow Processes

The first optimization problem is formulated for workflow processes which can be seen as a conceptual basis for business processes [44]. Workflow processes have also been formalized as a special class of Petri nets, more precisely as *workflow nets*, to validate them.

A *workflow process* [36] contains a number of information elements that are used as input or output elements of operations. An operation is a basic processing step; it has one or more input information elements and one output information element. An activity in a workflow process consists of one or more operations. The output of one of the operations can be the input of another operation of that activity. A workflow process is *valid* if (1) all operations occur at least once in an activity, and (2) if the input of one operation depends on the output of another operation, then the respective activities of which they are part of are ordered so that they respect this dependency. While constraint (1) ensures the completeness of the activity design, constraint (2) ensures the correctness of their ordering.

The *optimization problem* to be solved is to find a good clustering of information elements and operations into activities. A clustering with low coupling between activities and high cohesion within each activity is considered the best, since in this case the clusters can be well identified. In general, *coupling* measures the number of connections between the elements of a model [45]. In workflow processes, two activities are coupled if they share one or more common information elements. The *cohesion* metric for workflow processes in [36] measures the coherence within the activities of the process model. Similar to the coupling metric, this cohesion metric also focuses on the information processing in the process. The *clustering problem* can be formulated as a *multi-objective problem*, since we aim for low coupling and high cohesion. The validity constraints presented above formulate the *feasibility constraints.*

2.2 Multi-objective Optimization of Non-functional Properties

A business process has several non-functional properties of interest, such as cost, flow time, product quality, etc. Regularly, business processes are optimized with respect to these criteria. To be of practical importance, an optimization approach must consider multiple criteria simultaneously, e.g., minimizing cost and flow time while maximizing quality.

Vergidis et al. [46] propose a framework for the multi-objective evolutionary optimization of non-functional properties of business processes. They assume a set of *tasks* to be given, where for each task additional information is provided: A task comes with a set of *input resources* (it consumes), *output resources* (it produces), and values for attributes of interest (such as cost or duration). For a concrete optimization problem, a set of *input resources* and a set of *output resources* are given. The goal is to find a process (a subset of the given tasks) that produces the required output resources from the given input resources. That is, for a process to be *feasible*, the selected tasks must satisfy certain constraints: (1) each input resource of the optimization problem must be consumed by at least one selected task; (2) each output resource of the optimization problem must be produced by at least one selected task; and (3) the selected tasks must lead to a connected process diagram. The process should be optimal with respect to selected non-functional properties that are defined by objective functions. The proposed framework is generic in the sense that it is not restricted to specific non-functional properties. It is simply assumed that the objective functions can be computed by aggregating the attribute values of the selected tasks.

The framework in [46] uses multi objective evolutionary algorithms (MOEAs) to search for optimal business processes. Different representations of a solution are used for different operators of such an algorithm. Basically, a business process is encoded as an array containing the tasks that make up the process. Standard variation operators (crossover and mutation) are applied to these arrays. The fitness of a solution with respect to the multiple objectives can be computed by aggregating the values of selected tasks for the respective attributes; the necessary information for this is stored in a matrix. To check the feasibility constraints, [46] develops a *Process Composition Algorithm* (PCA) that assembles the selected processes into a process diagram (repairing certain constraint violations on the way). The resulting process design can then be checked for feasibility; if constraint violations remain, their severity is computed in a *Degree of Infeasibility*.

2.3 Parallelization of Tasks

The third optimization problem concerns business process optimization for processes described using the standardized workflow modeling language BPMN [34]. Durán and Salaün present an automated approach to optimizing *BPMN models that are enriched with a description of the execution time and resources associated with tasks* [12]. These enriched business process models take into account not only behavioral but also quantitative aspects.

The *optimization problem* aims at finding a reorganized enriched BPMN model with reduced execution time. Possible reorganizations of tasks within the BPMN model are described using *refactorings*. These reorganizations take into account the resources used by each task. The main idea for reducing the execution time is to increase parallelism between tasks as much as possible. The refactorings must take into account *specific constraints*. For example, tasks can only run in parallel if they do not compete for the same resources. Also, causal dependencies between tasks may need to be preserved when adding parallelism. These constraints represent *feasibility constraints* that must be satisfied for a solution to the optimization problem to be valid.

2.4 Summary of Optimization Problems

We briefly summarize the similarities and differences of the three selected optimization problems. All three optimization problems can be formulated for *models of business processes*. In the clustering problem, these models describe workflow processes; for the optimization of non-functional properties, the models are currently encoded as arrays of tasks in the business process. In the parallelization problem, the standard modeling language BPMN is already used. The *objective* of the clustering problem is of a structural kind, while the objectives of the other two problems are more behavioral, since they both focus on optimizing non-functional properties. In the parallelization problem, models even must be simulated in order to evaluate the objective function. All three problems share the fact that structural constraints must hold for a solution to be *feasible*. In particular, in the parallelization problem, it becomes clear that behavior preservation comes in addition to the preservation of structural constraints.

3 Applying Model-Driven Optimization to Business Processes

In this section, we outline how MDO can be used to solve business process optimization problems and illustrate our ideas at the selected problems just recalled. We also discuss where MDO offers promising solutions to these optimization problems and where the problems present challenges that could trigger interesting research in MDO.

MDO for Business Processes. As explained in the introduction, MDO denotes an approach to metaheuristic optimization in which models are crucial artefacts.[1] Thus, applying MDO to business processes amounts to developing model transformations in business process modelling languages such as BPMN. These

[1] Typically, models or model transformation sequences constitute the search space, and searching means modifying models or model transformation sequences. For feature model configuration [21,30], search operators are designed and verified based on models; however, for the actual search, models and operators are translated into more machine-oriented representations to increase efficiency.

model transformation rules can then be used as search operators in optimization processes, where the search space consists of models (or model transformation sequences). While rules tailored to a specific optimization problem and a specific optimization algorithm promise the best results, MDO offers a certain degree of genericity. A set of rules that specifies basic modifications of business process models can constitute the search operators for different types of optimization algorithms (such as evolutionary algorithms, hill climbing, or simulated annealing) and for different optimization problems (i.e., different objective functions with respect to which models are evaluated).

For the three optimization problems presented, MDO would mean the following. For the clustering of information elements (in Sect. 2.1), (coupling and cohesion) metrics have been defined that can be used to estimate the quality of a given business process. However, no automated approach has been proposed to optimize business processes with respect to these quality criteria. MDO serves as such an approach. Coupling and cohesion, as defined in [36], are the objectives; they can be combined into a single objective function or be kept separate and multi-objective optimization is employed. For example, given a suitable set of transformation rules and a business process, evolutionary search can be used to find a clustering of the information elements that has low coupling and high cohesion.

For the optimization of non-functional properties (in Sect. 2.2), multi-objective optimization via evolutionary algorithms is already performed to find business processes of high quality. Applying MDO here means to not encode the models into arrays for the search but to use them directly. Below, we discuss the benefits we expect from this change.

For the parallelization of tasks (in Sect. 2.3), the authors already use models as crucial artefacts and the search is performed by model transformation. Thus, the work [12] can be considered as an instance of MDO. However, instead of using metaheuristic search, they explore their search space completely or via a handcrafted heuristic, which is not feasible for larger search spaces or specific to their problem at hand. The framework of MDO here broadens the perspective to use the transformation rules suggested in [12] also for other optimization problems on business models, to try out other algorithms for the problem tackled in [12], or to complement the objective addressed in [12] by further ones, making the problem multi-objective.

Expected Benefits. One of the promises of MDO is to make optimization accessible to domain experts without deep technical expertise. Domain experts need only interact with models they are already familiar with, and may even be able to design domain-specific search operators (i.e. transformation rules) that are well suited to the search. While this hope has yet to be empirically verified, testing it for business process optimization is appealing because process design is typically a domain in which domain experts without technical expertise are involved.

On the technical side, the main promise of MDO lies in the strong formal foundation that model transformation has in graph transformation [15]. It can be used to make the search for *feasible* solutions easier. Realistic optimization problems often come with constraints that the solutions must satisfy. As seen above, this is also the case for all three optimization problems considered here. For the optimization of non-functional properties [46], the authors even explicitly mention that obtaining feasible process models during evolutionary search is a major challenge; most constructed solutions are infeasible. For graph transformation rules, there is ample experience with regard to the treatment of constraints (expressed in different kinds of logics). Such rules can be analyzed for preserving the validity of given constraints, e.g., [13,27,40], be equipped with application conditions that prevent rule applications that would introduce constraint violations, e.g., [19,32], or, for certain types of constraints, even be adapted so that the constraints are preserved, e.g., [7,21,25]. In addition, there is a research focus on repairing graphs and models with respect to constraints, which is another way to make infeasible mutation and crossover results feasible, e.g., [29,33,37,39].

MDO has begun to make use of these formal results. There is empirical evidence that evolutionary search on models benefits from transformation rules that preserve the given constraints [7,21,23]. For certain types of constraints (multiplicities), transformation rules that preserve them can be automatically derived from a meta-model [7], i.e., for a given modeling language. So we are convinced that MDO can be successfully used to optimize business processes for various purposes.

Raised Research Challenges. Above, we argued that basing (evolutionary) search on model transformation provides a means to address the problem of *structural constraints* that are expected to hold for solutions. However, when optimizing business processes, one usually needs to consider *behavioral constraints* as well: Mostly, the optimized process should still exhibit the same behavior as the original one. For example, the optimization of non-functional properties requires that a feasible process uses the given resources and produces the required resources. This constraint can be expressed as a formula on the graph structure and be treated as described above. However, behavioral equivalence of the optimized process with the original one is often expressed as a simulation or bisimulation.

There are techniques that allow one to check that the input and the output of a graph (or model) transformation are behaviorally equivalent, e.g., are in a (bi)simulation relation (see, e.g., [4,14,17,31,35]). However, there seems to be less research on this topic than for structural constraints. Furthermore, we are not aware of any research in MDO on preserving behavioral equivalence during search. There is recent work on formalizing the BPMN execution semantics using graph transformation, which facilitates behavioral property checking [28], which could serve as a starting point for this line of research. All in all, we expect that business process optimization can stimulate new research on preserving behavioral equivalence during model transformation, thus enriching the set of techniques that are used in MDO.

Another challenge is to develop an appropriate crossover operator for business process models. A crossover operator typically mixes information from two parent solutions to compute (one or two) child solutions that resemble their parents. In general, evolutionary search can benefit from using both crossover and mutation, rather than just mutation alone [11,42]. We have started to develop a generic (i.e., domain-agnostic) crossover operator for use in MDO [24,26]. It unifies many crossover operators that have been proposed for specific models or graphs. Initial experiments show some increase in search effectiveness compared to using mutation alone, but the experiments also seem to indicate that this generic operator suffers from producing too many infeasible solutions, i.e., from introducing constraint violations. We combined this operator with ad-hoc repair of the computed solutions [24]. In applying MDO to business processes, we expect this effect to occur as well, so the research on the development of constraint-preserving crossover operators [43] needs to be continued. In addition, research is needed on how to concretize the generic crossover operator for the business process domain.

4 Conclusion

Business process optimization is an important task for increasing the efficiency of workflows. To make optimization accessible to domain experts, an approach is needed that allows optimization tasks to be specified and executed without deep technical expertise. In this paper, we have argued that it would be promising to tailor model-driven optimization to business processes. We expect that this would make it easier for process designers to apply metaheuristic optimization such as evolutionary search to their optimization problems. We have recalled three selected business process optimization problems and sketched how they could be tackled and benefit from the use of MDO principles and techniques.

The application of MDO to business process optimization also poses new challenges for MDO. MDO has mostly been considered to address problems in software engineering with objectives of a structural nature such as modularization. Since business processes describe behavior, it must also be shown that the optimized processes comply with behavioral constraints. Finally, the development of domain-specific crossover operators seems to be another relevant research goal. We have discussed how existing research results on model and graph transformation can support these lines of research in order to successfully apply MDO to business processes.

References

1. Abdeen, H., et al.: Multi-objective optimization in rule-based design space exploration. In: Crnkovic, I., Chechik, M., Grünbacher, P. (eds.) ACM/IEEE International Conference on Automated Software Engineering, ASE 2014, Vasteras, Sweden, 15–19 September 2014, pp. 289–300. ACM (2014). https://doi.org/10.1145/2642937.2643005
2. Bagnall, A.J., Rayward-Smith, V.J., Whittley, I.M.: The next release problem. Inf. Softw. Technol. **43**(14), 883–890 (2001). https://doi.org/10.1016/S0950-5849(01)00194-X

3. Bill, R., Fleck, M., Troya, J., Mayerhofer, T., Wimmer, M.: A local and global tour on MOMoT. Softw. Syst. Model. **18**(2), 1017–1046 (2019). https://doi.org/10.1007/s10270-017-0644-3
4. Bisztray, D., Heckel, R., Ehrig, H.: Compositional verification of architectural refactorings. In: de Lemos, R., Fabre, J.-C., Gacek, C., Gadducci, F., ter Beek, M. (eds.) WADS 2008. LNCS, vol. 5835, pp. 308–333. Springer, Heidelberg (2009). https://doi.org/10.1007/978-3-642-10248-6_13
5. Blum, C., et al.: Evolutionary optimization. In: Chiong, R., Weise, T., Michalewicz, Z. (eds.) Variants of Evolutionary Algorithms for Real-World Applications, pp. 1–29. Springer, Heidelberg (2012). https://doi.org/10.1007/978-3-642-23424-8_1
6. Bowman, M., Briand, L.C., Labiche, Y.: Solving the class responsibility assignment problem in object-oriented analysis with multi-objective genetic algorithms. IEEE Trans. Softw. Eng. **36**(6), 817–837 (2010). https://doi.org/10.1109/TSE.2010.70
7. Burdusel, A., Zschaler, S., John, S.: Automatic generation of atomic multiplicity-preserving search operators for search-based model engineering. Softw. Syst. Model. **20**(6), 1857–1887 (2021). https://doi.org/10.1007/s10270-021-00914-w
8. Burdusel, A., Zschaler, S., Strüber, D.: MDEOptimiser: a search based model engineering tool. In: Babur, Ö., et al. (eds.) Proceedings of the 21st ACM/IEEE International Conference on Model Driven Engineering Languages and Systems: Companion Proceedings, MODELS 2018, Copenhagen, Denmark, 14–19 October 2018, pp. 12–16. ACM (2018). https://doi.org/10.1145/3270112.3270130
9. Burton, F.R., Poulding, S.M.: Complementing metaheuristic search with higher abstraction techniques. In: Paige, R.F., Harman, M., Williams, J.R. (eds.) 1st International Workshop on Combining Modelling and Search-Based Software Engineering, CMSBSE@ICSE 2013, San Francisco, CA, USA, 20 May 2013, pp. 45–48. IEEE Computer Society (2013). https://doi.org/10.1109/CMSBSE.2013.6604436
10. Deb, K., Agrawal, S., Pratap, A., Meyarivan, T.: A fast and elitist multiobjective genetic algorithm: NSGA-II. IEEE Trans. Evol. Comput. **6**(2), 182–197 (2002). https://doi.org/10.1109/4235.996017
11. Doerr, B., Happ, E., Klein, C.: Crossover can provably be useful in evolutionary computation. Theor. Comput. Sci. **425**, 17–33 (2012). https://doi.org/10.1016/j.tcs.2010.10.035
12. Durán, F., Salaün, G.: Optimization of BPMN processes via automated refactoring. In: Troya, J., Medjahed, B., Piattini, M., Yao, L., Fernández, P., Ruiz-Cortés, A. (eds.) ICSOC 2022. LNCS, vol. 13740, pp. 3–18. Springer, Cham (2022). https://doi.org/10.1007/978-3-031-20984-0_1
13. Dyck, J., Giese, H.: k-inductive invariant checking for graph transformation systems. In: de Lara, J., Plump, D. (eds.) ICGT 2017. LNCS, vol. 10373, pp. 142–158. Springer, Cham (2017). https://doi.org/10.1007/978-3-319-61470-0_9
14. Dyck, J., Giese, H., Lambers, L.: Automatic verification of behavior preservation at the transformation level for relational model transformation. Softw. Syst. Model. **18**(5), 2937–2972 (2019). https://doi.org/10.1007/s10270-018-00706-9
15. Ehrig, H., Ehrig, K., Prange, U., Taentzer, G.: Fundamentals of Algebraic Graph Transformation. Monographs in Theoretical Computer Science. Springer, Heidelberg (2006). https://doi.org/10.1007/3-540-31188-2
16. Eiben, A.E., Smith, J.E.: Introduction to Evolutionary Computing. Natural Computing Series, 2nd edn. Springer, Heidelberg (2015). https://doi.org/10.1007/978-3-662-44874-8
17. Engels, G., Kleppe, A., Rensink, A., Semenyak, M., Soltenborn, C., Wehrheim, H.: From UML activities to TAAL - towards behaviour-preserving model transformations. In: Schieferdecker, I., Hartman, A. (eds.) ECMDA-FA 2008. LNCS, vol.

5095, pp. 94–109. Springer, Heidelberg (2008). https://doi.org/10.1007/978-3-540-69100-6_7

18. Fleck, M., Troya, J., Wimmer, M.: Marrying search-based optimization and model transformation technology. In: Proceedings of the First North American Search Based Software Engineering Symposium. Elsevier (2015). http://publik.tuwien.ac.at/files/PubDat_237899.pdf. Accessed 07 Dec 2022
19. Habel, A., Pennemann, K.: Correctness of high-level transformation systems relative to nested conditions. Math. Struct. Comput. Sci. **19**(2), 245–296 (2009). https://doi.org/10.1017/S0960129508007202
20. Harman, M., Mansouri, S.A., Zhang, Y.: Search-based software engineering: trends, techniques and applications. ACM Comput. Surv. **45**(1), 11:1–11:61 (2012). https://doi.org/10.1145/2379776.2379787
21. Horcas, J.M., Strüber, D., Burdusel, A., Martinez, J., Zschaler, S.: We're not gonna break it! Consistency-preserving operators for efficient product line configuration. IEEE Trans. Softw. Eng. **49**(3), 1102–1117 (2023). https://doi.org/10.1109/TSE.2022.3171404
22. John, S., et al.: Searching for optimal models: comparing two encoding approaches. J. Object Technol. **18**(3), 6:1–22 (2019). https://doi.org/10.5381/jot.2019.18.3.a6
23. John, S., Kosiol, J., Lambers, L., Taentzer, G.: A graph-based framework for model-driven optimization facilitating impact analysis of mutation operator properties. Softw. Syst. Model. **22**(4), 1281–1318 (2023). https://doi.org/10.1007/s10270-022-01078-x
24. John, S., Kosiol, J., Taentzer, G.: Towards a configurable crossover operator for model-driven optimization. In: Kühn, T., Sousa, V. (eds.) Proceedings of the 25th International Conference on Model Driven Engineering Languages and Systems: Companion Proceedings, MODELS 2022, Montreal, Quebec, Canada, 23–28 October 2022, pp. 388–395. ACM (2022). https://doi.org/10.1145/3550356.3561603
25. Kosiol, J., Fritsche, L., Nassar, N., Schürr, A., Taentzer, G.: Constructing constraint-preserving interaction schemes in adhesive categories. In: Fiadeiro, J.L., Tutu, I. (eds.) WADT 2018. LNCS, vol. 11563, pp. 139–153. Springer, Cham (2019). https://doi.org/10.1007/978-3-030-23220-7_8
26. Kosiol, J., John, S., Taentzer, G.: A generic construction for crossovers of graph-like structures and its realization in the eclipse modeling framework. J. Log. Algebraic Methods Program. **136**, 100909 (2024). https://doi.org/10.1016/j.jlamp.2023.100909
27. Kosiol, J., Strüber, D., Taentzer, G., Zschaler, S.: Sustaining and improving graduated graph consistency: a static analysis of graph transformations. Sci. Comput. Program. **214**, 102729 (2022). https://doi.org/10.1016/j.scico.2021.102729
28. Kräuter, T., Rutle, A., König, H., Lamo, Y.: Formalization and analysis of BPMN using graph transformation systems. In: Fernández, M., Poskitt, C.M. (eds.) ICGT 2023. LNCS, vol. 13961, pp. 204–222. Springer, Cham (2023). https://doi.org/10.1007/978-3-031-36709-0_11
29. Lauer, A., Kosiol, J., Taentzer, G.: Empowering model repair: a rule-based approach to graph repair without side effects. In: ACM/IEEE International Conference on Model Driven Engineering Languages and Systems, MODELS 2023 Companion, Västerås, Sweden, 1–6 October 2023, pp. 831–840. IEEE (2023). https://doi.org/10.1109/MODELS-C59198.2023.00132
30. Martinez, J., Strüber, D., Horcas, J.M., Burdusel, A., Zschaler, S.: Acapulco: an extensible tool for identifying optimal and consistent feature model configurations. In: Felfernig, A., et al. (eds.) SPLC 2022: 26th ACM International Systems and

Software Product Line Conference, Graz, Austria, 12–16 September 2022, vol. B, pp. 50–53. ACM (2022). https://doi.org/10.1145/3503229.3547067
31. Narayanan, A., Karsai, G.: Towards verifying model transformations. In: Bruni, R., Varró, D. (eds.) Proceedings of the Fifth International Workshop on Graph Transformation and Visual Modeling Techniques, GT-VMT@ETAPS 2006, Vienna, Austria, 1–2 April 2006. Electronic Notes in Theoretical Computer Science, vol. 211, pp. 191–200. Elsevier (2006). https://doi.org/10.1016/J.ENTCS.2008.04.041
32. Nassar, N., Kosiol, J., Arendt, T., Taentzer, G.: Constructing optimized constraint-preserving application conditions for model transformation rules. J. Log. Algebraic Methods Program. **114**, 100564 (2020). https://doi.org/10.1016/j.jlamp.2020.100564
33. Nassar, N., Kosiol, J., Radke, H.: Rule-based repair of EMF models: formalization and correctness proof. In: Graph Computation Models (GCM 2017). Electronic Pre-Proceedings (2017). pages.di.unipi.it/corradini/Workshops/GCM2017/papers/Nassar-Kosiol-Radke-GCM2017.pdf
34. OMG: Business process model and notation. version 2.0 (2011). http://www.omg.org/spec/BPMN/2.0/
35. Rangel, G., Lambers, L., König, B., Ehrig, H., Baldan, P.: Behavior preservation in model refactoring using DPO transformations with borrowed contexts. In: Ehrig, H., Heckel, R., Rozenberg, G., Taentzer, G. (eds.) ICGT 2008. LNCS, vol. 5214, pp. 242–256. Springer, Heidelberg (2008). https://doi.org/10.1007/978-3-540-87405-8_17
36. Reijers, H.A., Vanderfeesten, I.T.P.: Cohesion and coupling metrics for workflow process design. In: Desel, J., Pernici, B., Weske, M. (eds.) BPM 2004. LNCS, vol. 3080, pp. 290–305. Springer, Heidelberg (2004). https://doi.org/10.1007/978-3-540-25970-1_19
37. Sandmann, C., Habel, A.: Rule-based graph repair. In: Echahed, R., Plump, D. (eds.) Proceedings Tenth International Workshop on Graph Computation Models, GCM@STAF 2019, Eindhoven, The Netherlands, 17 July 2019. EPTCS, vol. 309, pp. 87–104 (2019). https://doi.org/10.4204/EPTCS.309.5
38. Schmidt, D.C.: Guest editor's introduction: model-driven engineering. Computer **39**(2), 25–31 (2006). https://doi.org/10.1109/MC.2006.58
39. Schneider, S., Lambers, L., Orejas, F.: A logic-based incremental approach to graph repair featuring delta preservation. Int. J. Softw. Tools Technol. Transf. **23**(3), 369–410 (2021). https://doi.org/10.1007/s10009-020-00584-x
40. Schneider, S., Maximova, M., Giese, H.: Invariant analysis for multi-agent graph transformation systems using k-induction. In: Behr, N., Strüber, D. (eds.) ICGT 2022. LNCS, vol. 13349, pp. 173–192. Springer, Cham (2022). https://doi.org/10.1007/978-3-031-09843-7_10
41. Seada, H., Deb, K.: A unified evolutionary optimization procedure for single, multiple, and many objectives. IEEE Trans. Evol. Comput. **20**(3), 358–369 (2016). https://doi.org/10.1109/TEVC.2015.2459718
42. Sudholt, D.: How crossover speeds up building block assembly in genetic algorithms. Evol. Comput. **25**(2), 237–274 (2017). https://doi.org/10.1162/EVCO_a_00171
43. Thölke, H., Kosiol, J.: A multiplicity-preserving crossover operator on graphs. In: Kühn, T., Sousa, V. (eds.) Proceedings of the 25th International Conference on Model Driven Engineering Languages and Systems: Companion Proceedings, MODELS 2022, Montreal, Quebec, Canada, 23–28 October 2022, pp. 588–597. ACM (2022). https://doi.org/10.1145/3550356.3561587

44. Aalst, W.M.P.: Business process management demystified: a tutorial on models, systems and standards for workflow management. In: Desel, J., Reisig, W., Rozenberg, G. (eds.) ACPN 2003. LNCS, vol. 3098, pp. 1–65. Springer, Heidelberg (2004). https://doi.org/10.1007/978-3-540-27755-2_1
45. Vanderfeesten, I., Cardoso, J., Mendling, J., Reijers, H.A., Van der Aalst, W.: Quality metrics for business process models. In: BPM and Workflow Handbook, vol. 144, no. 2007, pp. 179–190 (2007)
46. Vergidis, K., Saxena, D.K., Tiwari, A.: An evolutionary multi-objective framework for business process optimisation. Appl. Soft Comput. **12**(8), 2638–2653 (2012). https://doi.org/10.1016/j.asoc.2012.04.009
47. Vergidis, K., Tiwari, A., Majeed, B.: Business process analysis and optimization: beyond reengineering. IEEE Trans. Syst. Man Cybern. Part C **38**(1), 69–82 (2008). https://doi.org/10.1109/TSMCC.2007.905812
48. Zschaler, S., Mandow, L.: Towards model-based optimisation: using domain knowledge explicitly. In: Milazzo, P., Varró, D., Wimmer, M. (eds.) STAF 2016. LNCS, vol. 9946, pp. 317–329. Springer, Cham (2016). https://doi.org/10.1007/978-3-319-50230-4_24

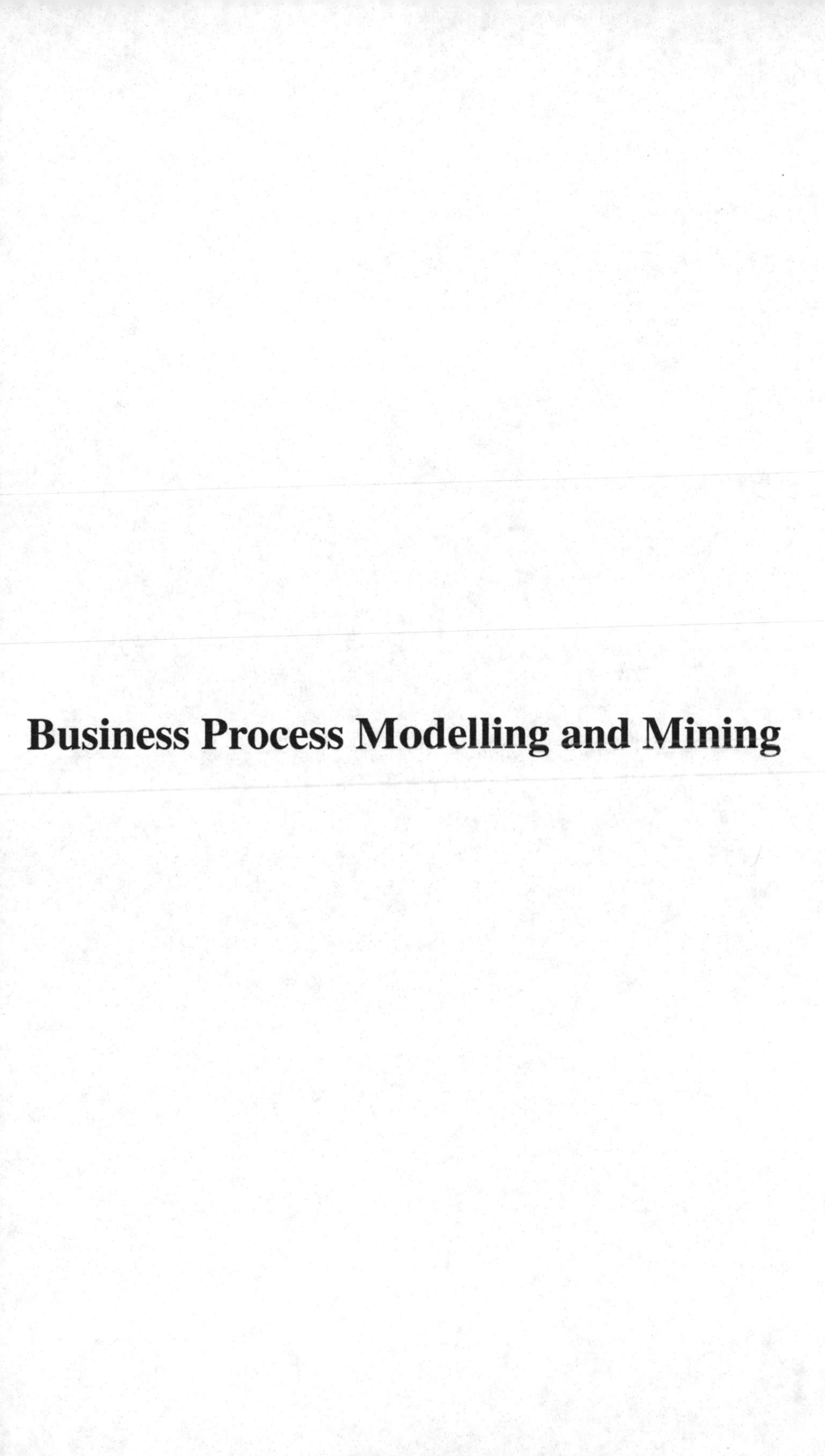

Business Process Modelling and Mining

Concurrent Context-Free Grammar for Parsing Business Processes with Iterated Shuffles

Akio Watanabe(✉), Ayumi Araragi, Hiroki Ikeuchi, and Yousuke Takahashi

NTT Network Service Systems Laboratories, NTT Corporation, Tokyo, Japan
{akio.watanabe,ayumi.araragi,hiroki.ikeuchi,yousuke.takahashi}@ntt.com

Abstract. Trace parsing, a technique for obtaining the correspondence between a trace, which is a string of activities, and a process model, forms the basis of process mining. Conventional trace parsing methods have not considered process models with concurrent loops (CLOOPs), in which a certain sub-process is executed in indefinite numbers of concurrent executions, despite their importance. In this paper, we propose a new formal grammar, a Concurrent Context-Free Grammar (CCFG), which can handle CLOOPs. CCFGs are a generalization of context-free grammars, which can handle concurrent strings allowing parallelism among strings. This simple generalization adds a new operator corresponding to a CLOOP to process trees, a type of process model, and greatly extends the representational capability of process trees. This paper also introduces a trace parsing method for a CCFG. This allows the CCFG to verify whether a trace can be derived from the process model, and the CCFG can be used for conformance checking.

Keywords: Conformance checking · Formal grammar · Iterated shuffle

1 Introduction

Conformance checking is an important task that forms the basis of process mining. Given a process model that represents a business process and an event log that records the execution trace of the business, conformance checking compares the process model and the event log to measure the amount of deviation. By finding traces that do not follow the process model, an audit can be performed to measure the appropriateness of the business. Moreover, if the conformance is quantified, the validity of the process model obtained by process discovery becomes clear. In fact, there is a method for discovering process models with optimal conformance in the process discovery method [1]. Conformance checking is not only a useful stand-alone task, but also the basis for other tasks in process mining.

Conformance checking starts with understanding whether the process model corresponds to the trace, which is the execution history of the business in each case. Correspondence means that the trace is included in the language of the process model. We call the task of understanding correspondence as *trace parsing*.

L. M. Kristensen and J. M. van der Werf (Eds.): PETRI NETS 2024, LNCS 14628, pp. 39–60, 2024.
https://doi.org/10.1007/978-3-031-61433-0_3

If the process model is a Petri net, this task equals to the reachability problem to the final markings [2], and if the model is a process tree, it equals to the membership problem (of a formal language [3] or an automaton [4]). By checking whether a trace corresponds to a process model, the conformance of the event log to the process model can be quantified. Almost all recent conformance checking methods are implemented according to the language of the process model (or the probability of each trace in the language of the process model). Therefore, conformance checking is basically realized by trace parsing.

However, current trace parsing methods cannot be applied to arbitrary business processes. An example of a business process that cannot be handled by existing trace parsing methods is a language called MIX [7]. MIX consists of three activities, which can be executed in any order. However, the number of executions of the three activities is always equal in the traces included in MIX;

$$\mathrm{MIX} = \{w \in \{a, b, c\}^*; |w|_a = |w|_b = |w|_c\}.$$

MIX cannot even be represented in a regular process tree because it is not included in the regular language class. On the other hand, Petri nets can represent MIX as shown in Fig. 1. However, this Petri net is not bounded because an unbounded number of tokens can be added to a place p_1 by an unobserved transition t_1. The unbounded Petri net is quite difficult to solve the reachability problem to the final marking.

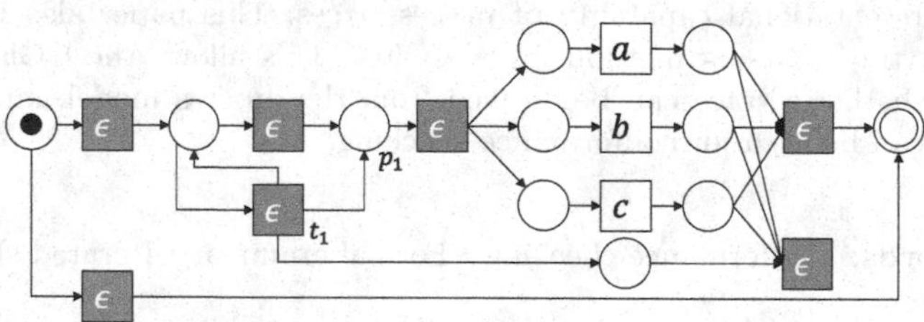

Fig. 1. Petri net representing MIX.

MIX is a structure that appears in many business processes. Suppose a hamburger store offers three dishes (hamburger, fries, and drink) to each customer. However, the order in which the dishes are prepared is irrelevant. In this case, the total number of dishes is not fixed because it depends on the customer's order, but the number of dishes made is always the same. Therefore, if a, b, and c are tasks to prepare each dish, the business process of this store is represented by MIX. Since business processes are often executed according to the number of customer orders, structures such as MIX that execute an indefinite number of specific tasks in parallel are frequently used.

We call an indefinite number of concurrent structures such as MIX a concurrent LOOP (CLOOP). A CLOOP is expressed in formal languages by an operator called *iterated shuffle*, but it is very difficult to handle because the iterated shuffle operation is not closed to regular languages [8].

This paper proposes a novel process model, Concurrent Context-Free Grammar (CCFG), which can represent business processes with an indefinite number of concurrent structures. The CCFG is a novel formal grammar that generates strings with concurrent structures represented by tuples called concurrent strings. CCFGs define the languages of traces by deriving concurrent strings by production rules and serializing them to represent the mapping from concurrent strings to traces. CCFGs can be easily obtained from process trees that has been extended to have CLOOPs. Thus, combined with any process discovery method that acquires process trees, conformance checking of process trees with CLOOPs can be performed [1,9].

We also propose an algorithm for trace parsing that determines whether any given trace is in the language of a CCFG. This method extends the parsing algorithm so that it can be applied to production rules with concurrency.

This paper is organized as follows. Section 2 refers to related studies on trace parsing. Section 3 describes the preliminaries of our study. Section 4 defines the process tree with CLOOP added. Section 5 describes concurrent strings and a CCFG, the subject of this paper. Section 6 describes the trace parsing algorithm for CCFGs. Section 7 presents an example of trace parsing with a CCFG. Section 8 presents conclusions and future work.

2 Related Work

A CLOOP is treated as an iterated shuffle in formal languages. This section describes how iterated shuffles are currently handled in Petri nets, process trees, and formal grammars, which are process models often used in trace parsing. Although declarative process models are widely included in trace parsing, they are outside the scope of this paper because their language is quite different from others and we have not found any studies that discuss iterated shuffle.

Adriansyah et al. proposed a method for measuring alignment distance by creating Petri nets that simultaneously represent transitions in the trace and transitions in the process model, and determining the reachability of the final marking on the Petri net [10]. Leemans et al. proposed a method to search for runs on Petri nets with ϵ transitions for stochastic labeled Petri nets [11]. Although these methods simultaneously measure the distance between the trace and the process model as well as trace parsing, the token-based approach is faster if one only wants to know the reachability of the end marking on the Petri net. Rozinat et al.'s token-based method finds the run to a final marking by moving a token along the trace on the Petri net. Unfortunately, these methods assume that the Petri net is bounded, i.e., the number of markings is finite. As we have already seen in the case of MIX, Petri nets are almost always not bounded when there are iterated shuffles. Therefore, these methods do not guarantee that a trace parsing can be performed on a process model with iterated shuffles. In fact, it is not easy to determine how many times to fire the transition t_0 when moving tokens on a Petri net with MIX in Fig. 1, and no explicit realization method has yet been proposed.

Recently, a polynomial-time trace parsing method for process trees has been proposed by Rocha et al. [4]. This method assumes that the language of the target process tree is included in the regular language. Watanabe et al. proposed PGPM, which replaces process trees with probabilistic context-free grammars (CFGs). In PGPM, the subtrees under the PARALLEL operator of the process tree are converted to a regular grammar (RG). This conversion procedure to RL is similar to the location automata described by Broda et al. [12]. Broda et al. proposed the concept of location, which represents the state of analysis of concurrent elements, to handle shuffle operators in regular expressions. However, although both methods deal with RGs or CFGs, as will be shown later, most languages with iterated shuffle cannot be expressed in CFGs. Thus, business processes with iterated shuffle cannot be handled by the usual process tree-based approach.

A technique that may be able to handle iterated shuffle is the Multiple CFG (MCFG) [13]. The m-MCFG generalizes the CFG so that a non-terminal symbol is a function whose input is at most m strings. For example, it is known that MIX can be represented by 2-MCFG [7]. However, even for MIX, which is relatively simple, the grammar of the MCFG becomes extremely complicated. For this reason, there is no known method for converting Petri nets and process trees to MCFGs.

In conclusion, business processes with iterated shuffles, i.e., CLOOPs, cannot be properly parsed using existing methods.

3 Preliminary

We first define the notation of strings used in this paper. Given some symbol set Σ, $w \in \Sigma^*$ is called a string or a serial string. In particular, given an activity set $\mathcal{A}$, a string $\sigma \in \mathcal{A}^*$ is called a *trace*. The σ_l denotes the l-th symbol of σ. $\sigma_{:l}$ denotes the l-th substring of σ and $\sigma_{l:}$ denotes the l-th symbol or later substrings of σ. In this paper, concatenation of strings is simply written as $\sigma_{:l-1}\sigma_{l:} = \sigma$. However, the concatenation of two sets U, V is expressed as $U \cdot V = \{uv | u \in U, v \in V\}$ using $\cdot$. Kleene star $*$ is treated in the same way as in common formal languages for both strings and sets.

3.1 Shuffle and Iterated Shuffle

Shuffle is an operation on two sets of strings, and iterated shuffle is an operation on one set of strings. In this paper, we use the same definitions and notations of shuffle and iterated shuffle as Jantzen et al. [8].

Definition 1 (shuffle $\sqcup\!\sqcup$). $U \sqcup\!\sqcup V := \{w | w = u_1v_1u_2v_2 \dots u_nv_n, u_1u_2 \dots u_n \in U, v_1v_2 \dots v_n \in V\}$.

Example 1. $\{ab\} \sqcup\!\sqcup \{c, de\} = \{abc, acb, cab, abde, dabe, adbe, deab, daeb, adeb\}$.

Definition 2 (Iterated shuffle $\otimes, \oplus$). $V^{\otimes} := \bigcup_{i \geq 0} V_i$, $V^{\oplus} := \bigcup_{i \geq 1} V_i$, *where* $V_0 = \{\epsilon\}, V_{i+1} = V_i \sqcup\!\sqcup V$.

Example 2. $\{abc, abcabc, aabbcc, aabcbc, aaabcbcbc, abababababcccc\} \subset \{abc\}^{\oplus}$.

3.2 Process Tree

In this study, a process tree, an excellent process model that guarantees soundness, is used for trace parsing [14]. The process tree is defined recursively as follows [5];

1. Let $\mathcal{A}$ be the set of activities. If $a \in \mathcal{A} \cup \{\epsilon\}$, then a is a process tree.
2. If $Q_1, \ldots, Q_n$ $(n \geq 1)$ is a process tree and o belongs to $\{\rightarrow, \times, \circlearrowleft, +\}$, then $o(Q_1, \ldots, Q_n)$ is also a process tree.

For example, when $\mathcal{A} = \{a, b, c, d, e\}$, $Q'_1 = a$, $Q'_2 = \times(a, b, c)$, $Q'_3 = \rightarrow (a, \circlearrowleft (\times(b, c, \epsilon), d), +(\rightarrow (e, f), g))$ are all process trees.

Language of Process Trees. Let $\mathcal{L}(Q) \subseteq \mathcal{A}^*$ denote the language of a process tree Q. The language of the process tree is defined as follows;

- $\mathcal{L}(a) := \{a\}$ $(a \in \mathcal{A} \cup \{\epsilon\})$,
- $\mathcal{L}(\rightarrow (Q_1, \ldots, Q_n)) := \mathcal{L}(Q_1) \cdot \cdots \cdot \mathcal{L}(Q_n)$,
- $\mathcal{L}(\times(Q_1, \ldots, Q_n)) := \mathcal{L}(Q_1) \cup \cdots \cup \mathcal{L}(Q_n) = \bigcup_{i=1}^{n} \mathcal{L}(Q_i)$,
- $\mathcal{L}(\circlearrowleft (Q_1, \ldots, Q_n)) := (\mathcal{L}(Q_1) \cdot \bigcup_{i=2}^{n} \mathcal{L}(Q_i))^* \cdot \mathcal{L}(Q_1) = \mathcal{L}(Q_1) \cdot (\bigcup_{i=2}^{n} \mathcal{L}(Q_i) \cdot \mathcal{L}(Q_1))^*$,
- $\mathcal{L}(+(Q_1, \ldots, Q_n)) := \mathcal{L}(Q_1) \sqcup\!\sqcup \cdots \sqcup\!\sqcup \mathcal{L}(Q_n)$.

$\mathcal{L}(Q)$ indicates the set of strings that can be derived from the process tree. For example, $\mathcal{L}(Q'_1) = \{a\}$, $\mathcal{L}(Q'_2) = \{a, b, c\}$, $\mathcal{L}(Q'_3) = \{abdea, aaed, abcbeda, \ldots\}$.

Since the language of the process tree consists of a concatenation $(\cdot)$, a union set $(\cup)$, and a Kleene closure $(*)$, and since shuffle operators can be converted to regular expressions [12], the language class of the process tree (PTL) is a subset of the regular language class (RL) (PTL $\subseteq$ RL). Conversely, given a regular expression, we can construct a process tree with equal language from any regular expression by mapping its concatenation to $\rightarrow$, its union set to $\times$, and its Kleene closure to $\circlearrowleft$ (RL $\subseteq$ PTL.) Thus, the process tree and the regular expression are equal (PTL = RL.)

The trace parsing task of this paper is to verify whether a trace is included in the language of the process tree, given a trace and a process tree as input.

3.3 Iterated Shuffle Cannot Be Expressed by Any (Common) Process Tree

Most languages containing iterated shuffle do not belong to the context-free language class (CFL). As a typical example, $(\{abc\})^{\oplus}$ and MIX do not belong to the CFL. This can be shown from the pumping lemma for context-free languages [15].

Lemma 1 (Pumping lemma for context-free language). *If L is a context-free language, there exists some pumping length $p \geq 1$, any trace of length p or more in L can be written as $\sigma = uvxyz$, where u, v, x, y, z satisfy the following conditions;.*

$$|vy| \geq 1, \quad |vxy| \leq p, \quad \forall i \geq 0, uv^i xy^i z \in L.$$

Using the pumping lemma, it can be proven that languages containing iterated shuffle are not in the CFL.

Lemma 2. *Neither $L_1 = (\{abc\})^{\oplus}$ nor MIX is in the CFL.*

Proof. Assume that L_1 and MIX are both context-free languages. Then, for each of L_1 and MIX, there exists a certain number p satisfying the condition of Lemma 1. Here, considering the trace $\sigma_1 = a^p b^p c^p$, $\sigma_1 \in L_1$ and $\sigma_1 \in$ MIX. However, considering the splitting of σ_1 into $uvxyz$, $|vy| \geq 1$ and $|vxy| \leq p$, vxy is always composed of two or less characters among a, b, and c. Therefore, uv^ixy^iz is $uv^ixy^iz \notin L_1$ and $uv^ixy^iz \notin$ MIX because the numbers of a, b, and c are not equal when $i \geq 2$. This inconsistent with Lemma 1. Thus, neither L_1 nor MIX is in the CFL.

Since the RL is a true subset of the CFL, L_1 and MIX are not included in the RL either. In other words, the process tree can represent neither L_1 nor MIX, and lacks the capability to represent actual business processes.

4 Process Tree with Concurrent Loop Operator

In Sect. 3.2, we defined a common process tree. Our interest is to handle indefinite concurrent structures, i.e., iterated shuffles, in the process tree. To treat iterated shuffles in process trees, we define a new operator called concurrent loop (CLOOP). CLOOPs are denoted by $\downdownarrows$ in the process tree.

To add CLOOPs, modify the process tree definition as follows;

1. Let $\mathcal{A}$ be the set of activities. If $a \in \mathcal{A} \cup \{\epsilon\}$, then a is a process tree.
2. If $Q_1, \ldots, Q_n$ $(n \geq 1)$ is a process tree and o belongs to $\{\rightarrow, \times, \circlearrowleft, + \downdownarrows\}$, then $o(Q_1, \ldots, Q_n)$ is also a process tree.

Also, add the following to the definition of language in the process tree;

- $\mathcal{L}(\downdownarrows (Q_1, \ldots, Q_n)) := (\mathcal{L}(Q_1) \cdot \cdots \cdot \mathcal{L}(Q_n))^{\oplus}$.

In the following, the process tree is defined as the one with the above extensions.

Using a CLOOP, $\downdownarrows (a, b, c)$ represents a process tree with L_1 as the language. Also, MIX is equivalent to the language of the process tree $\times(\downdownarrows (+(a, b, c)), \epsilon)$.

In the definition of a CLOOP in this paper, it is assumed that the process is always executed at least once, and $\oplus$ is used. Zero iterations would be also allowed by defining a CLOOP with $\otimes$ instead of $\oplus$. Actually, since $\otimes$ can be expressed by $\times(\epsilon, \downdownarrows (\cdot))$, there is no difference in expressive capability.

5 Concurrent Language

In Sect. 4, we defined a new process tree with CLOOP operators. Our goal is to realize trace parsing using this process tree. In this section, we define concurrent strings as a new system of strings with the addition of tuple, which expresses the concurrency of strings, so that iterated shuffle and shuffle can be handled. We then define a formal grammar, CCFG, from which concurrent strings can be derived, and describe how to obtain normal serial strings from concurrent strings.

5.1 Concurrent String

The set of $d(\geq 0)$ tuples in a set V of symbols is denoted by $V^{\times d}$. In this paper, tuples are denoted by brackets. E.g., when $V_0 = \{a, b\}$, $V_0^{\times 2} = \{(a, a), (a, b), (b, a), (b, b)\}$. Note that $V^{\times d}$ and V^d are different, as $V_0^2 = \{aa, ab, ba, bb\}$.

Let $V^{*\times*} := \bigcup_{d=0}^{\infty}(V^*)^{\times d}$. $V^{*\times*}$ represents a set of tuples consisting of an arbitrary number of strings. E.g., $\{(), (a), (aa), (a, b), (ab), (ab, baa), (bb, a, \epsilon, a)\} \subseteq V_0^{*\times*}$. Furthermore, we can consider $(V^{*\times*})^{*\times*}$, a set of tuples whose elements are strings containing an arbitrary number of tuples. E.g., $\{(), (aa(b, a)), (aa(a, b), b), (a, b, a, ab), (a(\epsilon, b), b, b)\} \subseteq (V_0^{*\times*})^{*\times*}$.

Repeating this, $V^{(0)} := V$ and $V^{(i)} := (V^{(i-1)})^{*\times*} (i > 0)$ can be defined. Then, define $V^{\oslash} := (V^{(\infty)} \cup V)^*$. Since $V^{(\infty)}$ is a set of arbitrary tuples, $V^{\oslash}$ is a set of strings composed of any tuples and strings. The elements of $V^{\oslash}$ are called *concurrent strings* over V. The reason why we call it a concurrent string is that two tuples (a, b) and (b, a) are equivalent in the serialization described below in Sect. 5.4, and the tuple elements can be considered to be unordered.

5.2 Concurrent Context-Free Grammar

Here we extend the common CFG to define a concurrent context-free grammar (CCFG), which can represent a set of concurrent strings as a concurrent language. A CCFG G is defined as (M, E, R, S). M, E, and S are the same as in common CFGs, and are the set of non-terminal symbols, the set of terminal symbols, and the starting symbol ($S \in M$), respectively. However, R in a common CFG is a set of production rules with the symbol sequence $(M \cup E)^*$ on the right side, whereas R in a CCFG is a set of concurrent production rules from which concurrent strings are derived. In CCFGs, $R \subseteq M \times (M \cup E)^{\oslash}$, and a concurrent production rule $r \in R$ is represented using "$\Rightarrow$" as $A \Rightarrow \alpha$ ($A \in M, \alpha \in (M \cup E)^{\oslash}$). That is, the right side of $r \in R$ is a concurrent string that can contain tuples.

Example 3. $G_1 = (M_1, E_1, R_1, S_1) = (\{S, A, B\}, \{a, b\}, \{S \Rightarrow (A, B), A \Rightarrow (A, a), A \Rightarrow a, B \Rightarrow bB, B \Rightarrow \epsilon\}, S)$ is a CCFG.

5.3 Concurrent Language by CCFG

If $A \Rightarrow \beta \in R$, there is a concurrent string w in any $(M \cup E)^{\oslash}$ where A is an element and a concurrent string w' with A in w replaced by $beta$, we say that w directly derives w' and denoted by $w \Longrightarrow w'$. If w'' is obtained by repeated direct derivation from w, we say that w derives w'' and denoted by $w \Longrightarrow^* w''$.

Given a CCFG $G = (M, E, R, S)$, the set of all concurrent strings on E which can be derived from S is called the *concurrent language* of G and denoted by $C(G)$.

$$C(G) := \{c \in E^{\oslash} | S \Longrightarrow^* c\}$$

A CCFG is a formal grammar that can define a concurrent language.

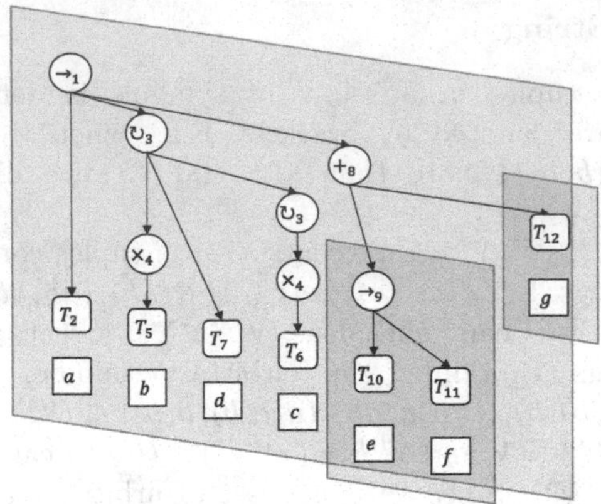

Fig. 2. Derivation example of $abdc(ef,g)$ on CCFG

Figure 2 illustrates the derivation process of a concurrent string $abdc(ef,g)$ by the CCFG. The derivation process of strings in a normal CFG is represented by a tree structure called a syntax tree. The derivation process of a CCFG can also be represented by a syntax tree, but in a CCFG, strings in tuples have no order relation to each other. Therefore, the branches corresponding to tuples are aligned horizontally on the syntax tree as shown in Fig. 2. The serialization described in the next section gives an order to each symbol in a concurrent string and converts it into a serial string. In other words, a CCFG generates a trace by deriving a concurrent string and assigning an order to the concurrent string.

5.4 Serialization

We have now defined concurrent strings and described a CCFG, a formal grammar for concurrent strings. However, our goal is not to represent a concurrent language, but to obtain a language of serial strings (without tuples). Therefore, in this section, we define the *serialization* process to obtain a set of serial strings from a concurrent string. By serialization, a CCFG can represent a language of serial strings. Especially, the key point of the proposal is that the serialization is defined as shuffling strings in a tuple. This allows CCFGs to represent languages with shuffle and iterated shuffle.

Definition 3 (Serialization $s(c)$). *The function $s : \mathcal{A}^{\oslash} \to 2^{\mathcal{A}^*}$ that obtains a set of serial strings from a concurrent string $c \in \mathcal{A}^{\oslash}$ is called a serialization function and is defined recursively as follows;*

- *If c is a string containing no tuples, then $s(c) := \{c\}$.*
- *If c is a tuple $c = ()$ with zero elements, then $s(c) := \{\epsilon\}$.*
- *If c is a tuple with one element denoted by $c = (c_1)$, then $s(c) := s(c_1)$.*
- *If c is a tuple represented as $c = (c_1, \ldots, c_n)$ $(n \geq 2)$, then $s(c) := s(c_1) \sqcup\!\sqcup \cdots \sqcup\!\sqcup\, s(c_n)$.*

– *Otherwise (c is a string containing a tuple), then c is represented as* $c = c_1 t_1 c_2 t_2 \ldots c_n t_n (n \geq 0)$, *where* $c_1, \ldots, c_n$ *is a non-tuple concurrent string and* $t_1, \ldots, t_n$ *is a tuple, as in* $s(c) := s(c_1) \cdot s(t_1) \cdot \cdots \cdot s(c_n) \cdot s(t_n)$.

Serialization is a function that yields the set of all serial strings, i.e., the language, obtained by ordering all concurrent elements in a concurrent string.

Example 4. The serialization for the concurrent string $a(bdc)(e, f, e)()(g, (h, i))$ is shown below.

$$
\begin{aligned}
& s(\ a(bdc)(e, f, e)()(g, (h, i))\ ) \\
&= s(a) \cdot s((bdc)) \cdot s((e, f, e)) \cdot s(()) \cdot s((g, (h, i))) \\
&= \{a\} \cdot s(bdc) \cdot (s(e) \sqcup\!\sqcup s(f) \sqcup\!\sqcup s(e)) \cdot \{\epsilon\} \cdot (s(g) \sqcup\!\sqcup s((h, i))) \\
&= \{a\} \cdot \{bdc\} \cdot \{fee, efe, eef\} \cdot (\{g\} \sqcup\!\sqcup (s(h) \sqcup\!\sqcup s(i))) \\
&= \{abdc\} \cdot \{fee, efe, eef\} \cdot (\{g\} \sqcup\!\sqcup \{hi, ih\}) \\
&= \{abdcfee, abdcefe, abdceef\} \cdot \{ghi, hgi, hig, gih, igh, ihg\} \\
&= \{abdcfeeghi, abdcfeehgi, abdcfeehig, abdcfeegih, abdcfeeigh, abdcfeeihg, \\
&\quad abdcefeghi, abdcefehgi, abdcefehig, abdcefegih, abdcefeigh, abdcefeihg, \\
&\quad abdceefghi, abdceefhgi, abdceefhig, abdceefgih, abdceefigh, abdceefihg\}.
\end{aligned}
$$

In the resulting language string, the e, f, and e contained in one tuple are arranged in random order. Also, g, h, and i are in random order. On the other hand, symbols that are not contained in the same tuple in a concurrent string maintain their order. For example, in the case of bdc and g, g always appears after bdc. The string obtained by serialization is ordered by the elements contained in the same tuple in the concurrent string.

If the concurrent string is a tuple, the serialization function applies the shuffle operator to the concurrent elements of the tuple. Conversely, if the concurrent string is not a tuple, it simply concatenates the strings. Therefore, languages that require shuffle can be expressed by corresponding PARALLEL(+) and CLOOP(⇊) to tuples, which are expressed using the shuffle operator.

Using serialization, we can obtain a set of corresponding strings from any concurrent string. In other words, any concurrent string has a corresponding language. The $s(c)$ corresponding to a concurrent string c is called the *serial language* of c.

A CCFG G has a concurrent language $C(G)$. Moreover, since $c \in C(G)$ has $s(c)$, a CCFG has a set of serial strings from the union set of serial languages. We call $\mathcal{L}(G) := \bigcup_{c \in C(G)} s(c)$ *the serial language* of G, or simply *the language* of G. Trace parsing for G is the task of checking whether $\sigma \in \mathcal{L}(G)$ or $\sigma \notin \mathcal{L}(G)$ for the trace $\sigma \in E^*$. The method of trace parsing for given G and σ is described in Sect. 6.

5.5 Conversion from Process Tree to CCFG

In this paper, we assume that the process model given in trace parsing is a process tree. In this section, we describe how to obtain a CCFG with an equivalent language for a process tree containing CLOOPs.

The procedure for converting a process tree to a CCFG is simple. First, the activity set $\mathcal{A}$ is simply mapped one-to-one to the set of terminal symbols E in the CCFG. Next, any node in the process tree is mapped one-to-one to a non-terminal symbol. Let $\mathcal{V}$ be the set of nodes in the process tree. A node in $\mathcal{V}$ can be either an operator or an activity. The operator of each $v \in \mathcal{V}$ is $o(v) : \mathcal{V} \to \{\to, \times, \circlearrowleft, +, \downdownarrows, T\}$, where $o(v) = T$ if v corresponds to an activity. When $o(v) = T$, the corresponding activity is obtained by l_v ($l_v = \epsilon$ if the activity is ϵ). The non-terminal symbol set is $M = \{o(v)_v | v \in \mathcal{V}\}$. When v_0 is a root node, the corresponding non-terminal symbol $o(v_0)_{v0}$ is the initial symbol S.

Table 1. Concurrent production rules to be added to R for each $o(v)$ operator

$o(v)$	Concurrent production rules	$o(v)$	Concurrent production rules
T	$T_v \Rightarrow l_v$	$\circlearrowleft$	$\circlearrowleft_v \Rightarrow C_{v(1)}$
$\to$	$\to_v \Rightarrow C_{v(1)} \cdots C_{v(\|v\|)}$		$\circlearrowleft_v \Rightarrow C_{v(1)} C_{v(2)} \circlearrowleft_v$
$\times$	$\times_v \Rightarrow C_{v(1)}$		$\cdots$
	$\cdots$		$\circlearrowleft_v \Rightarrow C_{v(1)} C_{v(\|v\|)} \circlearrowleft_v$
	$\times_v \Rightarrow C_{v(\|v\|)}$	$\downdownarrows$	$\downdownarrows_v \Rightarrow (C_{v(1)} \ldots C_{v(\|v\|)})$
$+$	$+_v \Rightarrow (C_{v(1)}, \ldots, C_{v(\|v\|)})$		$\downdownarrows_v \Rightarrow (C_{v(1)} \cdots C_{v(\|v\|)}, \downdownarrows_v)$

A set of concurrent production rules R can be constructed by combining the non-terminal symbols of each node $v \in \mathcal{V}$ of the process tree and the non-terminal symbols of the child nodes of v. Let $|v|$ be the number of child nodes of node v, $v(i)$ be the i-th child node of v, and $C_{v(i)}$ $(1 \leq i \leq |v|)$ be the corresponding non-terminal symbol. Then, for every $v \in \mathcal{V}$, the production rules shown in Table 1 are added to R, depending on the operator $o(v)$.

Theorem 1. *The CCFG G_Q, which obtained the concurrent production rules according to Table 1, has the same language as the original process tree Q, i.e. $\mathcal{L}(G_Q) = \mathcal{L}(Q)$.*

Proof. Let Q_v be a process tree with Q node v as the root node. When $o(v) = T$, obviously $\mathcal{L}(G_{Qv}) = \mathcal{L}(Q_v) = \{l_v\}$.

In this time, at least one node $v' \in \mathcal{V}$ of Q, $o(j) = T$ at every child node $j \in \{v(1), \ldots v(|v|)\}$, i.e. $\mathcal{L}(G_{Qj}) = \mathcal{L}(Q_j)$. If it can be proved that $\mathcal{L}(G_{Qv'}) = \mathcal{L}(Q_{v'})$ for this node v', the same can be proved for nodes with v' as a child node, and recursively $\mathcal{L}(G_{Qv}) = \mathcal{L}(Q_v)$ for all nodes $v \in \mathcal{V}$. That is, $\mathcal{L}(G_Q) = \mathcal{L}(Q)$ for Q. In the following, we will prove that $\mathcal{L}(G_{Qv}) = \mathcal{L}(Q_v)$ when $\forall j \in \{v(1), \ldots v(|v|)\}, \mathcal{L}(G_{Qj}) = \mathcal{L}(Q_j)$ in each case where $o(v)$ is $\to$, $\times$, $\circlearrowleft$, $+$, and $\downdownarrows$.

In case $o(v) = \rightarrow$: The only concurrent production rule applicable to the initial symbol $\rightarrow_v$ is $\rightarrow_v \Rightarrow C_1 \ldots C_{|v|}$. Since the string derived from each $C_i (1 \geq i \geq |v|)$ is $\mathcal{L}(G_{Qv(i)})$, thus $\mathcal{L}(G_{Qv}) = \mathcal{L}(G_{Qv(1)}) \cdot \cdots \cdot \mathcal{L}(G_{Qv(|v|)}) = \mathcal{L}(Q_{v(1)}) \cdot \cdots \cdot \mathcal{L}(Q_{v(|v|)}) = \mathcal{L}(Q_v)$.

In case $o(v) = \times$: Since the symbol sequence directly derived from the initial symbol $\times_v$ is one of $C_1, \ldots, C_n$, thus $\mathcal{L}(G_{Qv}) = \mathcal{L}(G_{Qv(1)}) \cup \cdots \cup \mathcal{L}(G_{Qv(|v|)}) = \mathcal{L}(Q_{v(1)}) \cup \cdots \cup \mathcal{L}(Q_{v(|v|)}) = \mathcal{L}(Q_v)$.

In case $o(v) = \circlearrowleft$: The symbol sequence directly derived from the initial symbol $\circlearrowleft_v$ is either C_1 or $C_1 C_i \circlearrowleft_v$ $(2 \leq i \leq |v|)$. Even the latter case, the symbol sequence derived from $\circlearrowleft_v$ is the same. Therefore, the symbol sequences derived from $\circlearrowleft_v$ are expressed as $(\{C_1\} \cdot \bigcup_{i=2}^{|v|} \{C_i\})^* \cdot \{C_1\}$. Thus, $\mathcal{L}(G_{Qv}) = (\mathcal{L}(G_{Qv(1)}) \cdot \bigcup_{i=2}^{|v|} \mathcal{L}(G_{Qv(i)}))^* \cdot \mathcal{L}(G_{Qv(1)}) = (\mathcal{L}(Q_{v(1)}) \cdot \bigcup_{i=2}^{|v|} \mathcal{L}(Q_{v(i)}))^* \cdot \mathcal{L}(Q_{v(1)}) = \mathcal{L}(Q_v)$.

In case $o(v) = +$: First, consider the concurrent language $C(G_{Qv})$ of G_{Qv}. The only concurrent production rule applicable to the initial symbol $+_v$ is $+_v \Rightarrow (C_1, \ldots, C_{|v|})$. Let $c_i \in C(G_{Qv(i)})$ be any concurrent string derived from each $C_i (1 \geq i \geq |v|)$, then

$$C(G_{Qv}) = \{(c_1, \ldots, c_{|v|}) | c_1 \in C(G_{Qv(1)}), \ldots, c_{|v|} \in C(G_{Qv(|v|)})\}.$$

At this time, since $\mathcal{L}(G_{Qv}) = \bigcup_{c \in C(G_{Qv})} s(c)$, thus

$$\begin{aligned}
\mathcal{L}(G_{Qv}) &= \bigcup_{(c_1,\ldots,c_{|v|}) \in C(G_{Qv})} s((c_1, \ldots, c_{|v|})) \\
&= \bigcup_{(c_1,\ldots,c_{|v|}) \in C(G_{Qv})} s(c_1) \sqcup\!\sqcup \cdots \sqcup\!\sqcup s(c_{|v|}) \\
&= (\bigcup_{c_1 \in C(G_{Qv(1)})} s(c_1)) \sqcup\!\sqcup \cdots \sqcup\!\sqcup (\bigcup_{c_1 \in C(G_{Qv(|v|)})} s(c_{|v|})) \\
&= \mathcal{L}(G_{Qv(1)}) \sqcup\!\sqcup \cdots \sqcup\!\sqcup \mathcal{L}(G_{Qv(|v|)}) \\
&= \mathcal{L}(Q_{v(1)}) \sqcup\!\sqcup \cdots \sqcup\!\sqcup \mathcal{L}(Q_{v(|v|)}) = \mathcal{L}(Q_v).
\end{aligned}$$

In case $o(v) = \Downarrow$: Let $\alpha = C_1 \ldots C_{|v|}$. Then the concurrent strings directly derived from the initial symbol $\Downarrow_v$ are either (α) or $(\alpha, \Downarrow_v)$. Even the latter case, the concurrent strings derived from $\Downarrow_v$ are the same. Thus, the set of strings derived from $\Downarrow_v$ can be expressed as an infinite sequence such as (α), $(\alpha, (\alpha))$, $(\alpha, (\alpha, (\alpha)))$,... where $(\alpha, \bullet)$ is added one by one to the outside. Among the concurrent strings derived from $\Downarrow_v$, let γ_i denote the concurrent string that contains α for i times. $\mathcal{L}(G_{Qv})$ is $\bigcup_{i \geq 1} s(\gamma_i)$. Also, $s(\gamma_1) = s(\alpha) = s(C_1 \ldots C_{|v|}) = s(C_1) \cdot \cdots \cdot s(C_{|v|}) = \mathcal{L}(G_{Qv(1)}) \cdot \cdots \cdot \mathcal{L}(G_{Qv(|v|)}) = \mathcal{L}(Q_{v(1)}) \cdot \cdots \cdot \mathcal{L}(Q_{v(|v|)})$. Moreover, since $\gamma_i = (\alpha, \gamma_{i-1})$ for any $i > 1$, $s(\gamma_i) = s(\alpha) \sqcup\!\sqcup s(\gamma_{i-1})$. Therefore, the following holds;

$$\begin{aligned}
\mathcal{L}(G_{Qv}) &= \bigcup_{i \geq 1} s(\gamma_i) = s(\alpha) \cup (s(\alpha) \sqcup\!\sqcup s(\alpha)) \cup (s(\alpha) \sqcup\!\sqcup s(\alpha) \sqcup\!\sqcup s(\alpha)) \cup \ldots \\
&= (\mathcal{L}(Q_{v(1)}) \cdot \cdots \cdot \mathcal{L}(Q_{v(|v|)}))^{\oplus} = \mathcal{L}(Q_v).
\end{aligned}$$

6 Trace Parsing by CCFG

The definition of the CCFG and the method of obtaining a CCFG were described in Sect. 5. The purpose of this study is trace parsing, i.e., to determine whether a given trace σ can be derived from a CCFG G. We propose a parsing method, Top-Down Trace Parsing (TDTP), which searches for the derivation process of σ in G.

The TDTP searches for a string that can be derived from the initial symbol and stores the parsed state in a memory called a *chart*. This is similar to top-down left-to-right chart parsing, but with three modifications. First, the definition of the left-most symbol has been modified. In CFG, the leftmost symbol in a string is simply called the leftmost symbol. In contrast, in CCFGs, multiple non-terminal symbols can be left-most because some elements are not ordered by tuples. The second modification is that the TDTP preacquires the left-most terminal symbol for any given symbol. This modification is not new, but is already defined in LR parsing as what is called the *first set* [16]. Third, a chart stores the entire tree under analysis as a single state. In ordinary chart parsing, the elements that correspond to the nodes of the tree called items are stored in the chart. However, in CCFGs, there can be multiple left-most non-terminal symbols in a parsing tree, so it is necessary to store the entire tree in the chart in order to know which is the left-most symbol. This modification requires a large amount of memory because all possible parsing trees must be kept, but it has the advantage that the parsing tree becomes a syntax tree directly at the end of parsing.

Since the TDTP is just an extension of the common parsing method to handle concurrent strings, it is expected to be able to parse CCFGs from arbitrary process trees. In our investigations, we did not find any cases of misjudged trace parsing. However, the validity of the TDTP has not yet been proven, and this is a future work.

The computational efficiency of the algorithm is out of scope of this paper. For example, Tomita's method [17], one of the generalized LR methods, is an efficient algorithm that manages multiple parsing trees together in a single stack and works similarly to the TDTP, but is expected to be more memory and computationally efficient. The ability to handle concurrent strings containing iterated shuffles is the main contribution of the TDTP.

Two constraints should be added to CCFGs provided to the TDTP. First, the TDTP handles only concurrent production rules whose right side is a tuple. For this reason, the concurrent production rule $A \Rightarrow \alpha$ is modified to $A \Rightarrow (\alpha)$. Since $S((\alpha)) = S(\alpha)$, this modification does not affect the languages of CCFGs. Second, the right side of a concurrent production rule is assumed to be in $(\alpha) \in (M \cup E)^{*\times *}$, not $(\alpha) \in (M \cup E)^{\oslash}$, i.e., there is no tuple within a tuple. This constraint is not a problem because a concurrent production rule $A \Rightarrow ((\alpha_1), (\alpha_2))$ can be decomposed into three concurrent production rules $A \Rightarrow (B, C)$, $B \Rightarrow (\alpha_1)$, and $C \Rightarrow (\alpha_2)$. In addition, tuples do not always contain tuples in the concurrent production rules obtained by the method in Sect. 5.5.

6.1 Definition of TDTP Component Concepts

First, we will define concepts related to the TDTP. In the following, $A, B, C, D \in M$ and $\alpha_i, \beta_i, \gamma_i \in (M \cup E)^*$ $(i \in \mathbb{N})$.

Item. Let $q = A \Rightarrow (\alpha_1 \bullet \beta_1, \ldots, \alpha_K \bullet \beta_K)$ be called an *item*. In item, the position of $\bullet$ indicates how far the analysis of the sequence has progressed. Sequences to the left of $\bullet$ are symbols that have already been analyzed, and sequences to the right of $\bullet$ are symbols that are expected to appear in the future.

Parsing Tree. The TDTP uses a parsing tree, which consists of components called nodes, to represent the progress of the parsing. A parsing tree is denoted by $t = [v_1, \ldots, v_{|t|}] = [(q_1, i_1, d_1), \ldots, (q_{|t|}, i_{|t|}, d_{|t|})]$ where $v_k = (q_k, i_k, d_k)$ is a node, i_k is the index of the parent node, and d_k is the index of elements in the tuple of the parent to which v_k belongs in. If a node is the root node, then $i_k = -1$ and $d_k = -1$. When $q_k = A \Rightarrow (\alpha_1 \bullet \beta_1, \ldots, \alpha_K \bullet \beta_K)$, $b_{ki} = \alpha_i \bullet \beta_i$ is called the i-th *branch* of v_k $(i \in 1, \ldots, K)$.

Completion of Nodes and Branches. Suppose that $q_k = A \Rightarrow (\alpha_1 \bullet \beta_1, \ldots, \alpha_K \bullet \beta_K)$ at node $v_k \in t$. A node v_K is said to be *complete* if $\beta_i = \epsilon$ for any $i \in 1, \ldots, K$. If not, the node is *incomplete*.

In the TDTP, when the i-th branch of node v_k is completed, a new parsing tree is created by replacing $b_{ki} = (\alpha_i \bullet B_i\beta_i)$ to $(\alpha_i B_i \bullet \beta_i)$. If v_k is completed as a result, the same process is performed recursively on the parent nodes of v_k. Therefore, all parents of incomplete nodes are maintained as being incomplete in the parsing tree in the chart.

Left-Most Incomplete Branch (lmib). Node v_k is the *left-most incomplete node* if $v_k \in t$ is incomplete and no other incomplete node in t has v_k as its parent. When v_k is the leftmost incomplete node, all branches $b_{ki} = \alpha_i \bullet \beta_i$ that are $\beta_i \neq \epsilon$ are called *left-most incomplete branches (lmib)*.

From the definition of lmib, a procedure lmib(t) that performs the following steps yields the set of indices of all lmib of any parsing tree t (the lmib set).

- For $k = 1, \ldots, |t|$, if $\beta_i \neq \epsilon$ at any branch $b_{ki} = \alpha_i \bullet \beta_i (i = 1, \ldots, |q_k|)$ of $v_k = (q_k, j_k, d_k)$, add j_k to the incomplete parent set $\mathcal{N}$ and (k, i) to the incomplete branch set $\mathcal{B}$.
- The left-most incomplete branch set is $\{(k, i) \in \mathcal{B} | k \notin \mathcal{N}\}$.

In this procedure, the first step searches for incomplete branches and acquires both incomplete branches and nodes. In the second step, all incomplete branches that are not parents of another incomplete branch are acquired. Note that in a CFG, there is only one lmib for any parsing tree, but in case of a CCFG, there can be multiple lmibs.

6.2 Related Function Definitions

In the following, we define three functions that are useful in realizing the TDTP.

first(A). If A is the symbol corresponding to the leftmost incomplete branch lmib(t) of the parsing tree t obtained for $\sigma_{:l-1}$, the TDTP checks whether A is capable of deriving a string prefixed by the next terminal symbol σ_l. To do this, for any symbol A, we need a way to obtain all possible terminal symbols that could come at the beginning of a terminal symbol sequence that can be derived from A.

Therefore, we define the set first(A) as follows;

$$\text{first}(A) := \{w \in E | w\beta \in s(\delta), A \Longrightarrow^* \delta\}.$$

first(A) indicates the set of terminal symbols located at the head of any terminal symbol sequence derived from any symbol A. In case of a concurrent production rule, if $A \Longrightarrow^* (bc, de, fg)$, then b, d, and f, the heads of each, are all in first(A).

first(A) can be obtained by the following procedure, starting from the symbol corresponding to the leaf node of the process tree.

- For all $w \in E$, $\text{first}(w) = \{w\}$.
- For all production rule $A \Rightarrow (B_1\gamma_1, \ldots, B_K\gamma_K) \in R$, $\text{first}(A) \leftarrow \text{first}(A) \cup \bigcup_{i=1}^{K} \text{first}(B_i)$.

Note that the first set is defined recursively by other first sets. In practice, the entire sets can be obtained by executing the procedure from the leaf nodes to the root node of a process tree. Also, since the first sets depend on only R, the procedure only needs to be computed once for a given process tree.

wait(q, d). wait(q, d) returns the next symbol at the d-th branch of item q. $\text{wait}(A \Rightarrow (\alpha_1 \bullet \beta_1, \ldots, \alpha_d \bullet B_d\beta_d, \ldots, \alpha_K \bullet \beta_K), d) = B_d$.

step(q, d). step(q, d) advances the analysis of the d-th branch of item q by one and returns the item with $\bullet$ moved to the right by one. $\text{step}(A \Rightarrow (\alpha_1 \bullet \beta_1, \ldots, \alpha_d \bullet B_d\beta_d, \ldots, \alpha_K \bullet \beta_K), d) = A \Rightarrow (\alpha_1 \bullet \beta_1, \ldots \alpha_d B_d \bullet \beta_d, \ldots, \alpha_K \bullet \beta_K)$.

6.3 Top-Down Trace Parsing

This section describes the Top-Down Trace Parsing (TDTP) algorithm for parsing traces by CCFGs. The TDTP is an extension of the parsing approach that accumulates parsing states in a chart by creating a parsing tree top-down from an initial symbol, so that it can be applied to CCFGs. Since a common CFG is considered to be a special case of CCFGs in which tuples of concurrent production rules are limited to have at most one concurrent element, the TDTP can also parse CFGs.

The TDTP is shown in Algorithm 1. The TDTP parse the trace σ sequentially. Starting from $l = 1$, parsing trees in which the string $\sigma_{:l}$ appears in the prefix are accumulated into chart[l]. The TDTP first performs a forward procedure to extend the current lmib set until it reaches the given terminal symbol

σ_l. When the target terminal symbol is reached, a backward procedure completes the branch in the opposite direction. Growing any branch of lmibs in this manner yields a parsing tree. The forward and backward processes are shown in Algorithm 2 and 3, respectively.

Algorithm 1. topdownTraceParsing(σ, G)

Require: σ: a trace, $G = (M, E, R, S)$: a CCFG.

Find first(A) for any symbol $A \in M \cup E$ from R.
For all $S \Rightarrow (\gamma_1, \dots, \gamma_{|r|}) \in R$, add $[(S \Rightarrow (\bullet\gamma_1, \dots, \bullet\gamma_{|r|}), -1, -1]$ to chart[0].
for $l = 1, \dots, |\sigma|$ **do**
    $w \leftarrow \sigma_l$
    for $t \in$ chart$[l-1]$ **do**
        for $(k, i) \in$ lmib(t) **do**
            $(q, j, d) \leftarrow t_k$
            $A \leftarrow$ wait(q, i)
            if $w \in$ first(A) **then**
                forward(l, t, k, i, w)
            end if
        end for
    end for
end for
return whether a parsing tree with all nodes completed is in chart$[|\sigma|]$ or not.

If chart$[|\sigma|]$ has a parsing tree with all nodes completed, the TDTP returns true because the trace σ can be derived from the target CCFG. If all items $A \Rightarrow (\gamma_1\bullet, \dots, \gamma_K\bullet)$ of the completed parsing tree are converted to $A \Rightarrow (\gamma_1, \dots, \gamma_K)$, the syntax tree can be obtained at the same time.

7 Trace Parsing Example

In this section, the operation of the TDTP is explained using an example of trace parsing for a process tree $\rightarrow(a, +(\rightarrow(b, c), d), \Downarrow(e, f))$ and a trace $\sigma = adbceeff$. Let $G_1 = (M_1, E_1, R_1, S_1)$ be the CCFG obtained from the process tree according to the procedure in Sect. 5.5. $M_1 = \{\rightarrow_1, T_2, +_3, \rightarrow_4, T_5, T_6, T_7, \Downarrow_8, T_9, T_{10}\}$, $E_1 = \{a, b, c, d, e, f\}$, $S_1 = \rightarrow_1$, and

$$\begin{aligned}R_1 = \{&\rightarrow_1 \Rightarrow T_2 +_3 \Downarrow_8,\ \ T_2 \Rightarrow a,\ \ +_3 \Rightarrow (\rightarrow_4, T_7),\ \ \rightarrow_4 \Rightarrow T_5 T_6,\\ &T_5 \Rightarrow b,\ \ T_6 \Rightarrow c,\ \ T_7 \Rightarrow d,\ \ \Downarrow_8 \Rightarrow (T_9 T_{10}, \Downarrow_8),\ \ \Downarrow_8 \Rightarrow (T_9 T_{10}),\\ &T_9 \Rightarrow e,\ \ T_{10} \Rightarrow f\}.\end{aligned}$$

Algorithm 2. forward(l, t, j, d, w)

Require: l: the current parsing position in the trace, t: parsing tree, j: index of the node to be analysed at t, d: index of the branch to be analysed at t_j, w: terminal symbol of σ_l.

```
(q, j', d') ← t_j
A ← wait(q, d)
if A = w then
    q_new ← step(q, d)
    t_new is an array of j-th element t_j = (q, j', d') of t replaced to (q_new, j', d').
    if q_new is complete then
        backward(l, t_new, j)
    else {q_new is incomplete}
        Add t_new to chart[l].
    end if
else
    for r = A ⇒ (γ_1, ..., γ_|r|) ∈ R do
        if w ∈ first(A) then
            q_child ← A ⇒ (•γ_1, ..., •γ_|r|)
            t_new is an array of (q_child, j, d) added to the end of t.
            for d_child = 1, ..., |q_child| do
                B ← wait(q_child, d_child)
                if A ≠ B and w ∈ first(B) then
                    forward(l, t_new, j_child, d_child, w)
                end if
            end for
        end if
    end for
end if
```

Algorithm 3. backward(l, t, j)

Require: l: The current parsing position in the trace, t: parsing tree, j: Index of the node at t

```
(q, j_parent, d_parent) ← t_j
if q is incomplete or q is root (j_parent = −1) then
    Add t to chart[l].
else
    (q_parent, j_gparent, d_gparent) ← t_{j_parent}
    q_new ← step(q_parent, d_parent)
    t_new is an array of j_parent-th element t_{j_parent} of t replaced to
    (q_new, j_gparent, d_gparent).
    backward(l, t_new, j_parent)
end if
```

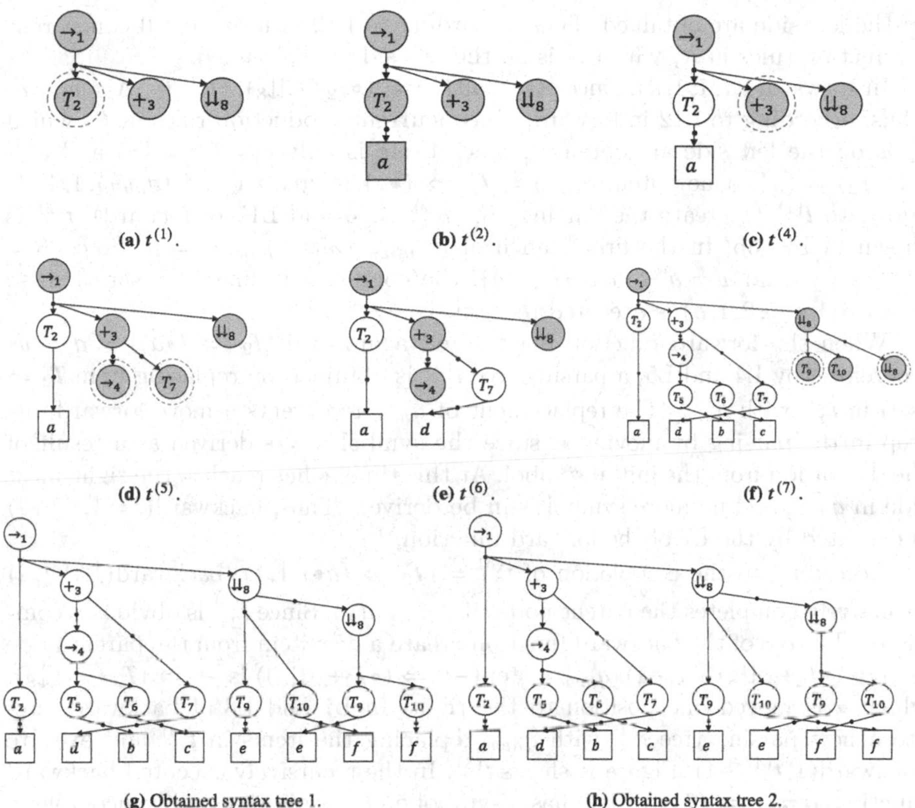

Fig. 3. Parsing trees

Initialization. When topdownTraceParsing is started, according to L2, the parsing tree $t^{(1)} = [(\rightarrow_1 \Rightarrow (\bullet T_2 +_3 \Downarrow_8), -1, -1)]$ corresponding to the derivation from the initial symbol S_1 as shown in Fig. 3a is added to chart[0]. This tree means that the string $T_2 +_3 \Downarrow_8$ may appear next. In Fig. 3, completed nodes are shown in white and incomplete nodes are shown in gray. The node with the lmib is circled by a red dashed line.

Parsing Process at $l = 1$. Considering the only parsing tree $t^{(1)}$ in chart[0] at $l = 1$, the only incomplete branch is $((\rightarrow_1 \Rightarrow (\bullet T_2 +_3 \Downarrow_8), -1, -1)$. Thus, $\text{lmib}(t_1) = \{(1,1)\}$. Since $\text{wait}(\rightarrow_1 \Rightarrow (\bullet T_2 +_3 \Downarrow_8), 1) = T_2$, the L9 of topdownTraceParsing checks whether $w = \sigma_1 = a$ is in $\text{first}(T_2)$. Since $T_2 \in \text{first}(T_2) = \{T_2, a\}$, $\text{forward}(1, t^{(1)}, k = 1, i = 1, a)$ in L10 of topdownTraceParsing is executed.

Thus, according to L12 in forward, all concurrent production rules in R_1 which T_2 is on the left side are obtained. Thus, according to L12 in forward, all concurrent production rules in R_1 which T_2 is on the left side are obtained. Thus, according to L12 in forward, all concurrent production rules in R_1 which T_2 is on the left side are obtained. Thus, according to L12 in forward, all concurrent production rules in R_1 which T_2 is on the left side are obtained.

In forward$(1, t, 1, 1, a)$, since A = wait$(\rightarrow\Rightarrow (\bullet T_2 +_3 \Downarrow_8), 1) = T_2$, $A \neq w = a$. Thus, according to L12 in forward, all concurrent production rules in R_1 which T_2 is on the left side are obtained. Since there is only $r = T_2 \Rightarrow (a)$ and $w \in$ first$(T_2) = \{a\}$, a new item $q_{child} = T_2 \Rightarrow (\bullet a)$ is created and $(q_{child}, 1, 1)$ is added to $t^{(1)}$ to create the parsing tree $t^{(2)}$ (L14 and L15 of forward). $t^{(2)}$ is shown in Fig. 3b. In the first branch of q_{child}, wait$(q_{child}, 1) = a$. Since $B = a \neq T_2 = A$ and $w = a \in$ first$(a) = \{a\}$, the condition in line 18 is satisfied, so forward$(1, t^{(2)}, 2, 1, a)$ is executed recursively.

When the forward function is executed again, wait$(T_2 \Rightarrow (\bullet a)) = a = w$. Therefore, by L4 and L5, a parsing tree $t^{(3)}$ is obtained by replacing item $T_2 \Rightarrow (\bullet a)$ in $t_2^{(2)}$ with q_{new}. The replacement of q_{new} represents a move forward one step in the parsing by moving •, since the symbol a was derived as a result of the derivation from the initial symbol. At this time, • has reached the right-most side in q_{new}, and no more symbols can be derived. Thus, backward$(l = 1, t^{(3)}, 2)$ is executed by the L7 of the forward function.

According to the completion of $t_2^{(3)} = (T_2 \Rightarrow (a\bullet), 1, 1)$, backward$(1, t^{(3)}, 2)$ recursively completes the parent node (if necessary). Since $t_2^{(3)}$ is obviously completed, L5 to 8 of the backward function create a new item from the parent node $(\rightarrow_1\Rightarrow (\bullet T_2 +_3 \Downarrow_8), -1, -1)$. q_{new} = step$(\rightarrow_1\Rightarrow (\bullet T_2 +_3 \Downarrow_8))$ is $\rightarrow_1\Rightarrow (T_2 \bullet +_3 \Downarrow_8)$, which • is moved one position to the right. In L7 and L8 of backward, create a new parsing tree $t^{(4)}$ with q_{new} replacing the items in $t^{(3)}$ and execute backward$(1, t^{(4)}, -1)$. Figure 3c shows $t^{(4)}$. In the recursively executed backward function, $q = \rightarrow_1\Rightarrow (T_2 \bullet +_3 \Downarrow_8)$ has a symbol on the right of • and is incomplete. Therefore, add $t^{(4)}$ to chart[1] by the third line and terminate the processing of the backward function.

This is the procedure at $l = 1$. It may seem complicated. In reality, however, the forward function generates nodes according to the generation rules from the initial symbol until $\sigma_1 = a$, and once a is reached, the backward function simply completes the node. Up to this point, there are no operations that are caused by the concurrent structure. Therefore, TDTP is almost all the same with common chart parsing except that the entire tree is stored in the chart and the non-terminal symbols derived from any symbol are known by the first function.

Parsing Process at $l = 2$. Next, topdownTraceParsing parses by the symbol $\sigma_2 = d$ at $l = 2$. In the parsing tree $t^{(4)}$ in chart[1], lmib$(t^{(4)}) = \{(1, 1)\}$, that is, the left-most incomplete branch of $t^{(4)}$ is $b_{11} = (T_2 \bullet +_3 \Downarrow_8)$ (which is shown in Fig. 3c as the node enclosed by the red dashed line). At this time, wait$(\rightarrow\Rightarrow (T_2 \bullet +_3 \Downarrow_8), 1) = +_3$, and first$(+_3) = \{b, d\}$. From $d \in$ first$(+_3)$, forward$(2, t^{(4)}, 2, 1, d)$ is executed according to line 10 of topdownTraceParsing.

In forward$(2, t^{(4)}, 2, 1, d)$, $q = t_1^{(4)} = \rightarrow\Rightarrow (T_2 \bullet +_3 \Downarrow_8)$ is considered. The forward function creates a parsing tree $T^{(5)}$ with $(q_{child}, 2, 1)$ added to $t^{(4)}$ as $q_{child} = +_3 \Rightarrow (\bullet \rightarrow_4, \bullet T_7)$ (L14 and L15 of forward). This is the first example of a concurrent string affecting the parsing behavior, with q_{child} having two branches. In this case, $t^{(5)}$ is represented as in Fig. 3d. In the figures in this paper, multiple edges are drawn on one node only for tuples with multiple branches. Note that $\rightarrow_1$ in Fig. 3d outputs one edge (with three heads), while $+_3$ outputs two edges (each with one head). $\text{wait}(q_{child}, 1) = \rightarrow_4$, $\text{wait}(q_{child}, 1) = T_7$, $d \notin \text{first}(\rightarrow_4) = \{b\}$, and $d \in \text{first}(T_7) = \{d\}$, L13 of forward is true only in case $A = T_7$, and $q_{child} = T_7 \Rightarrow (\bullet d)$ also satisfies the condition in L18. Therefore, $\text{forward}(2, t^{(5)}, 4, 2, d)$ is performed recursively.

In $\text{forward}(2, t^{(5)}, 4, 2, d)$, $(T_7 \Rightarrow (\bullet d), 4, 2)$ is added and further replaced to $(T_7 \Rightarrow (d\bullet), 4, 2)$. Since this additional node is completed, $(+_3 \Rightarrow (\bullet \rightarrow_4, \bullet T_7), 2, 1)$ is replaced to $(+_3 \Rightarrow (\bullet \rightarrow_4, T_7\bullet), 2, 1)$ by backward, and the parsing tree $t^{(6)}$ shown in Fig. 3e is created. $+_3 \Rightarrow (\bullet \rightarrow_4, T_7\bullet)$ is incomplete because the second branch is completed but the first branch is not. Therefore, $t^{(6)}$ is added to chart[2] by the third line in backward, and the analysis at $l = 2$ is finished.

Parsing Process at $l = 5$. The analysis process at $l \geq 3$ is omitted, but the procedure at e when $l = 5$ is explained because it helps to understand that the proposed method prevents infinite loops due to CLOOPs. When $l = 5$, the forward function adds a node with the item $\downarrow\downarrow_8 \Rightarrow (\bullet T_9 T_{10}, \bullet \downarrow\downarrow_8)$ to a parsing tree to create $t^{(7)}$ in Fig. 3f. In $t^{(7)}$, T_9 and $\downarrow\downarrow_8$ are the symbols corresponding to the left-most incomplete branches, and both have e in the first$(\cdot)$ sets. If there is no $A \neq B$ condition at L18 of the forward, $\downarrow\downarrow_8$ recursively executes the forward, generating an infinite number of parsing trees with $\downarrow\downarrow_8$ incremented. In fact, since successive $\downarrow\downarrow_8$ do not invoke forward as $A = B = \downarrow\downarrow_8$, only one $\downarrow\downarrow_8$ is created at most in any l. On the other hand, at $l = 6$, only one more $\downarrow\downarrow_8$ can be generated again because the forward is again executed on the generated $\downarrow\downarrow_8$.

The $A \neq B$ condition in the TDTP narrows the range of possible parsing trees. It is possible that e in $\downarrow\downarrow_8$ is actually derived from the second CLOOP ($\downarrow\downarrow_8$). However, the TDTP is unable to find a parsing tree that maps the first e to the second $\downarrow\downarrow_8$. Since the purpose of this paper is to determine only whether a trace is in the language of the process tree, it is not important which CLOOP the activity was generated from. However, this limitation should be carefully considered in cases where all possible syntax trees are required (e.g., to calculate the probability that a given trace is generated from a CCFG).

Parsing Result. Table 2 shows the chart at the end of the parsing. chart[8] contains four parsing trees, two of which have all nodes completed. Since the completed parsing tree is at the end of chart, the target trace $adbceeff$ is in $\mathcal{L}(G_1)$, i.e. $adbceeff$ conforms to the process tree $\rightarrow (a, +(\rightarrow (b, c), d), \downarrow\downarrow (e, f))$. Figure 3g and 3h represent two completed parsing trees. They differ in which of the two f corresponds to the two e, respectively.

Table 2. Chart when parsing *adbceeff* with G_1.

l	chart$[l]$
0	$[(\rightarrow_1 \Rightarrow (\bullet T_2 +_3 \downdownarrows_8), -1, -1)]$
1	$[(\rightarrow_1 \Rightarrow (T_2 \bullet +_3 \downdownarrows_8), -1, -1), (T_2 \Rightarrow (a\bullet), 1, 1)]$
2	$[(\rightarrow_1 \Rightarrow (T_2 \bullet +_3 \downdownarrows_8), -1, -1), (T_2 \Rightarrow (a\bullet), 1, 1), (+_3 \Rightarrow (\bullet \rightarrow_4, T_7\bullet), 1, 1), (T_7 \Rightarrow (d\bullet), 3, 2)]$
3	$[(\rightarrow_1 \Rightarrow (T_2 \bullet +_3 \downdownarrows_8), -1, -1), (T_2 \Rightarrow (a\bullet), 1, 1), (+_3 \Rightarrow (\bullet \rightarrow_4, T_7\bullet), 1, 1), (T_7 \Rightarrow (d\bullet), 3, 2),$ $(\rightarrow_4 \Rightarrow (T_5 \bullet T_6), 3, 1), (T_5 \Rightarrow (b\bullet), 5, 1)]$
4	$[(\rightarrow_1 \Rightarrow (T_2 +_3 \bullet \downdownarrows_8), -1, -1), (T_2 \Rightarrow (a\bullet), 1, 1), (+_3 \Rightarrow (\rightarrow_4 \bullet, T_7\bullet), 1, 1), (T_7 \Rightarrow (d\bullet), 3, 2),$ $(\rightarrow_4 \Rightarrow (T_5T_6\bullet), 3, 1), (T_5 \Rightarrow (b\bullet), 5, 1), (T_6 \Rightarrow (c\bullet), 5, 1)]$
5	$[(\rightarrow_1 \Rightarrow (T_2 +_3 \bullet \downdownarrows_8), -1, -1), (T_2 \Rightarrow (a\bullet), 1, 1), (+_3 \Rightarrow (\rightarrow_4 \bullet, T_7\bullet), 1, 1), (T_7 \Rightarrow (d\bullet), 3, 2),$ $(\rightarrow_4 \Rightarrow (T_5T_6\bullet), 3, 1), (T_5 \Rightarrow (b\bullet), 5, 1), (T_6 \Rightarrow (c\bullet), 5, 1), (\downdownarrows_8 \Rightarrow (T_9 \bullet T_{10}), 1, 1), (T_9 \Rightarrow (e\bullet), 8, 1)],$ $[(\rightarrow_1 \Rightarrow (T_2 +_3 \bullet \downdownarrows_8), -1, -1), (T_2 \Rightarrow (a\bullet), 1, 1), (+_3 \Rightarrow (\rightarrow_4 \bullet, T_7\bullet), 1, 1), (T_7 \Rightarrow (d\bullet), 3, 2),$ $(\rightarrow_4 \Rightarrow (T_5T_6\bullet), 3, 1), (T_5 \Rightarrow (b\bullet), 5, 1), (T_6 \Rightarrow (c\bullet), 5, 1), (\downdownarrows_8 \Rightarrow (T_9 \bullet T_{10}, \bullet \downdownarrows_8), 1, 1), (T_9 \Rightarrow (e\bullet), 8, 1)]$
6	$[(\rightarrow_1 \Rightarrow (T_2 +_3 \bullet \downdownarrows_8), -1, -1), (T_2 \Rightarrow (a\bullet), 1, 1), (+_3 \Rightarrow (\rightarrow_4 \bullet, T_7\bullet), 1, 1), (T_7 \Rightarrow (d\bullet), 3, 2),$ $(\rightarrow_4 \Rightarrow (T_5T_6\bullet), 3, 1), (T_5 \Rightarrow (b\bullet), 5, 1), (T_6 \Rightarrow (c\bullet), 5, 1), (\downdownarrows_8 \Rightarrow (T_9 \bullet T_{10}, \bullet \downdownarrows_8), 1, 1), (T_9 \Rightarrow (e\bullet), 8, 1),$ $(\downdownarrows_8 \Rightarrow (T_9 \bullet T_{10}), 8, 2), (T_9 \Rightarrow (e\bullet), 10, 1)],$ $[(\rightarrow_1 \Rightarrow (T_2 +_3 \bullet \downdownarrows_8), -1, -1), (T_2 \Rightarrow (a\bullet), 1, 1), (+_3 \Rightarrow (\rightarrow_4 \bullet, T_7\bullet), 1, 1), (T_7 \Rightarrow (d\bullet), 3, 2),$ $(\rightarrow_4 \Rightarrow (T_5T_6\bullet), 3, 1), (T_5 \Rightarrow (b\bullet), 5, 1), (T_6 \Rightarrow (c\bullet), 5, 1), (\downdownarrows_8 \Rightarrow (T_9 \bullet T_{10}, \bullet \downdownarrows_8), 1, 1), (T_9 \Rightarrow (e\bullet), 8, 1),$ $(\downdownarrows_8 \Rightarrow (T_9 \bullet T_{10}, \bullet \downdownarrows_8), 8, 2), (T_9 \Rightarrow (e\bullet), 10, 1)]$
7	$[(\rightarrow_1 \Rightarrow (T_2 +_3 \bullet \downdownarrows_8), -1, -1), (T_2 \Rightarrow (a\bullet), 1, 1), (+_3 \Rightarrow (\rightarrow_4 \bullet, T_7\bullet), 1, 1), (T_7 \Rightarrow (d\bullet), 3, 2),$ $(\rightarrow_4 \Rightarrow (T_5T_6\bullet), 3, 1), (T_5 \Rightarrow (b\bullet), 5, 1), (T_6 \Rightarrow (c\bullet), 5, 1), (\downdownarrows_8 \Rightarrow (T_9T_{10}\bullet, \bullet \downdownarrows_8), 1, 1), (T_9 \Rightarrow (e\bullet), 8, 1),$ $(\downdownarrows_8 \Rightarrow (T_9 \bullet T_{10}), 8, 2), (T_9 \Rightarrow (e\bullet), 10, 1), (T_{10} \Rightarrow (f\bullet), 8, 1)],$ $[(\rightarrow_1 \Rightarrow (T_2 +_3 \bullet \downdownarrows_8), -1, -1), (T_2 \Rightarrow (a\bullet), 1, 1), (+_3 \Rightarrow (\rightarrow_4 \bullet, T_7\bullet), 1, 1), (T_7 \Rightarrow (d\bullet), 3, 2),$ $(\rightarrow_4 \Rightarrow (T_5T_6\bullet), 3, 1), (T_5 \Rightarrow (b\bullet), 5, 1), (T_6 \Rightarrow (c\bullet), 5, 1), (\downdownarrows_8 \Rightarrow (T_9 \bullet T_{10}, \downdownarrows_8 \bullet), 1, 1), (T_9 \Rightarrow (e\bullet), 8, 1),$ $(\downdownarrows_8 \Rightarrow (T_9T_{10}\bullet), 8, 2), (T_9 \Rightarrow (e\bullet), 10, 1), (T_{10} \Rightarrow (f\bullet), 10, 1)],$ $[(\rightarrow_1 \Rightarrow (T_2 +_3 \bullet \downdownarrows_8), -1, -1), (T_2 \Rightarrow (a\bullet), 1, 1), (+_3 \Rightarrow (\rightarrow_4 \bullet, T_7\bullet), 1, 1), (T_7 \Rightarrow (d\bullet), 3, 2),$ $(\rightarrow_4 \Rightarrow (T_5T_6\bullet), 3, 1), (T_5 \Rightarrow (b\bullet), 5, 1), (T_6 \Rightarrow (c\bullet), 5, 1), (\downdownarrows_8 \Rightarrow (T_9T_{10}\bullet, \bullet \downdownarrows_8), 1, 1), (T_9 \Rightarrow (e\bullet), 8, 1),$ $(\downdownarrows_8 \Rightarrow (T_9 \bullet T_{10}, \bullet \downdownarrows_8), 8, 2), (T_9 \Rightarrow (e\bullet), 10, 1), (T_{10} \Rightarrow (f\bullet), 8, 1)],$ $[(\rightarrow_1 \Rightarrow (T_2 +_3 \bullet \downdownarrows_8), -1, -1), (T_2 \Rightarrow (a\bullet), 1, 1), (+_3 \Rightarrow (\rightarrow_4 \bullet, T_7\bullet), 1, 1), (T_7 \Rightarrow (d\bullet), 3, 2),$ $(\rightarrow_4 \Rightarrow (T_5T_6\bullet), 3, 1), (T_5 \Rightarrow (b\bullet), 5, 1), (T_6 \Rightarrow (c\bullet), 5, 1), (\downdownarrows_8 \Rightarrow (T_9 \bullet T_{10}, \bullet \downdownarrows_8), 1, 1), (T_9 \Rightarrow (e\bullet), 8, 1),$ $(\downdownarrows_8 \Rightarrow (T_9T_{10}\bullet, \bullet \downdownarrows_8), 8, 2), (T_9 \Rightarrow (e\bullet), 10, 1), (T_{10} \Rightarrow (f\bullet), 10, 1)]$
8	$[(\rightarrow_1 \Rightarrow (T_2 +_3 \downdownarrows_8 \bullet), -1, -1), (T_2 \Rightarrow (a\bullet), 1, 1), (+_3 \Rightarrow (\rightarrow_4 \bullet, T_7\bullet), 1, 1), (T_7 \Rightarrow (d\bullet), 3, 2),$ $(\rightarrow_4 \Rightarrow (T_5T_6\bullet), 3, 1), (T_5 \Rightarrow (b\bullet), 5, 1), (T_6 \Rightarrow (c\bullet), 5, 1), (\downdownarrows_8 \Rightarrow (T_9T_{10}\bullet, \downdownarrows_8 \bullet), 1, 1), (T_9 \Rightarrow (e\bullet), 8, 1),$ $(\downdownarrows_8 \Rightarrow (T_9T_{10}\bullet), 8, 2), (T_9 \Rightarrow (e\bullet), 10, 1), (T_{10} \Rightarrow (f\bullet), 8, 1), (T_{10} \Rightarrow (f\bullet), 10, 1)],$ $[(\rightarrow_1 \Rightarrow (T_2 +_3 \downdownarrows_8 \bullet), -1, -1), (T_2 \Rightarrow (a\bullet), 1, 1), (+_3 \Rightarrow (\rightarrow_4 \bullet, T_7\bullet), 1, 1), (T_7 \Rightarrow (d\bullet), 3, 2),$ $(\rightarrow_4 \Rightarrow (T_5T_6\bullet), 3, 1), (T_5 \Rightarrow (b\bullet), 5, 1), (T_6 \Rightarrow (c\bullet), 5, 1), (\downdownarrows_8 \Rightarrow (T_9T_{10}\bullet, \downdownarrows_8 \bullet), 1, 1), (T_9 \Rightarrow (e\bullet), 8, 1),$ $(\downdownarrows_8 \Rightarrow (T_9T_{10}\bullet), 8, 2), (T_9 \Rightarrow (e\bullet), 10, 1), (T_{10} \Rightarrow (f\bullet), 10, 1), (T_{10} \Rightarrow (f\bullet), 8, 1)],$ $[(\rightarrow_1 \Rightarrow (T_2 +_3 \bullet \downdownarrows_8), -1, -1), (T_2 \Rightarrow (a\bullet), 1, 1), (+_3 \Rightarrow (\rightarrow_4 \bullet, T_7\bullet), 1, 1), (T_7 \Rightarrow (d\bullet), 3, 2),$ $(\rightarrow_4 \Rightarrow (T_5T_6\bullet), 3, 1), (T_5 \Rightarrow (b\bullet), 5, 1), (T_6 \Rightarrow (c\bullet), 5, 1), (\downdownarrows_8 \Rightarrow (T_9T_{10}\bullet, \bullet \downdownarrows_8), 1, 1), (T_9 \Rightarrow (e\bullet), 8, 1),$ $(\downdownarrows_8 \Rightarrow (T_9T_{10}\bullet, \bullet \downdownarrows_8), 8, 2), (T_9 \Rightarrow (e\bullet), 10, 1), (T_{10} \Rightarrow (f\bullet), 8, 1), (T_{10} \Rightarrow (f\bullet), 10, 1)],$ $[(\rightarrow_1 \Rightarrow (T_2 +_3 \bullet \downdownarrows_8), -1, -1), (T_2 \Rightarrow (a\bullet), 1, 1), (+_3 \Rightarrow (\rightarrow_4 \bullet, T_7\bullet), 1, 1), (T_7 \Rightarrow (d\bullet), 3, 2),$ $(\rightarrow_4 \Rightarrow (T_5T_6\bullet), 3, 1), (T_5 \Rightarrow (b\bullet), 5, 1), (T_6 \Rightarrow (c\bullet), 5, 1), (\downdownarrows_8 \Rightarrow (T_9T_{10}\bullet, \bullet \downdownarrows_8), 1, 1), (T_9 \Rightarrow (e\bullet), 8, 1),$ $(\downdownarrows_8 \Rightarrow (T_9T_{10}\bullet, \bullet \downdownarrows_8), 8, 2), (T_9 \Rightarrow (e\bullet), 10, 1), (T_{10} \Rightarrow (f\bullet), 10, 1), (T_{10} \Rightarrow (f\bullet), 8, 1)]$

8 Conclusion

In this paper, we propose a novel formal grammar, a CCFG, for representing business processes with CLOOPs. We also proved that a CCFG can be easily obtained from a process tree, and that the obtained CCFG has the same language as the original process tree. A CCFG can represent a language containing iterated shuffles by a combination of concurrent string derivation and serialization. Therefore, the CCFG extends the expressive power of process trees to real business processes.

In this paper, we also proposed the TDTP, an algorithm for trace parsing using a CCFG. The TDTP acquires a syntax tree corresponding to a given trace by growing a parsing tree constructed from initial symbols according to the given trace. However, the validity and computational efficiency of the TDTP algorithm were not verified in this paper, and this is a future work.

Most of the linguistic properties of the CCFG are still unknown, and this is also a future work. In particular, the relations between the CCFG and automata, including Petri nets, are unknown, although this is a very interesting topic. The relations with other formal grammars, such as multiple context free grammars, also deserve attention.

Probabilistic CCFG, which adds the concept of probability to a CCFG, is also a topic of practical interest. While the modeling of probabilities for concurrent strings is expected to be analogous in case of a PCFG for a common CFG, the modeling of probabilities for the serialization of concurrent strings is expected to be neither simple nor unique. We intend to study the probability modeling of the CCFG in the future.

References

1. Buijs, J.C., van Dongen, B.F., van der Aalst, W.M.P.: Quality dimensions in process discovery: the importance of fitness, precision, generalization and simplicity. Int. J. Cooper. Inf. Syst. **23**(01) (2014)
2. Rozinat, A., van der Aalst, W.M.P.: Conformance checking of processes based on monitoring real behavior. Inf. Syst. **33**(1), 64–95 (2008)
3. Watanabe, A., Takahashi, Y., Ikeuchi, H., Matsuda, K.: Grammar-based process model representation for probabilistic conformance checking. In: 2022 4th International Conference on Process Mining, pp. 88–95. IEEE (2022)
4. Goulart Rocha, E., van der Aalst, W.M.P.: Polynomial-time conformance checking for process trees. In: Di Francescomarino, C., Burattin, A., Janiesch, C., Sadiq, S. (eds.) BPM 2023. LNCS, vol. 14159, pp. 109–125. Springer, Cham (2023). https://doi.org/10.1007/978-3-031-41620-0_7
5. van der Aalst, W.M.P.: Process Mining: Data Science in Action, 2nd edn. Springer, Heidelberg (2016)
6. Leemans, S.J.J., Polyvyanyy, A.: Stochastic-aware conformance checking: an entropy-based approach. In: Dustdar, S., Yu, E., Salinesi, C., Rieu, D., Pant, V. (eds.) CAiSE 2020. LNCS, vol. 12127, pp. 217–233. Springer, Cham (2020). https://doi.org/10.1007/978-3-030-49435-3_14

7. Salvati, S.: MIX is a 2-MCFL and the word problem in $\mathbb{Z}^2$ is captured by the IO and the OI hierarchies. J. Comput. Syst. Sci. **81**(7), 1252–1277 (2015)
8. Jantzen, M.: Extending regular expressions with iterated shuffle. Theor. Comput. Sci. **38**, 223–247 (1985)
9. Leemans, S.J.J., Fahland, D., van der Aalst, W.M.P.: Discovering block-structured process models from event logs - a constructive approach. In: Colom, J.-M., Desel, J. (eds.) PETRI NETS 2013. LNCS, vol. 7927, pp. 311–329. Springer, Heidelberg (2013). https://doi.org/10.1007/978-3-642-38697-8_17
10. Adriansyah, A., Sidorova, N., van Dongen, B.F.: Cost-based fitness in conformance checking. In: 2011 11th International Conference on Application of Concurrency to System Design, pp. 57–66. IEEE (2011)
11. Leemans, S.J., Maggi, F.M., Montali, M.: Enjoy the silence: analysis of stochastic Petri nets with silent transitions. arXiv preprint arXiv:2306.06376 (2023)
12. Broda, S., Machiavelo, A., Moreira, N., Reis, R.: Location automata for regular expressions with shuffle and intersection. Inf. Comput. **295** (2023)
13. Seki, H., Matsumura, T., Fujii, M., Kasami, T.: On multiple context-free grammars. Theor. Comput. Sci. **88**(2), 191–229 (1991)
14. van der Aalst, W., Buijs, J., van Dongen, B.: Towards improving the representational bias of process mining. In: Aberer, K., Damiani, E., Dillon, T. (eds.) SIMPDA 2011. LNBIP, vol. 116, pp. 39–54. Springer, Heidelberg (2012). https://doi.org/10.1007/978-3-642-34044-4_3
15. Hopcroft, E.J., Motwani, R., Ullman, D.J.: The pumping lemma for context free languages. In: Introduction to Automata Theory, Languages, and Computation, 3rd edn., pp. 279–287. Addison Wesley (2006)
16. Hegerle, B.: Parsing Transformative LR (1) Languages. arXiv preprint cs/0605104 (2006)
17. Tomita, M.: An efficient augmented-context-free parsing algorithm. Comput. Linguist. **13**, 31–46 (1987)

Conformance Checking with Model Projections

Rethinking Log-Model Alignments for Processes with Interacting Objects

Dominique Sommers(✉), Natalia Sidorova, and Boudewijn van Dongen

Department of Mathematics and Computer Science, Eindhoven University of Technology, Eindhoven, The Netherlands
{d.sommers,n.sidorova,b.f.v.dongen}@tue.nl

Abstract. Alignments are a well-established conformance checking technique that serve to reconcile system logs with normative process models. For processes involving multiple entities, such as objects and resources performing different tasks, the interaction of these entities must be taken into account in the alignments. In traditional approaches, it is achieved by considering a log event as matching a model event when *all* the entities registered in the logged event fully match the ones required by the model, thus considering events as unbreakable process atoms. We relax this requirement to deal with partial agreements between logged and modeled events and define *relaxed alignments*, aiming at maximizing the synchronized interactions between entities. Our approach is based on the use of projections of the log and model on individual objects, to deal with partial agreements. The optimality criterion is based on the cost function taking into account the deviating moves (events with non-matching event labels), the degree of (partial) event matches, and the correlations between non matching parts of events involved in the process execution. Illustrating with a running example, we demonstrate that our approach yields alignments that better capture deviations in the context of complex, multi-object processes.

Keywords: Petri nets · Conformance checking · Interacting objects · Projections

1 Introduction

Real-world processes exhibit complex behavior involving various activities executed by system resources (including both humans and machines) for different objects following certain rules. The executed activities get recorded in event logs, which serve as input data for numerous process mining algorithms. Conformance checking is one of the branches of process mining, focusing on checking the process behavior registered in an event log against a normative process model representing the expected process behavior.

Many approaches consider process instances (cases) in isolation from other instances [1]. However, the execution of a process instance is often affected by the execution of other instances, e.g. because of shared resources, case batching, or interaction between objects [3,10,12]. Event logs and process models that include information

L. M. Kristensen and J. M. van der Werf (Eds.): PETRI NETS 2024, LNCS 14628, pp. 61–82, 2024.
https://doi.org/10.1007/978-3-031-61433-0_4

about object interaction and constraints enable the use of conformance checking for processes with interacting objects. Such analysis goes beyond the single instance control-flow perspective, checking whether and where the process behavior recorded in an event log deviates from the interaction constraints prescribed by the process model [20].

Alignments is one of the popular conformance checking techniques, exposing where the behavior recorded in a log and the model agree, which activities prescribed by the model are missing in the log, and which log activities should not have been performed according to the model [2,6]. Usually alignments focus on the control flow of a process, with more advanced techniques additionally incorporating data and/or resource information [4,7,14,15]. Even in these cases, they still operate on a case-by-case basis, providing optimal alignments considering individual cases, but failing to synchronize interactions between objects and expose deviations concerning inter-object dependencies. In our previous work, we proposed system alignments, working on multiple objects simultaneously and taking into account inter-object dependencies [20,21].

One limitation of the existing methods, which we address in this paper, concerns their binary approach to matching logged activities with modeled ones. A logged activity is deemed a match if and only if the correct task is executed on the right objects at the right moment in time. For instance, a successful match within a package delivery process might require a package to be delivered at a predetermined depot by the deliverer who picked up the package. Any deviation from this exact combination—such as a different delivery location (at another depot or at home), or by a different deliverer—would be considered a mismatch. In fact, an alignment including such partially deviating activities, would receive a lower score than an alignment in which the entire delivery step is skipped, because of being penalized both for an unwanted activity execution (deviating delivery) and for skipping the required activity (matching delivery). This binary approach oversimplifies the matching process and may lead to penalizing deviations that are almost compliant with the overall model requirements.

To overcome this limitation, we propose relaxed alignments that can accommodate partially compliant behavior by loosening constraints with respect to certain perspectives, represented by projections of the event log and the process model on corresponding objects. We show that relaxed system alignments incorporate optimal trace alignments while maximizing that match of object interactions imposed by log and model.

This paper is organized as follows. In Sect. 2 we introduce basic concepts of poset theory, event logs, and Petri nets. In Sect. 3, we explain the limitations of standard alignment mechanisms using a running example. In Sect. 4, we introduce and formalize the notion of projections on objects for event logs, process models, and alignments. Section 5 presents the concept of relaxed alignments, using these projections, enabling partial compliance of L and M. We discuss implications of our work in Sect. 6.

2 Preliminaries

In this section, we start with definitions and notation of multisets and partially ordered sets. Then we briefly introduce some definitions for logged and modeled behavior of processes with interacting objects, coming from related work. For more detailed explanations, we refer the reader to the respective sources.

2.1 Multisets, Posets, and Sequences

Definition 1 *(Multiset). A* multiset *m over a set X is $m : X \rightarrow \mathbb{N}$. $X^{\oplus}$ denotes the set of all multisets over X. The support $supp(m)$ of a multiset m is the set $\{x \in X \mid m(x) > 0\}$. We list elements of the multiset as $[m(x) \cdot x \mid x \in X]$.*

For two multisets m_1, m_2 over X, we write $m_1 \leq m_2$ if $\forall_{x \in X} m_1(x) \leq m_2(x)$, and $m_1 < m_2$ if $m_1 \leq m_2 \wedge m_1 \neq m_2$.

Definition 2 *(Partial order, Covering relation, Partially ordered set). A* partially ordered set *(*poset*) $X = (\bar{X}, \prec_X)$ is a pair of a set $\bar{X}$ and a* partial order *$\prec_X \subseteq X \times X$ (irreflexive, antisymmetric, and transitive). We overload the notation and write $x \in X$ if $x \in \bar{X}$. For $x, y \in X$, we write $x \|_X y$ if $x \not\prec y \wedge y \not\prec x$, and $x \preceq y$ if $x \prec y \vee x = y$.*

The covering relation *$\lessdot \subset \prec$ is the transitive reduction of partial order $\prec$ which is the smallest subset of $\prec$ with $\lessdot^+ = \prec$, i.e., $\lessdot = \{(x, y) \in \prec \mid \forall_{(x,z) \in \prec} z \not\prec y\}$.*

We define $X^<$ as the set of all sequences with elements from $\bar{X}$ that respects the partial order, i.e., for a $\sigma = \langle \sigma_1, \ldots, \sigma_{|\bar{X}|} \rangle \in X^<$ we have $\{x \in \sigma\} = \bar{X}$ and $\prec_X \subseteq <_\sigma$.

Definition 3 *(Poset projection). For a poset $X = (\bar{X}, \prec_X)$ and a subset $Y \subseteq \bar{X}$, we define the* projection *$X\restriction_Y$ of X on Y as $X\restriction_Y = (\bar{X} \cap Y, \prec_X \cap (Y \times Y))$.*

Definition 4 *(Sequence). A* sequence *σ over a set X of length $n \in \mathbb{N}$ is X^n. With $n > 0$, we write $\sigma = \langle \sigma_1, \ldots, \sigma_n \rangle$, and denote n by $|\sigma|$. $\langle\rangle$ denotes the empty sequence where $n = 0$. The set of all finite sequences over X is denoted by X^*. A projection of a sequence σ on a set Y is defined inductively by $\langle\rangle\restriction_Y = \langle\rangle$ and $(\langle x \rangle \cdot \sigma)\restriction_Y = \begin{cases} \langle x \rangle \cdot \sigma\restriction_Y & \text{if } x \in Y \\ \sigma\restriction_Y & \text{otherwise} \end{cases}$, where $x \in X$ and $\cdot$ is the sequence concatenation operator.*

2.2 Recorded Behavior: System Logs

We consider processes with interacting objects and we assume that each object belongs to exactly one object type, known beforehand. This object type corresponds in fact to a role played by objects of this type in the process. Note that while in reality, every object is identifiable in a way, and $\mathcal{O}$ could be considered a set rather than a multiset, we allow for abstractions of objects. Take for example warehouse objects, where a warehouse has space for two packages, without distinguishing between the two spots.

Definition 5 *(Objects, Object roles). We assume a finite multiset $\mathcal{O}$ of* object names *and a finite set $\mathcal{R}$ of* object roles. *Each* object name *$o \in \mathcal{O}$ has a single role* role$(o) \in \mathcal{R}$ *assigned to it.*

Events are atomic activity executions that occurred at a specific time and involved objects and actors. In a system log, we abstract away from the events' actual timestamps and only use the relative ordering between them. An event log can be transformed to *traces*, defined as projections e.g., on the object names.

Definition 6 *(Event, System log, Trace). Let Σ be a set of activities. An* event *e is a tuple $(a, t, O)^{id}$, with a unique identifier id, an* activity *$a \in \Sigma$, timestamp t and a non-empty multiset of object names $O \leq \mathcal{O}$, indicating objects involved in it.*

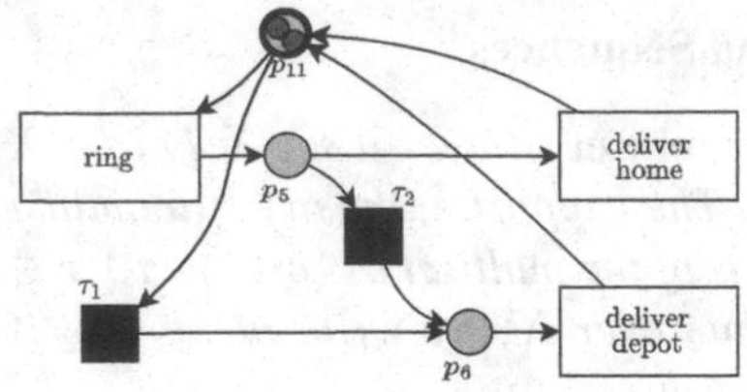

Fig. 1. Example: labeled Petri net N_1 with marking m such that $m(p_{11}) = 2$.

A system log $L = (\bar{L}, \prec_L)$ is a set of recorded events $\bar{L}$ with partial order $\prec_L$ inferred from the recording mechanism(s).

A trace L_o of object $o \in \mathcal{O}$ is defined as $L_o = (\{(a,t,O)^{id} \in L \,|\, o \in O\}, \prec_{L_o})$ with $\prec_{L_o} = \prec_L \cap (\overline{L_o} \times \overline{L_o})$.

We write e instead of e^{id} when it is clear that e refers to an identifiable event.

2.3 Modeled Behavior: Process Models

Petri nets [19] are process models used to describe and reason about the execution of a process.

Definition 7 *(Labeled Petri net).* A labeled Petri net *[18] is a tuple $N = (P, T, \mathcal{F}, \ell)$, with sets of places and transitions P and T, respectively, such that $P \cap T = \emptyset$, and a multiset of arcs $\mathcal{F} : (P \times T) \cup (T \times P) \rightarrow \mathbb{N}$ defining the flow of the net. $\ell : T \rightarrow \Sigma^\tau = \Sigma \cup \{\tau\}$ is a* labeling *function, assigning each transition t a label from alphabet Σ or label τ for silent transitions.*

Given an $x \in P \cup T$, its pre- and post-set, *${}^\bullet x$ and $x^\bullet$, are sets defined by ${}^\bullet x = \{y \in P \cup T \mid \mathcal{F}((y,x)) \geq 1\}$ and $x^\bullet = \{y \in P \cup T \mid \mathcal{F}((x,y)) \geq 1\}$ respectively.*

The Petri net $N_1 = (T_1, P_1, \mathcal{F}_1, \ell_1)$ in Fig. 1 is a simple example of a model for a package delivery process. A deliverer rings a doorbell to deliver a package at home or at a depot, depending on whether the door is answered. Alternatively, the package can be delivered at a depot immediately. $\tau_1 \in T_1$ and $\tau_2 \in T_1$ are the silent transitions in N_1 ($\ell_1(\tau_1) = \ell_1(\tau_2) = \tau$), denoted as black boxes.

Definition 8 *(Marking, Enabling and firing of transitions, Reachable markings). The state of a labeled Petri net $N = (P, T, \mathcal{F}, \ell)$ is defined by its* marking *$m \in P^\oplus$ that describes how many tokens each place contains.*

A transition $t \in T$ is enabled *in marking m iff $m \geq {}^\bullet t$. An enabled transition can* fire, *resulting in a state change, denoted by $m \xrightarrow{t} m'$, where the resulting marking m' is defined by $m' = m - {}^\bullet t + t^\bullet$. We say that m' is reachable from m and write $m \xrightarrow{*} m'$ if there is some sequence of transition firings $\sigma \in T^*$ such that $m \xrightarrow{\sigma_1} \cdots \xrightarrow{\sigma_{|\sigma|}} m'$.*

Certain types of Petri nets allow to model differentiation between object roles and their interaction. For the technique we propose in this paper, any formalism of process models can be used as long as its language can be defined in terms of partial orders.

Formalisms from [11,13,16,17,20] are examples of that. In previous work [20,21], we proposed alignment methods for resource-constrained ν-nets, which trivially extend to typed Jackson nets and easy sound t-PNIDs. Note that these formalisms incorporate multiple interacting processes with only one-to-one interactions. For many-to-many relations, i.e., with variable number of interacting objects, one should resort to formalisms like Object-centric nets [3] or synchronizing proclets [10]. In this paper we use typed Petri nets with identifiers (t-PNIDS) [22], defined as follows:

Definition 9 *(Typed variables, Typed Petri net with identifiers [22]).* $\mathcal{V} = \biguplus_{r\in\mathcal{R}} \mathcal{V}_r$ *is the set of variables with for every* $r \in \mathcal{R}$*:* $\text{role}(v) = r$ *for* $v \in \mathcal{V}_r$*. We write* $\mathcal{V}_R = \biguplus_{r\in R} \mathcal{V}_r$ *with* $R \subseteq \mathcal{R}$*. A* typed Petri net with identifiers *(t-PNID) is a tuple* $N = (P, T, F, \ell, \alpha, \beta)$*, with* $F = supp(\mathcal{F})$*, where:*

- $(P, T, \mathcal{F}, \ell)$ *is a labeled Petri net (c.f., Definition 7);*
- $\alpha : P \to \mathcal{R}^*$ *is the* place typing function*;*
- $\beta : F \to (\mathcal{V}^*)^\oplus$ *defines for each flow a multiset of variable vectors such that the variable types correspond with the place types:* $a(p) = (\text{role}(\mathbf{v}_1), \dots, \text{role}(\mathbf{v}_{|\mathbf{v}|}))$ *for any* $\mathbf{v} \in supp(\beta((p,t))$ *and* $a(p') = (\text{role}(\mathbf{v}'_1), \dots, \text{role}(\mathbf{v}'_{|\mathbf{v}'|}))$ *for any* $\mathbf{v}' \in supp(\beta((t,p'))$ *where* $t \in T, p \in^\bullet t, p' \in t^\bullet$*.* $Var(t) = In(t) \cup Out(t)$ *with* $In(t) = \{\mathbf{v}_i \mid p \in^\bullet t, \mathbf{v} \in supp(\beta((p,t))), 1 \le i \le |\mathbf{v}|\}$ *and* $Out(t) = \{\mathbf{v}_i \mid p \in t^\bullet, \mathbf{v} \in supp(\beta((t,p))), 1 \le i \le |\mathbf{v}|\}$*.*

$\mathcal{R}_p$ *and* $\mathcal{R}_t$ *are the sets of roles related to places and transitions in* N *according to* α*, defined as* $\mathcal{R}_p = \{\alpha(p)_i \mid 1 \le i \le |\alpha(p)|\}$ *and* $\mathcal{R}_t = \bigcup_{p\in(^\bullet t\cup t^\bullet)} \mathcal{R}_p$*.*

A state of a t-PNID is defined by its marking, which can change by firing enabled transitions in the net, as defined in Definition 10:

Definition 10 *(Marking, Enabling and firing of transitions). Let* $N = (P, T, F, \alpha, \beta)$ *be a t-PNID. A* marking *of* N *is a function* $m : P \to (supp(\mathcal{O})^*)^\oplus$ *such that* $\alpha(p) = (\text{role}(\mathbf{o}_1), \dots, \text{role}(\mathbf{o}_k))$ *for any* $(\mathbf{o}_1, \dots, \mathbf{o}_k) \in m(p)$*. For the set of identifiers in marking* m*, we write* $Id(m) = \{\mathbf{o}_i \mid p \in P, \mathbf{o} \in supp(m(p)), 1 \le i \le |\mathbf{o}|\}$*.*

A mode *μ of a transition t is an injection* $\mu : \mathcal{V} \to supp(\mathcal{O})$ *such that* $\text{role}(v) = \text{role}(\mu(v))$*. We overload the notation of* μ *to* $(\mathcal{V}^*)^\oplus \to (\mathcal{O}^*)^\oplus$ *and write* $\mu(\beta(a)) = [\beta(a)(\mathbf{v}) \cdot (\mu(\mathbf{v}_1), \dots, \mu(\mathbf{v}_{|\mathbf{v}|})) \mid \mathbf{v} supp(\beta(a))]$ *for any* $a \in (P \times T) \cup (T \times P)$*.*

A transition firing *t^{id}_μ denotes a firing of transition t with mode μ and identifier id and defines a state change in N:* $m \xrightarrow{t^{id}_\mu} m'$*.* t *is enabled with mode* μ *if* $\mu(\beta((p,t))) \le m(p)$ *for all* $p \in P$ *and* $\mu(v) \notin Id(m)$ *for each* $v \in Out(t) \setminus In(t)$*, i.e., fresh variables are truly fresh. The marking* m' *reached after firing of* t *with mode* μ *is defined as:*

$$m'(p) = m(p) - \mu(\beta((p,t))) + \mu(\beta((t,p))) \text{ for all } p \in P \tag{1}$$

$\text{objects}(t^{id}_\mu)$ *denotes the multiset of involved objects in* t^{id}_μ*, i.e., the consumed objects and newly created ones, defined as follows:*

$$\text{objects}(t^{id}_\mu) = [\beta((p,t))(\mathbf{v}) \cdot \mu(\mathbf{v}_i) \mid p \in^\bullet t, \mathbf{v} \in supp(\beta((p,t))), 1 \le i \le |\mathbf{v}|] + [\beta((t,p))(\mathbf{v}) \cdot \mu(\mathbf{v}_i) \mid p \in t^\bullet, \mathbf{v} \in supp(\beta((t,p))), 1 \le i \le |\mathbf{v}|, \mathbf{v}_i \in Out(t) \setminus In(t)] \tag{2}$$

T_μ *is the set of all possible transition firings.*

We write t_μ instead of t_μ^{id} when t_μ clearly refers to an identifiable transition firing.

Definition 11 *(t-PNID Process model, Execution poset and sequence, Run, Language). Let $M = (N, m_i, m_f)$ be a t-PNID process model, where N is a t-PNID and m_i and m_f are respectively the initial and final states of M. An* execution sequence *in M is a firing sequence σ reaching m_f from m_i, denoted as $m_i \xrightarrow{\sigma} m_f$.*

An execution poset *(also called a* run*) $\varphi = (\bar{\varphi}, \prec_\varphi)$ is a poset of transition firings such that each sequence $\sigma \in \varphi^>$ respecting the partial order is a firing sequence, i.e., $m_i \xrightarrow{\varphi} m_f \iff \forall_{\sigma \in \varphi^>} m_i \xrightarrow{\sigma} m_f$.*

The language *$\mathcal{L}(M)$ of a model M is the set of all execution posets in M.*

3 Rethinking Alignments for Processes with Interacting Objects

We use a running example to describe system alignments as introduced in previous works and then show some limitations of such alignments related to the binary interpretation of the deviations, thus explaining the need for partial synchronization.

Our running example extends the package delivery process from Fig. 1 by adding activities for ordering, depot registration, and collection. Figure 2a shows its recorded system log L containing the delivery of three packages by one deliverer (d_1) in the same street. d_1 delivers package p_1, which fits in the mailbox, and then proceeds to ring the doorbells to deliver package p_2, which remains unanswered, and package p_3. In the meantime, d_1 receives a notification that no ring activity has been recorded for the first package, so the deliverer logs it, while package p_2 is registered for depot delivery. The example ends with d_1 handing over p_3 and delivering p_2 to the warehouse w_1; p_3 is collected afterwards. Figure 2b shows the t-PNID process model $M = (N, m_i, m_f)$ modeling this process' behavior, with correlations between deliverers (blue) and packages (white) in places p_5 and p_6, and between warehouses (red) and packages in places p_7 and p_9. Note that the log deviates from the model as the model requires ringing before delivery and delivering the package at the depot before moving on to the next package delivery, if the delivery was unsuccessful, while the deliverer optimizes this business process as shown in the log.

Alignments between logged and modeled behavior consist of *moves* belonging to one of three types: log moves, model moves, and synchronous moves. *Log moves* indicate that an event from the log cannot be mimicked by the process model, and *model moves* show that the model requires the execution of an activity that has no matching recorded event in the log at the corresponding position. Unlike these *deviating moves*, *synchronous moves* are *conforming* moves and signify that the observed and modeled behavior agree on the event. Figure 3 shows these moves for the running example by layering transition firings (purple) from a run in the process model M and the events (yellow) from the system log L. The synchronous moves are thus the ones combining the yellow and the purple visual elements (e.g., $\substack{\text{order}\\\text{home}}[p_1]$), model moves are the elements in purple (e.g., $\substack{\text{order}\\\text{depot}}[p_2]$) and log moves are the yellow ones (e.g., $\substack{\text{order}\\\text{home}}[p_2]$). Formally, we define these moves as follows:

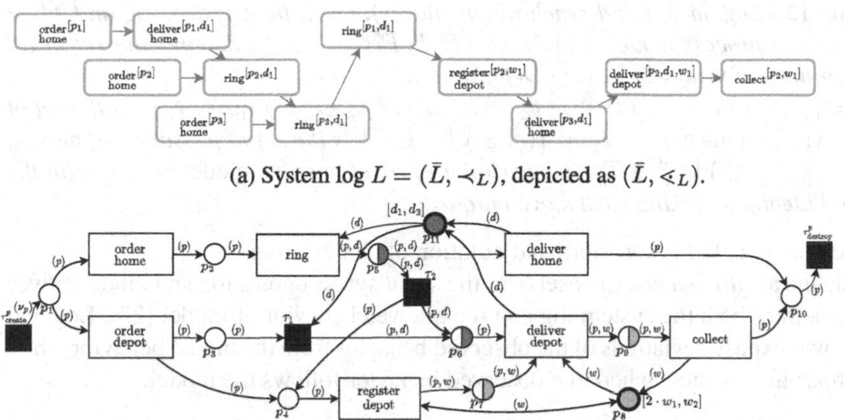

(a) System log $L = (\bar{L}, \prec_L)$, depicted as $(\bar{L}, \lessdot_L)$.

(b) t-PNID process model $M = (N, m_i, m_f)$ with initial marking m_i depicted by multisets for p_{11} and p_8 and final marking $m_f = m_i$.

Fig. 2. Running example of a delivery process with object roles $\mathcal{R} = \{p, d, w\}$. (Color figure online)

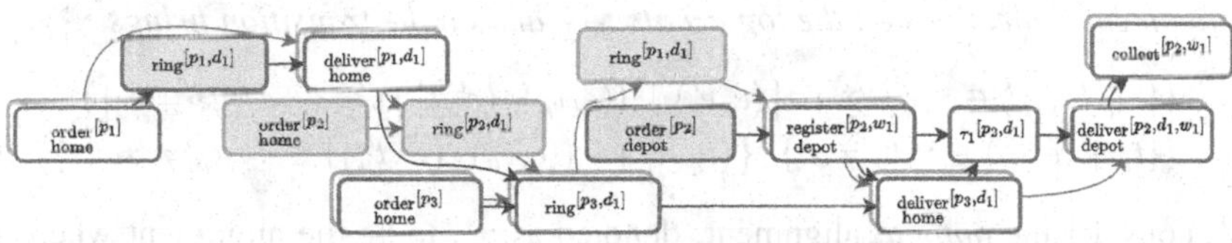

Fig. 3. An optimal system alignment γ^*, where events and transition firings are depicted as a^O (yellow) and $t^{\mathtt{objects}(t_\mu)}$ (purple) respectively, with their respective covering relations. Conforming and deviating moves are annotated in white and red respectively. (Color figure online)

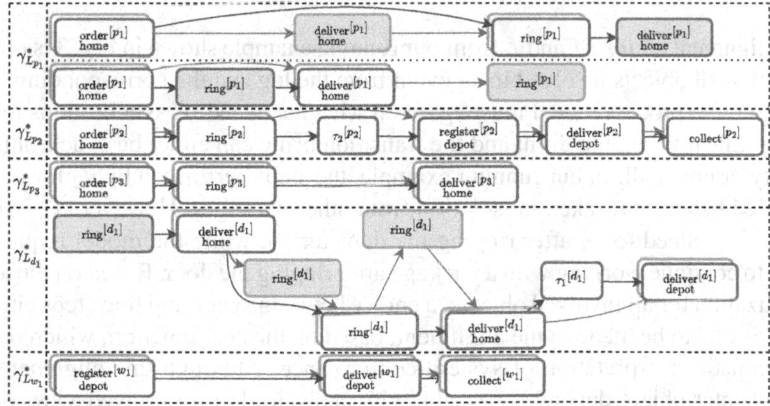

Fig. 4. Trace alignments $\gamma^*_{L_o}$ for $o \in \{p_1, p_2, p_3, d_1, w_1\}$, with conforming and deviating moves are annotated in white and red respectively. (Color figure online)

Definition 12 *(Log, model and synchronous moves). Let L be a system log and $M = (N, m_i, m_f)$ a process model with $N = (P, T, F, \ell, \alpha, \beta)$. T_μ denotes the set of all possible transition firings in M (see Definition 8).*

$\Gamma_s = \{(e, t_\mu) \mid e = (a, O) \in L, t_\mu \in T_\mu, a = \ell(t), O = \text{objects}(t_\mu)\}$ *is the set of possible* synchronous moves, $\Gamma_l = \{(e, \gg) \mid e \in L\}$ *is the set of possible* log moves, *and* $\Gamma_m = \{(\gg, t_\mu^{id}) \mid t_\mu^{id} \in T_\mu\}$ *is the (infinite) set of possible* model moves, *with the additional identifier making each move unique.*

We write $\Gamma_{lm} = \Gamma_l \cup \Gamma_m$ for shorthand notation of all deviating moves.

We define an *alignment* as a poset over the set of synchronous, log and model moves that incorporates both the system log and the allowed behavior of model [20]. Log and model moves expose deviations of the observed behavior from the model behavior while synchronous moves show where the observed behavior follows the model.

Definition 13 *(Alignment). Let $L = (\bar{L}, \prec_L)$ be a system log and $M = (N, m_i, m_f)$ be a process model. An* alignment $\gamma = \text{align}(N_{M,L})$ *is a poset* $\gamma = (\bar{\gamma}, \prec_\gamma)$, *where* $\bar{\gamma} \subseteq (\Gamma_l \cup \Gamma_s \cup \Gamma_m)$, *having the following properties:*

1. $\overline{\gamma\restriction_L} = \bar{L}$ *and* $\prec_L \subseteq \prec_{\gamma\restriction_L}$
2. $m_i \xrightarrow{\gamma\restriction_T} m_f$, *i.e.,* $\forall_{\sigma \in (\gamma\restriction_T)^<}, m_i \xrightarrow{\sigma} m_f$

with alignment projections on the log events $\gamma\restriction_L$ and on the transition firings $\gamma\restriction_T$:

$$\gamma\restriction_L = (\{e \mid (e, t_\mu) \in \bar{\gamma}, e \neq \gg\}, \{(e, e') \mid ((e, t_\mu), (e', t'_\mu)) \in \prec_\gamma, e \neq \gg \neq e'\}) \quad (3)$$

$$\gamma\restriction_T = (\{t_\mu \mid (e, t_\mu) \in \bar{\gamma}, t_\mu \neq \gg\}, \{(t_\mu, t'_\mu) \mid ((e, t_\mu), (e', t'_\mu)) \in \prec_\gamma, t_\mu \neq \gg t'_\mu\}) \quad (4)$$

We consider the *optimal* alignment, denoted as γ^*, to be the alignment where the number of deviating moves is minimized, which is achieved by minimizing the standard cost function c:

$$c((e, t_\mu)) = \begin{cases} 1 & \text{if } (e, t_\mu) \in \Gamma_{lm} \wedge \ell(t) \neq \tau \\ 0 & \text{otherwise} \end{cases} \quad (5)$$

The alignment γ^* for M and L from our running example shown in Fig. 3 is optimal.

Note that all objects involved in an event from the log and the corresponding transition firing in the process model must agree on what happened. A synchronous move is defined atomically, i.e., the event and the transition firing can either be synchronized in its entirety or not at all. In our running example, the steps performed by d_2 for p_2 do fit the modeled behavior if taken in isolation from other packages. However, d_1 switched to an activity related to p_3 after ringing the door for p_2, while the model requires the deliverer to continue working with a package after ringing the door. Based on the atomic synchronization for all involved objects, a choice between synchronizing steps either for p_2 or for p_3 had to be made in the alignment, based on the cost function, which resulted in a pessimistic interpretation of system conformance. Although this alignment does respect all inter-object dependencies, this ultimately leads to a skewed view of how well the individual objects fit the process model.

We further illustrate this idea using optimal trace alignments for individual objects, i.e., for packages p_1, p_2 and p_3, deliverer d_1, and warehouse w_1 (see Fig. 4). Here, any interaction between different objects is ignored, assuming we have a process model for each object role that models the behavior of individual objects. Therefore, such trace alignments are defined on these corresponding models. We highlight that the log trace for package p_2 is aligned completely synchronously with the model, while the optimal alignment γ^* from Fig. 3 shows that the recorded event $\text{ring}^{[p_2,d_1]}$ is not synchronized with the model. During the interval in which deliverer d_1 is recorded to be involved in activities for package p_2, d_1 is also involved in other activities with package p_3, causing the deviation in γ^*. An intuitive way of resolving this is by changing the $\substack{\text{order}\\\text{home}}^{[p_2]}$ event from the log to an $\substack{\text{order}\\\text{depot}}^{[p_2]}$ transition firing in the process model. This shows that the problem caused by d_1 is cascaded to activity executions only involving p_2. While there are other optimal alignments for M and L, the propagation of this deviation caused by d_1 to package from their interaction is inevitable.

Therefore there is a need for relaxing system alignments by allowing partial synchronization of events and transition firings. We explore the properties of such alignments concerning the incorporated individual trace alignments.

4 Object Projections for Alignments

The regular alignment denotes the worst-case alignment considering all objects simultaneously, providing a pessimistic view on the alignment behaviors from the perspective of individual objects or subsets of objects. We exploit traces, defined as log projections (c.f., Definition 6), to reveal these perspectives in the recorded behavior. We define projections for the modeled behavior as well, and subsequently, use them in the system alignments.

4.1 Process Model Projections

A projection of a process model captures the behavior for a subset of the objects involved in the process. Within a t-PNID, every object follows the behavior specified by the corresponding role, which is defined through its place typing function. We can use this property to define the projection on objects of t-PNIDs, as well as their markings and transition firings as follows:

Definition 14 *(Object projection of t-PNIDs and its markings and firings). Given a t-PNID $N = (P, T, F, \ell, \alpha, \beta)$ and submultiset of objects $O \leq \mathcal{O}$. Let $R = \{\text{role}(o) \mid o \in supp(O)\} \subseteq \mathcal{R}$ be the roles associated with O. The projection of N on O is $N{\restriction_O} = (P_R, T_R, F_R, \ell_R, \alpha_R, \beta_R)$ with*

$$P_R = \{p_{R_p} \mid p \in P, R_p = \mathcal{R}_p \cap R \neq \emptyset\} \text{ with } \alpha_R(p_{R_p}) = \alpha(p){\restriction_R} \quad (6)$$

$$T_R = \{t_V \mid t \in T, V \subseteq (Var(t) \cap \mathcal{V}_R)\} \text{ with } \ell_R(t_V) = \ell(t) \quad (7)$$

$$F_R = \{(p_{R_p}, t_V) \in (P_R \times T_R) \mid (p,t) \in F\} \cup \{(t_V, p_{R_p}) \in (T_R \times P_R) \mid (p,t) \in F\}$$
$$\text{with } \beta_R((p_{R_p}, t_V)) = [\beta((p,t))(\mathbf{v}) \cdot \mathbf{v}{\restriction_V} \mid \mathbf{v} \in supp(\beta((p,t)))] \quad (8)$$

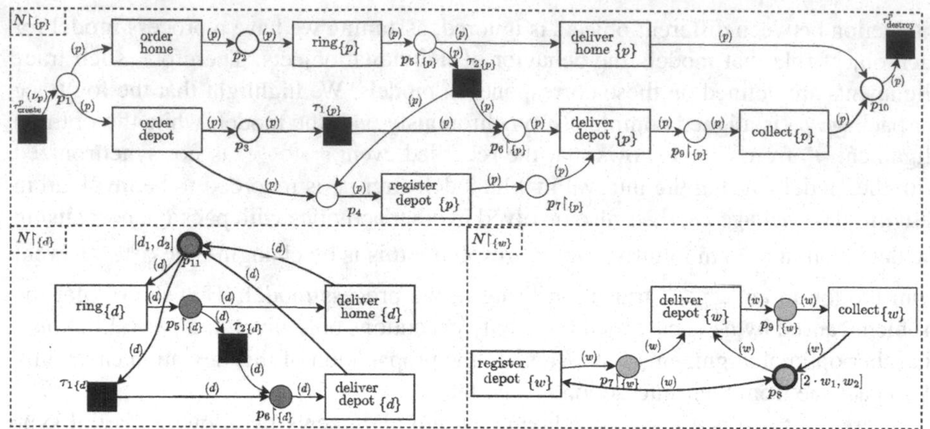

Fig. 5. Projections of N on object roles p, d, and w.

Let m be a marking in N, i.e., $m : P \to (supp(\mathcal{O})^*)^{\oplus}$. Its projection on objects O is the submarking $m\restriction_O : P_R \to (supp(O)^*)^{\oplus}$ defined for each $p \in P_R$ as follows: $m\restriction_O(p) = [m(p)(\mathbf{o}) \cdot \mathbf{o}\restriction_O \mid \mathbf{o} \in supp(m(p))]$.

Let t_μ be a transition firing in N; its projection on objects O is $t_\mu\restriction_O = (t\restriction_O)_{\mu\restriction_O}$ with $t\restriction_O = t_V \in T_R$ such that there exists a bijection $\{(o, v) \in (supp(O) \times V) \mid \text{role}(o) = \text{role}(v)\}$ and $\mu\restriction_O = \mu\restriction_V$.

For the projection on a set of roles $R \subseteq \mathcal{R}$, we also write $N\restriction_R = N\restriction_O$ for any $O \leq \mathcal{O}$ such that $R = \{\text{role}(o) \mid o \in supp(O)\}$. For the sake of readability, we write $N\restriction_o$ instead of $N\restriction_{[o]}$ and $N\restriction_r$ instead of $N\restriction_{\{r\}}$, for $o \in supp(\mathcal{O})$ and $r \in \mathcal{R}$.

In order to allow individual objects to traverse the net, the projection of transitions is applied to the variables of corresponding types.

For the process model M in Fig. 2b with $\mathcal{R} = \{p, d, w\}$, $M\restriction_r$ is shown in Fig. 5 for each $r \in \mathcal{R}$.

We show that the projections of t-PNIDs, of their markings and transition firings maintain the properties of the original t-PNIDs, their markings and firings, and hence a projection is also a process model.

Theorem 1 *(A projected (t-PNID) process model is a (t-PNID) process model). Let $M = (N, m_i, m_f)$ be a process model with t-PNID N and $O \leq \mathcal{O}$ be a multiset of objects with roles $R = \{\text{role}(o) \mid o \in supp(O)\}$. $M\restriction_O = (N\restriction_O, m_i\restriction_O, m_f\restriction_O)$ is also a process model, i.e., $N\restriction_O$ is a t-PNID, $m_i\restriction_O$ and $m_f\restriction_O$ are markings in $N\restriction_O$, and $t_\mu\restriction_O$ is a transition firing in $N\restriction_O$ for any $t \in T_R$.*

Proof. $N\restriction_O$ is a t-PNID as it has the three properties from Definition 9:

- $(P, T, \mathcal{F}, \ell)$ with any $\mathcal{F} \in F^{\oplus}$ is a labeled Petri net by definition of t-PNIDs. $(P_R, T_R, \mathcal{F}_R, \ell_R)$ is essentially a subnet of $(P, T, \mathcal{F}, \ell)$ with the same labeling function and therefore a labeled Petri net;
- $\alpha_R : P_R \to R^*$ by definition of vector type projection, and since $R \subseteq \mathcal{R}$, we have $\alpha_R : P_R \to \mathcal{R}^*$;

- Again, by definition of vector type projection, $\beta_R : F_R \to (\mathcal{V}_R^*)^{\oplus}$, hence $\beta_R : F_R \to (\mathcal{V}^*)^{\oplus}$. Furthermore, for any $p_{R_p} \in P_R$, $|\alpha_R(p_{R_p})| = |\alpha(p)\restriction_R| = |R_p|$, and for all $t_V \in T_R$ such that $(p_{R_p}, t_V) \in F_R$, $|\beta_R((p_{R_p}, t_V))| = |\beta((p,t))\restriction_R| = |R_p|$. Let $\mathbf{v} \in \mathit{supp}(\beta((p_{R_p}, t_V)))$, and let $\text{roles}(\mathbf{v}) = (\text{role}(\mathbf{v}_1), \ldots, \text{role}(\mathbf{v}_{|\mathbf{v}|}))$. $\alpha(p) = \text{role}(\mathbf{v})$ by definition of t-PNIDs, and therefore we have $\alpha_R(p_{R_p}) = \alpha(p)\restriction_R = \text{roles}(\mathbf{v})\restriction_R = (\text{role}((\mathbf{v}\restriction_R)_1), \ldots, \text{role}((\mathbf{v}\restriction_R)_{|R_p|}))$.
  Similarly, for all $t_V \in T_R$ such that $(t_V, p_{R_p}) \in F_R$ and $\mathbf{v} \in \mathit{supp}(\beta((t_V, p_{R_p})))$, $|\beta_R((t_V, p_{R_p}))| = |\beta((a,b))\restriction_R| = |R_p|$ and $\alpha_R(p_{R_p}) = \alpha(p)\restriction_R = \text{roles}(\mathbf{v})\restriction_R = (\text{role}((\mathbf{v}\restriction_R)_1), \ldots, \text{role}((\mathbf{v}\restriction_R)_{|R_p|}))$.

By definition, $m\restriction_O$ is only defined on P_R. Furthermore, for a place $p_{R_p} \in P_R$, its marking $m\restriction_O(p) = [m(p)\mathbf{o} \cdot \mathbf{o}\restriction_O \mid \mathbf{o} \in \mathit{supp}(m(p))]$, where $\text{role}((\mathbf{o}\restriction_O)_i) \in R$ for $1 \leq i \leq |\mathbf{o}\restriction_O|$, i.e., it only contains objects from O. Hence, $m_i\restriction_O$ and $m_f\restriction_O$ are markings in $N\restriction_O$.

The projection of transition firing t_μ on O from marking m to m' is $t_\mu\restriction_O$, which is a firing in $N\restriction_O$, i.e., $m\restriction_O \xrightarrow{t_\mu\restriction_O} m'\restriction_O$. □

Linking Elements in Process Model Projections. With Definition 14, projections of t-PNIDs are defined so that its elements, i.e., places, transitions, and arcs, are linked when their properties are the same. This is especially useful when combining multiple projection nets, which we do in the next section.

Definition 15 *(Linking projection elements). Let $N = (P, T, F, \ell, \alpha, \beta)$ be a t-PNID, $O \leq \mathcal{O}$ a multiset of objects, and $N\restriction_O = (P_R, T_R, F_R, \ell_R, \alpha_R, \beta_R)$ the projection of N on O, with $R = \{\text{role}(o) \mid o \in \mathit{supp}(O)\} \subseteq \mathcal{R}$.*

An element $x \in (P \cup T \cup F)$ and its corresponding projected element $x\restriction_O \in (P_R \cup T_R \cup F_R)$ are linked, *denoted $x = x'$, if and only if they reason over the same objects. For places $p \in P$ and $p_{R_p} \in P_R$, $p = p_{R_p}$ iff $\mathcal{R}_p = \mathcal{R}_{p_{R_p}} = R_p$. For transitions $t \in T$ and $t_V \in T_R$, $t = t_V$ iff $Var(t) = Var(t_V) = V$. For arcs $(p,t) \in F$ and $(p_{R_p}, t_V) \in F_R$ (and $(t,p) \in F$ and $(t_V, p_{R_p}) \in F_R$), $(p,t) = (p_{R_p}, t_V)$ ($(t,p) = (t_V, p_{R_p})$) iff $p = p_{R_p}$ and $t = t_V$.*

By Definition 14, in all three cases, the element and its corresponding projection have the same properties when they are linked. For linked places p and p_{P_R}, $\alpha(p) = \alpha_R(p_{R_p})$, for linked transitions t and t_V, $\ell(t) = \ell_R(t_V)$, and for linked arcs (p,t) and (p_{R_p}, t_V) (and (t,p) and (t_V, p_{R_p})), $\beta((p,t)) = \beta_R((p_{R_p}, t_V))$ ($\beta((t,p)) = \beta_R((t_V, p_{R_p}))$).

Figure 6 depicts the union of several projection nets regarding the transition $\substack{\text{deliver}\\\text{depot}}$, showing how and which elements are linked. On top of Fig. 6, we see that $N = N\restriction_{\{p,d,w\}}$, since $\{p,d,w\}$ covers all object roles in N. This is trivially true for any net and its projection on all involved object roles, i.e., $N\restriction_R = N$ iff $(\bigcup_{p \in P} \mathcal{R}_p) \subseteq R$.

4.2 Alignment Projections

The definition of alignment projections follows directly from the definition of poset projections (c.f., Definition 3): only the moves involving the projected objects are kept.

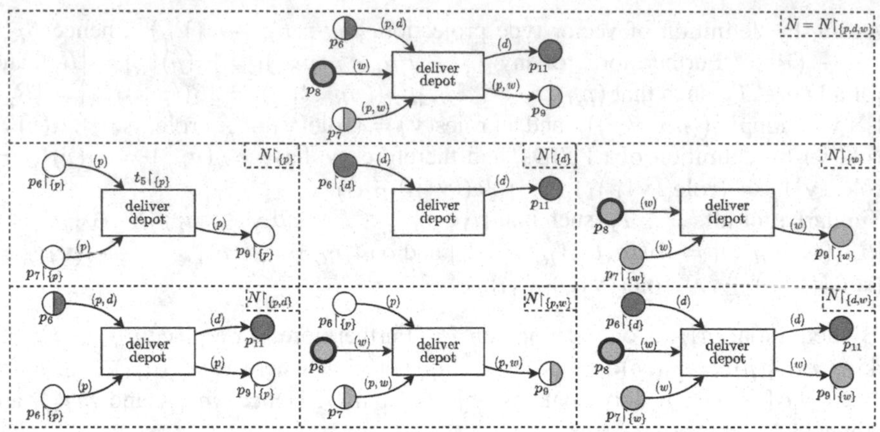

Fig. 6. All projections of N for the transition $^{\text{deliver}}_{\text{depot}}$, i.e., $N\restriction_R$ for all $R \in \mathcal{P}(\{p, d, w\})$.

Let γ be an alignment between system log L and process model $M = (N, m_i, m_f)$, and $O \leq \mathcal{O}$ be a multiset of objects, the projection of γ on O, denoted as $\gamma\restriction_O$ and is defined by $\gamma\restriction_O = (\{t_\mu\restriction_O \mid t_\mu \in \gamma, \text{objects}(t_\mu) \cap \mathit{supp}(O) \neq \emptyset\}, \prec_{\gamma\restriction_O})$ with $\prec_{\gamma\restriction_O} = \prec_\gamma \cap (\prec_{\gamma\restriction_O} \times \prec_{\gamma\restriction_O})$. In Lemma 1, we show that the projected alignment is an alignment on the projected log and model.

Lemma 1 *($\gamma\restriction_O$ is an alignment between $L\restriction_O$ and $M\restriction_O$). Let γ be an alignment between event log L and process model $M = (N, m_i, m_f)$, and let $O \leq \mathcal{O}$ be a multiset of objects. The projection of γ on objects O, i.e., $\gamma\restriction_O$, is an alignment between the projections of L and M on objects O, i.e., $\overline{\gamma\restriction_O\restriction_L} = \bar{L}$ and $\prec_L \subseteq \prec_{\gamma\restriction_O\restriction_L}$, and $m_i\restriction_O \xrightarrow{\gamma\restriction_O\restriction_T} m_f\restriction_O$.*

Proof. This follows directly from the fact that projections of transition firings are transition firings in the projected net, as proved in Theorem 1. □

A *projected trace alignment* is a projected alignment on a single object $o \in \mathcal{O}$, denoted as $\gamma\restriction_o$ (shorthand notation for $\gamma\restriction_{\{o\}}$). It relates directly to the respective trace L_o in the log (c.f., Definition 6).

5 Partial Synchronization of Objects

In this section, we use projection nets to partially synchronize events and transition firings. We modify the synchronous product model by adding all possible projections to create a relaxed alignment. We discuss and provide a cost function that optimizes criteria to ensure that projected trace alignments correspond to optimal trace alignments while synchronizing as much shared behavior as possible.

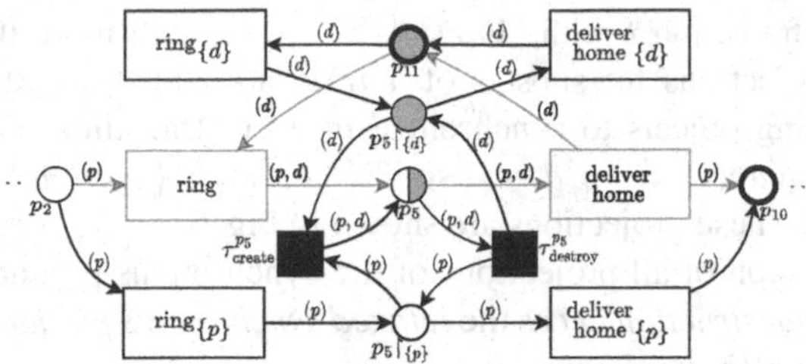

Fig. 7. Example fraction of the union of projection nets and correlation creation/destruction net N^C, i.e., a fraction of $(\bigcup_{R \in \mathcal{P}(\mathcal{R})} N\lceil_R) \cup N^C$.

5.1 Partial Synchronization in Alignments via Projections

As described in Sect. 3, a system alignment suffers from the problem that its projected trace alignments are not maximally conforming as they inherit deviating moves from their interacting objects. Take for example p_2 from alignments shown in Figs. 3 and 4. $\gamma^* \lceil_{p_2}$ has higher cost than $\gamma^*_{L_{p_2}}$, because of the alignment for its interacting object d_1. While this alignment technique clearly exposes the violations of inter-object dependencies, it obscures how well the recorded behavior of p_2 matches with its projected modeled behavior.

In this paper, our goal is to get to the other side of the spectrum where projected trace alignments are optimal regarding their projected behavior, while still synchronizing the interactions as much as possible. We have shown that optimal trace alignment can be achieved by aligning a trace, i.e., a projection from the log, to the corresponding projection of the model. As a basis for computing the alignment, we use the synchronous product model of the process model and a representation of L in the same formalism as M, denoted as $M^L = (N^L, m_i^L, m_f^L)$ [20]. M^S incorporates M, M^L and *synchronous transitions* added for each pair of a transition t^L from N^L a transition $t \in T$ of net N having the same label. Synchronous transitions inherit the incoming and outgoing places from both t^L and t. Log, model, and synchronous moves relate directly to the firing of these log, model, and synchronous transitions, and a run $\gamma \in \mathcal{L}(M^S)$ is an alignment of M and L. The optimal trace alignments can also be computed on projections of this synchronous product model immediately. Using this fact, here we propose an alignment method that allows for partial synchronization of the behavior of interacting objects by incorporating projections in the synchronous product model.

To allow for synchronized as well as projected moves, we modify the synchronous product model by adding all projections directly to it. Figure 7 depicts an example of the union of projection nets regarding the transition for ring and $\substack{\text{deliver}\\\text{home}}$ from the running example. Note that this union includes additional silent transitions $\tau^{p_5}_{\text{create}}$ and $\tau^{p_5}_{\text{destroy}}$. Without these, it is not possible to fire a projected transition when a correlation is established. In other words, we can only stay in one projection net throughout the interaction of objects. To break this up, we introduce the correlation creation/destruction net which can create and destroy new and existing correlations.

Note that for a transition t with $|Var(t)| > 2$, i.e., with more than two involved objects, also the projections for subsets of $Var(t)$ are added, to allow for any combination of interacting objects to synchronize in every transition. For example, with $t = \frac{\text{deliver}}{\text{depot}}$, we have $Var(t) = \{p, d, w\}$, so $|\mathcal{P}(Var(t)) \setminus \{\{p, d, w\}, \emptyset\}| = 6$ projections of t are added. These projections are shown in Fig. 6.

We denote the union of all projections of the synchronous product model together with the *correlation/destruction net* as the *relaxed synchronous product model*, of which we denote a run as a *relaxed alignment*:

Definition 16 *(Relaxed synchronous product net, Correlation creation/destruction net, Relaxed alignment). With t-PNID synchronous product model $M^S = (N^S, m_i^S, m_f^S)$, the fully* relaxed synchronous product model *is $\tilde{M}^S = (\tilde{N}^S, m_i^S, m_f^S)$ where $\tilde{N}^S = (\bigcup_{O \in \mathcal{P}(\mathcal{O})} N^S\restriction_O) \cup N^C$, with $N^C = (P^S, T^C, F^C, \ell^C, \alpha^S, \beta^C)$ the* correlation creation/destruction net*. For every $p \in P$ with $|\mathcal{R}_p| > 1$, there is for every partition $B_{\mathcal{R}_p}$ of $\mathcal{R}_p$, with $|B_{\mathcal{R}_p}| > 1$, a create correlation transition $\tau_{create}^{p,B_{\mathcal{R}_p}}$ and a destroy correlation transition $\tau_{destroy}^{p,B_{\mathcal{R}_p}}$. N^C contains arcs from p to $\tau_{create}^{p,B_{\mathcal{R}_p}}$ (and from $\tau_{destroy}^{p,B_{\mathcal{R}_p}}$ to p) and from $\tau_{create}^{p,B_{\mathcal{R}_p}}$ to p_R for every $R \in B_{\mathcal{R}_p}$ (from p_R to $\tau_{destroy}^{p,B_{\mathcal{R}_p}}$) with β^C such that a single variable with the corresponding type is assigned to each arc.*

A relaxed alignment *$\tilde{\gamma}$ is an alignment where partial matching is allowed, which is a run in $\tilde{M}^S$, i.e., $\tilde{\gamma} \in \mathcal{L}(\tilde{M}^S)$.*

We show that easy soundness (the reachability of the final marking from the initial marking) of the relaxed synchronous product model is preserved, since any run in the original synchronous product model is also a run in the relaxed one. Therefore, relaxed alignments can be computed using the same methods as for regular alignments.

Lemma 2 *(Any run in M^S is also a run in $\tilde{M}^S$). Let $M^S = (N^S, m_i^S, m_f^S)$ be a synchronous product model and $\tilde{M}^S = (\tilde{N}^S, \tilde{m}_i^S, \tilde{m}_f^S)$ be its relaxed version. Any run $\varphi \in \mathcal{L}(M^S)$ is a run $\varphi \in \mathcal{L}(\tilde{M}^S)$ in $\tilde{M}^S$, i.e., $m_i \xrightarrow{\varphi} m_f \implies \tilde{m}_i \xrightarrow{\varphi} \tilde{m}_f$.*

Proof. Let $\varphi \in \mathcal{L}(M)$ be a run in M. With T and $\tilde{T}$ the transitions of M and $\tilde{M}$ respectively, we have by definition $T \subseteq \tilde{T}$, so for every transition firing $t_\mu \in \varphi$, we have $t \in \tilde{T}$. Furthermore, no new conditions are introduced in $\tilde{M}$, i.e., the pre-sets (and post-sets) of transitions in T remain unchanged in $\tilde{M}$, hence $\varphi \in \mathcal{L}(M) \implies \varphi \in \mathcal{L}(\tilde{M})$. □

Corollary 1 *(Easy soundness is preserved after relaxing a synchronous product net). With $M^S = (N^S, m_i^S, m_f^S)$ and $\tilde{M}^S = (\tilde{N}^S, \tilde{m}_i^S, \tilde{m}_f^S)$ a synchronous product t-PNID model and the corresponding relaxed model. $\tilde{M}^S$ is easy sound if M^S is easy sound, i.e., $m_i \xrightarrow{*} m_f \implies \tilde{m}_i \xrightarrow{*} \tilde{m}_f$.*

Proof. From Lemma 2, any run $\varphi \in \mathcal{L}(M^S)$ is also a run in M^S, i.e., $\varphi \in \mathcal{L}(\tilde{M}^S)$ and therefore $\tilde{m}_i \xrightarrow{*} \tilde{m}_f$. □

As we show in Lemma 1, the relaxed alignment has the property that for any $o \in \mathcal{O}$, its projected trace alignment $\tilde{\gamma}\restriction_o'$ is also a valid trace alignment γ_{L_o}, i.e., it is an alignment of the trace L_o onto the corresponding projection process model $M\restriction_o$.

Lemma 3 *(Projected trace alignments are trace alignments). Let $\tilde{\gamma}$ be a relaxed alignment. Its projected trace alignment $\tilde{\gamma}\restriction'_o$, for each $o \in \mathcal{O}$, is an alignment for $L\restriction_o$ on the corresponding projection net $M\restriction_o$, where $\tilde{\gamma}\restriction'_o = \tilde{\gamma}\restriction_o \setminus T^C_\mu$, i.e., without create/destroy correlation moves.*

Proof. Let us write $\tilde{\gamma}_o$ and L_o for abbreviation of $\tilde{\gamma}\restriction'_o$ and $L\restriction_o$. By definition of projections, we have $\tilde{M}\restriction_o = M\restriction_o$ and $\tilde{M}^L\restriction_o = M^L\restriction_o = M^{L_o}$. Hence, $\tilde{M}^S\restriction_o = M^S\restriction_o = M^{S_o}$ and

(1) $\overline{\tilde{\gamma}_o\restriction_L} = \overline{L_o}$ and $\prec_{L_o} \subseteq \prec_{\tilde{\gamma}_o\restriction_L}$;
(2) $\tilde{\gamma}_o\restriction_T \in \mathcal{L}(M\restriction_o)$, i.e., $m_i\restriction_o \xrightarrow{\tilde{\gamma}_o\restriction_T} m_f\restriction_o$.

This implies that $\tilde{\gamma}_o \in \mathcal{L}(M^S_o)$, and therefore we can write $\tilde{\gamma}_o = \gamma_{L\restriction_o, M\restriction_o}$. □

5.2 Optimal Relaxed Alignments: Three Criteria

The aim for the optimal relaxed alignment is to synchronize as much behavior as possible for as many objects simultaneously as possible, balancing it with optimization of alignments on individual objects. This requires a cost function with the following criteria:

The first-order criterium, based on $k_1 : \mathcal{L}(\tilde{M}^S) \to \mathbb{N}$, is similar to regular alignments. We want to minimize the number of deviating moves, and more specifically, the number of distinguishable objects involved in deviating moves. Formally,

$$k_1(\tilde{\gamma}) = \sum_{t_\mu \in \tilde{\gamma} \cap \Gamma_{lm}, \ell(t) \neq \tau} |supp(\text{objects}(t_\mu))| \tag{9}$$

should be minimized. Later we show that this criterium ensures optimal projected trace alignments.

The second-order criterium, based on $k_2 : \mathcal{L}(\tilde{M}^S) \to \mathbb{N}$, synchronizes behavior between objects where possible, such that a projected move indicates that only the involved objects allow for the behavior. In other words, the alignment is minimally relaxed. Formally,

$$k_2(\tilde{\gamma}) = \sum_{(t_V)_\mu \in \tilde{\gamma}} |Var(t)| - |Var(t_V)| \tag{10}$$

should be minimized.

The third-order criterium, based on $k_3 : \mathcal{L}(\tilde{M}^S) \to \mathbb{N}$, is to minimize the creation (and destruction) of correlations, so that they are only used either to decrease k_1 and/or k_2, or to create (destroy) correlations as efficiently as possible. For example, if we want to destroy the correlation of three objects with roles x, y and z, we prefer the firing of transition $\tau^{p,\{\{x\},\{y\},\{z\}\}}_{\text{destroy}}$ over the firings of transitions $\tau^{p,\{\{x,y\},\{z\}\}}_{\text{destroy}}$ and $\tau^{p\restriction_{\{x,y\}},\{\{x\},\{y\}\}}_{\text{destroy}}$, even though they achieve the same result. Formally,

$$k_3(\tilde{\gamma}) = |\{t_\mu \in \tilde{\gamma} \mid t \in T^C\}| \tag{11}$$

should be minimized.

We define our cost function and show that it respects criteria k_1 to k_3 for certain choices for parameter ϵ used in the definition. For any $(t_V^{(e)}, t_V) \in \tilde{T}^S$:

$$c((t_V^{(e)}, t_V)_\mu) = \begin{cases} |V| + (|\mathit{Var}(t)| - |V|)\epsilon & \text{if } \gg \in \{t^{(e)}, t\} \wedge \ell(t) \neq \tau \\ \epsilon^2 & \text{if } t_V \in T^C \\ (|\mathit{Var}(t)| - |V|)\epsilon & \text{otherwise} \end{cases} \tag{12}$$

We overload the notation to denote the cost of a complete alignment $\tilde{\gamma} \in \mathcal{L}(\tilde{M}^S)$:

$$c(\tilde{\gamma}) = \sum_{(t_V^{(e)}, t_V)_\mu \in \tilde{\gamma}} c((t_V^{(e)}, t_V)_\mu) \tag{13}$$

which we rewrite to decompose it into components of c_1, c_2, and c_3.

$$\begin{aligned} c(\tilde{\gamma}) = & \sum_{(t_V^{(e)}, t_V)_\mu \in \tilde{\gamma} \cap \Gamma_{lm}, \ell(t) \neq \tau} |V| + (|\mathit{Var}(t)| - |V|)\epsilon + \\ & \sum_{(t_V^{(e)}, t_V)_\mu \in \tilde{\gamma} \cap \Gamma_s} (|\mathit{Var}(t)| - |V|)\epsilon + \sum_{t_\mu \in \tilde{\gamma}, t \in T^C} \epsilon^2 \\ = & \sum_{(t_V^{(e)}, t_V)_\mu \in \tilde{\gamma} \cap \Gamma_{lm}, \ell(t) \neq \tau} |V| + \sum_{(t_V^{(e)}, t_V)_\mu \in \tilde{\gamma}, t \notin T^C} (|\mathit{Var}(t)| - |V|)\epsilon + \sum_{t_\mu \in \tilde{\gamma}, t \in T^C} \epsilon^2 \\ = & \; c_1(\tilde{\gamma}) + c_2(\tilde{\gamma}) + c_3(\tilde{\gamma}) \end{aligned} \tag{14}$$

with $c_1(\tilde{\gamma}) = \sum_{(t_V^{(e)}, t_V)_\mu \in \tilde{\gamma} \cap \Gamma_{lm}, \ell(t) \neq \tau} |V|$, $c_2(\tilde{\gamma}) = \sum_{(t_V^{(e)}, t_V)_\mu \in \tilde{\gamma}, t \notin T^C} (|\mathit{Var}(t)| - |V|)\epsilon$, $c_3(\tilde{\gamma}) = \sum_{t_\mu \in \tilde{\gamma}, t \in T^C} \epsilon^2$.

Minimizing this cost function results in an optimal relaxed alignment $\tilde{\gamma}^*$ with the following properties:

Lemma 4. *There is a value for ϵ such that $\tilde{\gamma}^*$ has minimum number of deviating involved objects, i.e. $k_1(\tilde{\gamma})$ is minimized.*

Proof. We have, by definition, for any (relaxed) transition firing $(t_V^{(e)}, t_V)_\mu \in \Gamma$:

$$|\mathit{supp}(\text{objects}((t_V^{(e)}, t_V)_\mu))| = |\mathit{Var}((t_V^{(e)}, t_V))| = |V| \tag{15}$$

Therefore for any relaxed alignment $\tilde{\gamma} \in \tilde{M}^S$, w: $c_1(\tilde{\gamma}) = k_1(\tilde{\gamma})$.

Now suppose there is a relaxed alignment $\tilde{\gamma}'$ with $k_1(\tilde{\gamma}') < k_1(\tilde{\gamma}^*)$, then $k_1(\tilde{\gamma}^*) \geq k_1(\tilde{\gamma}') + 1$.

$$\begin{aligned} c(\tilde{\gamma}^*) &= c_1(\tilde{\gamma}^*) + c_2(\tilde{\gamma}^*) + c_3(\tilde{\gamma}^*) = k_1(\tilde{\gamma}^*) + c_2(\tilde{\gamma}^*) + c_3(\tilde{\gamma}^*) \\ &\geq k_1(\tilde{\gamma}') + 1 + c_2(\tilde{\gamma}^*) + c_3(\tilde{\gamma}^*) \\ &= c(\tilde{\gamma}') - c_2(\tilde{\gamma}') - c_3(\tilde{\gamma}') + 1 + c_2(\tilde{\gamma}^*) + c_3(\tilde{\gamma}^*) \\ &> c(\tilde{\gamma}') \text{ iff } c_2(\tilde{\gamma}^*) + c_3(\tilde{\gamma}^*) - c_2(\tilde{\gamma}') - c_3(\tilde{\gamma}') < 1 \end{aligned} \tag{16}$$

which holds if for both $\tilde{\gamma} \in \{\tilde{\gamma}^*, \tilde{\gamma}'\}$ we have $c_2(\tilde{\gamma}) + c_3(\tilde{\gamma}) < 1$. This is achieved with ϵ such that

$$0 < \epsilon < 1/\max(\sum_{(t_V^{(e)}, t_V)_\mu \in \tilde{\gamma}, t \notin T^C} (|Var(t)| - |V|), |\{t_\mu \in \tilde{\gamma} \mid t \in T^C\}|) \quad (17)$$

Then $c(\tilde{\gamma}^*) > c(\tilde{\gamma}')$ is a contradiction, since $\tilde{\gamma}^*$ is optimal.

Lemma 5 *($\tilde{\gamma}^*$ is minimally relaxed as second-order criterium). Number of relaxations $k_2 = \sum_{(t_V)_\mu \in \tilde{\gamma}}(|Var(t)| - |Var(t_V)|)$.*

Proof. By definition, $|Var(t_V)| = |V|$, so $c_2(\tilde{\gamma}) = k_2(\tilde{\gamma})\epsilon$.

Suppose there is a relaxed alignment $\tilde{\gamma}'$ with $k_1(\tilde{\gamma}^*) = k_1(\tilde{\gamma}')$ and has $k_2(\tilde{\gamma}') < k_2(\tilde{\gamma}^*)$, then $k_2(\tilde{\gamma}^*) \geq k_2(\tilde{\gamma}') + 1$.

From Lemma 4, we know that $c_1(\tilde{\gamma}) = k_1(\tilde{\gamma})$ for any $\tilde{\gamma} \in \mathcal{L}(\tilde{M}^S)$, so we can rewrite the cost function as:

$$\begin{aligned} c(\tilde{\gamma}^*) &= c_1(\tilde{\gamma}^*) + c_2(\tilde{\gamma}^*) + c_3(\tilde{\gamma}^*) = c_1(\tilde{\gamma}^*) + k_2(\tilde{\gamma}^*)\epsilon + c_3(\tilde{\gamma}^*) \\ &\geq c_1(\tilde{\gamma}^*) + (k_2(\tilde{\gamma}') + 1)\epsilon + c_3(\tilde{\gamma}^*) \\ &= c_1(\tilde{\gamma}^*) + c(\tilde{\gamma}') - c_1(\tilde{\gamma}') - c_3(\tilde{\gamma}') + \epsilon + c_3(\tilde{\gamma}^*) \\ &> c(\tilde{\gamma}') \text{ iff } c_3(\tilde{\gamma}^*) - c_3(\tilde{\gamma}') < \epsilon \end{aligned} \quad (18)$$

which is the case if $0 < \epsilon < 1$ ($\epsilon^2 < \epsilon \iff \epsilon < 1$), and then again $c(\tilde{\gamma}^*) > c(\tilde{\gamma}')$ is a contradiction, since $\tilde{\gamma}^*$ is optimal.

Correlations are *intact* when they are created and destroyed throughout their original creation and destruction.

Lemma 6. *$\tilde{\gamma}^*$ has maximum intact correlations, satisfying the third-order criterium.*

Proof. By definition we have $c_3(\tilde{\gamma}) = k_3(\tilde{\gamma})\epsilon^2$. Now suppose there is a relaxed alignment $\tilde{\gamma}'$ with $k_1(\tilde{\gamma}^*) = k_1(\tilde{\gamma}')$, $k_2(\tilde{\gamma}^*) = k_2(\tilde{\gamma}')$ and has $k_3(\tilde{\gamma}') < k_3(\tilde{\gamma}^*)$, then $k_3(\tilde{\gamma}^*) \geq k_3(\tilde{\gamma}') + 1$.

From Lemma 4 and 5, we know that $c_1(\tilde{\gamma}) = k_1(\tilde{\gamma})$ and $c_2(\tilde{\gamma}) = k_2(\tilde{\gamma})\epsilon$ for any $\tilde{\gamma} \in \mathcal{L}(\tilde{M}^S)$, so we can rewrite the cost function as:

$$\begin{aligned} c(\tilde{\gamma}^*) &= c_1(\tilde{\gamma}^*) + c_2(\tilde{\gamma}^*) + c_3(\tilde{\gamma}^*) = c_1(\tilde{\gamma}^*) + c_2(\tilde{\gamma}^*)\epsilon + k_3(\tilde{\gamma}^*)\epsilon^2 \\ &\geq c_1(\tilde{\gamma}^*) + c_2(\tilde{\gamma}^*)\epsilon + (k_3(\tilde{\gamma}^*) + 1)\epsilon^2 \\ &= c_1(\tilde{\gamma}^*) + c_2(\tilde{\gamma}^*)\epsilon + c(\tilde{\gamma}') - c_1(\tilde{\gamma}') - c_2(\tilde{\gamma}') + \epsilon^2 \\ &> c(\tilde{\gamma}') + \epsilon^2 \end{aligned} \quad (19)$$

which is the case if $\epsilon > 0$, and then again $c(\tilde{\gamma}^*) > c(\tilde{\gamma}')$ is a contradiction, since $\tilde{\gamma}^*$ is optimal.

So if ϵ is chosen correctly, i.e., $0 < \epsilon < 1$ and respecting Eq. 17, minimizing cost function c results in the optimal relaxed alignment having the desired properties.

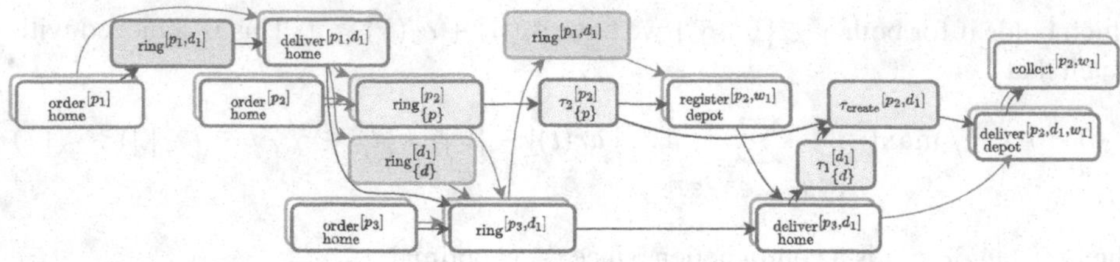

Fig. 8. Optimal relaxed alignment $\tilde{\gamma}^*$ with conforming, partial conforming, deviating, partial deviating, and create/destroy correlation moves annotated in white, blue, red, purple, and grey respectively. (Color figure online)

The optimal relaxed alignment for the running example is shown in Fig. 8. From $\tilde{\gamma}^*$, we see that the deviations for package p_1 are for all involved objects since it does not contain any projections. The deviations discussed in Sect. 3 for package p_2, caused by the deliverer d_1, are only contained in the projection of d_1. From the move $\text{ring}^{[p_2]}_{\{p\}}$, it is evident that otherwise required objects are missing. $\tilde{\gamma}^*$ shows the individually optimal alignment objects, while maximizing the synchronization of interaction where possible, i.e., if you project $\tilde{\gamma}^*$ back onto the individual objects, you get the optimal trace alignments shown in Fig. 4. For package p_1, having two optimal trace alignments, the one is chosen that maximizes the interaction with d_1. Next, we show that the optimal relaxed alignment always has these properties.

5.3 Other Properties Regarding Projected Trace Alignments

Following from the criteria k_1 to k_3 defined above, and the fact that the cost function c from Eq. 12 respects these criteria, we show two other properties of any relaxed alignment regarding its projected trace alignments, through Lemmas 7 and 8. Finally, we use these lemmas to show, in Theorem 2, that the projected trace alignments of the optimal relaxed alignment are also optimal trace alignments. Firstly, the difference between the total cost of a relaxed alignment and of the corresponding trace alignments is smaller than 1:

Lemma 7 *($c(\tilde{\gamma}) - \sum_{o \in \mathcal{O}} c(\gamma_{L_o}) < 1$ for some $0 < \epsilon < 1$). Let c be as defined in Eq. 12, $\tilde{\gamma} \in \mathcal{L}(\tilde{M}^S)$ be a relaxed alignment and let γ_{L_o}, $o \in \mathcal{O}$, be the trace alignment such that $\gamma_{L_o} = \tilde{\gamma}\restriction'_o$, i.e., γ_{L_o} is linked to the projected trace alignment without create/destroy correlation moves. There exists an ϵ, such that $0 < \epsilon < 1$ and $c(\tilde{\gamma}) - \sum_{o \in \mathcal{O}} c(\gamma_{L_o}) < 1$.*

Proof. We have $c(\tilde{\gamma}) = c_1(\tilde{\gamma}) + c_2(\tilde{\gamma}) + c_3(\tilde{\gamma}) = k_1(\tilde{\gamma}) + k_2(\tilde{\gamma})\epsilon + k_3(\tilde{\gamma})\epsilon^2$ with $k_1(\tilde{\gamma}) = \sum_{o \in \mathcal{O}} c(\tilde{\gamma}\restriction'_o) = \sum_{o \in \mathcal{O}} c(\gamma_{L_o})$ by definition and construction of γ_{L_o}. Furthermore, we know from Lemmas 4 to 6 that $k_2(\tilde{\gamma})\epsilon + k_3(\tilde{\gamma})\epsilon^2 < 1$ with any ϵ respecting Eq. 17. Therefore, $c(\tilde{\gamma}) - \sum_{o \in \mathcal{O}} c(\gamma_{L_o}) = k_2(\tilde{\gamma})\epsilon + k_3(\tilde{\gamma})\epsilon^2 < 1$. □

Secondly, we can construct a relaxed alignment such that the projected trace alignments are optimal trace alignments:

Lemma 8 *(There exists an alignment with optimal projected trace alignments). Let $\tilde{M}^S$ be the relaxed product model of M^S. There is a relaxed alignment $\tilde{\gamma} \in \mathcal{L}(\tilde{M}^S)$ such that for every $o \in \mathcal{O}$, $\tilde{\gamma}\restriction'_o = \gamma^*_{L_o, M\restriction_o}$.*

Proof. We show that combining optimal trace alignments, i.e., $\gamma = \bigcup_{o \in \mathcal{O}} \gamma^*_{L_o}$ with $\gamma^*_{L_o}$ any optimal alignment for $o \in \mathcal{O}$ on $M\restriction_o$, results in a relaxed alignment without synchronization. Additionally, $\prec_\gamma$ is extended with additional relations from L.

By construction we have $\overline{\gamma\restriction_o} = \overline{\gamma^*_{L_o}}$ and $\prec_{\gamma^*_{L_o}} \subseteq \prec_{\gamma\restriction_o}$, so $\gamma\restriction_o \in \mathcal{L}(M\restriction_o)$. Since $M\restriction_o \subseteq \tilde{M}$ by definition of $\tilde{M}$, we have $\gamma\restriction_o \in \mathcal{L}(\tilde{M})$. Since there is no synchronization, there are no inter-object dependencies to adhere to from M, hence $\gamma \in \mathcal{L}(\tilde{M})$. □

Using the properties above, we show that the optimal relaxed alignment incorporates projected trace alignments that are optimal as well:

Theorem 2. *Projected trace alignments of $\tilde{\gamma}^*$ are optimal trace alignments.*

Let $\tilde{\gamma}^* \in \mathcal{L}(\tilde{M}^S)$ be an optimal relaxed alignment. For every $o \in \mathcal{O}$, $\tilde{\gamma}^*\restriction'_o$ is an optimal trace alignment, i.e., $\tilde{\gamma}^*\restriction'_o = \gamma^*_{L_o}$.

Proof. We prove the property of optimal projected trace alignments by contradiction. Let us assume that there is a projected trace alignment in $\tilde{\gamma}^*$ which is not optimal, i.e., $\exists_{o \in \mathcal{O}} \tilde{\gamma}^*\restriction'_o = \gamma_{L_o}$. From Lemma 1, we know that every projected trace alignment is a trace alignment, i.e., $\tilde{\gamma}^*\restriction'_o = \gamma^{(*)}_{L_o}$ (i.e., regardless of optimality).

From Lemma 8, we know that there exists a relaxed alignment with optimal projected trace alignment, i.e., $\tilde{\gamma} \in \mathcal{L}(\tilde{M}^S)$ with $\forall_{o \in \mathcal{O}} \tilde{\gamma}\restriction_o = \gamma^*_{L_o}$. Furthermore, with an appropriate choice of ϵ, i.e., respecting Eq. 17, we know from Lemma 7 that for both $\tilde{\gamma}^*$ and $\tilde{\gamma}$, we have

$$c(\tilde{\gamma}^*) - \sum_{o' \in \mathcal{O} \setminus \{o\}} c(\gamma^*_{L_{o'}}) - c(\gamma_{L_o}) = k_2(\tilde{\gamma}^*)\epsilon + k_3(\tilde{\gamma}^*)\epsilon^2 < 1 \tag{20}$$

$$c(\tilde{\gamma}) - \sum_{o' \in \mathcal{O} \setminus \{o\}} c(\gamma^*_{L_{o'}}) - c(\gamma^*_{L_o}) = k_2(\tilde{\gamma})\epsilon + k_3(\tilde{\gamma})\epsilon^2 < 1 \tag{21}$$

We know from optimal regular alignment that $c(\gamma_{L_o}) \geq c(\gamma^*_{L_o}) + 1$, so we can rewrite the cost of $\tilde{\gamma}^*$ such that

$$\begin{aligned} c(\tilde{\gamma}^*) &= k_2(\tilde{\gamma}^*)\epsilon + k_3(\tilde{\gamma}^*)\epsilon^2 + \sum_{o' \in \mathcal{O} \setminus \{o\}} c(\gamma^*_{L_{o'}}) + c(\gamma_{L_o}) \\ &\geq k_2(\tilde{\gamma}^*)\epsilon + k_3(\tilde{\gamma}^*)\epsilon^2 + \sum_{o' \in \mathcal{O} \setminus \{o\}} c(\gamma^*_{L_{o'}}) + c(\gamma^*_{L_o}) + 1 \\ &= c(\tilde{\gamma}) + 1 + (k_2(\tilde{\gamma}^*)\epsilon + k_3(\tilde{\gamma}^*)\epsilon^2) - (k_2(\tilde{\gamma})\epsilon + k_3(\tilde{\gamma})\epsilon^2) \\ &> c(\tilde{\gamma}) \text{ (by Eqs. 20 and 21)} \end{aligned} \tag{22}$$

which is a contradiction since $\tilde{\gamma}^*$ is optimal. □

Lemmas 4 to 6 and Theorem 2 prove that the optimal trace alignment combines objects maximally with the least number of objects deviating and maximum intact correlations, while the projected trace alignments represent optimal trace alignments. Hence, the goal described in Sect. 3 is accomplished.

5.4 Computational Complexity

The complexity of computing the optimal alignment is based on the size of the synchronous product model. In the relaxed alignment problem, the synchronous product model grows combinatorially in the number of interacting objects. However, the combinatorial growth happens locally for the projections of transitions in the synchronous product model. In practice, we expect only a low number of interacting objects in individual activities, which greatly mitigates the combinatorial explosion.

Furthermore, the search strategy (e.g., A^*) can be enhanced by providing an upper bound for the total cost of the relaxed alignment, to prune non-optimal solutions in an early stage. From Theorem 2 and Lemma 7, we know that the optimal relaxed alignment contains optimal trace alignments and its cost never exceeds that of the optimal trace alignments by more than 1. The optimal trace alignments can be computed first, and these computations do not suffer from the increased complexity. After that we can compute the upper bound for the optimal relaxed alignment, namely $\sum_{o \in \mathcal{O}} c(\gamma^*_{L_o}) + 1$.

Lastly, the relaxed synchronous product may exhibit structural patterns that favor other methods for computing the optimal alignment [9], e.g., using SAT encodings [5] or automated planning [8]. Further investigation into such structures is needed to verify this conjecture.

6 Conclusion

In real-life processes, different objects interact with each other when executing different tasks. Alignments aim at reconciling a system log and a normative process model, providing a richer representation of the process by exposing deviations in the objects' recorded and modeled behaviors, and violations of inter-object dependencies.

Due to the nature of object interactions specified in the model, traditional system alignments provide a pessimistic view on the conformance for individual objects. We exploit the multiple perspectives of the system by adding relaxations before generating alignments, thus allowing for partial matching of activities in the logged and modeled behaviors. Relaxed alignments address this problem in a general sense, relaxing constraints for some objects and ensuring that the optimal relaxed system alignment contains optimal trace alignments on the individual level.

In a real-life setting, it is interesting to incorporate domain knowledge into the definition of an appropriate cost function for relaxed moves. This might both reduce the computational complexity and enhancing the interpretability of the relaxed alignments.

Acknowledgements. This work is done within the project "Certification of production process quality through Artificial Intelligence (CERTIF-AI)", funded by NWO (project number: 17998).

References

1. van der Aalst, W.M.P.: The application of Petri nets to workflow management. J. Circuits Syst. Comput. **8**(01), 21–66 (1998)
2. van der Aalst, W.M.P., Adriansyah, A., van Dongen, B.F.: Replaying history on process models for conformance checking and performance analysis. Wiley Interdisc. Rev. Data Min. Knowl. Discov. **2**(2), 182–192 (2012)
3. van der Aalst, W.M.P., Berti, A.: Discovering object-centric Petri nets. Fundam. Inform. **175**(1–4), 1–40 (2020)
4. Alizadeh, M., Lu, X., Fahland, D., Zannone, N., van der Aalst, W.M.P.: Linking data and process perspectives for conformance analysis. Comput. Secur. **73**, 172–193 (2018)
5. Boltenhagen, M., Chatain, T., Carmona, J.: Optimized sat encoding of conformance checking artefacts. Computing **103**(1), 29–50 (2021)
6. Carmona, J., van Dongen, B.F., Solti, A., Weidlich, M.: Conformance Checking - Relating Processes and Models. Springer, Cham (2018). https://doi.org/10.1007/978-3-319-99414-7
7. de Leoni, M., van der Aalst, W.M.P.: Aligning event logs and process models for multi-perspective conformance checking: an approach based on integer linear programming. In: Daniel, F., Wang, J., Weber, B. (eds.) BPM 2013. LNCS, vol. 8094, pp. 113–129. Springer, Heidelberg (2013). https://doi.org/10.1007/978-3-642-40176-3_10
8. de Leoni, M., Marrella, A.: Aligning real process executions and prescriptive process models through automated planning. Expert Syst. Appl. **82**, 162–183 (2017)
9. Dongen, B.F.: Efficiently computing alignments. In: Weske, M., Montali, M., Weber, I., vom Brocke, J. (eds.) BPM 2018. LNCS, vol. 11080, pp. 197–214. Springer, Cham (2018). https://doi.org/10.1007/978-3-319-98648-7_12
10. Fahland, D.: Describing behavior of processes with many-to-many interactions. In: Donatelli, S., Haar, S. (eds.) PETRI NETS 2019. LNCS, vol. 11522, pp. 3–24. Springer, Cham (2019). https://doi.org/10.1007/978-3-030-21571-2_1
11. Ghilardi, S., Gianola, A., Montali, M., Rivkin, A.: Petri net-based object-centric processes with read-only data. Inf. Syst. **107**, 102011 (2022)
12. van Hee, K., Sidorova, N., Voorhoeve, M.: Soundness and separability of workflow nets in the stepwise refinement approach. In: van der Aalst, W.M.P., Best, E. (eds.) ICATPN 2003. LNCS, vol. 2679, pp. 337–356. Springer, Heidelberg (2003). https://doi.org/10.1007/3-540-44919-1_22
13. van Hee, K., Sidorova, N., Voorhoeve, M.: Resource-constrained workflow nets. Fundam. Inform. **71**(2,3), 243–257 (2006)
14. Mannhardt, F., De Leoni, M., Reijers, H.A., van der Aalst, W.M.P.: Balanced multi-perspective checking of process conformance. Computing **98**(4), 407–437 (2016)
15. Mozafari Mehr, A.S., de Carvalho, R.M., van Dongen, B.: Detecting privacy, data and control-flow deviations in business processes. In: Nurcan, S., Korthaus, A. (eds.) CAiSE 2021. LNBIP, vol. 424, pp. 82–91. Springer, Cham (2021). https://doi.org/10.1007/978-3-030-79108-7_10
16. Montali, M., Rivkin, A.: Model checking Petri nets with names using data-centric dynamic systems. Formal Aspects Comput. **28**(4), 615–641 (2016)
17. Montali, M., Rivkin, A.: DB-nets: on the marriage of colored Petri nets and relational databases. In: Koutny, M., Kleijn, J., Penczek, W. (eds.) Transactions on Petri Nets and Other Models of Concurrency XII. LNCS, vol. 10470, pp. 91–118. Springer, Heidelberg (2017). https://doi.org/10.1007/978-3-662-55862-1_5
18. Murata, T.: Petri nets: properties, analysis and applications. Proc. IEEE **77**(4), 541–580 (1989)

19. Peterson, J.L.: Petri Net Theory and the Modeling of Systems. Prentice Hall PTR, Hoboken (1981)
20. Sommers, D., Sidorova, N., van Dongen, B.F.: Aligning event logs to resource-constrained petri nets. In: Bernardinello, L., Petrucci, L. (eds.) PETRI NETS 2022. LNCS, vol. 13288, pp. 325–345. Springer, Cham (2022). https://doi.org/10.1007/978-3-031-06653-5_17
21. Sommers, D., Sidorova, N., van Dongen, B.: Exact and approximated log alignments for processes with inter-case dependencies. In: Gomes, L., Lorenz, R. (eds.) PETRI NETS 2023. LNCS, vol. 13929, pp. 99–119. Springer, Cham (2023). https://doi.org/10.1007/978-3-031-33620-1_6
22. van der Werf, J.M.E.M., Rivkin, A., Polyvyanyy, A., Montali, M.: Data and process resonance: identifier soundness for models of information systems. In: Bernardinello, L., Petrucci, L. (eds.) PETRI NETS 2022. LNCS, vol. 13288, pp. 369–392. Springer, Cham (2022). https://doi.org/10.1007/978-3-031-06653-5_19

Process Comparison Using Petri Net Decomposition

Tobias Brockhoff[1(✉)], Moritz Nicolas Gose[2], Merih Seran Uysal[1], and Wil M.P. van der Aalst[1]

[1] Chair of Process and Data Science, RWTH Aachen University, Aachen, Germany
{brockhoff,uysal,wvdaalst}@pads.rwth-aachen.de
[2] RWTH Aachen University, Aachen, Germany
moritz.gose@rwth-aachen.de

Abstract. Business processes drive the value creation at companies requiring them to constantly monitor and improve the former. The field of *Process Comparison* (PC) offers promising approaches to gain insight into differences between variants of a process that one can leverage to improve the latter. For example, one might consider the same process at different points in time or at different sites. Recent PC methods consider event logs containing data on real-life process executions the single source of truth. However, there often exist additional specifications that can be represented as Petri nets. In this paper, we propose an approach that leverages a given Petri net to compare two event logs in a hierarchical manner. To this end, we decompose the provided net into subprocesses and extract data on their executions from the event logs. Based on these executions, we exemplify how one can flexibly assess different aspects of a process (e.g., control flow, performance, or conformance). Using statistical tests, we eventually detect differences between subprocesses with respect to a selected aspect. Despite the approach is mostly agnostic to the decomposition applied, we present a decomposition strategy that we deem particularly suitable for PC. For this purpose, we consider the ned Process Structure Tree of a Petri net and propose a novel preprocessing approach to improve the final decomposition. We implemented the approach in ProM and evaluate it in a real-life case study.

Keywords: Process Mining · Process Comparison · Process Variant Analysis · Business Process Intelligence

1 Introduction

Modern information systems record increasing amounts of data on business process executions. *Event data* are a special type of these data where each data point is an *event* comprising a *timestamp*, an *activity* name, and a *case id* related to a business case. *Process mining* methods are concerned with the analysis of event data and the implementation of event-data-to-knowledge pipelines. Ultimately,

L. M. Kristensen and J. M. van der Werf (Eds.): PETRI NETS 2024, LNCS 14628, pp. 83–105, 2024.
https://doi.org/10.1007/978-3-031-61433-0_5

process mining aims to provide insight to improve the process. *Process Comparison* (PC) approaches take this idea one step further revealing differences between two (or more) process variants. Differences can, for example, concern the process' control flow (e.g., decision likelihoods) or performance (e.g., cycle times). Thereby, PC provides insight into the effect of changes, comparing process executions before and after a change, or into particularities of a process' environment, comparing implementations at different sites.

Recent PC approaches often consider event data as the single source of truth and, therefore, only take event logs as inputs [8,10,25]. Yet, this neglects process models (e.g., BPMN models) as another valuable source of information. In practice, BPMN models are fundamental for the design and re-design of enterprise information systems [3] making them an essential part of a process' documentation. For analytical purposes, these models can be seamlessly converted into Petri nets. The advantages of incorporating process models into PC are manifold: (i) process models help to *manage the complexity of processes.* PC approaches solely based on event data often rely on follows relations between activities. However, example traces cannot properly represent these relations for large and concurrent processes (e.g., production processes). Besides, process models allow to explicitly consider duplicated transitions and to model and analyze interesting relations. For example, in [4], the authors add user-defined places to measure times for larger subprocesses within a single process variant. (ii) Submodels often define *logical units* enabling us to analyze differences at different levels of granularity. (iii) Finally, models provide a *natural context* to present the results of PC. They are the common ground in a projected view of the differences. Despite the advantages of model-based analysis, merely considering models usually results in a loss of information. For instance, model notations such as Petri nets do not support modeling frequencies. Besides, the actual process executions can deviate from the model or exhibit dependencies not represented by the model.

In this paper, we propose a hybrid approach, which takes two event logs and a shared process model as an input. Figure 1 shows the main concepts of the approach. In contrast to existing approaches, we compare the event logs in a hierarchical manner enabling process analysts to conduct an analysis at different levels of granularity. For example, in Fig. 1, we detect a cycle time difference for a large subprocess on the first level of the hierarchy. By drilling down into the subprocess, one can refine this knowledge: (i) there is no significant difference regarding the time it takes to execute the redo part and (ii) one concurrent branch is even executed faster in the other process variant.

In this work, we use process models, transformed in Petri nets, to enable a hierarchical analysis, but we discover differences based on the event data. To this end, we decompose the Petri net into a hierarchy of subprocesses and compute subprocess executions relating subprocesses and event data. Based on these executions, we define measurements that assess different aspects of a process. In doing so, we not only consider the control-flow and performance perspective, but also propose to compare how process variants deviate from the model.

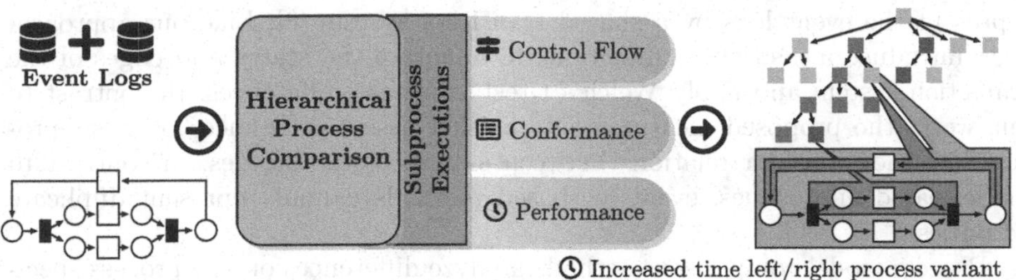

Fig. 1. Hierarchical Process Comparison on two event logs using a shared process model. First, we decompose the Petri net and relate the resulting subprocesses with the event data. Based on these subprocess executions, we compare the two process variants with respect to various perspectives. Ultimately, each vertex in the decomposition shows whether a difference was discovered for the respective subprocess. The color is chosen based on the effect size of the difference. Selecting a subprocess, it is highlighted in the original model.

While the approach is mostly agnostic to the decomposition applied, it is usually preferable that the nodes in the decomposition represent logically coherent subprocesses. We therefore propose a decomposition strategy that leverages the *Refined Process Structure Tree* (RPST) of a model [27]. In the RPST, each node corresponds to a subprocess that is entered and exited via a single node, respectively. However, as illustrated in Sect. 4.3, relaxing this requirement can improve the decomposition (e.g., to handle long-term dependencies). To this end, we propose an additional preprocessing step preceding the RPST computation, where we use the event log to remove places to improve the block-structuredness of the Petri net and, consecutively, the decomposition.

The remainder of this paper is structured as follows: we discuss related works and preliminary concepts in Sects. 2 and 3, respectively. In Sects. 4.1 and 4.3, we discuss the decomposition approach and illustrate its application to PC in Sect. 4.2. In Sect. 5, we evaluate the method in a real-world case study. Finally, we give our conclusion and directions for future work in Sect. 6.

2 Related Work

Our approach is directly related to works on PC and Petri net decomposition as well as to approaches for decomposed conformance checking.

For a comprehensive survey on PC, we refer the reader to [26]. Despite not being limited to the control flow or performance perspective [22], most PC approaches consider these. One can distinguish PC approaches that consider event logs as input [7,8,10,22,25,28], compare process models [6,14,18], or require a process model and two (or more) event logs as input [1,12,17,23,30]. Log-based approaches often represent follows relation between activities by means of a graph [7,8,22,25,28]. Then, a statistical test is used to detect frequency differences [8,22,25]. For example, in their seminal approach, Bolt et al.

represent the event logs by a shared transition system [8]. Like our approach, they introduce measurement functions to annotate the states and edges of the transition system and apply Welch's t-test to detect differences. In contrast to our work, the proposed measurements neither assess differences on a subprocess level nor consider relations between subprocesses. Besides, in contrast to model-based approaches, event log-based methods cannot represent duplicate activities.

Process model comparison methods analyze differences on the process specification level [6,14,18]. If event data is available, only considering models would discard the event data as a valuable source of information providing insight into decision likelihoods and time. Therefore, approaches for data-driven PC using process models generally enrich the latter with information obtained from the event data. Thereby, one can investigate aspects like execution times and de facto path frequencies in the model. For example, in [17], the authors merge two process models into a difference model which is further enriched with instance traffic information. For each edge, they then detect frequency differences. In [30], Wynn et al. explicitly consider Petri nets. Like our approach, they use alignments [5] to relate the model with the (potentially slightly deviating) data. Compared to our method, their approach provides more detailed results on waiting times between transitions. However, they only consider pairs of transitions rather than larger subprocesses and their relations. Moreover, they do not validate the statistical significance of the differences returned. In general, to the best of our knowledge, there currently exists no *hierarchical*, model-based PC approach.

Petri net decomposition techniques simplify models to (heuristically) solve problems on subnets. For different applications, various approaches have been proposed [2,9,13,16,27,32]. In this work, we compare process variants with respect to coherent sub-workflows (i.e., subprocesses) of the original model. To this end, we build on the notion of *single-entry, single-exit* (SESE)-fragments, which have originally been proposed in [27] as a unique, hierarchical decomposition of a process model into self-contained subnets. This notion expands upon an earlier definition by Johnson et al. [15]. The resulting, more granular decomposition coined the term *Refined Process Structure Tree* (RPST). Further improvements and a more efficient way to compute the RPST have been proposed in [24]. In this work, we propose an additional pre-processing step that can improve the decomposition for the sake of PC but weakens the structural guarantees.

An application of SESE fragments related to our approach is decomposed conformance checking [20]. While we also consider conformance in the context of RPSTs, we use fragments to aggregate the diagnostics, similar to [21], rather than improving the computational efficiency.

3 Petri Nets and Process Mining Concepts

We denote sets by capital letters. Given a set S, its powerset is denoted by $\mathcal{P}(S)$, and the set of all multisets is denoted by $\mathcal{B}(S)$. For a multiset $m \in \mathcal{B}(S)$, the multiplicity and an element $s \in S$ is $m(s) \in \mathbb{N}$. Given two multisets $m_1, m_2 \in \mathcal{B}(S)$, we write $m_1 + m_2$ ($m_1 - m_2$) to denote their sum (difference). Besides, we write $m_1 \leq m_2$ if $m_1(s) \leq m_2(s)$ holds for all $s \in S$. In an abuse of notation, we also apply these operators to pairs of sets and multisets.

The Kleene star (S^*) represents the set of all finite sequences over a (countable infinite) alphabet S. Given a sequence $\boldsymbol{\sigma} = \langle \sigma_1, \dots, \sigma_n \rangle \in S^*$, we refer to its ith element as $\boldsymbol{\sigma}_i$. The length of $\boldsymbol{\sigma}$ is denoted by $|\boldsymbol{\sigma}|$. Let $I = \{i_1, \dots, i_m\} \subseteq \{1, \dots, |\boldsymbol{\sigma}|\}$ be a set of indices. Assuming the order $i_1 < \cdots < i_m$, the index set I induces the subsequence $\boldsymbol{\sigma}_{[I]} = \langle \boldsymbol{\sigma}_{i_1}, \dots, \boldsymbol{\sigma}_{i_m} \rangle$.

Petri Nets. Let $\mathcal{A}$ denote the universe of activity labels.

Definition 1 (Labeled Petri Net). *Let $\tau \notin \mathcal{A}$ be a special silent label. A labeled Petri net is a tuple $N = (P, T, F, l)$, where (i) P is a finite set of places, (ii) T is a finite set of transitions, (iii) $F \subseteq (P \times T) \cup (T \times P)$ is a flow relation, and (iv) $l \colon T \to \mathcal{A} \cup \{\tau\}$ is a labeling function.*

Let $N = (P, T, F, l)$ be a Petri net. For an element $x \in P \cup T$, its preset (postset) are defined as $^\bullet x := \{y | (y, x) \in F\}$ ($x^\bullet := \{y | (x, y) \in F\}$). Given a set of edges $F' \subseteq F$, we denote the set of adjacent places by $P_{F'} := \{p \in P \exists t \in T\, ((p, t) \in F' \vee (t, p) \in F')\}$. Likewise, for the set $T_{F'}$ of adjacent transitions. The edge-induced subnet $N_{F'}$ is the Petri net $(P_{F'}, T_{F'}, F', l \restriction_{T_{F'}})$, where $l \restriction_{T_{F'}}$ denotes the restriction of l on the adjacent transitions. The semantics of Petri nets are determined by marking places, firing (sequences of) transitions.

Definition 2 (Marking, Firing). *Let $N = (P, T, F, l)$ be a labeled Petri net. A marking $m \in \mathcal{B}(P)$ of N is a finite multiset of places. A transition $t \in T$ is* enabled *in m if $^\bullet t \leq m$. If t is* enabled, *firing t in m results in the marking $m' = (m - {^\bullet t}) + t^\bullet$, written as $m[t\rangle_N m'$.*

Definition 3 (Firing Sequence). *Let $N = (P, T, F, l)$ be a labeled Petri net and $m_I, m_F \in \mathcal{B}(P)$ be two markings. A sequence $\boldsymbol{\sigma} = \langle t_1, \dots, t_n \rangle \in T^*$ is a* valid firing sequence *from m_I to m_F of N, written $m_I \xrightarrow{\boldsymbol{\sigma}}_N m_F$, if there exist markings $m_1, \dots, m_{n+1}$ such that (i) $m_1 = m_I$, (ii) $m_{n+1} = m_F$, and (ii) for $1 \leq i \leq n$ we have $m_i[t_i\rangle_N m_{i+1}$.*

In process mining, a commonly considered class of Petri nets are workflow nets (WF-nets). A WF-net has a clear start and end, and all elements are on a directed path from the start to the end.

Definition 4 (Workflow Net (WF-net)). *A labeled Petri net $N = (P, T, F, l)$ is a labeled workflow net (WF-net) if (i) there is a unique source place p_I (i.e., $\{p_I\} = \{p \in P | {^\bullet p} = \emptyset\}$), (ii) there is a unique sink place p_F (i.e., $\{p_F\} = \{p \in P | p^\bullet = \emptyset\}$), and (iii) every node is on a path from p_I to p_F.*

The workflow system net (WF-sytem net) explicitly establishes a connection between a WF-net and its semantics (i.e., its initial and final marking).

Definition 5 (Workflow System Net (WF-sytem net)). *A workflow system net (WF-sytem net)* $\text{SN} = (N, m_I, m_F)$ *comprises a labeled WF-net* $N = (P, T, F, l)$*, the initial marking* $m_I = [p_I]$*, and the final marking* $m_F = [p_F]$*.*

The *language* of a WF-sytem net comprises all sequences of visible activity labels of its valid firing sequences. Finally, a WF-sytem net is safe if we cannot reach a marking where a place contains multiple tokens.

Definition 6 (Safeness). *A WF-sytem net* $\text{SN} = ((P, T, F, l), m_I, m_F)$ *is safe if there is no valid firing sequence* $\boldsymbol{\sigma} \in T^*$ *reaching a marking* $m' \in \mathcal{B}(P)$ *from* m_I *(i.e.,* $m_I \xrightarrow{\boldsymbol{\sigma}}_N m'$*) with* $m'(p) > 1$ *for some* $p \in P$*.*

Event Data. We leverage Petri nets to compare two process variants based on data of their real-life executions. Each process execution corresponds to a business case and comprises information on the activities executed for this particular case. For each activity execution, we record the activity's name and a timestamp. While additional attributes are possible, they are not considered in this work. An *event log* collects multiple process executions.

Definition 7 (Event Log). *Let* $\mathcal{A}$ *and* $\mathcal{T}$ *denote (countable infinite) universes of activity names and timestamps. The set of all activity executions is defined as* $\mathcal{E} := \mathcal{A} \times \mathcal{T}$*. A* trace $\boldsymbol{\sigma} = \langle (a_1, t_1), \ldots, (a_n, t_n) \rangle \in \mathcal{E}^*$ *is a finite sequence of activity executions respecting time—that is* $t_i \leq t_j$ *for all* $1 \leq i < j \leq n$*. An* event log $L \in \mathcal{B}(\mathcal{E}^*)$ *is a finite multiset of traces.*

Dealing with event data, we frequently use two types of projection. The projection $\pi_{S'} : S^* \to (S')^*$ of a sequence over a set S on a subset $S' \subseteq S$ only keeps the elements contained in S'. Consider the trace $\boldsymbol{\sigma} = \langle (a, 1), (b, 2), (c, 3) \rangle$. Projecting $\boldsymbol{\sigma}$ onto all executions of a and b yields the trace $\pi_{\{a,b\} \times \mathcal{T}}(\boldsymbol{\sigma}) = \langle (a, 1), (b, 2) \rangle$. Moreover, we write $\pi_{time} : \mathcal{E}^* \to \mathcal{T}^*$ ($\pi_{act} : \mathcal{E}^* \to \mathcal{A}^*$) to denote the projection of traces onto the associated timestamps (activities). For example, for $\boldsymbol{\sigma}$, we obtain the sequence $\pi_{time}(\boldsymbol{\sigma}) = \langle 1, 2, 3 \rangle$. In a slight abuse of notation, we also apply π^{time} and π^{act} to individual activity executions.

Alignments. In real life processes, the traces recorded might not perfectly match with the prescribed Petri net model. Introducing a dedicated skip symbol, we additionally require that columns either contain a single skip symbol or that the trace's activity matches the transition's label. Thereby, an alignment represents a joined, synchronized execution of the trace and the model.

Definition 8 (Alignment [2]). *Let* $\boldsymbol{\sigma} \in \mathcal{E}^*$ *be a trace and* $\text{SN} = (N, m_I, m_F)$*,* $N = (P, T, F, l)$*,* $l : T \to \mathcal{A}$ *be a WF-sytem net. Let* $\gg \notin \mathcal{A} \cup T$ *denote a special no-move symbol. We define the sets of synchronous, log, model, and all moves as* $M_{\text{SYNC}} := \{(e, t) | t \in T, l(t) \neq \tau, e \in \mathcal{E}, \pi^{act}(e) = l(t)\}$*,* $M_{\text{LM}} := \mathcal{E} \times \{\gg\}$*,* $M_{\text{MM}} := \{\gg\} \times T$*, and* $M_{\text{ALL}} := M_{\text{SYNC}} \cup M_{\text{LM}} \cup M_{\text{MM}}$*, respectively. An alignment of* $\boldsymbol{\sigma}$ *and* N *is a sequence of moves* $\boldsymbol{\gamma} \in M_{\text{ALL}}^*$ *such that*

(i) projection $\pi^1(\boldsymbol{\gamma})$ *on the first element, ignoring* $\gg$*, yields* $\boldsymbol{\sigma}$*—that is,* $\pi_{\not\gg}(\pi^1(\boldsymbol{\gamma})) = \boldsymbol{\sigma}$*—and*

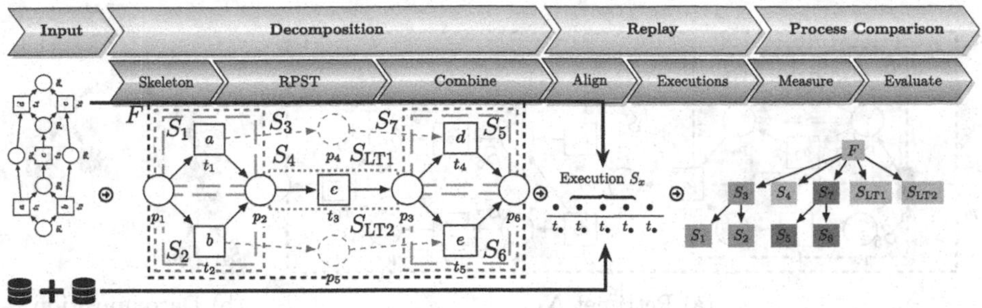

Fig. 2. Overview of our three-stage approach for comparing two process variants. First, we hierarchically decompose the model into subprocesses. Second, we replay the event data to relate the subprocesses with the data. Third, we compare the variants with respect to the subprocesses from various perspectives. Eventually, we project the differences onto the decomposition.

(ii) projection $\pi^2(\gamma)$ on the second element, ignoring $\gg$, yields a valid firing sequence from m_I to m_F of N—that is, $m_I \xrightarrow{\pi_{\not\gg}(\pi^2(\gamma))}_N m_F$,

where $\pi_{\not\gg} := \pi_{(\mathcal{E}\cup T)\setminus\{\gg\}}$ denotes the projection that removes skips.

The set of best alignments is typically determined by a cost function minimizing the number of log and visible model moves. In this work, we assume that a single best alignment is given, which can make our approach non-deterministic. Finally, given an alignment γ of a trace σ and WF-sytem net $\text{SN} = ((P, T, F, l), m_I, m_F)$, the marking reached after executing the first $1 \leq k \leq |\gamma|$ steps is $m^{al}_\gamma(k)$ with

$$m_I \xrightarrow{\pi_{\not\gg}(\pi^2(\gamma_{[\{1,\dots,k\}]}))}_N m^{al}_\gamma(k). \quad (1)$$

4 Hierarchical Process Comparison

This section presents our approach for comparing the executions of two process variants based on a shared process model. To avoid boundary cases (e.g., empty alignments), we assume that event logs do not contain empty traces and that WF-sytem nets have at least one transition. Figure 2 shows an overview of our approach that has three stages: (i) the *model decomposition* stage, where we hierarchically decompose the model into subprocesses; (ii) the *replay stage*, where we relate the subprocesses with the event data; and (iii) the *comparison stage*, where we compare the process variants with respect to different perspectives.

4.1 Hierarchical Decomposition

Existing Petri net decomposition approaches [2,9,13,16,27] focus on the semantic relation between the original Petri net and the sub-nets. For example, the

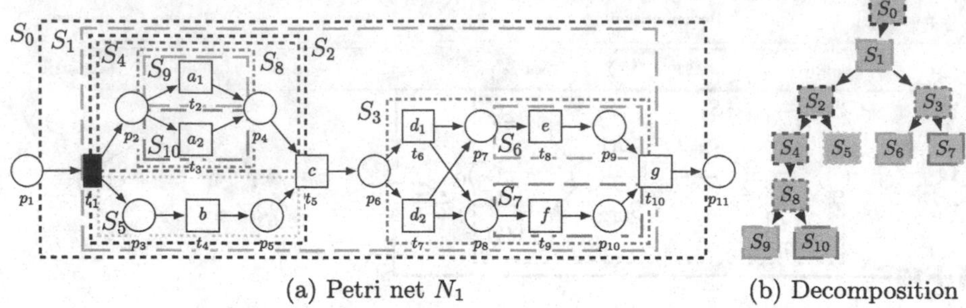

Fig. 3. Illustration of a hierarchical WF-net decomposition. Subfigure (a) shows a WF-net and ten ($S_0, \ldots, S_9$) (*single-entry, single-exit* (SESE)) subprocesses. Each subprocess contains the edges inside the illustrated rectangle. A hierarchy of these subprocesses is depicted in Subfigure (b).

decomposition proposed in [2] is motivated by the idea of stitching together *valid firing sequences* of the sub-nets to an (over-optimistic) execution of the original net. In contrast, as illustrated in Fig. 2, we align the log with the original net. Thereby, we gain more freedom to decompose the net in a hierarchical manner without facing problems when relating process executions to sub-nets. Yet, computing alignments can be computationally costly. In this work, we employ a generic hierarchical decomposition based on the edges of a WF-net.

Definition 9 (Hierarchical Decomposition). *Let* $N = (P, T, F, l)$ *be a WF-net. A hierarchical decomposition of* N *is a tree* $H = (\mathbb{S}, E), \mathbb{S} \subseteq \mathcal{P}(F), E \subseteq \mathbb{S} \times \mathbb{S}$ *rooted at a vertex* $v_0 \in \mathbb{S}$ *and downward-pointing edges such that* $v_0 = F$*; for* $(v_1, v_2) \in E$*, we have* $v_1 \supset v_2$*; and for* $S \in \mathbb{S}$*, the induced subnet* N_S *is connected.*

Intuitively, each vertex in the hierarchy corresponds to an edge induced subnet of the original net. Under this interpretation, the original net is at the root, and each non-root vertex's net is a subnet of its parent's net—that is, it comprises a subset of its parent's net's places, transitions, and edges. We require connectedness of subprocesses to later define the semantics of a *subprocess execution.*

Figure 3b shows a decomposition of the WF-net in Fig. 3a. Note that, given the ten subprocesses in Fig. 3a, Definition 9 does not enforce a unique hierarchy. A decomposition where S_0 is the root and all remaining subprocesses are S_0's children would also be valid. While a deeper hierarchy is usually preferable, there can be exceptions. For example, an analysis that investigates parent-child relations can benefit from this additional freedom (cf. Eq. (10)).

In general, Definition 9 does not impose strong restrictions on the decomposition applied. To relate event data to transitions of a Petri net, we use alignments where we consider the complete trace and net. In contrast to relating individual subprocesses independently of each other, this guarantees a globally consistent assignment without additional considerations (e.g., an event cannot be assigned to different transitions having the same label). Therefore, the measurements proposed in Sect. 4.2 are mostly independent of the decomposition.

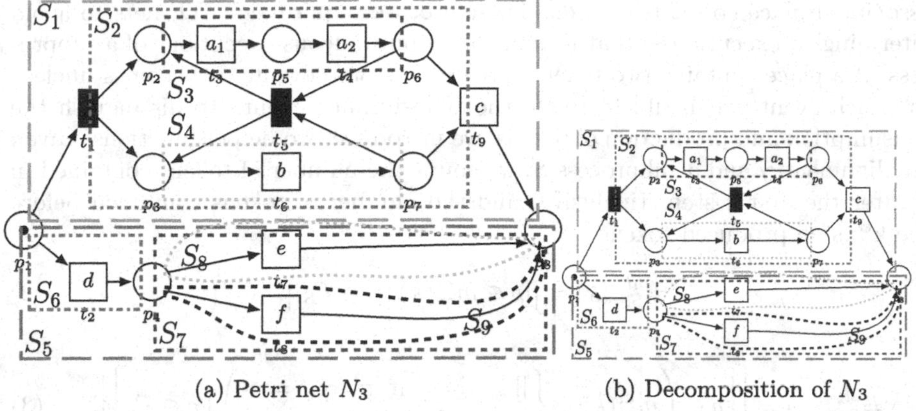

(a) Petri net N_3 (b) Decomposition of N_3

Fig. 4. WF-net $N_3 = (P_3, T_3, F_3, l)$ illustrating hierarchical PC. The edges in the colored box in (a) depict the canonical fragments of the net. Subfigure (b) depicts a decomposition of N_3.

Despite our approach being mostly agnostic to the decomposition applied, structural subprocess properties can play a role when interpreting the results. In particular, we distinguish subprocesses—so-called *fragments* [27]—where control flow enters and exits through a single node, respectively.

Definition 10 (*Single-entry, single-exit* (SESE) Subprocess). *Let $N = (P, T, F, l)$ be a WF-net and $S \subseteq F$ be a set of edges such that N_S is connected. A vertex $v \in P_S \cup T_S$ is a* boundary *vertex of N_S if it is the source or the sink of N or if v is incident to edges $e_1 \in S$ and $e_2 \notin S$; otherwise v is an* internal *vertex. The subprocess S is a SESE subprocess if there exists entry and exit vertices $v_i, v_e \in P_S \cup T_S$ such that (i) v_i (v_e) are boundary vertices; (ii) no incoming edge of v_i is in S or all outgoing edges of v_i are in S; (iii) no outgoing edge of v_e is in S or all incoming edges of v_e are in S; (iv) there is no other boundary vertex $v \in (P_S \cup T_S) \setminus \{v_i, v_e\}$.*

All subprocesses illustrated in Fig. 3a are SESE subprocesses.

4.2 Measuring Differences

To compare event logs using the decomposition, we first relate the subprocesses to the event data. We then define various measurements that, for example, assess differences in the control flow or performance of subprocesses. Finally, we use hypothesis tests to detect differences with respect to the measurements. We illustrate the following concepts and a few interesting measurements on the WF-net shown in Fig. 4a. Figure 4b shows its decomposition.

Subprocess Executions. In contrast to works that project the data onto the subprocesses [2,20], we extract subprocess executions from alignments. Replaying the alignment, we consider a subprocess being under execution as long as an associated place contains a token. Therefore, we require *safe* WF-nets to avoid intermingled executions—that is, multiple, simultaneous executions of a subprocess. If a place contains two tokens "created" by two events, it becomes unclear to which event we should to relate the "consuming" event. To distinguish the consumption and production of tokens, we introduce two helper functions. Given an alignment γ and a subprocess S, we count the number of tokens contained in S after the first k steps (inclusive) and the number of tokens contained before the kth step produced tokens:

$$\#^{al}_{S,\gamma}(k) = \left|[p \in m^{al}_{\gamma}(k) \mid p \in P_S]\right|, \tag{2}$$

$$\#^{al,-}_{S,\gamma}(k) = \left|\left[p \in \left(m^{al}_{\gamma}(k) - \begin{cases} [] & \text{if } \pi^2(\gamma_k) = \gg \\ \pi^2(\gamma_k)^{\bullet} & \text{else} \end{cases}\right) \middle| p \in P_S\right]\right|. \tag{3}$$

Next, we define the intervals during which a subprocess is under executions.

Definition 11 (Subprocess Execution). *Let* $\mathrm{SN} = ((P,T,F,l), m_I, m_F)$ *be a WF-sytem net,* $S \subseteq F$ *be a set of edges,* $\boldsymbol{\sigma} \in \mathcal{E}^*$ *be a trace, and* $\boldsymbol{\gamma} \in M^*_{\mathrm{ALL}}$ *be an alignment of* $\boldsymbol{\sigma}$ *and* SN. *The* partial execution intervals *of* S *given* γ *are*

$$\begin{aligned}
E^{part}_{\mathrm{SN},\gamma,S} = \{\{i,\dots,j\} | 1 \le i \le j \le |\gamma| & \qquad \textit{(Intervals)} \\
\wedge\, \forall i \le k < j \left(\#^{al}_{S,\gamma}(k) \ge 1\right) & \qquad \textit{(Token contained)} \\
\wedge\, \exists i \le k \le j \left(\pi^2(\gamma_k) \in T_S\right) & \qquad \textit{(Transition fired)} \\
\wedge \forall i < k < j \left(\#^{al,-}_{S,\gamma}(k) = 0 \rightarrow \pi^2(\gamma_k) \in T_S\right)\Big\}. & \qquad \textit{(No short-circuiting loops)}
\end{aligned} \tag{4}$$

The complete execution intervals *of* S *given* γ *are the maximal partial execution intervals*

$$E^{exec}_{\mathrm{SN},\gamma,S} = \left\{\mathcal{I}_1 \in E^{part}_{\mathrm{SN},\gamma,S} | \forall \mathcal{I}_2 \in E^{part}_{\mathrm{SN},\gamma,S} \left(\mathcal{I}_1 \subseteq \mathcal{I}_2 \rightarrow \mathcal{I}_1 = \mathcal{I}_2\right)\right\}. \tag{5}$$

Equation (4) gives the conditions for a subprocess to be considered under executions. First, the subprocess must contain a token. Second, at least one transition of the subprocess must fire as for place-bounded subprocesses a token can pass without entering the subprocess. For example, consider p_1 in Fig. 4a and the subprocesses S_1 and S_5. Despite both subprocesses contain p_1, a trace can only enter one. Third, there is no outside short-circuiting loop (i.e., a loop containing a single transition that is not adjacent to the subprocess) that consumes the last token and produces a new token in it. For example, consider the subprocess S_3 in Fig. 4a and the following alignment γ_2:

$$\begin{array}{r}
\#^{al,-}_{S_3,\gamma_2}(k) \;\; 0 \;\; 0 \;\; 1 \;\; 0 \;\; 0 \;\;\; 0 \;\;\; 0 \;\; 0 \;\; 1 \;\; 0 \\
\#^{al}_{S_3,\gamma_2}(k) \;\; (1 \;\; 1 \;\; 1 \;\; 1 \;\; 1 \;\; \{1) \;\; 1 \;\; 1 \;\; 1 \;\; 0\} \\
\gamma_2 = \begin{array}{|cccccccccc|}\hline \gg & a_1 & b & a_2 & a_2 & \gg & a_1 & a_2 & b & c \\ \hline t_1 & t_3 & t_6 & t_4 & \gg & t_5 & t_3 & t_4 & t_6 & t_9 \\ \hline \end{array}
\end{array}. \tag{6}$$

The token counts of S_3, displayed on top of the alignment, show that t_5 consumes the last token before it produces a new token in S_3. Yet, t_5 is not contained in S_3. Finally, Eq. (5) defines a subprocess' executions as the longest intervals during which the process is considered being under execution. In Eq. (6), the parentheses and braces indicate the resulting executions.

Subprocess Measurements. Based on the subprocess executions, we can compare processes with respect to various aspects on different levels of granularity.

Definition 12 (Trace Measurement). *Let* $\text{SN} = ((P, T, F, l), m_I, m_F)$ *be a WF-sytem net,* $S \subseteq F$ *be a set of edges,* $\boldsymbol{\sigma} \in \mathcal{E}^*$ *be a trace, and* $\gamma \in M^*_{\text{ALL}}$ *be an alignment of* $\boldsymbol{\sigma}$ *and* SN. *The universe of perspectives is* $\mathcal{F}$. *The subprocess trace measurement with respect to a perspective* $p \in \mathcal{F}$ *is a function* $\mu^p_{\text{SN},S} : M^*_{\text{ALL}} \to \mathbb{R} \cup \{\bot\}$ *where* $\bot$ *denotes the absence of a measurement.*

In the following, we exemplify five measurements to illustrate how the decomposition and subprocess executions facilitate the detection of process variant differences. In particular, we consider the control flow, performance, and conformance perspective. To this end, we assume the following context for the remainder of this section: Let $\text{SN} = (N, m_I, m_F), N = (P, T, F, l)$ be a WF-sytem net, $\mathcal{H} = (\mathbb{S}, E)$, $\mathbb{S} \subseteq \mathcal{P}(F)$ be a hierarchical decomposition of SN, $S \in \mathbb{S}$ be a subprocess, $\boldsymbol{\sigma} \in \mathcal{E}^*$ be a trace, and $\gamma \in M^*_{\text{ALL}}$ be an alignment of $\boldsymbol{\sigma}$ and SN.

Control Flow. First, we can measure whether a subprocess was activated, and how often it was executed:

$$\mu^{act}_{\text{SN},S}(\gamma) = \begin{cases} 1 & \text{if } \left|E^{exec}_{\text{SN},\gamma,S}\right| > 0 \\ 0 & \text{else} \end{cases}, \quad \mu^{freq}_{\text{SN},S}(\gamma) = \left|E^{exec}_{\text{SN},\gamma,S}\right|. \tag{7}$$

These measurements show differences with respect to the frequency of branching decisions and the number of repetitions.

In addition, the hierarchical nature of the decomposition facilitates the analysis of conditional decisions. Consider the following two event logs (not showing time for simplicity) that perfectly fit the Petri net N_3:

$$L_1 = \left[\langle a_1, a_2, b, c\rangle^{60}, \langle d, e\rangle^{14}, \langle d, f\rangle^{26}\right], \tag{8}$$

$$L_2 = \left[\langle a_1, a_2, b, c\rangle^{40}, \langle d, e\rangle^{27}, \langle d, f\rangle^{33}\right]. \tag{9}$$

In L_2, the activities e and f occur more frequently. Accordingly, the control flow measurements in Eq. 7 show that, in L_2, executing S_8 and S_9 is more likely. However, if we respect the initial choice, the likelihood of observing f given that d was initially chosen is higher for L_1 (65% vs 55%). We can incorporate this into our frequency measurement by considering the parent subprocess S_7. We only consider whether S_8 and S_9 were activated if S_7 was activated. Assuming that S is not the root of $\mathcal{H}$, let $\bar{S}$ be the parent of S (i.e., $(\bar{S}, S) \in E$). We define the *conditional subprocess activation* measurement

$$\mu^{c.act}_{\text{SN},S|\bar{S}}(\gamma) = \begin{cases} \mu^{act}_{\text{SN},S}(\gamma) & \text{if } \mu^{act}_{\text{SN},\bar{S}}(\gamma) = 1 \\ \bot & \text{else} \end{cases}. \tag{10}$$

Conformance. Process executions do not always comply with the process model. Thus, process variants might not only differ in how frequently they activate model elements but also in how they deviate from the prescribed control flow. An example is the *occurrence of log moves* measurement. For the subprocess S, we consider log moves on activities that are among the labels of the transitions T_S. In particular, we count log moves that occur outside the subprocess' executions:

$$\mu^{cc}_{\text{SN},S}(\gamma) = \left|\left\{i \in \{1,\dots,|\gamma|\} \mid \gamma_i \in M_{\text{LM}} \wedge \exists t \in T_S\ l(t) = \pi^{act}\left(\pi^1\left(\gamma_i\right)\right) \right.\right. \\ \left.\left. \wedge \forall I \in E^{exec}_{\text{SN},\gamma,S}\ i \notin I\right\}\right|. \quad (11)$$

Differences with respect to this measurement show that, in one process variant, transitions of S are more likely to occur at unexpected positions. However, this measurement also illustrates a major challenge dealing with log moves—namely, duplicate transition labels. In case multiple transitions have the same label, we count log moves multiple times.

Performance. Performance differences between process variants are often of major interest. Using our notion of process executions, one can define various performance measurements. In the following, we exemplify two complementary measurements and discuss their limitations. Given a complete execution interval $I \in E^{exec}_{\text{SN},\gamma,S}$ of the subprocess S, we first consider the time series

$$\gamma^{sync\_t}_{S,[I]} = \pi^{time}\left(\pi^1\left(\pi_{M_{\text{SYNC}} \cap (\mathcal{E} \times T_S)}\left(\gamma_{[I]}\right)\right)\right) \quad (12)$$

of synchronously executed subprocess transitions in I. Next, we define the *synchronous subprocess execution duration* as the time difference between the first and last synchronously executed transition that is adjacent to the subprocess:

$$\mu^{sync\_t}_{\text{SN},S}(\gamma) = \begin{cases} \bot & \text{if } |\gamma^{sync\_t}_{S,[I]}| = 0 \text{ for all } I \in E^{exec}_{\text{SN},\gamma,S} \\ \text{avg}\left(\left\{ \begin{array}{c} \max(\gamma^{sync\_t}_{S,[I]}) \\ -\min(\gamma^{sync\_t}_{S,[I]}) \end{array} \middle| \begin{array}{c} I \in E^{exec}_{\text{SN},\gamma,S} \\ \wedge |\gamma^{sync\_t}_{S,[I]}| > 0 \end{array} \right\}\right) & \text{else} \end{cases} \quad (13)$$

While this measurement provides insight into differences in the duration during which a subprocess was active, it does not account for delays prior or after the execution. For example, consider the subprocess S_3 in N_3 and the alignment

$$\gamma_3 = \frac{\gg\ \overset{1}{b}\ \overset{10}{a_1}\ \overset{11}{a_2}\ \overset{20}{c}}{t_1\ t_6\ t_3\ t_4\ t_9}, \quad (14)$$

where the activity executions' timestamps are depicted on top of the activities. Despite the first move marks S_3, the first activity is executed at time step 10. In contrast, $\mu^{sync\_t}_{N_3,S_3}(\gamma_3) = 1$ suggests a fast execution of S_3. To compare initial delays, we can consider the *elapsed time since case start* measurement

$$\mu^{elap\_t}_{\text{SN},S}(\gamma, \boldsymbol{\sigma}) = \min_{I \in E^{exec}_{\text{SN},\gamma,S}}\left(\min(\gamma^{sync\_t}_{S,[I]})\right) - \pi^{act}(\boldsymbol{\sigma}_0) \quad (15)$$

that extracts the time until the very first synchronous execution of a subprocess' transition. However, even if we consider both measurements, there are four major limitations: (i) for loops, we only consider the first delayed start. (ii) The *elapsed time since case start* measurement is monotonically increasing (e.g., $\mu^{elap\_t}_{N_3,(t_4,p_6)}(\gamma_3) = \mu^{elap\_t}_{N_3,\bar{S_3}}(\gamma_3)+(11-10)$). Thus, the analyst needs to consider that differences can propagate to later parts of the process. (iii) The *synchronous subprocess execution duration* of subprocesses containing a single transition is either zero or undefined. (iv) For non-SESE subprocesses, there can be additional externally caused delays even after a subprocess' execution started. Therefore, one generally needs to consider multiple performance measurements depending on the performance aspect of interest and the subprocess under consideration. For example, to address the third limitation, we might consider the causal predecessors and successors of a transition in the run of the Petri net. For S_8, we would compute the time between firing t_2 and t_7. Yet, for the concurrent subprocesses S_3 and S_4, this means that their duration would be determined by t_1 and possibly t_9 if we consider the transition that removes the last token. In this case, we would get the same measurement value for both subprocesses.

Hypothesis Testing. Applying the presented trace measurements, we extract zero ($\perp$) or one value per alignment. As we aim to compare subprocesses based on their executions, we first discard the irrelevant measurements (i.e., measurements having the value $\perp$). For each subprocess and perspective, we thereby obtain a population of real-valued measurements. To detect statistically significant differences between populations, we apply hypothesis testing under the null hypothesis that there is no significant difference. In doing so, we implicitly assume that the measurements are independent of each other. In practice, this assumption might not always hold (e.g., if a single resource handles multiple cases simultaneously). Nevertheless, case independence is a common assumption in the field of PC [8,10,25]. Moreover, we additionally assume that the measurements are approximately normally distributed; yet the populations might have different means and variances. Based on these assumptions, we apply Welch's t-test [29] with a p-value of 0.05 to test if two populations' mean values differ.

Besides the significance of a difference, the *effect size* assesses its strength. For large populations, even small differences in their means can become statistically significant. Therefore, we employ Cohen's d [11] to quantify a difference's effect size. Eventually, we determine the color of each subprocess in the decomposition based on whether there is a significant difference, whether the mean is larger for the left or right process variant, and the Cohen's d value.

4.3 Strategies for Decomposition

While our notion of WF-net decomposition is very generic, it is usually desirable that vertices in the hierarchy define logically coherent and independent subprocesses. In the literature, a structural characterization of such a subprocess is to require that it has a single entry and a single exit vertex [27]. Thereby, each

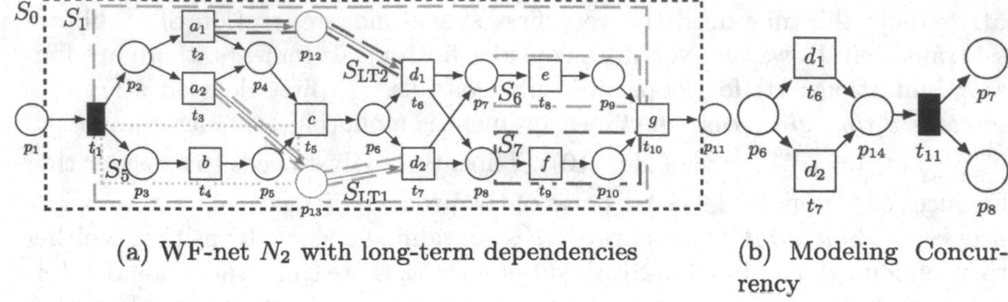

(a) WF-net N_2 with long-term dependencies (b) Modeling Concurrency

Fig. 5. Limitations of a SESE-based decomposition. The WF-net shown in (a) extends the WF-net in Fig. 3a by an additional long-term dependency (red). The colored polygons depict the non-trivial SESE subprocesses that constitute the RPST of this model. Using a SESE-based decomposition, we cannot analyze the subprocesses between t_1 and t_5 (i.e., S_2 in Fig. 3a) and p_6 and t_{10} (formerly S_3). The language preserving transformation depicted in (b) would allow to further decompose S_3 in N_1. (Color figure online)

subprocess is self-contained interacting with the remaining net at its entry and exit node. This characterization serves as a basis for the RPST of a WF-net [27]. The RPST is a decomposition comprising a maximal set of vertices satisfying the following conditions: (i) each vertex is a fragment (i.e., a SESE subprocess); (ii) each fragment is *canonical*—that is, there exists no overlapping fragment (not necessarily contained in the RPST) that neither is a proper super- nor subset; and (iii) each vertex is a child of its smallest superset. These conditions imply that the edges of a net are the leaves of its RPST. Such an edge fragment is also called *trivial* (i.e., a fragment of size one). For example, excluding trivial fragments, Fig. 3b shows the RPST of the WF-net in Fig. 3a. For further details on (computing) RPSTs, we refer the reader to [24,27].

The property of defining self-contained subprocesses makes the RPST a promising decomposition technique for hierarchical PC. Therefore, in our implementation, we leverage an adapted version of a WF-net's RPST. We propose to adapt the RPST computation for the sake of PC because the strict SESE requirement makes the decomposition sensitive. For example, consider the WF-net N_2 depicted in Fig. 5a which extends N_1 (Fig. 3a) by two additional places. The highlighted places p_{12} and p_{13} couple the choices between a_1 and a_2 and between d_1 and d_2. This significantly changes the set of canonical fragments. Compared to N_1, the added places induce two additional canonical fragments S_{LT1} and S_{LT2} but violate the SESE property of six of the original canonical fragments. Thereby, the resulting RPST would neither allow us to explicitly analyze the choice between a_1 and a_2 nor could we distinguish the two main blocks of the workflow (i.e., the process before and after c).

WF-net Skeleton. As shown in Fig. 2, we propose an additional preprocessing step of the WF-net to create more SESE subprocesses. The idea is to remove places that connect structurally distant parts of the model preventing a more fine granular decomposition. As places constrain the behavior of the WF-net, removing places instead of transitions yields a net describing a relaxed process. In contrast, removing transitions can result in Petri nets without valid firing sequences. Thus, the RPST obtained by removing places describes a relaxed baseline process that we consider the *skeleton process* of the original model.

In order to later compute its RPST, we need to ensure that the *skeleton process* is a WF-net. Intuitively, this means that we can only remove constraints (i.e., places) that do not break the workflow. Accordingly, we can neither remove the source nor the sink. Moreover, we might only remove a place if, in the resulting net, each node is on a path between the source and the sink. For example, in N_2 (Fig. 5a), the only candidates for removal are p_4, p_{12}, and p_{13}.

While removing p_{12} and p_{13} results in the original net that has a fine granular decomposition, this is not the case for p_4. After removing p_4, no further place can be removed resulting in a decomposition similar to the one before ($S_{\mathrm{LT}1}$ and $S_{\mathrm{LT}2}$ would be extended up to p_2). What distinguishes the places p_{12} and p_{13} from p_4 is that latter place constraints the *local* order of activities. In each firing sequence of the net that contains t_2, the number of steps in which p_4 is marked is less than for p_{12}. Therefore, we propose to preferably remove places that are marked for many steps.

Definition 13 (Place Marking Interval). *Let $N = (P, T, F, l)$ be a safe WF-net and $\mathbb{T} \in \mathcal{B}(T^*)$ be a multiset of valid firing sequences of N. Given a firing sequence $\sigma = \langle t_1, \ldots, t_n \rangle \in \mathbb{T}$, the* token intervals *of a place $p \in P$ are*

$$\mathrm{ti}_\sigma(p) = \{(i,j) | 1 \leq i < j \leq n, p \in t_i{}^\bullet, p \in {}^\bullet t_j, \forall k (i < k < j) \rightarrow p \notin {}^\bullet t_k \cup t_k{}^\bullet\} \,. \tag{16}$$

The average number of steps a place $p \in P$ is marked is

$$\mathrm{tt}_\mathbb{T}(p) = \mathrm{avg}([j - i | (i,j) \in \biguplus_{\sigma \in \mathbb{T}} \mathrm{ti}_\sigma(p)]) \,. \tag{17}$$

This idea is inspired by an ILP Miner [31] variant that prefers adding places to a net where the token is consumed quickly. The safeness of the WF-net is crucial to uniquely match the transitions that produced and consumed a token. In contrast to structural place removal conditions, Definition 13 is not affected by implicit places (i.e., places that do not affect the valid firing sequences). For the sake of efficiency, in our implementation, we re-use the alignments of the event logs to evaluate Eq. (17) rather than sampling the net.

Finally, we propose the following decomposition strategy for hierarchical PC: (i) sort the places according to Definition 13 in descending order, (ii) iterate over the sorted places removing places as long as the resulting net remains a WF-net (i.e., remains connected), (iii) compute the RPST of the resulting WF-net skeleton, and (iv) add the edges of the removed places as subprocesses under the root. Consider a set of firing sequences for N_2 (Fig. 5a) such that p_{12} and p_{13} are marked at least once. Depending on the position of t_4 in the firing sequence, the places p_{12} and p_{13} are marked for two or three steps. Each time one of these places is marked, the place p_4 is marked for at least one step less. Therefore, one can show that we have $\text{tt}_S(p_4) < \text{tt}_S(p_{12})$ or $\text{tt}_S(p_4) < \text{tt}_S(p_{13})$ (not necessarily both). Removing any other place would violate the WF-net constraint. Thus, we first remove p_{12} or p_{13} after which p_4 cannot be removed anymore.

While there are various ways to combine the RPST and the places removed, we simply add the associated sets of edges under the root. In contrast, one could add them under the smallest subprocess that contains the place's adjacent transitions. Adding the edges under the root has the advantage that the remaining tree is a proper RPST for the block skeleton. Assuming that the set of edges removed is usually small, we expect this design decision to have a minor impact.

Our place removal approach might change the language of the WF-sytem net. Fortunately, this does not affect the analysis since we compute the alignments with respect to the original net. Besides, there are other language-preserving transformations that might improve a net's SESE decomposability. Figure 5b shows an example using a silent transition as a concurrency split. Since this does not change the model in terms of its language, we currently leave this to the modeler. Besides, approaches that transform other model notations into Petri nets often natively use silent transitions to create SESE subprocesses.

5 Case Study

We implemented our approach as a ProM plugin, available in the ProM nightly builds[1], and evaluate it in a case study on the real-life *Road Traffic Fine Management* (RTFM) event log [19]. To the best of our knowledge, from the model-based approaches discussed in Sect. 2, only Wynn et al. [30] provide the publicly available implementation *Profiler 3d*. Therefore, we qualitatively compare the results to *Profiler 3d*. Like existing works that consider this log [8,10,25], we split it into low fine cases—that is, the initial fine is less than €50—and high fine cases—that is, the fine is larger or equal to €50.

Based on the original log, we created a highly fitting BPMN model and transformed it into the Petri net depicted in Fig. 6a. On a high level, a fine is first created (CF) and then either paid (P) or sent (SF). In the latter case, the offender may either pay it, or the case enters a subprocess concerned with additional penalization (AP) or appealing (IDA2P). We duplicated the payment transition (P) to precisely represent highly frequent variants where the fine was paid. Thereby, we can explicitly analyze the different ways of paying a fine. Finally, we add an implicit place (i.e., a place that does not constrain the behavior), highlighted red in Fig. 6a. It creates additional boundary nodes in the additional penalization subprocess. Thereby, no proper fragment can contain this subprocess making it difficult to decompose. Despite the place being implicit, it is well-suited to investigate the time between inserting the fine notification (IFN) and paying (P) or collecting (SCC) it.

[1] https://promtools.org/prom-6-nightly-builds/.

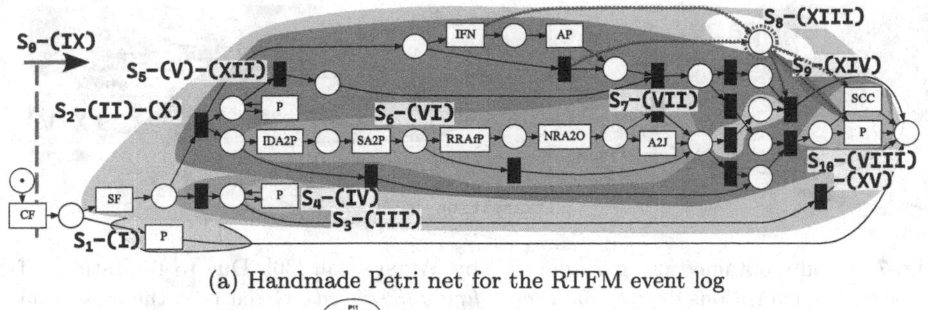

(a) Handmade Petri net for the RTFM event log

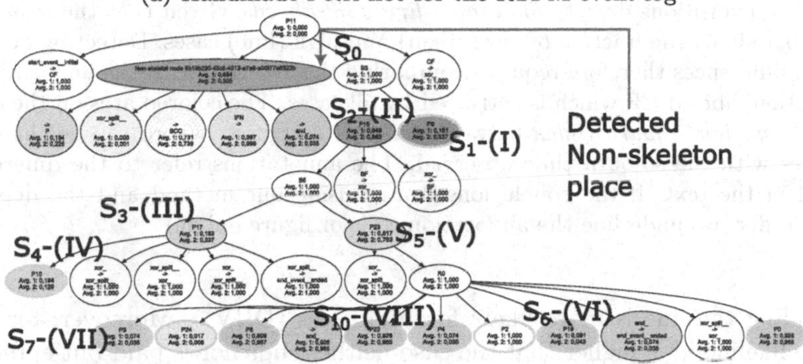

(b) Decomposition of the Petri net in (a) enhanced by differences in a subprocess' conditional activation likelihood

Fig. 6. Comparing low and high fine cases in terms of differences in the control flow. Subfigure (a) depicts a Petri net of the process. The decomposition depicted in (b) shows the differences detected using the *conditional subprocess activation* measurement. For significant differences, the shade depicts the effect size, and the color illustrates whether subprocesses are activated more likely for low (blue) or high (red) fines. Besides, we collapsed uninteresting subtrees (yellow outlines). The red annotations relate the nodes in the decomposition of the Petri net and to the descriptions in the text. (Color figure online)

5.1 Results

Figure 6b shows the hierarchical decomposition of the Petri net in Fig. 6a, and Fig. 7 depicts the output of *Profiler 3d*. Our proposed decomposition strategy correctly identifies the additionally introduced place as a non-skeleton place. Next, we investigate control flow differences. Since the Petri net in Fig. 6a contains a considerable number of choices, we employ the *conditional subprocess activation* measurement. Thereby, we can analyze subprocesses proceeding a choice irrespective of the choice's likelihood. Figure 6b shows that, for low fines, it is considerably more likely that we observe an immediate payment (**D**iagnostic D(I)). *Profiler 3d* also detect this difference. In contrast, high fines are often first sent (SF) to the offender (D(II)). After a fine was sent, an offender might pay, or—an interesting, frequent option present in the event log—the case may end. While latter option is more probable for low fines (D(III)), receiving a payment

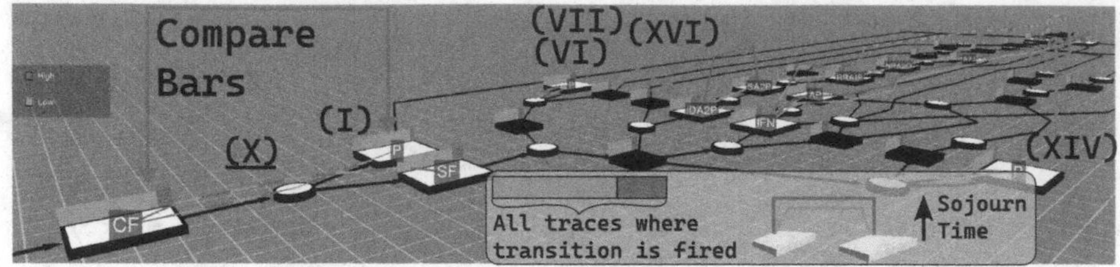

Fig. 7. Results obtained using *Profiler 3d* by Wynn et al. [30]. Due to limitations of the tool, the transitions depict *conditional firing likelihoods.* Given that the transition fires, the bar shows the fraction of low (green) and high (fine) cases. Detecting decision likelihood differences therefore requires comparing this fraction with the annotation of the transition labeled CF which is activated by all cases. The colored arcs on the edges depict the *median sojourn times* between pairs of activities, where the height of an edge scales with the sojourn time observed. The annotations refer to the differences described in the text. If the conclusions drawn using our method and the depicted approach differ, we underline the annotation. (Color figure online)

after sending the fine is more likely for high fines (D(IV)). Moreover, for high fines, the likelihood is higher that the case neither ends nor is paid but enters an extended subprocess (D(V)). This subprocess includes three concurrent strands of work. On the prefecture's site, the delayed payment leads to a notification (IFN) and an additional penalty (AP). Moreover, the prefecture might receive an appeal (IDA2P). On the offenders' side, they might decide to start paying the fine (possibly in multiple steps) even though an additional penalty will already be added due to the delay. For this subprocess, we observe two main differences. First, appeals are more likely for high fines (D(VI)). Second, given that an appeal was made, it is more probable to be successful for high fine cases (D(VII)). In particular, this difference might be quite interesting for stakeholders. Consequently, for low fine cases, it is slightly more likely that one eventually observes a payment or credit collection (D(VIII)). From the preceding differences, D(VI) and D(VII) are most noticeable in Fig. 7. For the remaining differences, the fact that one needs to visually compare fractions of cases makes them very difficult to detect.

Performance. Analyzing performance differences, we investigate the *synchronous subprocess execution duration* as well as the *elapsed time since case start.* Figure 8 depicts the projection of measurement differences onto the decomposition. First, high fine cases tend to have a longer overall cycle time (D(IX)). The average duration between a fine's creation and its last observed activity is approximately a year, while it is 263 days for low fine cases. Next, is noteworthy that, on average, low fines are sent 18 days earlier (D(X))—that is, 72 versus 90 days. Since there is no difference regarding the *elapsed time since case start* for CF, we would expect that Fig. 7 shows a sojourn time difference for the edge (CF, SF). Yet, this is not the case. Moreover, one can attribute all significant differences with

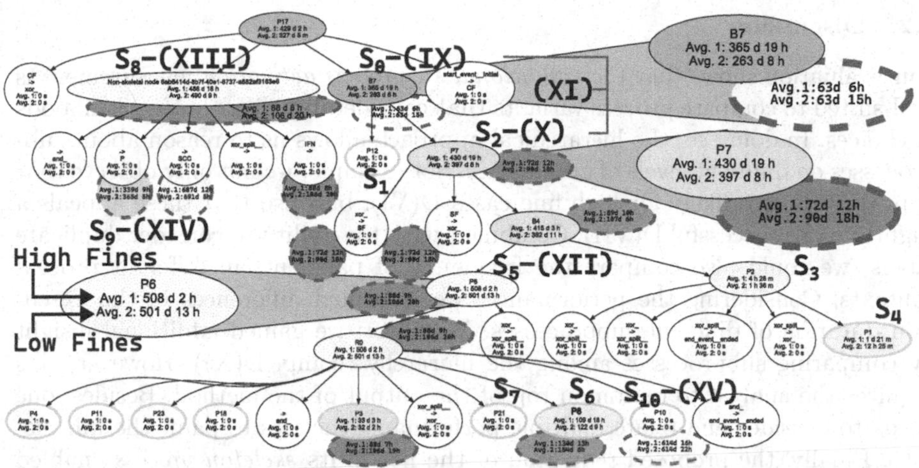

Fig. 8. Hierarchical comparison of low (avg. 2) and high fine (avg. 1) cases in terms of performance. In this decomposition of the Petri net in Fig. 6a, we collapsed uninteresting subtrees (yellow outlines). For each subprocess (i.e., vertex), we compare its *synchronous subprocess execution duration*. For selected subprocesses (red dashed), we also depict the *elapsed time since case start*. The annotations depict the corresponding subprocesses in Fig. 6a as well as labels for the diagnostics. We further enlarged important subprocesses (gray). (Color figure online)

respect to the elapsed time in the subtree rooted at S_2 to D(X). However, this requires an additional analysis of the average values. Due to the hierarchical decomposition, we can also easily see that the difference in the average duration of the subprocess S_2, entered by sending the fine, is smaller than for S_0—that is, 33 days vs 98 days (D(XI)). This suggests that the time savings for low fines are due to immediate payments (S_1). In fact, there is almost no difference regarding the duration of the appealing and additional fining subprocess S_5 (D(XII)). Finally, the introduced non-skeleton place gives additional insight into the time between inserting the notification (IFN) and the final payment (P) or credit collection (SCC). It shows that this duration does not differ significantly (D(XIII)). However, low fine cases that mark this place (i.e., cases that enter S_5) tend to be paid slightly earlier shown by a shorter *elapsed time since case start* (D(XIV)). Nevertheless, the overall time until the last payment is made or is collected forcefully, does not differ significantly (D(XV)). Finally, *Profiler 3d* detects increased sojourn times between IDA2P and SA2P for high fines. Similar to D(XIV), it also indicates that after a fine is inserted (IFN), the final payment is made slightly earlier for low fines (D(XVI)).

5.2 Discussion

Our evaluation shows that the *conditional subprocess activation* measurement is well-suited to compare process variants that considerably differ in the likelihoods of choices. In doing so, the hierarchical approach allows us to reason about subprocesses on different levels of granularity. For example, we not only observe that appeals are more likely for high fine cases D(VI), but also that these appeals a slightly more successful D(VII). Combined with the ability to consider duplicate labels, we could also compare the frequency of payments in different process contexts. Considering the performance, we identified differences in the execution duration of different subprocesses. Moreover, we gained additional insight by comparing subprocesses among the hierarchy (comp. D(X)). However, this requires the analyst to reason on top of the output of the method. Besides, one needs to consider and relate multiple performance metrics to paint the full picture. Finally, the proposed reduction of the net to its *skeleton process* enabled us to add an implicit place without negatively affecting the decomposition. This place could then be used to investigate a specific aspect of the process.

Compared to *Profiler 3d* [30], our approach shows differences more clearly. Moreover, the proposed method allows to reason about larger subprocesses and automatically analyzes subprocesses on different levels of granularity. While *Profiler 3d* also supports hierarchical Petri nets as input, it requires that the hierarchy is specified upfront.

6 Conclusion

In this paper, we leverage a shared Petri net to compare two event logs in a hierarchical manner. To this end, we decompose the model into a hierarchy of subprocesses. For each subprocess and case in the event logs, we then extract intervals during which the subprocess is considered being under execution. Based on these intervals, measurements that assess different aspects of the process can be defined. In this paper, we exemplify measurements that assess differences in the control flow, performance, and conformance. Furthermore, we propose a decomposition strategy based on Refined Process Structure Trees. In doing so, we introduce a new preprocessing step to improve the decomposability of the net. The case study shows how the proposed method allows to reason on larger subprocesses but also to drill down on interesting details.

For future work, we plan to investigate process-aware measurements that do not consider each case in isolation. The guidance provided by a good model might help to alleviate this limitation of existing process comparison approaches. Moreover, we intend to make the approach more interactive by allowing the user to assess performance metrics of flexibly defined sections during runtime.

Acknowledgments. Funded by the Deutsche Forschungsgemeinschaft (DFG, German Research Foundation) under Germany's Excellence Strategy-EXC-2023 Internet of Production-390621612. We also thank the Alexander von Humboldt (AvH) Stiftung for supporting our research.

References

1. van der Aalst, W.M.P., de Medeiros, A.K.A., Weijters, A.J.M.M.: Process equivalence: comparing two process models based on observed behavior. In: Dustdar, S., Fiadeiro, J.L., Sheth, A.P. (eds.) BPM 2006. LNCS, vol. 4102, pp. 129–144. Springer, Heidelberg (2006). https://doi.org/10.1007/11841760_10
2. van der Aalst, W.M.P.: Decomposing Petri nets for process mining: a generic approach. Distrib. Parallel Databases **31**(4), 471–507 (2013). https://doi.org/10.1007/s10619-013-7127-5
3. van der Aalst, W.M.P., Stahl, C.: Information systems. In: Modeling Business Processes: A Petri Net-Oriented Approach (2011). https://doi.org/10.7551/mitpress/8811.003.0003
4. van der Aalst, W.M.P., Tacke Genannt Unterberg, D., Denisov, V., Fahland, D.: Visualizing token flows using interactive performance spectra. In: Janicki, R., Sidorova, N., Chatain, T. (eds.) PETRI NETS 2020. LNCS, vol. 12152, pp. 369–380. Springer, Cham (2020). https://doi.org/10.1007/978-3-030-51831-8_18
5. Adriansyah, A., van Dongen, B.F., van der Aalst, W.M.P.: Towards robust conformance checking. In: zur Muehlen, M., Su, J. (eds.) BPM 2010. LNBIP, vol. 66, pp. 122–133. Springer, Heidelberg (2011). https://doi.org/10.1007/978-3-642-20511-8_11
6. Andrews, K., Wohlfahrt, M., Wurzinger, G.: Visual graph comparison. In: 13th International Conference Information Visualisation, pp. 62–67 (2009). https://doi.org/10.1109/IV.2009.108
7. van Beest, N.R.T.P., Dumas, M., García-Bañuelos, L., La Rosa, M.: Log delta analysis: interpretable differencing of business process event logs. In: Motahari-Nezhad, H.R., Recker, J., Weidlich, M. (eds.) BPM 2015. LNCS, vol. 9253, pp. 386–405. Springer, Cham (2015). https://doi.org/10.1007/978-3-319-23063-4_26
8. Bolt, A., de Leoni, M., van der Aalst, W.M.P.: A visual approach to spot statistically-significant differences in event logs based on process metrics. In: Nurcan, S., Soffer, P., Bajec, M., Eder, J. (eds.) CAiSE 2016. LNCS, vol. 9694, pp. 151–166. Springer, Cham (2016). https://doi.org/10.1007/978-3-319-39696-5_10
9. Bouvier, P., Garavel, H., Ponce-de-León, H.: Automatic decomposition of Petri nets into automata networks – a synthetic account. In: Janicki, R., Sidorova, N., Chatain, T. (eds.) PETRI NETS 2020. LNCS, vol. 12152, pp. 3–23. Springer, Cham (2020). https://doi.org/10.1007/978-3-030-51831-8_1
10. Cecconi, A., Augusto, A., Di Ciccio, C.: Detection of statistically significant differences between process variants through declarative rules. In: Polyvyanyy, A., Wynn, M.T., Van Looy, A., Reichert, M. (eds.) BPM 2021. LNBIP, vol. 427, pp. 73–91. Springer, Cham (2021). https://doi.org/10.1007/978-3-030-85440-9_5
11. Cohen, J.: Statistical Power Analysis for the Behavioral Sciences, chap. 2, 2nd edn., pp. 20–27 (1988). https://doi.org/10.4324/9780203771587
12. Cordes, C., Vogelgesang, T., Appelrath, H.-J.: A generic approach for calculating and visualizing differences between process models in multidimensional process mining. In: Fournier, F., Mendling, J. (eds.) BPM 2014. LNBIP, vol. 202, pp. 383–394. Springer, Cham (2015). https://doi.org/10.1007/978-3-319-15895-2_32
13. Eshuis, R.: Translating safe Petri nets to statecharts in a structure-preserving way. In: Cavalcanti, A., Dams, D.R. (eds.) FM 2009. LNCS, vol. 5850, pp. 239–255. Springer, Heidelberg (2009). https://doi.org/10.1007/978-3-642-05089-3_16
14. Ivanov, S.Y., Kalenkova, A.A., van der Aalst, W.M.P.: BPMNDiffViz: a tool for BPMN models comparison. BPM reports **1507** (2015)

15. Johnson, R., Pearson, D., Pingali, K.: The program structure tree: computing control regions in linear time. In: Proceedings of the ACM SIGPLAN 1994 Conference on Programming Language Design and Implementation, pp. 171–185 (1994). https://doi.org/10.1145/773473.178258
16. Karatkevich, A., Andrzejewski, G.: Hierarchical decomposition of Petri nets for digital microsystems design. In: 2006 International Conference - Modern Problems of Radio Engineering, Telecommunications, and Computer Science, pp. 518–521 (2006). https://doi.org/10.1109/TCSET.2006.4404613
17. Kriglstein, S., Wallner, G., Rinderle-Ma, S.: A visualization approach for difference analysis of process models and instance traffic. In: Daniel, F., Wang, J., Weber, B. (eds.) BPM 2013. LNCS, vol. 8094, pp. 219–226. Springer, Heidelberg (2013). https://doi.org/10.1007/978-3-642-40176-3_18
18. Küster, J.M., Gerth, C., Förster, A., Engels, G.: Detecting and resolving process model differences in the absence of a change log. In: Dumas, M., Reichert, M., Shan, M.-C. (eds.) BPM 2008. LNCS, vol. 5240, pp. 244–260. Springer, Heidelberg (2008). https://doi.org/10.1007/978-3-540-85758-7_19
19. de Leoni, M., Mannhardt, F.: Road traffic fine management process (2015). https://doi.org/10.4121/uuid:270fd440-1057-4fb9-89a9-b699b47990f5
20. Munoz-Gama, J., Carmona, J., van der Aalst, W.M.P.: Hierarchical conformance checking of process models based on event logs. In: Colom, J.-M., Desel, J. (eds.) PETRI NETS 2013. LNCS, vol. 7927, pp. 291–310. Springer, Heidelberg (2013). https://doi.org/10.1007/978-3-642-38697-8_16
21. Munoz-Gama, J., Carmona, J., van der Aalst, W.M.P.: Single-entry single-exit decomposed conformance checking. Inf. Syst. **46**, 102–122 (2014). https://doi.org/10.1016/j.is.2014.04.003
22. Nguyen, H., Dumas, M., La Rosa, M., ter Hofstede, A.H.M.: Multi-perspective comparison of business process variants based on event logs. In: Trujillo, J.C., et al. (eds.) ER 2018. LNCS, vol. 11157, pp. 449–459. Springer, Cham (2018). https://doi.org/10.1007/978-3-030-00847-5_32
23. Pini, A., Brown, R., Wynn, M.T.: Process visualization techniques for multi-perspective process comparisons. In: Bae, J., Suriadi, S., Wen, L. (eds.) AP-BPM 2015. LNBIP, vol. 219, pp. 183–197. Springer, Cham (2015). https://doi.org/10.1007/978-3-319-19509-4_14
24. Polyvyanyy, A., Vanhatalo, J., Völzer, H.: Simplified computation and generalization of the refined process structure tree. In: Bravetti, M., Bultan, T. (eds.) WS-FM 2010. LNCS, vol. 6551, pp. 25–41. Springer, Heidelberg (2011). https://doi.org/10.1007/978-3-642-19589-1_2
25. Taymouri, F., La Rosa, M., Carmona, J.: Business process variant analysis based on mutual fingerprints of event logs. In: Dustdar, S., Yu, E., Salinesi, C., Rieu, D., Pant, V. (eds.) CAiSE 2020. LNCS, vol. 12127, pp. 299–318. Springer, Cham (2020). https://doi.org/10.1007/978-3-030-49435-3_19
26. Taymouri, F., Rosa, M.L., Dumas, M., Maggi, F.M.: Business process variant analysis: survey and classification. Knowl.-Based Syst. **211**, 106557 (2021). https://doi.org/10.1016/j.knosys.2020.106557
27. Vanhatalo, J., Völzer, H., Koehler, J.: The refined process structure tree. In: Dumas, M., Reichert, M., Shan, M.-C. (eds.) BPM 2008. LNCS, vol. 5240, pp. 100–115. Springer, Heidelberg (2008). https://doi.org/10.1007/978-3-540-85758-7_10
28. Vidgof, M., Djurica, D., Bala, S., Mendling, J.: Interactive log-delta analysis using multi-range filtering. Softw. Syst. Model. **21**, 847–868 (2022). https://doi.org/10.1007/s10270-021-00902-0

29. Welch, B.L.: The generalization of 'student's' problem when several different population variances are involved. Biometrika **34**(1–2), 28–35 (1947). https://doi.org/10.1093/biomet/34.1-2.28
30. Wynn, M.T., et al.: ProcessProfiler3D: a visualisation framework for log-based process performance comparison. Decis. Support Syst. **100**, 93–108 (2017). https://doi.org/10.1016/j.dss.2017.04.004
31. van Zelst, S.J., van Dongen, B.F., van der Aalst, W.M.P.: ILP-based process discovery using hybrid regions. In: Algorithms & Theories for the Analysis of Event Data, pp. 47–61. CEUR Workshop Proceedings (2015)
32. Zhong, C., He, W., Li, Z., Wu, N., Qu, T.: Deadlock analysis and control using petri net decomposition techniques. Inf. Sci. **482**, 440–456 (2019). https://doi.org/10.1016/j.ins.2019.01.029

Semantics and Complexity

On the Expressive Power of Transfinite Sequences for Continuous Petri Nets

Stefan Haar(✉) and Serge Haddad(✉)

Université Paris-Saclay, INRIA and LMF, ENS Paris-Saclay, CNRS, Gif-sur-Yvette, France
stefan.haar@inria.fr, shaddad@lmf.cnrs.fr

Abstract. Continuous Petri nets (CPNs) form a model of (uncountably infinite) dynamic systems that has been successfully explored for modelling and theoretical purposes. Here, we focus on the following topic. Let the *mode* of a marking be the set of transitions fireable in the future. Along a firing sequence, the sequence of different modes is non increasing, and forms what we call the *trajectory* of the sequence. The set of achievable trajectories is an important issue, for instance in the study of biological processes. In CPNs, a marking can be reachable by a finite sequence, or lim-reachable by an infinite convergent sequence. The set of trajectories (resp. markings) obtained via lim-reachability (sometimes strictly) includes the set of trajectories (resp. markings) obtained via reachability. Here, we introduce transfinite firing sequences over countable ordinals and establish several results: (1) while trans-reachability is equivalent to lim-reachability, the set of trajectories associated with trans-reachability may be strictly larger than the one associated with lim-reachability; (2) w.r.t. trajectories, transfinite sequences over ordinals smaller than ω^2 are enough; and (3) checking whether a trajectory is achievable is NP-complete.

We then turn to a more difficult problem: the specification, for all transfinite firing sequences, of their achievable *signatures*, i.e. the sequences of markings witnessing the changes of mode along the trajectory. In view of this goal, we define a finite symbolic reachability tree (SRT) that tracks the possible signatures of the system; in the SRT, a set of markings with same mode is associated with each vertex. We establish that, for bounded CPNs, reversibility holds inside the leaves of the SRT (which correspond to the long-run behaviours). This property is also crucial in the application domains that motivate this work, namely regulation, signaling, ecosystems and other biological networks, where all quantities are bounded in mass or energy. Finally, from an algorithmic point of view, we show how to build an effective representation of the SRT in exponential time, even when the CPN is unbounded.

1 Introduction

Petri Nets in Life Sciences. Over the last decades, life sciences have been increasingly benefitting from the increased application of formal methods involving discrete event system models. This is particularly true for boolean and

L. M. Kristensen and J. M. van der Werf (Eds.): PETRI NETS 2024, LNCS 14628, pp. 109–131, 2024.
https://doi.org/10.1007/978-3-031-61433-0_6

Thomas networks; but several authors (Heiner et al. [2,15], Chaouia et al. [4–6], and others) have successfully used and studied (discrete) Petri nets, specifically in systems biology for the modelling of metabolism, cellular regulation (including the first author [7,17]), signalling [19], or ecosystems [1]. For the relation between Boolean Networks and Petri nets, see [8]. Further application fields for Petri nets in the life sciences, such as ecology [1], are currently emerging.

A central objective in these efforts is to identify and study *attractors*, informally stated as the possible long run behaviours of the system. These objects are crucial in several domains. For instance, in the context of cellular regulation, attractors give exactly the *phenotypes* of the studied system, whereas the collapse or blossom of an ecosystem is characterized by the attractors it enters. Furthermore an attractor fulfills *reversibility*: all visited states in an attractor can be visited again.

The Limits of Discrete Models. The discrete and non-deterministic nature of Petri nets are a key asset in these applications in the sense that they provide exhaustive treatment at a reasonable complexity (PSPACE for the reachability problem of bounded nets), compared to the predominant ODE or Markovian models in life sciences thus far. In fact, some works have studied more permissive discrete firing rules, between discrete Petri nets [9] and Boolean Networks [7,10,11,18]. The key motivation here is to ensure a better state space coverage, to avoid false predictions, diagnoses, or therapies. On the formal side, a gap in the reachability coverage had been first reported for contextual Petri nets [9]; following this, extensions to the traditional spectrum of semantics was developed for Boolean networks [7,10,11,18], culminating in a *most permissive* or *MP* semantics [18]. This MPS provides a sound overapproximation of the natural, continuous behaviour of the system, paired with a low complexity for the discrete reachability problem (see [18] and the *zenodo* tool]. Faithful refinement of this overapproximation remains an important open problem. From the *continuous* end of the spectrum, several authors (e.g. Heiner et al. [16]) have introduced continuous and fuzzy Petri nets as models of biological systems, in order to account for uncertainties in the modeling and observation of natural systems. However, the theory necessary for exploiting such models in prediction is still missing.

Continuous Petri Nets (CPN) [3,12,14]**.** In a CPN, the marking can evolve by firing a real quantity of a transition thus leading to a place marking defined by a real. CPNs present several interests both from a theoretical and a modelling point of view:

- Infinite firing sequences can be "convergent" and reach at the limit some marking. This yields the notion of lim-reachability which is sometimes more appropriate for modelling biological systems;
- reachability, lim-reachability and some other interesting problems can be solved in polynomial time while more difficult problems like the deadlock-freeness problem are NP-complete. These results hold even in unbounded CPNs, a major advantage over Petri nets.

Our Contributions. Motivated by biological applications we focus on the following topic. Let the *mode* of a marking be the set of transitions fireable in the future. Along a firing sequence, the sequence of different modes is non increasing and forms what we call the *trajectory* of the sequence. The set of achievable trajectories is an important issue for instance in the study of biological processes. In CPNs, a marking can be reachable by a finite sequence or lim-reachable by an infinite convergent sequence. The set of signatures (resp. markings) obtained via lim-reachability (sometimes strictly) includes the set of trajectories (resp. markings) obtained via reachability. Here we introduce transfinite firing sequences over countable ordinals and establish several results:

- while trans-reachability is equivalent to lim-reachability, the set of trajectories associated with trans-reachability may be strictly larger than the one associated with lim-reachability;
- w.r.t. trajectories transfinite sequences over ordinals less than ω^2 are enough;
- checking whether a trajectory is achievable is a NP-complete problem.

Afterwards, we turn on a more difficult problem: the specification for all transfinite firing sequences of their *signature*, i.e. the sequence of markings witnessing the changes of mode in the trajectory. In view of this goal, we define a finite symbolic reachability tree (SRT) which tracks the possible signatures of the system and where a set of markings with same mode is associated with each vertex. From an algorithmic point of view, we show how to build an effective representation of the SRT in exponential time even. While all these results hold for possibly unbounded CPNs, we establish that for bounded CPNs inside the leaves of the SRT (which correspond to the long-run behaviours), the CPN is reversible. This corresponds to the presence of *attractors* in the case of bounded *discrete* models. Indeed, while in general Petri nets, attractors may not exist, as soon as the net is bounded, the reachability graph is finite and must have finite terminal strongly connected components, which correspond to attractors. Boundedness is a reasonable assumption in the application domains; indeed, regulation, signaling and other biological networks are mass/energy bounded. Note that an efficient method for exhaustive attractor search via unfolding prefixes of safe discrete Petri nets was presented in [7].

Organisation. In Sect. 2, we introduce continuous Petri nets and recall some of their theoretical properties needed for developing our results. Then in Sect. 3, we define transfinite firing sequences, their trajectory and signature and establish several related results including the NP-completeness of the trajectory problem. Afterwards in Sect. 4, we define the SRT and establish the reversibility inside its leafs and design the building of an effective representation. Finally in Sect. 5, we conclude and give some perspectives to this work. Omitted proofs can be found in the appendix of the technical report [13].

2 Continuous Petri Nets

We will follow the terminology and notations of [14] and for some notions the ones of [3,12]. The omitted proofs of the results presented in this section can be found in [12].

Notations. Denote $\mathbb{R}_+ \triangleq [0,\infty)$, $\mathbb{I} \triangleq [0,1]$, $\mathbf{0} \triangleq (0,\ldots,0) \in \mathbb{R}^n$. Let $\mathbf{v} \in \mathbb{R}_+^X$ where X is a finite set. Then $[[\mathbf{v}]] \triangleq \{x \in X \mid \mathbf{v}[x] > 0\}$ and will be called the *support* of $\mathbf{v}$.

2.1 Definitions and Previous Results

Syntactically, there is no difference between CPNs and ordinary Petri nets.

Definition 1 (Continuous Petri Net (CPN)). *A* net *is a tuple* $\mathcal{N} = (P, T, \mathsf{Pre}, \mathsf{Post})$, *where:*

- P *a finite nonempty set of* places;
- T *is a finite set of* transitions *with* $P \cap T = \emptyset$;
- Pre *(resp.* Post*) is the backward (resp. forward)* $P \times T$ *incidence matrix, whose entries belong to* $\mathbb{N}$.

Notations. The *incidence matrix* of $\mathcal{N}$ is the matrix $\mathbf{C} \triangleq \mathsf{Post} - \mathsf{Pre}$. For $p \in P$, set ${}^\bullet p \triangleq \{t \in T : \mathsf{Post}(p,t) > 0\}$ and $p^\bullet \triangleq \{t \in T : \mathsf{Pre}(p,t) > 0\}$. Dually, for $t \in T$, set ${}^\bullet t \triangleq \{p \in P : \mathsf{Pre}(p,t) > 0\}$ and $t^\bullet \triangleq \{p \in P : \mathsf{Post}(p,t) > 0\}$. If $x \in P \cup T$, write ${}^\bullet x^\bullet \triangleq {}^\bullet x \cup x^\bullet$. These notations are extended to sets of items: ${}^\bullet X \triangleq \bigcup_{x\in X} {}^\bullet x$, $X^\bullet \triangleq \bigcup_{x\in X} x^\bullet$ and ${}^\bullet X^\bullet \triangleq \bigcup_{x\in X} {}^\bullet x^\bullet$. The *reverse* CPN is defined by: $\mathcal{N}^{-1} \triangleq (P, T, \mathsf{Post}, \mathsf{Pre})$.

In a CPN, the marking of a place is a non negative real and a CPN allows a fraction of a transition firing, scaling the quantities to be consumed and produced accordingly.

Definition 2 (Marking, Marked CPN). *A* (continuous) marking $\mathbf{m}$ *of a CPN* $\mathcal{N}$ *is an item of* $\mathbb{R}_+^P$. *A* marked CPN *is a pair* $(\mathcal{N}, \mathbf{m}_0)$ *where* $\mathbf{m}_0$ *is an initial marking.*

Definition 3 (Enabling, Firing). *Let* $\mathcal{N}$ *be a CPN,* $\mathbf{m}$ *be a marking of* $\mathcal{N}$, *and* $t \in T$. *Then:*

1. $\mathsf{enab}(t,\mathbf{m})$, *the* enabling degree *of* t *in* $\mathbf{m}$, *is defined by* $\min_{p\in{}^\bullet t} \frac{\mathbf{m}(p)}{\mathsf{Pre}(p,t)}$ *when* ${}^\bullet t \neq \emptyset$, *and* ∞ *otherwise. If* $\mathsf{enab}(t,\mathbf{m}) > 0$, *one says that* t *is* enabled *in* $\mathbf{m}$.
2. *For every* t *and* $\alpha \in [0, \mathsf{enab}(t,\mathbf{m})] \cap \mathbb{R}$, t *can be* α-fired *in* $\mathbf{m}$, *leading to a new marking* $\mathbf{m}'$ *given by:* $\forall\ p \in P\ \mathbf{m}'(p) \triangleq \mathbf{m}(p) + \alpha \cdot \mathbf{C}(p,t)$, *and we write* $\mathbf{m} \xrightarrow{\alpha t}_{\mathcal{N}} \mathbf{m}'$ *(allowing null firings).*

Notation. By analogy with Petri nets, we sometimes rewrite $\mathbf{m} \xrightarrow{1t}_{\mathcal{N}} \mathbf{m}'$ as $\mathbf{m} \xrightarrow{t}_{\mathcal{N}} \mathbf{m}'$. $\mathcal{Z} \triangleq \mathbb{R}_+ \times T$ denotes the set of *firing steps.* We denote by ω the first infinite ordinal.

Definition 4. *Let* $\mathbf{m}_0$ *be a marking,* $n \in \mathbb{N}$ *and* $\sigma = (\alpha_i t_i)_{i \leq n}$ *be a finite sequence over* $\mathcal{Z}$. *Then* σ *is a* finite firing sequence *from* $\mathbf{m}_0$ *if there exists a finite sequence of markings* $(\mathbf{m}_i)_{i \leq n+1}$ *such that for all* $i \leq n$, $\mathbf{m}_i \xrightarrow{\alpha_i t_i} \mathbf{m}_{i+1}$. *In that case, write* $\mathbf{m}_0 \xrightarrow{\sigma}_{\mathcal{N}} \mathbf{m}_{n+1}$.

Let $\sigma = (\alpha_i t_i)_{i \in \mathbb{N}}$ *be an infinite sequence over* $\mathcal{Z}$ *with* $\sum_{i \in \mathbb{N}} \alpha_i < \infty$. *Then* σ *is an* infinite firing sequence *from* $\mathbf{m}_0$ *if there exists a infinite family of markings* $(\mathbf{m}_i)_{i \leq \omega}$ *such that (1) for all* $i \in \mathbb{N}$, $\mathbf{m}_i \xrightarrow{\alpha_i t_i} \mathbf{m}_{i+1}$ *and (2)* $\lim_{i \to \infty} \mathbf{m}_i = \mathbf{m}_\omega$. *In that case, write* $\mathbf{m}_0 \xrightarrow{\sigma}_{\mathcal{N}} \mathbf{m}_\omega$.

Observation and Notations. The finiteness of $\sum_{i \in \mathbb{N}} \alpha_i$ ensures the existence of $\lim_{i \to \infty} \mathbf{m}_i$. When there is no ambiguity about $\mathcal{N}$, $\mathbf{m} \xrightarrow{\sigma}_{\mathcal{N}} \mathbf{m}'$ will simply be denoted $\mathbf{m} \xrightarrow{\sigma} \mathbf{m}'$. Sometimes we omit the final marking and write $\mathbf{m} \xrightarrow{\sigma}_{\mathcal{N}}$ instead of $\mathbf{m} \xrightarrow{\sigma}_{\mathcal{N}} \mathbf{m}'$. Let $\sigma = (\alpha_i t_i)_{i \leq n}$ be a finite sequence (resp. $\sigma = (\alpha_i t_i)_{i \in \mathbb{N}}$ be an infinite sequence such that $\sum_{i \in \mathbb{N}} \alpha_i < \infty$). Then $\vec{\sigma} \in \mathbb{R}_+^T$, the *Parikh vector* of σ, is defined by $\vec{\sigma}[t] \triangleq \sum_{t_i = t} \alpha_i$. Let $\mathbf{m}_0 \xrightarrow{\sigma} \mathbf{m}$. Then the *state equation* $\mathbf{m} = \mathbf{m}_0 + \mathbf{C}\vec{\sigma}$ can be established by recurrence and possibly taking limits.

Definition 5. *Let* $(\mathcal{N}, \mathbf{m}_0)$ *be a marked CPN. Then:*
The reachability set *is defined by:*

$$\mathbf{RS}(\mathcal{N}, \mathbf{m}_0) \triangleq \{\mathbf{m} : \exists\, \sigma \in \mathcal{Z}^*\ \mathbf{m}_0 \xrightarrow{\sigma} \mathbf{m}\}.$$

The lim-reachability set *is defined by:*

$$\lim\!-\mathbf{RS}(\mathcal{N}, \mathbf{m}_0) \triangleq \{\mathbf{m} : \exists\, \sigma \in \mathcal{Z}^\infty\ \mathbf{m}_0 \xrightarrow{\sigma} \mathbf{m}\}.$$

Since the last step $\mathbf{m} \xrightarrow{\alpha t} \mathbf{m}'$ of a finite firing sequence can be mimicked by the infinite firing sequence $\mathbf{m} \xrightarrow{(2^{-n}\alpha t)_{n \geq 1}} \mathbf{m}'$, we have $\mathbf{RS}(\mathcal{N}, \mathbf{m}_0) \subseteq \lim\!-\mathbf{RS}(\mathcal{N}, \mathbf{m}_0)$. However generally these two sets are different.

The next definitions and propositions are related to the main topic of our study: given a marking $\mathbf{m}_0$, what are the transitions eventually firable in the future, starting from $\mathbf{m}_0$?

Definition 6 (Firing set). *The* firing set *of a marked CPN* $(\mathcal{N}, \mathbf{m}_0)$ $\mathbf{FS}(\mathcal{N}, \mathbf{m}_0) \subseteq 2^T$ *is defined by:* $\mathbf{FS}(\mathcal{N}, \mathbf{m}_0) \triangleq \{[[\vec{\sigma}]] \mid \exists\, \sigma \in \mathcal{Z}^*\ \exists\, \mathbf{m}\ \mathbf{m}_0 \xrightarrow{\sigma} \mathbf{m}\}$.

The next propositions summarize key results about firing sets and their close connexion with reachability and lim-reachability.

Proposition 1. *Let* $\mathcal{N}$ *be a CPN and* $\mathbf{m}$, $\mathbf{m}'$ *be markings of* $\mathcal{N}$.

- *If* $[[\mathbf{m}]] \subseteq [[\mathbf{m}']]$ *then* $\mathbf{FS}(\mathcal{N}, \mathbf{m}) \subseteq \mathbf{FS}(\mathcal{N}, \mathbf{m}')$;
- $\mathbf{FS}(\mathcal{N}, \mathbf{m})$ *is closed under union;*
- *if* $T' = \{t_1, \ldots, t_k\} \in \mathbf{FS}(\mathcal{N}, \mathbf{m}_0)$ *then there exists a sequence* $\mathbf{m}_0 \xrightarrow{\sigma} \mathbf{m}$ *with* $\sigma = \alpha_1 t_{\beta(1)} \ldots \alpha_k t_{\beta(k)}$, β *a permutation of* $\{1, \ldots, k\}$, $\alpha_i > 0$ *for all* i, *and* $[[\mathbf{m}]] = [[\mathbf{m}_0]] \cup {}^\bullet T'^\bullet$.

Definition 7. *Let $\mathcal{N}$ be a CPN and* $\mathbf{m}$ *be a marking. Then* $T_{\mathcal{N},\mathbf{m}}$*, the* mode *of* $\mathbf{m}$ *in* $\mathcal{N}$ *is defined by:* $T_{\mathcal{N},\mathbf{m}} \triangleq \{t \in T \mid \exists \mathbf{m} \xrightarrow{\sigma}_{\mathcal{N}} \text{ with } t \in [[\vec{\sigma}]]\}$.

Notations and Observations. The previous proposition implies that $T_{\mathcal{N},\mathbf{m}}$ is both the maximal item of $\mathbf{FS}(\mathcal{N},\mathbf{m})$ and the union of these items. When there is no ambiguity about $\mathcal{N}$, we will simply write $T_\mathbf{m}$ for $T_{\mathcal{N},\mathbf{m}}$. Since the firing set only depends on the support of the marking, $T_\mathbf{m}$ (resp. $T_{\mathcal{N},\mathbf{m}}$, $\mathbf{FS}(\mathcal{N},\mathbf{m})$) can also be denoted $T_{[[\mathbf{m}]]}$ (resp. $T_{\mathcal{N},[[\mathbf{m}]]}$, $\mathbf{FS}(\mathcal{N},[[\mathbf{m}]])$).

Proposition 2. *Let $\mathcal{N}$ be a CPN, $P' \subseteq P$ and $T' \subseteq T$. Then:*

- *one can check in polynomial time whether $T' \in \mathbf{FS}(\mathcal{N},P')$;*
- *one can compute in polynomial time $T_{\mathcal{N},P'}$.*

The next theorems now establish a characterization of reachability and lim-reachability. Observe that the latter one is obtained by dropping from the former one Condition (3).

Theorem 1. *Let $(\mathcal{N},\mathbf{m}_0)$ be a marked CPN and* $\mathbf{m}$ *be a marking.*
Then $\mathbf{m} \in \mathbf{RS}(\mathcal{N},\mathbf{m}_0)$ iff there exists $\mathbf{v} \in \mathbb{R}_+^T$ such that:
(1) $\mathbf{m} = \mathbf{m}_0 + \mathbf{C}\cdot\mathbf{v}$, (2) $[[\mathbf{v}]] \in \mathbf{FS}(\mathcal{N},\mathbf{m}_0)$, and (3) $[[\mathbf{v}]] \in \mathbf{FS}(\mathcal{N}^{-1},\mathbf{m})$.
When such a $\mathbf{v}$ *exists, there exists a finite σ with $\vec{\sigma} = \mathbf{v}$ and $\mathbf{m}_0 \xrightarrow{\sigma} \mathbf{m}$.*

The structure of an infinite firing sequence witnessing the membership in $\lim\!-\!\mathbf{RS}(\mathcal{N},\mathbf{m}_0)$ was established in the proof but not stated in the previous version of the following theorem.

Theorem 2. *Let $(\mathcal{N},\mathbf{m}_0)$ be a marked CPN and* $\mathbf{m}$ *be a marking.*
Then $\mathbf{m} \in \lim\!-\!\mathbf{RS}(\mathcal{N},\mathbf{m}_0)$ iff there exists $\mathbf{v} \in \mathbb{R}_+^T$ such that:
(1) $\mathbf{m} = \mathbf{m}_0 + \mathbf{C}\cdot\mathbf{v}$ and (2) $[[\mathbf{v}]] \in \mathbf{FS}(\mathcal{N},\mathbf{m}_0)$.
Furthermore if $\mathbf{m} \in \lim\!-\!\mathbf{RS}(\mathcal{N},\mathbf{m}_0)$, then there exist finite sequences σ_0,σ_1 such that $\mathbf{m}_0 \xrightarrow{\sigma_0} \mathbf{m}_1 \xrightarrow{(2^{-n}\sigma_1)_{n\geq 1}} \mathbf{m}$
with $[[\vec{\sigma}_0]] = [[\vec{\sigma}_1]] = [[\mathbf{v}]]$ and $[[\mathbf{m}_0]] \cup {}^\bullet[[\vec{\sigma}_1]]^\bullet = [[\mathbf{m}_1]]$.

Theorem 2 and its applications here motivate the following definition.

Definition 8. *Let $\mathcal{N}$ be a CPN,* $\mathbf{m}$*,* $\mathbf{m}'$ *be markings and σ be a finite sequence. Then $\sigma' = (2^{-n}\sigma)_{n\geq 1}$ is a* repetitive discounted *sequence from* $\mathbf{m}$ *to* $\mathbf{m}'$ *if* $\mathbf{m} \xrightarrow{\sigma'} \mathbf{m}'$ *and* ${}^\bullet[[\vec{\sigma}]]^\bullet \subseteq [[\mathbf{m}]]$.

The next lemma characterizes existence of repetitive discounted sequences.

Lemma 1. *Let $\mathcal{N}$ be a CPN and* $\mathbf{m}$*,* $\mathbf{m}'$ *be two markings. Then there exists a repetitive discounted sequence from* $\mathbf{m}$ *to* $\mathbf{m}'$ *iff there exists $\mathbf{v} \in \mathbb{R}_+^T$ such that:* $\mathbf{m}' = \mathbf{m} + \mathbf{C}\cdot\mathbf{v}$ *and* ${}^\bullet[[\mathbf{v}]]^\bullet \subseteq [[\mathbf{m}]]$.

Proof. The necessity of this condition follows immediately by defining $\mathbf{v} = \vec{\sigma}$ when $\sigma' = (2^{-n}\sigma)_{n\geq 1}$ is the repetitive discounted sequence.

Since $^\bullet[[\mathbf{v}]]^\bullet \subseteq [[\mathbf{m}]]$, for all $t \in [[\mathbf{v}]]$, $\{t\} \in T_{\mathcal{N},[[\mathbf{m}]]} \cap T_{\mathcal{N}^{-1},[[\mathbf{m}]]}$ and by union $[[\mathbf{v}]] \in T_{\mathcal{N},[[\mathbf{m}]]} \cap T_{\mathcal{N}^{-1},[[\mathbf{m}]]}$.

For all $n \geq 0$, let us introduce $\mathbf{m}_n = 2^{-n}\mathbf{m} + (1 - 2^{-n})\mathbf{m}'$. $\mathbf{m}_0 = \mathbf{m}$ and for all $n \geq 0$, $\mathbf{m}_{n+1} = \mathbf{m}_n + \mathbf{C} \cdot 2^{-(n+1)}\mathbf{v}$ and $[[\mathbf{m}_n]] = [[\mathbf{m}]]$.

Using Theorem 1, there exists σ'' such that $\mathbf{m} \xrightarrow{\sigma''} \mathbf{m}_1 = 2^{-1}\mathbf{m} + 2^{-1}\mathbf{m}'$ with $\vec{\sigma}'' = \frac{1}{2}\mathbf{v}$. Thus for all $n \geq 0$,
$2^{-n}\mathbf{m} \xrightarrow{2^{-n}\sigma''} 2^{-(n+1)}\mathbf{m} + 2^{-(n+1)}\mathbf{m}'$ implying in turn:
$\mathbf{m}_n = (1-2^{-n})\mathbf{m}' + 2^{-n}\mathbf{m} \xrightarrow{2^{-n}\sigma''} (1-2^{-n})\mathbf{m}' + 2^{-(n+1)}\mathbf{m} + 2^{-(n+1)}\mathbf{m}' = \mathbf{m}_{n+1}$

So, with a slight abuse of notation, denoting $\sigma = 2\sigma''$, one gets: $\mathbf{m} \xrightarrow{(2^{-n}\sigma)_{n\geq 1}} \mathbf{m}'$. □

2.2 Two Motivating Examples

Example 1. In figures, places are represented by circles with their initial markings inside, transitions by rectangles (each one with a label identifying it) and Pre (resp. Post) specified by weighted edges entering (resp. leaving) transitions. Consider the Petri net/CPN on the left hand side of Fig. 1. It can be thought of as describing an epidemics situation, in which persons are initially healthy (p_1) but prone to catch one of two mild diseases (p_2 or p_3). If carriers of both diseases meet, a new and highly contagious syndrome (p_4) may emerge, which can spread in the populations of both p_2 and p_3. The reachability graph with only three states is depicted in the center part of Fig. 1, where the modes are noted next to each node. The only terminal strongly connected components (TSCCs), also called *attractors* in this context, are $\{p_2\}$ and $\{p_3\}$. However using the CPN firing rules we obtain the much richer dynamics. It can be checked that While $\{p_2\}$ and $\{p_3\}$ remain attractors, a third one emerges in $\{p_4\}$, showing how the continuous dynamics may *increase* the set of attractors of a system (in other circumstances, attractors may lose that status, merge etc.).

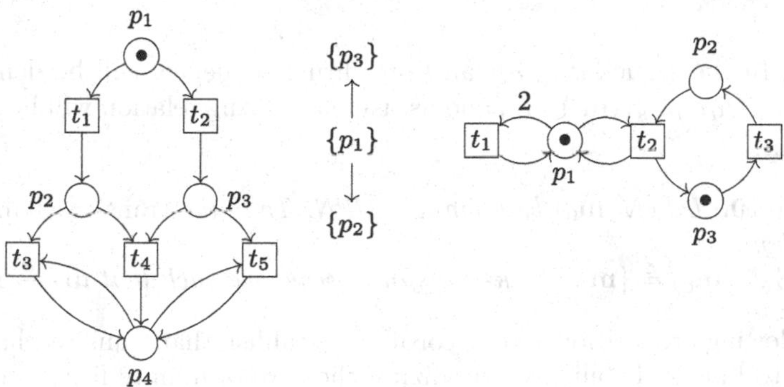

Fig. 1. A CPN/Petri net (left), its reachability graph (center) and another CPN (right).

Example 2. Let us examine the CPN on the right of Fig. 1. In natural systems, such a structure may correspond to a subsystem spanned by p_2 and p_3 that is reversible as long as some amount of ressource p_1 is available, but stops when p_1 is depleted by a decay process. Let us examine the sequence of modes visited by a firing sequence σ. If σ is a finite sequence then p_1 remains marked along the sequence; thus the single visited mode is T. For $\sigma = (2^{-n}t_1)_{n\geq 1}$, the marking reached by this infinite sequence is $1p_3$, and the corresponding sequence of modes is $\{t_3\}$. The marking $1p_2$ with associated mode $\emptyset$ is reachable from $1p_3$ in one step (by firing $1t_3$). It is possible to reach $1p_2$ with $\sigma = (2^{-n}t_1 2^{-n}t_3)_{n\geq 1}$. However the sequence of modes of this infinite sequence is $T\emptyset$. In fact, no finite or infinite sequence of firings from $\mathbf{m}_0$ will produce the sequence of modes $T\{t_3\}\emptyset$. However, if we introduce transfinite sequences (i.e. indexed by ordinals instead of integers) then the sequence $\sigma = (2^{-n}t_1)_{n\geq 1}1t_3$, where the last transition firing is indexed by ordinal ω, yields the sequence of modes $T\{t_3\}\emptyset$. This example motivates what will be introduced in the next section.

3 Signatures and Trajectories

In the following, we lift the representation level to abstract away from the individual continuous markings. The key step is to shift attention from a marking $\mathbf{m}$ to its mode $T_{\mathbf{m}}$ (i.e. the transitions still fireable in the future).

Notation. Let κ be a countable ordinal (i.e. an ordinal with countable cardinality). Then $\sigma = (\alpha_\iota t_\iota)_{\iota<\kappa}$, where for all ι, $(\alpha_\iota t_\iota) \in \mathcal{Z}$, is a *$\kappa$-transfinite sequence.* Let $\sigma = (\alpha_\iota t_\iota)_{\iota<\kappa}$ be a κ-transfinite sequence and $\iota < \iota' \leq \kappa$. Then $\sigma_{\iota,\iota'} = (\alpha_{\iota''} t_{\iota''})_{\iota\leq\iota''<\iota'}$.

Definition 9. *Let $\mathbf{m}_0$ be a marking, κ be a countable ordinal and $\sigma = (\alpha_\iota t_\iota)_{\iota<\kappa}$ be a κ-transfinite sequence over $\mathcal{Z}$. Then σ is a* firing sequence *from $\mathbf{m}_0$, denoted $\mathbf{m}_0 \xrightarrow{\sigma} \mathbf{m}_\kappa$, iff there exists a transfinite family of markings $(\mathbf{m}_\iota)_{\iota\leq\kappa}$ such that:*

- *for all $\iota < \kappa$, $\mathbf{m}_\iota \xrightarrow{\alpha_\iota t_\iota} \mathbf{m}_{\iota+1}$;*
- *for all limit ordinals $\kappa' \leq \kappa$, $\lim_{\iota<\kappa'} \mathbf{m}_\iota = \mathbf{m}_{\kappa'}$;*
- *$\sum_{\iota<\kappa} \alpha_\iota < \infty$.*

Notation. In the sequel, σ, a κ-transfinite firing sequence, will be denoted by the pair $\sigma = \langle(\alpha_\iota t_\iota)_{\iota<\kappa}, (\mathbf{m}_\iota)_{\iota\leq\kappa}\rangle$, and as usual the firing relation will be denoted by $\mathbf{m}_0 \xrightarrow{\sigma} \mathbf{m}_\kappa$.

Definition 10. *Let $(\mathcal{N}, \mathbf{m}_0)$ be a marked CPN. Then the* trans-reachability set *is defined by:*
trans$-$ $\mathbf{RS}(\mathcal{N}, \mathbf{m}_0) \triangleq \{\mathbf{m} \mid \exists\, \sigma$ *κ-transfinite sequence such that* $\mathbf{m}_0 \xrightarrow{\sigma} \mathbf{m}\}$.

The following proposition and its corollary establish that trans-reachability is equivalent to lim-reachability. Generalizing the case of infinite firing sequences, the Parikh vector of σ is defined by $\vec{\sigma}[t] \triangleq \sum_{t_\iota=t} \alpha_\iota$.

Proposition 3. *Let* $(\mathcal{N}, \mathbf{m}_0)$ *be a marked CPN and* $\sigma = \langle(\alpha_\iota t_\iota)_{\iota<\kappa}, (\mathbf{m}_\iota)_{\iota\leq\kappa}\rangle$ *be a* κ*-transfinite firing sequence. Then :*

$$(1)\ \mathbf{m}_\kappa = \mathbf{m}_0 + \mathbf{C}\cdot\vec{\sigma} \text{ and } (2)[[\vec{\sigma}]] \in \mathbf{FS}(\mathcal{N}, \mathbf{m}_0).$$

Proof. We proceed by induction on ordinals.

Case 1: $\kappa = \kappa' + 1$. Thus $\mathbf{m}_\kappa = \mathbf{m}_{\kappa'} + \alpha_{\kappa'}\mathbf{C}(t_{\kappa'})$. By induction, $\mathbf{m}_{\kappa'} = \mathbf{m}_0 + \mathbf{C}\cdot\vec{\sigma}_{0,\kappa'}$. Thus $\mathbf{m}_\kappa = \mathbf{m}_0 + \mathbf{C}\cdot\vec{\sigma}_{0,\kappa}$. Since $\vec{\sigma}_{0,\kappa'} \in \mathbf{FS}(\mathcal{N}, \mathbf{m}_0)$, applying Proposition 1, there exists a finite sequence $\mathbf{m}_0 \xrightarrow{\sigma'} \mathbf{m}'$ with $\vec{\sigma}' = \vec{\sigma}_{0,\kappa'}$ and $[[\mathbf{m}']] = [[\mathbf{m}_0]] \cup {}^\bullet\vec{\sigma}'^\bullet$ implying $[[\mathbf{m}_{\kappa'}]] \subseteq [[\mathbf{m}']]$. Thus there exists some $\alpha > 0$ such that $\mathbf{m}' \xrightarrow{\alpha t_{\kappa'}}$, which entails that $[[\vec{\sigma}_{0,\kappa}]] \in \mathbf{FS}(\mathcal{N}, \mathbf{m}_0)$.

Case 2: κ is a limit ordinal. Since T is finite, there exists an ordinal $\kappa' < \kappa$ with $[[\vec{\sigma}_{0,\kappa'}]] = [[\vec{\sigma}_{0,\kappa}]]$. Applying the induction hypothesis, $[[\vec{\sigma}_{0,\kappa}]] \in \mathbf{FS}(\mathcal{N}, \mathbf{m}_0)$.

Let $\varepsilon > 0$. Since $\sum_{\iota<\kappa}\alpha_\iota < \infty$, there exists some $\kappa_\varepsilon < \kappa$ such that for all $\kappa_\varepsilon \leq \kappa' < \kappa$ $\sum_{\kappa'\leq\iota<\kappa}\alpha_\iota \leq \varepsilon$. Let $B = \max(\mathsf{Pre}(p,t), \mathsf{Post}(p,t) \mid p \in P, t \in T)$. Then for all $\kappa_\varepsilon \leq \kappa' < \kappa$:

- by induction, $\mathbf{m}_{\kappa'} = \mathbf{m}_0 + \mathbf{C}\cdot\vec{\sigma}_{0,\kappa'}$;
- and so $\|\mathbf{m}_{\kappa'} - \mathbf{m}_0 + \mathbf{C}\cdot\vec{\sigma}_{0,\kappa}\| \leq B\varepsilon$.

Since $\mathbf{m}_\kappa = \lim_{\kappa'<\kappa}\mathbf{m}_{\kappa'}$, this implies that $\|\mathbf{m}_\kappa - \mathbf{m}_0 + \mathbf{C}\cdot\vec{\sigma}_{0,\kappa}\| \leq B\varepsilon$. Letting ε go to 0, one gets that $\mathbf{m}_\kappa = \mathbf{m}_0 + \mathbf{C}\cdot\vec{\sigma}_{0,\kappa}$. □

Combining Proposition 3 with Theorem 2, we obtain the following corollary which generalizes the case of infinite firing sequences.

Corollary 1. *Let* $(\mathcal{N}, \mathbf{m}_0)$ *be a marked CPN and* $\sigma = \langle(\alpha_\iota t_\iota)_{\iota<\kappa}, (\mathbf{m}_\iota)_{\iota\leq\kappa}\rangle$ *a* κ*-transfinite firing sequence. Then there exist finite sequences* σ_0, σ_1 *such that* $\mathbf{m}_0 \xrightarrow{\sigma_0} \mathbf{m}_1 \xrightarrow{(2^{-n}\sigma_1)_{n\geq1}} \mathbf{m}_\kappa$ *with* $[[\vec{\sigma}_0]] = [[\vec{\sigma}_1]] = [[\vec{\sigma}]]$ *and* $[[\mathbf{m}_0]] \cup {}^\bullet[[\vec{\sigma}_1]]^\bullet = [[\mathbf{m}_1]]$.

As shown by the next proposition, along a transfinite firing sequence the modes associated with the visited markings are non increasing.

Proposition 4. *Let* $(\mathcal{N}, \mathbf{m}_0)$ *be a marked CPN and* $\sigma = \langle(\alpha_\iota t_\iota)_{\iota<\kappa}, (\mathbf{m}_\iota)_{\iota\leq\kappa}\rangle$ *be a* κ*-transfinite firing sequence. Then for all* $\iota < \iota'$, $T_{\mathbf{m}_{\iota'}} \subseteq T_{\mathbf{m}_\iota}$.

Proof. We proceed by a transfinite induction with nothing to prove for $\kappa = 0$.

Case 1: $\kappa = \kappa' + 1$. Thus $\mathbf{m}_{\kappa'} \xrightarrow{\alpha_{\kappa'} t_{\kappa'}} \mathbf{m}_\kappa$ which implies that every transition eventually fireable from $\mathbf{m}_\kappa$ is also eventually fireable from $\mathbf{m}_{\kappa'}$. So $T_{\mathbf{m}_\kappa} \subseteq T_{\mathbf{m}_{\kappa'}}$.

Case 2: κ is a limit ordinal. Since $\mathbf{m}_\kappa = \lim_{\iota<\kappa}\mathbf{m}_\iota$, there exists some κ_0 such that for all $\kappa_0 \leq \kappa' < \kappa$, $[[\mathbf{m}_\kappa]] \subseteq [[\mathbf{m}_{\kappa'}]]$. So $T_{\mathbf{m}_\kappa} = T_{[[\mathbf{m}_\kappa]]} \subseteq T_{[[\mathbf{m}_{\kappa'}]]} = T_{\mathbf{m}_{\kappa'}}$. Thus given an arbitrary $\kappa'' < \kappa$, either $\kappa'' \geq \kappa_0$ and the result is established or $\kappa'' < \kappa_0$ which implies by induction that $T_{\mathbf{m}_{\kappa''}} \supseteq T_{\mathbf{m}_{\kappa_0}} \supseteq T_{\mathbf{m}_\kappa}$. □

So along a transfinite firing sequence the mode of the visited markings may only decrease a finite number of times. We aim at tracking these changes of mode and so we introduce some useful abstractions for a sequence.

Definition 11. *Let $(\mathcal{N}, \mathbf{m}_0)$ be a marked CPN and $\sigma = \langle(\alpha_\iota t_\iota)_{\iota<\kappa}, (\mathbf{m}_\iota)_{\iota\leq\kappa}\rangle$ be a κ-transfinite firing sequence.*

- *The* leaps *of σ are those ordinals $\iota_0 < \ldots < \iota_k$ with $k \leq |T|$ inductively defined by $\iota_0 = 0$ and if ι_ℓ exists and $T_{\mathbf{m}_{\iota_\ell}} \neq T_{\mathbf{m}_\kappa}$ then $\iota_{\ell+1}$ exists and $\iota_{\ell+1} = \min(\iota > \iota_\ell \mid T_{\mathbf{m}_\iota} \neq T_{\mathbf{m}_{\iota_\ell}})$;*
- *the* signature *of σ, $sig(\sigma)$, is the pair $((\mathbf{m}_{\iota_i})_{i\leq k}, \mathbf{m}_\kappa)$;*
- *the* abstract signature *of σ, $asig(\sigma)$, is the pair $((T_{\mathbf{m}_{\iota_i}})_{i\leq k}, \mathbf{m}_\kappa)$;*
- *the* trajectory *of σ, $traj(\sigma)$, is $(T_{\mathbf{m}_{\iota_i}})_{i\leq k}$.*

Notation. Let T be a set. Then the set of possible trajectories $Traj(T) \triangleq \{(T_i)_{i\leq k} \mid \forall i < k \; T_{i+1} \subsetneq T_i \subseteq T\}$. By a slight abuse of notations, the markings $\mathbf{m}_{\iota_\ell}$ will also be called leaps of σ.

The next proposition shows that given a signature of a transfinite firing sequence, there exists another transfinite firing sequence of ordinal less than ω^2 with same signature. Furthermore this sequence has a special shape.

Proposition 5. *Let $(\mathcal{N}, \mathbf{m}_0)$ be a marked CPN and $\sigma = \langle(\alpha_\iota t_\iota)_{\iota<\kappa}, (\mathbf{m}_\iota)_{\iota\leq\kappa}\rangle$ be a κ-transfinite firing sequence with $k+1$ leaps. Then there exist finite sequences $\sigma_{0,0}, \sigma_{0,1}, \ldots, \sigma_{k,0}, \sigma_{k,1}$ such that:*

- *for all $i \leq k$, $\mathbf{m}_{i,0} \xrightarrow{\sigma_{i,0}} \mathbf{m}_{i,1} \xrightarrow{(2^{-n}\sigma_{i,1})_{n\geq 1}} \mathbf{m}_{i+1,0}$ with $\mathbf{m}_{0,0} = \mathbf{m}_0$ and $\mathbf{m}_{k+1,0} = \mathbf{m}_\kappa$;*
- *$[[\vec{\sigma}_{i,0}]] = [[\vec{\sigma}_{i,1}]]$ and $[[\mathbf{m}_{i,0}]] \cup {}^\bullet[[\vec{\sigma}_{i,1}]]^\bullet = [[\mathbf{m}_{i,1}]]$;*
- *$sig(\sigma') = sig(\sigma)$ with $\sigma' = (\sigma_{i,0}(2^{-n}\sigma_{i,1})_{n\geq 1})_{i\leq k}$, being a $(k+1)\omega$-transfinite sequence;*
- *the leaps of σ' are $0, \omega, 2\omega, \ldots, k\omega$.*

Proof. We establish the result by recurrence on k.

Basis Case $k = 0$. This case corresponds to the situation where $\mathbf{m}_0 \xrightarrow{\sigma} \mathbf{m}$ with $T_\mathbf{m} = T_{\mathbf{m}_0}$. Applying Corollary 1, there exist finite sequences σ_0, σ_1 such that $\mathbf{m}_0 \xrightarrow{\sigma_0} \mathbf{m}_1 \xrightarrow{(2^{-n}\sigma_1)_{n\geq 1}} \mathbf{m}_\kappa$ with $[[\vec{\sigma}_0]] = [[\vec{\sigma}_1]] = [[\vec{\sigma}]]$ and $[[\mathbf{m}_0]] \cup {}^\bullet[[\vec{\sigma}_1]]^\bullet = [[\mathbf{m}_1]]$. Let $\sigma' = \sigma_0(2^{-n}\sigma_1)_{n\geq 1}$ and consider an arbitrary marking $\mathbf{m}'$ visited by σ'. Since the modes are non increasing $T_{\mathbf{m}'} \supseteq T_\mathbf{m}$ and so there is no leaps other than $\mathbf{m}_0$ in σ' which establishes this case.

Inductive Case. Let $\sigma = \langle(\alpha_\iota t_\iota)_{\iota<\kappa}, (\mathbf{m}_\iota)_{\iota\leq\kappa}\rangle$ be a transfinite firing sequence with signature $((\mathbf{m}_{\iota_i})_{i\leq k}, \mathbf{m}_\kappa)$. Let $\sigma = \sigma_1\sigma_2$ where σ_1 leads from $\mathbf{m}_0$ to $\mathbf{m}_{\iota_1}$ (the second leap of σ) and σ_2 leads from $\mathbf{m}_{\iota_1}$ to $\mathbf{m}$. By hypothesis of recurrence there exist finite sequences $\sigma_{1,0}, \sigma_{1,1}, \ldots, \sigma_{k,0}, \sigma_{k,1}$ fulfilling the properties of the proposition w.r.t. σ_2. Applying Corollary 1, there exist finite sequences σ'_0, σ'_1

such that $\mathbf{m}_0 \xrightarrow{\sigma_0'} \mathbf{m}_1 \xrightarrow{(2^{-n}\sigma_1')_{n\geq 1}} \mathbf{m}_{\iota_1}$ with $[[\vec{\sigma}_0]] = [[\vec{\sigma}_1']] = [[\vec{\sigma}_1]]$ and $[[\mathbf{m}_0]] \cup {}^\bullet[[\vec{\sigma}_1]]^\bullet = [[\mathbf{m}_1]]$. Let $\sigma' = \sigma_0'(2^{-n}\sigma_1')_{n\geq 1}$. Since $[[\mathbf{m}_0]] \subseteq [[\mathbf{m}_1]]$, $T_{\mathbf{m}_1} = T_{\mathbf{m}_0}$

For $i \geq 1$, let $\mathbf{m}_{1,i}$ be the marking reached by the sequence $\sigma_0'(2^{-n}\sigma_1')_{1\leq n\leq i}$. Observe that $\mathbf{m}_{1,i}$ is a convex combination of $\mathbf{m}_1$ and $\mathbf{m}_{\iota_1}$ with non null coefficients. So $[[\mathbf{m}_1]] \subseteq [[\mathbf{m}_{1,i}]]$ which implies that $T_{\mathbf{m}_{1,i}} = T_{\mathbf{m}_1} = T_{\mathbf{m}_0}$. For any arbitrary marking $\mathbf{m}' \neq \mathbf{m}_{\iota_1}$ visited by σ', there exists some $\mathbf{m}_{1,i}$ visited later and so $T_{\mathbf{m}'} \supseteq T_{\mathbf{m}_{1,i}} = T_{\mathbf{m}_0}$. So the only leaps of σ' are $\mathbf{m}_0$ and $\mathbf{m}_{\iota_1}$ which concludes the proof. □

If we only consider modes and omit visited markings that we will tackle in the next section, the existence of a trajectory is a central issue.

Definition 12. *The* trajectory problem *takes as input a marked CPN* $(\mathcal{N}, \mathbf{m}_0)$ *and a trajectory* $\tau \in Traj(T)$ *and checks whether there exists* σ *a transfinite firing sequence of* $(\mathcal{N}, \mathbf{m}_0)$ *such that* $traj(\sigma) = \tau$.

Proposition 6. *The* trajectory problem *is* NP*-complete.*

Proof. The proof of the hardness part is presented in the appendix. For the membership in NP, let us consider a marked CPN $(\mathcal{N}, \mathbf{m}_0)$ and a sequence $\tau = T_0 \ldots T_K \in Traj(T)$ with $T_0 = T_{\mathbf{m}_0}$.

The non deterministic procedure, denoted $\mathcal{A}$, relies on the existence of the special shape (provided by Proposition 5) of the possibly transfinite firing sequence with associated trajectory τ,

$$\mathbf{m}_0 = \mathbf{m}_{0,0} \xrightarrow{\sigma_{0,0}} \mathbf{m}_{0,1} \xrightarrow{(2^{-n}\sigma_{0,1})_{n\geq 1}} \mathbf{m}_{1,0} \xrightarrow{\sigma_{1,0}} \cdots$$

$$\cdots \xrightarrow{(2^{-n}\sigma_{K-2,1})_{n\geq 1}} \mathbf{m}_{K-1,0} \xrightarrow{\sigma_{K-1,0}} \mathbf{m}_{K-1,1} \xrightarrow{(2^{-n}\sigma_{K-1,1})_{n\geq 1}} \mathbf{m}_{K,0}$$

i.e., such that for all $i < K$:

- $[[\vec{\sigma}_{i,0}]] = [[\vec{\sigma}_{i,0}]]$ and $[[\mathbf{m}_{i,0}]] \cup {}^\bullet[[\vec{\sigma}_{i,1}]]^\bullet = [[\mathbf{m}_{i,1}]]$;
- $T_i = T_{[[\mathbf{m}_{i,0}]]} = T_{[[\mathbf{m}_{i,1}]]}$;
- and $T_K = T_{[[\mathbf{m}_{K,0}]]}$.
  - $\mathcal{A}$ first guesses (in polynomial time) a sequence of subset of transitions $(X_i)_{i<K}$ and a sequence of subsets of places $(P_{i,0}, P_{i,1})_{0<i<K} P_{K,0}$ with $P_{0,0} = [[\mathbf{m}_0]]$.
  - Then $\mathcal{A}$ checks whether for all $i < K$ $P_{i,0} \cup {}^\bullet X_i{}^\bullet = P_{i,1}$.
  - Afterwards $\mathcal{A}$ checks (in polynomial time due to Proposition 1) whether for all $i < K$, $T_{P_{i,0}} = T_{P_{i,1}} = T_i$, and $T_{P_{K,0}} = T_K$.
  - Then $\mathcal{A}$ checks the qualitative conditions of reachability and lim-reachability characterizations (see Theorems 1 and 2) : for all $i < K$, $X_i \in \mathbf{FS}(\mathcal{N}, P_{i,0}) \cap \mathbf{FS}(\mathcal{N}^{-1}, P_{i,1})$. Since $\mathbf{FS}(\mathcal{N}, P_{i,0}) \subseteq \mathbf{FS}(\mathcal{N}, P_{i,1})$, there is no need to check whether $X_i \in \mathbf{FS}(\mathcal{N}, P_{i,1})$. Again due to Proposition 1, these tests are performed in polynomial time.

- If the previous checks are successful, $\mathcal{A}$ builds the following linear program (including implicit strict inequalities due to the conditions about the supports) related to the quantitative conditions of reachability and lim-reachability characterizations where the (positive) variables are the components of the set of markings $\{\mathbf{m}_{i,j}\}_{i<K,j\in\{0,1\}} \cup \{\mathbf{m}_{K,0}\}$ and the set of Parikh vectors $\{\mathbf{v}_{i,j}\}_{i<K,j\in\{0,1\}}$:

$$\mathbf{m}_{0,0} = \mathbf{m}_0 \wedge \bigwedge_{0\leq i<K} \mathbf{m}_{i,1} = \mathbf{m}_{i,0} + \mathbf{C}\cdot\mathbf{v}_{i,0}$$

$$\wedge \bigwedge_{0\leq i<K} \mathbf{m}_{i+1,0} = \mathbf{m}_{i,1} + \mathbf{C}\cdot\mathbf{v}_{i,1}$$

$$\wedge \bigwedge_{0\leq i<K, j\in\{0,1\}} [[\mathbf{m}_{i,j}]] = P_{i,j} \wedge [[\mathbf{v}_{i,j}]] = X_i$$

Here and later on, an equation like $[[\mathbf{v}_{i,j}]] = X_i$ is an abbreviation for:

$$\bigwedge\nolimits_{t\in X_i} \mathbf{v}_{i,j}[t] > 0 \wedge \bigwedge\nolimits_{t\in T\setminus X_i} \mathbf{v}_{i,j}[t] = 0.$$

Then $\mathcal{A}$ checks in polynomial time if this linear program is satisfiable. □

4 A Symbolic Reachability Tree

In Examples 1 and 2 above, we have used an abstraction approach that lumps together markings having the same set of eventually firable transitions into *modes*. Here, we will formalize the associated semantics in the form of symbolic reachability trees, introduced in Subsect. 4.1. In Subsect. 4.2, the "reversibility" of the leaves of these trees will be established, using linear algebra theory. Finally we develop a construction of an effective representation of symbolic reachability trees in Subsect. 4.3.

4.1 Definition

The aim of the symbolic reachability tree (SRT) that we will build is to represent in an effective way all the abstract signatures of κ-transfinite sequences of a marked CPN. By effective we mean that we can check not only whether a potential abstract signature exists, but also for inclusion or equality of the sets of abstract signatures of two CPNs.

In order to define an appropriate SRT, we first introduce the *abstract reachability graph*. Since the mode of a marking only depends on its support, the abstract reachability graph tracks the possible evolution of the supports of reachable markings. Therefore, the vertices of this graph are the subsets of P. Let us summarize some of the results of the previous section that guide us for the construction of the edges of this graph:

- For $\mathbf{m} \xrightarrow{\sigma} \mathbf{m}'$, a transfinite firing sequence with $T_{\mathbf{m}'} = T_{\mathbf{m}}$, there exists an infinite firing sequence σ' with $\mathbf{m} \xrightarrow{\sigma'} \mathbf{m}'$ implying $[[\vec{\sigma}']] \in \mathbf{FS}(\mathcal{N}, [[\mathbf{m}]])$;
- For $\mathbf{m} \xrightarrow{\sigma} \mathbf{m}'$, a transfinite firing sequence whose only leaps are the initial and final markings, there exist a finite sequence σ_0 and a repeated discounted firing sequence $(2^{-n}\sigma_1)_{n\geq 1}$ with $\mathbf{m} \xrightarrow{\sigma_0} \mathbf{m}_1 \xrightarrow{(2^{-n}\sigma_1)_{n\geq 1}} \mathbf{m}'$, $[[\vec{\sigma}_0]] = [[\vec{\sigma}_1]] = [[\vec{\sigma}]]$ and $[[\mathbf{m}_1]] = [[\mathbf{m}_0]] \cup {}^\bullet[[\vec{\sigma}_1]]^\bullet$. The leaps of this alternative firing sequence are also the initial and final markings.

Thus there will be two kinds of edges from P' to $P'' \neq P'$ in our graph. When $T_{P''} = T_{P'}$, there is an *anonymous* edge from P' to P'' if there exists a set of transitions T' with $T' \in \mathbf{FS}(\mathcal{N}, P')$ and $P'' \subseteq P' \cup T'^\bullet$, since those are the only places that may be marked after the firing of an infinite sequence whose support is T' and $P' \setminus {}^\bullet T' \subseteq P''$ since these places remain marked after such a firing. When $P'' \subsetneq P$ and $T_{P''} \subsetneq T_{P'}$ there is an edge labelled by some $T' \subseteq T$ with (1) ${}^\bullet T'^\bullet \subseteq P'$ and (2) $P' \setminus {}^\bullet T' \subseteq P''$. Here, (1) is a necessary condition for the existence of a repeated discounted firing sequence described in Lemma 1, and (2) is a necessary condition ensuring that the support of the target marking is P''. This kind of edges is called *border* edges. Omitting labels for anonymous edges will be justified during the description of the SRT.

Definition 13 (Abstract reachability graph). *Let $\mathcal{N}$ be a CPN. Then its abstract reachability graph $ARG(\mathcal{N}) = (V, E)$ is defined as follows:*

- *$V = 2^P$ is its set of vertices;*
- *For all $P' \neq P'' \subseteq P$ with $T_{P'} = T_{P''}$, $P' \to P''$ is an edge of E iff there exists $T' \subseteq T$ such that:*
  (1) $T' \in \mathbf{FS}(\mathcal{N}, P')$ and (2) $P' \setminus {}^\bullet T' \subseteq P'' \subseteq P' \cup T'^\bullet$.
- *For all $T' \subseteq T$, $P'' \subsetneq P' \subseteq P$, $T_{P''} \subsetneq T_{P'}$, $P' \xrightarrow{T'} P''$ is an edge of E iff:*
  (1) ${}^\bullet T'^\bullet \subseteq P'$ and (2) $P' \setminus {}^\bullet T' \subseteq P''$.

Let $(\mathcal{N}, \mathbf{m}_0)$ be a marked CPN. Then its abstract reachability graph $ARG(\mathcal{N}, \mathbf{m}_0)$ is the restriction of $ARG(\mathcal{N})$ to the vertices reachable from $[[\mathbf{m}_0]]$.

Lemma 2. *The reflexive closure of $\to$, the anonymous relation of $ARG(\mathcal{N})$, denoted $\to^*$ is transitive.*

Proof. Let $P_1 \to P_2 \to P_3$ with $P_3 \neq P_1$. So for $i \in \{1, 2\}$, there exists T_i with $T_i \in \mathbf{FS}(\mathcal{N}, P_i)$ and $P_i \setminus {}^\bullet T_i \subseteq P_{i+1} \subseteq P_i \cup T_i{}^\bullet$. Let $T' = T_1 \cup T_2$. Then:
$P_1 \setminus {}^\bullet T' = (P_1 \setminus {}^\bullet T_1) \setminus {}^\bullet T_2 \subseteq P_2 \setminus {}^\bullet T_2 \subseteq P_3 \subseteq P_2 \cup T_2{}^\bullet \subseteq P_1 \cup T_1{}^\bullet \cup T_2{}^\bullet = P_1 \cup T'^\bullet$.
Pick an arbitrary marking $\mathbf{m}$ with $[[\mathbf{m}]] = P'$. Applying Proposition 1, there exists $\mathbf{m} \xrightarrow{\sigma} \mathbf{m}'$ with $[[\vec{\sigma}]] = T_1$ and $[[\mathbf{m}']] = [[\mathbf{m}]] \cup {}^\bullet T_1{}^\bullet$.
Thus $[[\mathbf{m}']] \supseteq P_2$. Since $T_2 \in \mathbf{FS}(\mathcal{N}, P_2)$ there exists $\mathbf{m}' \xrightarrow{\sigma'}$ with $[[\vec{\sigma}]] = T_2$.
So $\mathbf{m} \xrightarrow{\sigma\sigma'}$ implying $T' \in \mathbf{FS}(\mathcal{N}, P_1)$. Thus $P_1 \to P_3$. □

The next definitions show how to define the acceptance of a signature of a transfinite firing sequence by the ARG.

Definition 14. *Let* $(\mathcal{N}, \mathbf{m}_0)$ *be a marked CPN,* $s = (P_i^- \xrightarrow{X_i} P_i^+)_{0<i\leq k}$ *a sequence of border edges of* $ARG(\mathcal{N}, \mathbf{m}_0)$ *and* $P_f \subseteq P$*. Then the pair* (s, P_f) *is a* symbolic path *of* $ARG(\mathcal{N}, \mathbf{m}_0)$ *if* $[[\mathbf{m}_0]] \rightarrow^* P_f$ *when* $s = \varepsilon$*, and* $s \neq \varepsilon$ *implies*

- $[[\mathbf{m}_0]] \rightarrow^* P_1^-$ *and* $P_k^+ \rightarrow^* P_f$*;*
- *for all* $0 < i < k$*,* $P_i^+ \rightarrow^* P_{i+1}^-$*.*

Definition 15. *Let* $(\mathcal{N}, \mathbf{m}_0)$ *be a marked CPN,* (s, P_f) *be a symbolic path of* $ARG(\mathcal{N}, \mathbf{m}_0)$ *with* $s = (P_i^- \xrightarrow{X_i} P_i^+)_{0<i\leq k}$ *and* σ *be a transfinite firing sequence with* $sig(\sigma) = ((\mathbf{m}_i)_{i\leq k}, \mathbf{m}_f)$*. Then* $sig(\sigma)$ *is* accepted *by* (s, P_f) *if for all* $i \leq k$*,* $P_i^+ = [[\mathbf{m}_i]]$ *and* $P_f = [[\mathbf{m}_f]]$*.*

Example 3. Figure 2 depicts a marked CPN $(\mathcal{N}, \mathbf{m}_0)$ and its abstract reachability graph. The anonymous edges are represented by single lines, border edges by double lines. For border edges labelled by T', we omit the brackets defining T' (e.g., $\{t_1, t_2\}$ is shown as t_1, t_2). Let $\sigma = \mathbf{m}_0 \xrightarrow{t_1 t_2} \mathbf{0}$ a firing sequence with $sig(\sigma) = (\mathbf{m}_0\ 1p_2\ \mathbf{0}, \mathbf{0})$, and the following symbolic path in $ARG(\mathcal{N}, \mathbf{m}_0)$: $c = (\{p_1\} \xrightarrow{t_1} \{p_2\}\ \{p_2\} \xrightarrow{t_2} \emptyset, \emptyset)$; c accepts $sig(\sigma)$. For firing sequence $\sigma' = \mathbf{m}_0 \xrightarrow{(2^{-n}t_1 2^{-n}t_2)_{n\geq 1}} \mathbf{0}$ with $sig(\sigma') = (\mathbf{m}_0\ \mathbf{0}, \mathbf{0})$, take pair: $c' = (\{p_1, p_2\} \xrightarrow{\{t_1,t_2\}} \emptyset, \emptyset)$ in $ARG(\mathcal{N}, \mathbf{m}_0)$. Since $\{p_1\} \rightarrow^* \{p_1, p_2\}$, c' accepts $sig(\sigma')$. Figure 3 depicts another marked CPN $(\mathcal{N}', \mathbf{m}_0')$ and its abstract reachability graph. Consider the symbolic path in $ARG(\mathcal{N}', \mathbf{m}_0')$: $c'' = (\{p_1, p_2\} \xrightarrow{\{t_1,t_2\}} \emptyset, \emptyset)$. The abstract reachability graph does not depend on the exact weights of the net. Here, if $x = 1$, then no signature of a firing sequence will be accepted by c'' since for all lim-reachable marking $\mathbf{m}$s, $\|\mathbf{m}\|_1 \triangleq \mathbf{m}(p_1) + \mathbf{m}(p_2) = 1$. On the other hand, if $x = 2$, then $\sigma'' = \mathbf{m}_0 \xrightarrow{\frac{1}{4}t_1} \mathbf{m}_1 \xrightarrow{(2^{-n}(\frac{1}{2}t_1 1t_2 \frac{1}{4}t_1))_{n\geq 1}} \mathbf{0}$ with $\mathbf{m}_1 = \frac{1}{2}p_1 + \frac{1}{4}p_2$ fulfills $sig(\sigma'') = (\mathbf{m}_0\ \mathbf{0}, \mathbf{0})$.

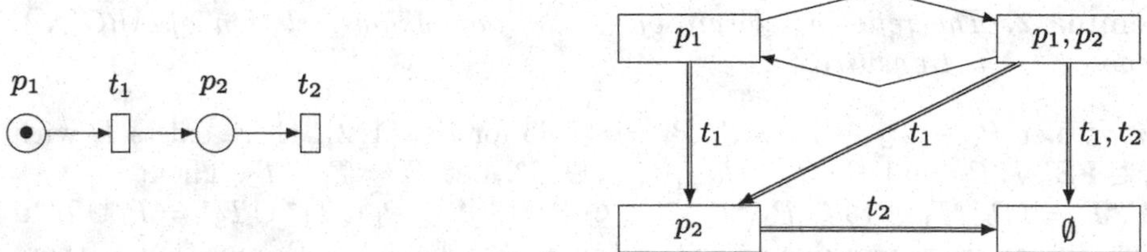

Fig. 2. A marked CPN and its ARG.

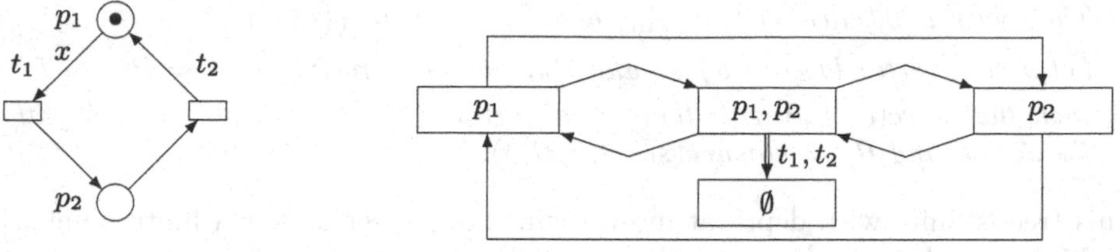

Fig. 3. Another marked CPN and its ARG.

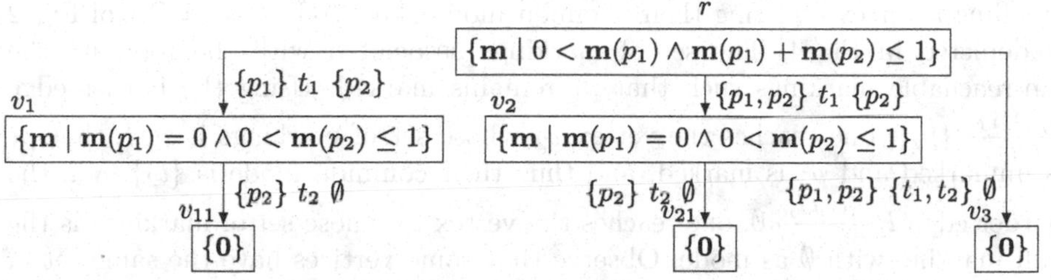

Fig. 4. The symbolic reachability tree of the marked CPN of Fig. 2.

The next proposition shows that the ARG accepts the signatures of all transfinite firing sequences (but possibly more as illustrated by the CPN of Fig. 3). Its proof is an immediate consequence of the definition of the edges as discussed above.

Proposition 7. *Let $(\mathcal{N}, \mathbf{m}_0)$ be a marked CPN and σ be a transfinite firing sequence for $(\mathcal{N}, \mathbf{m}_0)$. Then there exists a symbolic path (s, P_f) of $ARG(\mathcal{N}, \mathbf{m}_0)$ such that $sig(\sigma)$ is accepted by (s, P_f).*

In order to take into account the markings while building the SRT, we introduce two operators on sets of markings of CPN.

Definition 16. *Let $\mathcal{N}$ be a CPN and R be a set of markings. Then:*

$$closure(R) = \{\mathbf{m}' \mid \exists \mathbf{m} \in R\ [[\mathbf{m}]] \rightarrow^* [[\mathbf{m}']] \wedge \mathbf{m}' \in \lim\text{–}\mathbf{RS}(\mathcal{N}, \mathbf{m})\}.$$

Let $tr = P' \xrightarrow{T'} P''$ be a border edge of $ARG(\mathcal{N})$. Then $succ(tr, R)$ is:

$$\{\mathbf{m}' \mid \exists \mathbf{m} \xrightarrow{(2^{-n}\sigma)_{n \geq 1}} \mathbf{m}' \wedge \mathbf{m} \in R \wedge [[\mathbf{m}]] = P' \wedge [[\mathbf{m}']] = P'' \wedge [[\vec{\sigma}]] = T'\}.$$

We are now in a position to define the symbolic reachability tree which will keep track of the abstract signatures of a marked CPN.

Definition 17 (Symbolic reachability tree). *Let $(\mathcal{N}, \mathbf{m}_0)$ be a marked CPN. The* symbolic reachability tree *$SRT(\mathcal{N}, \mathbf{m}_0)$, of $(\mathcal{N}, \mathbf{m}_0)$ is a directed tree whose vertices v are labelled by a non empty set of markings R_v and their common mode T_v, and edges are labelled by border edges of $ARG(\mathcal{N}, \mathbf{m}_0)$, inductively defined by:*

- *The root r is labelled by $T_r = T_{\mathbf{m}_0}$ and $R_r = closure(\{\mathbf{m}_0\})$.*
- *Let v be a vertex labelled by T_v and R_v. For all border edge $tr = P' \xrightarrow{T'} P''$ such that $succ(tr, R_v) \neq \emptyset$ there is a vertex v' and an edge $v \xrightarrow{tr} v'$ with $T_{v'} = T_{P''}$ and $R_{v'} = closure(succ(tr, R_v))$.*

This tree is finite with depth at most T since every vertex has a finite number of children with their mode strictly included in the mode of their father.

Example 4. In the examples of SRTs we only represent the set of markings labelling a vertex omitting their common mode. The SRT of the CPN of Fig. 2 is depicted in Fig. 4. The set of markings associated with the root are the lim-reachable markings such that p_1 remains marked. Using the border edge $P \xrightarrow{\{t_1\}} \{p_2\}$, one reaches the vertex v_2 whose set of markings is such that p_1 is unmarked and p_2 is marked, and thus their common mode is $\{t_2\}$. Via the border edge $P \xrightarrow{\{t_1,t_2\}} \emptyset$, one reaches the vertex v_3 whose set of markings is the null marking with $\emptyset$ as mode. Observe that some vertices have the same set of markings and could be merged, thus producing a directed acyclic graph. The SRT of the CPN of Fig. 3 when $x = 2$ is depicted in Fig. 5. The set of markings associated with the root are those lim-reachable markings for which either p_1 or p_2 remains marked; hence their common mode is T. This set is characterized by the inequalities $0 < \mathbf{m}(p_1) + 2\mathbf{m}(p_2) \leq 1$. Using the border edge $P \xrightarrow{\{t_1,t_2\}} \emptyset$, one creates the vertex v_1, whose set of markings is the null marking, with $\emptyset$ as mode.

The next two propositions establish that the SRT exactly captures the signatures of the CPN.

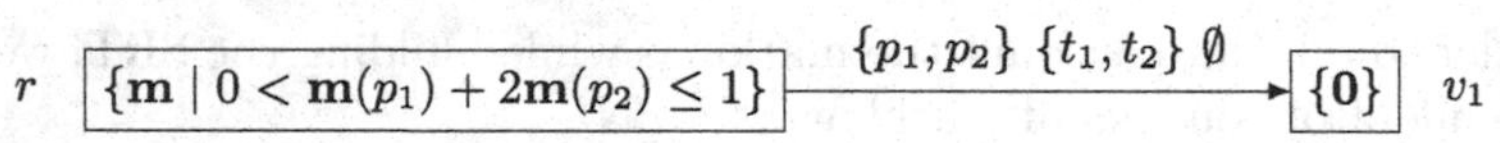

$$r \quad \boxed{\{\mathbf{m} \mid 0 < \mathbf{m}(p_1) + 2\mathbf{m}(p_2) \leq 1\}} \xrightarrow{\{p_1,p_2\}\ \{t_1,t_2\}\ \emptyset} \boxed{\{\mathbf{0}\}} \quad v_1$$

Fig. 5. The symbolic reachability tree of the marked CPN of Fig. 3 when $x = 2$.

Proposition 8. *Let $(\mathcal{N}, \mathbf{m}_0)$ be a marked CPN and σ be a κ-transfinite firing sequence with $sig(\sigma) = ((\mathbf{m}_i)_{i \leq k}, \mathbf{m}_\kappa)$. Then there exists a path $\rho = r \xrightarrow{tr_1} v_1 \xrightarrow{tr_2} \cdots \xrightarrow{tr_k} v_k$ in $SRT(\mathcal{N}, \mathbf{m}_0)$ such that: $\forall 1 \leq i \leq k\ \mathbf{m}_i \in R_{v_i} \wedge \mathbf{m}_\kappa \in R_{v_k}$.*

Proof. Due to Proposition 5, there exist finite (possibly null) sequences $\sigma_{0,0}$, $\sigma_{0,1}$, ..., $\sigma_{k,0}$, $\sigma_{k,1}$ with $0 \leq k \leq |T|$ such that:

- for all $i \leq k$, $\mathbf{m}_{i,0} \xrightarrow{\sigma_{i,0}} \mathbf{m}_{i,1} \xrightarrow{(2^{-n}\sigma_{i,1})_{n\geq 1}} \mathbf{m}_{i+1,0}$ with $\mathbf{m}_{0,0} = \mathbf{m}_0$ and $\mathbf{m}_{k+1,0} = \mathbf{m}_\kappa$;
- $[[\vec{\sigma}_{i,0}]] = [[\vec{\sigma}_{i,1}]]$ and $[[\mathbf{m}_{i,0}]] \cup {}^\bullet[[\vec{\sigma}_{i,1}]]^\bullet \subseteq [[\mathbf{m}_{i,1}]]$;

- $sig(\sigma') = sig(\sigma)$, where $\sigma' = (\sigma_{i,0}(2^{-n}\sigma_{i,1})_{n\geq 1})_{i\leq k}$ is a finite-length $(k+1)\omega$-transfinite sequence;
- the leaps of σ' are $0, \omega, 2\omega, \ldots, k\omega$.

When $T_{\mathbf{m}_\kappa} = T_{\mathbf{m}_0}$, by definition of $R_r = closure(\mathbf{m}_0)$, one has $\mathbf{m}_\kappa \in R_r$ since $\mathbf{m}_\kappa$ is lim-reachable from $\mathbf{m}_0$; thus ρ consists only of r.

Otherwise we build ρ in an inductive way. Note that $T_{\mathbf{m}_{0,1}} = T_{\mathbf{m}_0}$, and since $\mathbf{m}_{0,1}$ is reachable from $\mathbf{m}_0$, also $\mathbf{m}_{0,1} \in R_r$. Define $P' = [[\mathbf{m}_{0,1}]]$, $P'' = [[\mathbf{m}_{1,0}]]$ and $X = [[\sigma_{0,1}]]$. Note that $tr_1 = P' \xrightarrow{X} P''$ is a border edge of $ARG(\mathcal{N}, \mathbf{m}_0)$ and $\mathbf{m}_{1,0} \in succ(tr_1, R_r)$. Thus $succ(tr_1, R_r) \neq \emptyset$, and there is a vertex v_1 and a transition $r \xrightarrow{tr_1} v_1$ with $\mathbf{m}_{1,0} \in succ(tr_1, R_r)$ and $\mathbf{m}_{1,1} \in closure(succ(tr_1, R_r)) = R_{v_1}$.

By induction hypothesis, one gets a path $r \xrightarrow{tr_1} v_1 \cdots \xrightarrow{tr_k} v_k$ such that for all $1 \leq i \leq k$, $\mathbf{m}_{i,0} \in R_{v_i}$. Since $\mathbf{m}_\kappa = \mathbf{m}_{k+1,0}$ is lim-reachable from $\mathbf{m}_{k,0}$ and $T_{\mathbf{m}_\kappa} = T_{\mathbf{m}_{k,0}}$, we have $\mathbf{m}_\kappa \in R_{v_k}$, which concludes the proof. □

Proposition 9. *Let $(\mathcal{N}, \mathbf{m}_0)$ be a marked CPN, ρ be a path $r = v_0 \xrightarrow{tr_1} v_1 \xrightarrow{tr_2} \cdots \xrightarrow{tr_k} v_k$ in $SRT(\mathcal{N}, \mathbf{m}_0)$ and $\mathbf{m}_\kappa \in R_{v_k}$. Then there exists a transfinite firing sequence σ with $sig(\sigma) = ((\mathbf{m}_i)_{i\leq k}, \mathbf{m}_\kappa)$. such that: $\forall 1 \leq i \leq k$ $\mathbf{m}_i \in R_{v_i}$.*

Proof. We build σ by induction. When $k = 0$, $\mathbf{m}_\kappa \in R_{v_0}$. By definition of R_{v_0}, there exists a infinite firing sequence $\mathbf{m}_0 \xrightarrow{\sigma} \mathbf{m}_\kappa$ and $T_{\mathbf{m}_\kappa} = T_{\mathbf{m}_0}$ which implies that the signature of σ is $(\mathbf{m}_0, \mathbf{m}_\kappa)$.

Assume now that $k > 0$. Then there exists a border edge $tr_k = P' \xrightarrow{X} P''$ and a marking $\mathbf{m}_k \in succ(tr, R_{v_{k-1}})$ such that $[[\mathbf{m}_k]] = P''$ and $\mathbf{m}_\kappa$ is lim-reachable from $\mathbf{m}_k$ and $T_{\mathbf{m}_\kappa} = T_{\mathbf{m}_k}$.

Since $\mathbf{m}_k \in succ(tr, R_{v_{k-1}})$, there exists $\mathbf{m}_k^- \in R_{v_{k-1}}$ such that $[[\mathbf{m}_k^-]] = P'$, and σ is a repeated discounted sequence with $[[\vec{\sigma}]] = X$ and $\mathbf{m}_k^- \xrightarrow{\sigma} \mathbf{m}_k$.

By induction hypothesis, we obtain a transfinite firing sequence from $\mathbf{m}_1^- \in R_r$ to $\mathbf{m}_\kappa$, and thus a transfinite firing sequence from $\mathbf{m}_0$ to $\mathbf{m}_\kappa$ such that its leaps are $\mathbf{m}_0, \mathbf{m}_1, \ldots, \mathbf{m}_k$ as required by the proposition. □

4.2 Terminal Components of the SRT for Bounded CPNs

As for marked Petri nets, a marked CPN is *bounded* if there exists a bound $B \in \mathbb{N}$ such that for all reachable markings $\mathbf{m}$, $\|\mathbf{m}\|_\infty \leq B$. There is a useful characterization of (un)boundedness for marked CPN via linear programming.

Theorem 3 ([12]). *Let $(\mathcal{N}, \mathbf{m}_0)$ be a marked CPN. Then $(\mathcal{N}, \mathbf{m}_0)$ is unbounded iff there exists $\mathbf{v} \in \mathbb{R}_+^T$ such that (1) $\mathbf{C} \cdot \mathbf{v} \gneq 0$ and (2) $[[\mathbf{v}]] \subseteq T_{\mathbf{m}_0}$.*

Consider $\mathbf{m}$ a marking belonging to a strongly connected component (SCC) of the reachability graph of a bounded Petri net. Then for all $\mathbf{m}'$ reachable from $\mathbf{m}$, $\mathbf{m}$ is reachable from $\mathbf{m}'$. This property, which is called *reversibility* and whose proof follows from the definition of SCCs, is of crucial importance in biological applications (characterization of phenotypes, etc.).

Notation. A *terminal component* of the SRT is the set of markings associated with a leaf of the SRT. Observe that by definition of the SRT for all $\mathbf{m}$, $\mathbf{m}'$ in a terminal component, $T_{\mathbf{m}} = T_{\mathbf{m}'}$.

Here we establish reversibility inside terminal components of the SRT of a bounded marked CPN. In the context of CPNs, reversibility can be stated as follows: Let $\mathbf{m}$ be a marking of a terminal SRT component and $\mathbf{m}'$ be lim-reachable from $\mathbf{m}$, then $\mathbf{m}$ is lim-reachable from $\mathbf{m}'$. To prove this, let us recall the following proposition from linear programming theory about duality.

Proposition 10. *Let* $\mathbf{A}$ *be a real matrix of dimension* $K \times L$ *and* $1 \leq k \leq K$. *Then the following statements are equivalent:*

- $\exists \mathbf{v} \in \mathbb{R}_+^K\ \mathbf{v} \cdot \mathbf{A} \leq 0 \wedge \mathbf{v}[k] > 0$
- $\nexists \mathbf{w} \in \mathbb{R}_+^L\ \mathbf{A} \cdot \mathbf{w} \geq 0 \wedge (\mathbf{A} \cdot \mathbf{w})[k] > 0$

Let us consider $\mathbf{B} = -\mathbf{A}^t$ with dimension of $\mathbf{A}$ being now $L \times K$. Then we get another formulation for the duality property obtained by combining transposition and additive inversion.

Proposition 11. *Let* $\mathbf{B}$ *be a real matrix of dimension* $L \times K$ *and* $1 \leq k \leq K$. *Then the following statements are equivalent:*

- $\exists \mathbf{v} \in \mathbb{R}_+^K\ \mathbf{B} \cdot \mathbf{v} \geq 0 \wedge \mathbf{v}[k] > 0$
- $\nexists \mathbf{w} \in \mathbb{R}_+^L\ \mathbf{w} \cdot \mathbf{B} \leq 0 \wedge (\mathbf{w} \cdot \mathbf{B})[k] < 0$

Using Proposition 10, one can reformulate the characterization of the boundedness of a marked CPN stated by Theorem 3.

Notation. Let $\mathcal{N}$ be a CPN, $\mathbf{C}$ be its incidence matrix and $T' \subseteq T$. Then $\mathbf{C}_{T'}$ denotes $\mathbf{C}$ reduced to the columns of T'.

Theorem 4. *Let* $(\mathcal{N}, \mathbf{m}_0)$ *be a marked CPN. Then* $(\mathcal{N}, \mathbf{m}_0)$ *is bounded iff there exists* $\mathbf{w} \in \mathbb{R}_+^P$ *such that* $\mathbf{w} \cdot \mathbf{C}_{T_{\mathbf{m}_0}} \leq 0 \wedge [[\mathbf{w}]] = P$.

Proof. Let us recall the characterization of Theorem 3. $(\mathcal{N}, \mathbf{m}_0)$ is bounded iff there does not exist $\mathbf{v} \in \mathbb{R}_+^T$ such that $\mathbf{C} \cdot \mathbf{v} \gneq 0$ and $[[\mathbf{v}]] \subseteq T_{\mathbf{m}_0}$, which is equivalent to the assertion that there does not exist $\mathbf{v} \in \mathbb{R}_+^{T_{\mathbf{m}_0}}$ such that $\mathbf{C}_{T_{\mathbf{m}_0}} \cdot \mathbf{v} \gneq 0$. This is also equivalent to: "For all $p \in P$, there does not exist $\mathbf{v} \in \mathbb{R}_+^{T_{\mathbf{m}_0}}$ such that $\mathbf{C}_{T_{\mathbf{m}_0}} \cdot \mathbf{v} \geq 0 \wedge (\mathbf{C}_{T_{\mathbf{m}_0}} \cdot \mathbf{v})[p] > 0$."

Applying Proposition 10, this is also equivalent to the following statement: "For all $p \in P$, there exists $\mathbf{w}_p \in \mathbb{R}_+^P$ such that $\mathbf{w}_p \cdot \mathbf{C}_{T_{\mathbf{m}_0}} \leq 0 \wedge \mathbf{w}[p] > 0$." Setting $\mathbf{w} = \sum_{p \in P} \mathbf{w}_p$, this is equivalent to:
There exists $\mathbf{w} \in \mathbb{R}_+^P$ such that $\mathbf{w} \cdot \mathbf{C}_{T_{\mathbf{m}_0}} \leq 0 \wedge [[\mathbf{w}]] = P$. □

The next lemma implies that, starting from a marking of a terminal component, the inequality of Theorem 4 is in fact an equality; that will be stated by Theorem 5 below.

Lemma 3. *Let $(\mathcal{N}, \mathbf{m}_0)$ be a bounded marked CPN such that for all $\mathbf{m} \in \lim{-}\mathbf{RS}(\mathcal{N}, \mathbf{m}_0)$, $T_{\mathbf{m}} = T_{\mathbf{m}_0}$. There does not exist $\mathbf{w} \in \mathbb{R}_+^P$ such that:*

$$\mathbf{w} \cdot \mathbf{C}_{T_{\mathbf{m}_0}} \lneq 0.$$

Proof. Due to Proposition 1, there exists $\mathbf{m}_1 \in \mathbf{RS}(\mathcal{N}, \mathbf{m}_0)$ such that, for all $p \in {}^{\bullet}T_{\mathbf{m}_0}$, $\mathbf{m}_1(p) > 0$. Thus for all $t \in T_{\mathbf{m}_0}$, $\{t\} \in \mathbf{FS}(\mathcal{N}, \mathbf{m}_1)$. Since $\mathbf{FS}(\mathcal{N}, \mathbf{m}_1)$ is closed under union, any subset of $T_{\mathbf{m}_0}$ belongs to $\mathbf{FS}(\mathcal{N}, \mathbf{m}_1)$. Hence $\lim{-}\mathbf{RS}(\mathcal{N}, \mathbf{m}_1) = \{\mathbf{m} \in \mathbb{R}_+^P \mid \exists \mathbf{v} \in \mathbb{R}_+^{T_{\mathbf{m}_0}} \wedge \mathbf{m} = \mathbf{m}_1 + \mathbf{C}_{T_{\mathbf{m}_0}} \cdot \mathbf{v}\}$.

Assume by contradiction that there exists $\mathbf{w} \in \mathbb{R}_+^P$ and $t \in T_{\mathbf{m}_0}$ such that:

$$\mathbf{w} \cdot \mathbf{C}_{T_{\mathbf{m}_0}} \leq 0 \text{ and } (\mathbf{w} \cdot \mathbf{C}_{T_{\mathbf{m}_0}})[t] < 0.$$

Let $\alpha = \sup(\mathbf{v}[t] \mid \mathbf{v} \in \mathbb{R}_+^{T_{\mathbf{m}_0}} \wedge \mathbf{m}_1 + \mathbf{C}_{T_{\mathbf{m}_0}} \cdot \mathbf{v} \geq 0)$. Since $\mathbf{w} \cdot \mathbf{C}_{T_{\mathbf{m}_0}} \leq 0$ and $(\mathbf{w} \cdot \mathbf{C}_{T_{\mathbf{m}_0}})[t] < 0$, α is finite. Due to linear programming theory, there is some $\mathbf{v}$ such that $\mathbf{m}_1 + \mathbf{C}_{T_{\mathbf{m}_0}} \cdot \mathbf{v} \geq 0$ and $\mathbf{v}[t] = \alpha$. Let $\mathbf{m} = \mathbf{m}_1 + \mathbf{C}_{T_{\mathbf{m}_0}} \cdot \mathbf{v}$. Then $\mathbf{m} \in \lim{-}\mathbf{RS}(\mathcal{N}, \mathbf{m}_1) \subseteq \lim{-}\mathbf{RS}(\mathcal{N}, \mathbf{m}_0)$ and $t \notin T_{\mathbf{m}}$, a contradiction. □

Theorem 5. *Let $(\mathcal{N}, \mathbf{m}_0)$ be a bounded marked CPN such that $T_{\mathbf{m}} = T_{\mathbf{m}_0}$ for all $\mathbf{m} \in \lim{-}\mathbf{RS}(\mathcal{N}, \mathbf{m}_0)$. Then there exists $\mathbf{w} \in \mathbb{R}_+^P$: $\mathbf{w} \cdot \mathbf{C}_{T_{\mathbf{m}_0}} = 0 \wedge [[\mathbf{w}]] = P$.*

Proof. Since $(\mathcal{N}, \mathbf{m}_0)$ is bounded, there exists $\mathbf{w} \in \mathbb{R}_+^P$ such that $\mathbf{w} \cdot \mathbf{C}_{T_{\mathbf{m}_0}} \leq 0$ and $[[\mathbf{w}]] = P$. By Lemma 3, there is no $\mathbf{w}$ such that $\mathbf{w} \cdot \mathbf{C}_{T_{\mathbf{m}_0}} \lneq 0$. Hence $\mathbf{w} \cdot \mathbf{C}_{T_{\mathbf{m}_0}} = 0$. □

Let $\mathbf{m}_0$ be a marking of a terminal component of the SRT of a bounded marked CPN. While Theorem 5 states that, when restricted to $T_{\mathbf{m}_0}$, there is a place invariant whose support is P, one has by the following theorem that there is a transition invariant whose support is $T_{\mathbf{m}_0}$.

Theorem 6. *Let $(\mathcal{N}, \mathbf{m}_0)$ be a bounded marked CPN such that for all $\mathbf{m} \in \lim{-}\mathbf{RS}(\mathcal{N}, \mathbf{m}_0)$, $T_{\mathbf{m}} = T_{\mathbf{m}_0}$. Then there exists $\mathbf{v} \in \mathbb{R}_+^T$ such that $\mathbf{C}_{T_{\mathbf{m}_0}} \cdot \mathbf{v} = 0 \wedge [[\mathbf{v}]] = T_{\mathbf{m}_0}$.*

Proof. Let $t \in T_{\mathbf{m}_0}$. Lemma 3 implies that there does not exist $\mathbf{w} \in \mathbb{R}_+^P$ such that $\mathbf{w} \cdot \mathbf{C}_{T_{\mathbf{m}_0}} \leq 0$ and $(\mathbf{w} \cdot \mathbf{C}_{T_{\mathbf{m}_0}})[t] < 0$. Applying Proposition 11, there exists $\mathbf{v}_t$ such that $\mathbf{C}_{T_{\mathbf{m}_0}} \cdot \mathbf{v}_t \geq 0$ and $\mathbf{v}_t[t] > 0$. Let $\mathbf{v} = \sum_{t \in T_{\mathbf{m}_0}} \mathbf{v}_t$. Then $\mathbf{C}_{T_{\mathbf{m}_0}} \cdot \mathbf{v} \geq 0$ and $[[\mathbf{v}]] = T_{\mathbf{m}_0}$. If $\mathbf{C}_{T_{\mathbf{m}_0}} \cdot \mathbf{v} \gneq 0$ then, due to Theorem 3, $\mathbf{C}_{T_{\mathbf{m}_0}}$ would be unbounded, a contradiction. So $\mathbf{C}_{T_{\mathbf{m}_0}} \cdot \mathbf{v} = 0$. □

The next theorem establishes reversibility inside terminal components of the SRT of a bounded marked CPN.

Theorem 7. *Let $(\mathcal{N}, \mathbf{m}_0)$ be a bounded marked CPN such that for all $\mathbf{m} \in \lim{-}\mathbf{RS}(\mathcal{N}, \mathbf{m}_0)$, $T_{\mathbf{m}} = T_{\mathbf{m}_0}$. Then for all $\mathbf{m} \in \lim{-}\mathbf{RS}(\mathcal{N}, \mathbf{m}_0)$, $\lim{-}\mathbf{RS}(\mathcal{N}, \mathbf{m}) = \lim{-}\mathbf{RS}(\mathcal{N}, \mathbf{m}_0)$.*

Proof. Let $\mathbf{m}$ be an arbitrary marking in lim$-\mathbf{RS}(\mathcal{N}, \mathbf{m}_0)$, there exists $\mathbf{v} \in \mathbb{R}_+^{T_{\mathbf{m}_0}}$ such that $\mathbf{m} = \mathbf{m}_0 + \mathbf{C}_{T_{\mathbf{m}_0}} \cdot \mathbf{v}$. Due to Theorem 6, there exists $\mathbf{v}' \in \mathbb{R}_+^T$ such that $\mathbf{C}_{T_{\mathbf{m}_0}} \cdot \mathbf{v}' = 0 \wedge [[\mathbf{v}']] = T_{\mathbf{m}_0}$.

There exists some $n \in \mathbb{N}$ such that for all $t \in T_{\mathbf{m}_0}$, $n\mathbf{v}'[t] > \mathbf{v}[t]$. Thus $\mathbf{v}'' = n\mathbf{v}' - \mathbf{v} \geq 0$, $[[\mathbf{v}'']] = T_{\mathbf{m}_0}$ and $\mathbf{m}_0 = \mathbf{m} + \mathbf{C}_{T_{\mathbf{m}_0}} \cdot \mathbf{v}''$. Since $T_{\mathbf{m}_0}$ is the maximal element of $\mathbf{FS}(\mathcal{N}, \mathbf{m})$, then using Theorem 2, $\mathbf{m}_0 \in$ lim$-\mathbf{RS}(\mathcal{N}, \mathbf{m})$. □

One could ask whether this result could be strenghtened with reachability instead of lim-reachability. The next proposition shows that this is not the case.

Proposition 12. *There exists a bounded marked CPN such that for all* $\mathbf{m} \in \mathbf{RS}(\mathcal{N}, \mathbf{m}_0)$, $T_{\mathbf{m}} = T_{\mathbf{m}_0}$ *and there exists* $\mathbf{m}' \in \mathbf{RS}(\mathcal{N}, \mathbf{m})$ *with* $\mathbf{m} \notin \mathbf{RS}(\mathcal{N}, \mathbf{m}')$.

Proof. Consider the marked CPN of Fig. 3 with $x = 2$. Any reachable marking $\mathbf{m}$ can be written as $\mathbf{m} = ap_1 + bp_2$ with $0 < a + 2b \leq 1$, hence $T_{\mathbf{m}} = T$. Then $\mathbf{m} \xrightarrow{bt_2 \frac{a+b}{2} t_1} \mathbf{m}'$ with $\mathbf{m}' = \frac{a+b}{2} p_2$. Since the total number of tokens cannot increase, $\mathbf{m} \notin \mathbf{RS}(\mathcal{N}, \mathbf{m}')$. □

4.3 Building the Symbolic Reachability Tree

In order to build the SRT, we need to specify a finite representation of the sets R_v and the intermediate sets (see Sect. 4.1) that allows us to check emptyness. To do so, we introduce existential formulas of linear inequalities whose variables are either place markings denoted by $\mathbf{m}(p)$ for $p \in P$, or additional variables in a countable set X. Such a formula can be written as:

$$\varphi = \exists x_1 \ldots \exists x_n \bigvee_{i \leq m} \bigwedge_{j \leq n_i} \sum_{k \leq n} a_{i,j,k} x_k + \sum_{p \in P} a_{i,j,p} \mathbf{m}(p) \bowtie_{i,j} b_{i,j}$$

where for all i, j, k and p, $a_{i,j,k}, a_{i,j,p}, b_{i,j} \in \mathbb{Q}$ and $\bowtie_{i,j} \in \{\leq, <, \geq, >, =\}$.

$\Phi(\mathcal{N})$ is the set of such formulas. Given a formula φ, $[[\varphi]]$ denotes the set of markings that satisfy φ. The emptyness of $[[\varphi]]$ can be decided in polynomial time w.r.t. the size of φ by solving the m linear programs corresponding to the clauses of the external disjunction.

Following the definition of the SRT, one observes that given a formula φ, our problem boils down to compute a formula φ^* such that $[[\varphi^*]] = closure([[\varphi]])$ and given a border edge tr to compute a formula φ_{tr} such that $[[\varphi_{tr}]] = succ(tr, [[\varphi]])$.

Proposition 13. *Let* $\mathcal{N}$ *be a CPN and* $\varphi \in \Phi(\mathcal{N})$. *Then one can compute a formula* $\varphi^* \in \Phi(\mathcal{N})$ *such that* $[[\varphi^*]] = closure([[\varphi]])$ *and given a border edge* tr *a formula* $\varphi_{tr} \in \Phi(\mathcal{N})$ *such that* $[[\varphi_{tr}]] = succ(tr, [[\varphi]])$.

Proof. Let $\varphi \in \Phi(\mathcal{N})$. Then φ^* is defined by:

$$\exists \mathbf{m}' \in \mathbb{R}_+^P \ \varphi[\mathbf{m}'/\mathbf{m}] \wedge \exists \mathbf{v} \in \mathbb{R}_+^T \ \mathbf{m} = \mathbf{m}' + \mathbf{C} \cdot \mathbf{v}$$
$$\wedge \bigvee_{P' \to^* P''} \ \bigvee_{T' \in \mathbf{FS}(\mathcal{N}, P')} [[\mathbf{m}']] = P' \wedge [[\mathbf{m}]] = P'' \wedge [[\mathbf{v}]] = T'$$

Here the new variables are $\mathbf{m}'$ that must fulfill $\mathbf{m}' \in [[\varphi]]$ and the Parikh vector $\mathbf{v}$ of an infinite sequence from $\mathbf{m}'$ to $\mathbf{m}$. The second line ensures that $T_{\mathbf{m}} = T_{\mathbf{m}'}$ and that combined with the state equation of the first line such an infinite sequence exists (by the characterization of lim-reachability).

Let $\varphi \in \Phi(\mathcal{N})$ and $tr = P' \xrightarrow{T'} P''$ be a border edge. Then φ_{tr} is defined by:

$$\exists \mathbf{m}' \in \mathbb{R}_+^P \ \varphi[\mathbf{m}'/\mathbf{m}] \wedge \exists \mathbf{v} \in \mathbb{R}_+^T \ \mathbf{m} = \mathbf{m}' + \mathbf{C} \cdot \mathbf{v}$$
$$\wedge [[\mathbf{m}']] = P' \wedge [[\mathbf{v}]] = T' \wedge \mathbf{m} = \wedge [[\mathbf{m}]] = P''$$

The new variables are $\mathbf{m}'$ that must fulfill $\mathbf{m}' \in [[\varphi]]$ and the Parikh vector $\mathbf{v}$ of a repeated discounted sequence from $\mathbf{m}'$ to $\mathbf{m}$. The second line combined with the state equation of the first line ensures the existence of such a repetitive discounted sequence (due to their characterization). □

Using Proposition 13, we can associate with every vertex v a satisfiable formula $\varphi_v \in \Phi(\mathcal{N})$ such that $R_v = [[\varphi_v]]$. The whole construction is performed in exponential time for two reasons. First the number of vertices of the SRT may be exponential. Then, due to the disjunctions indexed over some subset of places P', P'' and over $\mathbf{FS}(\mathcal{N}, P')$, the size of the formulas may also be exponential. Fortunately these two factors are independent yielding a single exponential bound.

5 Conclusion

In order to analyze the qualitative behaviour of CPNs, we have focused on the mode of a marking: i.e., the subset of transitions fireable in the future. To do so, we have introduced transfinite firing sequences and two finite abstractions: *trajectories* (sequences of decreasing modes) and *signatures* (trajectories enlarged by witnessing markings). We have shown that w.r.t. these abstractions, transfinite sequences generate more behaviours than infinite sequences, but within transfinite sequences, those of ordinal less than ω^2 were expressive enough.

The symbolic reachability tree (SRT) we have introduced captures all possible signatures of a CPN. We have established that the set of markings associated with the leafs of the SRT satisfy *reversibility*, a desirable property corresponding e.g. to attractors of biological systems.

From an algorithmic point of view, we have shown that the trajectory problem is NP-complete. In addition, we have designed an exponential time building of an effective representation of the SRT.

Several lines of future work remain to be explored. From a theoretical point of view, we plan to study transfinite sequences with infinite length (as opposed to those studied here): investigation of *fairness* properties, transfinite sequences with multiple accumulation points, etc. On a more abstract level, the relations between CPN models with other existing dynamic models for biological networks are a major issue; we also hope to gain new perspectives for the *control* of the long-term behaviour (e.g. in *cellular reprogramming* [17] and medical therapies).

References

1. Aguirre-Samboní, G.K., Gaucherel, C., Haar, S., Pommereau, F.: Reset Petri net unfolding semantics for ecosystem hypergraphs. In: PNSE 2022. CEUR, Bergen, Norway, vol. 3170, pp. 213–214 (2022)
2. Baldan, P., Bocci, M., Brigolin, D., Cocco, N., Heiner, M., Simeoni, M.: Petri nets for modelling and analysing trophic networks. Fundam. Informaticae **160**(1–2), 27–52 (2018)
3. Blondin, M., Finkel, A., Haase, C., Haddad, S.: The logical view on continuous Petri nets. ACM Trans. Comput. Log. **18**(3), 24:1–24:28 (2017)
4. Chaouiya, C., Klaudel, H., Pommereau, F.: A modular, qualitative modeling of regulatory networks using Petri nets. Comput. Biol. **16**, 253–279 (2016)
5. Chaouiya, C., Naldi, A., Remy, E., Thieffry, D.: Petri net representation of multi-valued logical regulatory graphs. Nat. Comput. **10**(2), 727–750 (2011)
6. Chaouiya, C., Remy, E., Thieffry, D.: Petri net modelling of biological regulatory networks. J. Discrete Algorithms **6**(2), 165–177 (2008)
7. Chatain, T., Haar, S., Jezequel, L., Paulevé, L., Schwoon, S.: Characterization of reachable attractors using Petri net unfoldings. In: Mendes, P., Dada, J.O., Smallbone, K. (eds.) CMSB 2014. LNCS, vol. 8859, pp. 129–142. Springer, Cham (2014). https://doi.org/10.1007/978-3-319-12982-2_10
8. Chatain, T., Haar, S., Kolcák, J., Paulevé, L., Thakkar, A.: Concurrency in Boolean networks. Nat. Comput. **19**, 91–109 (2020)
9. Chatain, T., Haar, S., Koutny, M., Schwoon, S.: Non-atomic transition firing in contextual nets. In: Devillers, R., Valmari, A. (eds.) PETRI NETS 2015. LNCS, vol. 9115, pp. 117–136. Springer, Cham (2015). https://doi.org/10.1007/978-3-319-19488-2_6
10. Chatain, T., Haar, S., Paulevé, L.: Boolean networks: beyond generalized asynchronicity. In: Baetens, J.M., Kutrib, M. (eds.) AUTOMATA 2018. LNCS, vol. 10875, pp. 29–42. Springer, Cham (2018). https://doi.org/10.1007/978-3-319-92675-9_3
11. Chatain, T., Haar, S., Paulevé, L.: Most permissive semantics of Boolean networks. Research report 1808.10240, Computing Research Repository, p. 15 (2018). https://arxiv.org/abs/1808.10240
12. Fraca, E., Haddad, S.: Complexity analysis of continuous Petri nets. Fundam. Informaticae **137**(1), 1–28 (2015)
13. Haar, S., Haddad, S.: On the expressive power of transfinite sequences for continuous petri nets. Technical report, INRIA (2024). Long version of this paper; INRIA report hal-04514473v1
14. Júlvez, J., Recalde, L., Silva, M.: On reachability in autonomous continuous Petri net systems. In: van der Aalst, W.M.P., Best, E. (eds.) ICATPN 2003. LNCS, vol. 2679, pp. 221–240. Springer, Heidelberg (2003). https://doi.org/10.1007/3-540-44919-1_16
15. Liu, F., Assaf, G., Chen, M., Heiner, M.: A Petri nets-based framework for whole-cell modeling. Biosystems **210**, 104533 (2021)
16. Liu, F., Sun, W., Heiner, M., Gilbert, D.: Hybrid modelling of biological systems using fuzzy continuous Petri nets. Briefings Bioinform. **22**(1), 438–450 (2019)

17. Mandon, H., Su, C., Haar, S., Pang, J., Paulevé, L.: Sequential reprogramming of Boolean networks made practical. In: Bortolussi, L., Sanguinetti, G. (eds.) CMSB 2019. LNCS, vol. 11773, pp. 3–19. Springer, Cham (2019). https://doi.org/10.1007/978-3-030-31304-3_1
18. Paulevé, L., Kolcák, J., Chatain, T., Haar, S.: Reconciling qualitative, abstract, and scalable modeling of biological networks. Nat. Commun. **11**(1), 4256 (2020)
19. Takai-Igarashi, T.: Ontology based standardization of Petri net modeling for signaling pathways. In Silico Biol. **5**(5–6), 529–536 (2005)

Hilbert Composition of Multilabelled Events

Elvio G. Amparore(✉), Susanna Donatelli, and Lea Terracini

Università degli Studi di Torino, Turin, Italy
{amparore,susi}@di.unito.it, lea.terracini@unito.it

Abstract. Any language for modelling concurrent and distributed systems features some sorts of composition and synchronization. This is usually beneficial in the design and verification of complex models. The focus of the paper is on events, like events in Discrete Events Dynamic Systems, or transitions in Petri nets, in which events are labelled with multisets of (conjugated) symbols.

We propose a novel synchronization approach that is based on a well-grounded mathematical theory. Using a simple and intuitive pairwise composition with regular labels, we show that the synchronization generates a set of events that is equivalent to a set of Hilbert bases of polyhedral convex cones. Such connection with the theory of Hilbert bases allows us to prove several useful properties of the composition, as well as an effective algorithm to compute such synchronizations. Finally a calculus of events, named *Hilbert Calculus of Events*, is formulated, for which basic properties are proved.

Keywords: Event synchronization · Petri Box Calculus · Graver basis · Hilbert basis

1 Introduction

Composition is a central aspect of Discrete Events Dynamic Systems (DEDS) modelling. Whether it is realized through a formally defined calculus and associated properties, or whether it is just a graphical composition in a tool (or, better, both), whether the underlying formalism are Petri nets or process algebras, it is often essential to model and verify large systems.

Synchronization of transitions in Petri nets has often been inspired by CSP [9], as in [5], or by the action/co-action paradigm of CCS [11], as in the *Petri Box Calculus* (PBC) [3]. PBC actually goes beyond the standard CCS synchronization by allowing transition labels to be multisets of (conjugated) symbols, for reasons that we shall review in Sect. 1.1. PBC transition synchronization is given in terms of a pairwise composition rule that, under some conditions,generates infinite nets. A later attempt to produce a finite set is presented in [2]. The produced set is indeed finite, but it may not produce some relevant synchronization (see [1], pag. 13, for an example).

In this paper we address the following research question:

L. M. Kristensen and J. M. van der Werf (Eds.): PETRI NETS 2024, LNCS 14628, pp. 132–152, 2024.
https://doi.org/10.1007/978-3-031-61433-0_7

Is it possible to define a synchronization of events labelled with multisets of symbols that produces a finite set of events and for which we can prove essential properties (like commutativity and associativity) and which is meaningful from a modelling point of view?

The main contribution of this paper is to define such synchronization through a bounded pairwise composition of events, provide an algorithm for the computation of the synchronized events and to prove the finiteness and uniqueness of the computed set through a parallel with the computation of the Hilbert basis of a monoid.

To do so, the paper first reviews some previous work (Sect. 1.1) and gives the basic definitions used throughout the paper (Sect. 2), and then defines the synchronization operator hc (Sect. 3). The hc operator adds to a set of events a set of synchronized events and Sect. 4 provides a finite procedure to compute them, and proves its relationship with the Graver basis of a monoid of events. Section 4 also compares the proposed hc operator with the synchronization operators of CCS and PBC. Section 5 defines the Hilbert Calculus of Events (HCE), which includes parallel composition and restriction and we prove a few basic properties.

1.1 Previous Work

Although HCE deals with events independently of the high-level formalism for defining them, our long-term goal is to define a calculus of Petri nets, therefore this section on previous work, and the comparison in the paper, concentrates mainly on Petri nets. Moreover we concentrate on the works for Petri net synchronization that are directly connected with the presented approach. A wider review of the net algebras proposed in the literature, also for colored nets, and an identification of the Petri net tools that support compositionality can be found, for example, in [1].

A first idea of algebra for Petri nets was presented in [10], based on composition of "head" and "tail" places. No transition synchronization is provided, but a transition-based refinement is defined, in which a transition is substituted by an expression of nets. The fundamental work on CSP [9] and CCS [11] has had a large impact on the study of net algebras. An early proposal for CSP-like transition synchronization can be found in [5], while the main proposal of CCS-like net algebra is PBC [4].

PBC is a complete algebra that operates on Petri net components, known as *Petri boxes*, which are nets where places are labeled as either *enter*, *exit* or internal and transitions are labelled by a *multiset* of actions and co-actions. Place and transition composition operators are distinct. Places can be composed as *sequences* or *choices*. Transitions are composed by *synchronization* over an action a and its co-action $\hat{a}$. Synchronization of an action a with its co-action $\hat{a}$ produces the silent action τ. The operator is called sy: performing $\mathcal{B}$ sy a adds to the Petri box $\mathcal{B}$ a new transition for each pair of transitions that include action a and, respectively, $\hat{a}$ in their labels. Multiple nets can be composed by *parallel composition*, followed by a synchronization.

The choice of multiset labelling in PBC is motivated by the need to achieve "multi-handshake" (or multi-way) synchronization. This synchronization is relevant from a modelling point of view, as discussed in [4, Sec. 2.8 and Chap. 9], for example to model an atomic operation on multiple variables. Although it is enough to label transitions with a *set* of actions, to model multi-way synchronization among three transitions requires that the label of the resulting transition is the sum (and not the union) of the labels of the transitions that have been synchronized, which naturally gives rise to *multiset* labelling. As discussed in [4, p. 21], if this is not the case commutativity of the synchronization is not guaranteed, and $(\mathcal{B}\,\mathtt{sy}\,a)\,\mathtt{sy}\,b$ may be different from $(\mathcal{B}\,\mathtt{sy}\,b)\,\mathtt{sy}\,a$.

Unfortunately, PBC synchronization can generate an infinite number of new transitions, as in [3, Sec. 4.5], and, although only a finite subset of them can be enabled for a given initial marking, this is clearly impractical. The work in [2] proposes a different definition for $\mathtt{sy}\,a$, which is guaranteed to be finite as it is shown to be equivalent to the computation of the minimal semiflows of the matrix that describes the labels of the transitions. This computation, however, may result in generating only a subset of the possible synchronizations (particularly, not all τ labelled transitions are generated). An example of this problem is presented and discussed in [1, page 13].

This paper builds on our previous work [1], which proposes an open framework for composition of Petri nets that can be instantiated to achieve various patterns of composition, as, for example, CSP-like composition [5], and synchronization of transitions as in PBC. Unlike PBC, in [1] the same composition criteria is used for composing transitions and composing places. The composition algorithm in [1] exploits the ideas of [2], of modifying the Fourier-Motzkin algorithm for minimal semiflows: proceeding by pairwise sums as PBC it ensures, unlike the proposal in [2], that all τ interactions are generated. Unfortunately, the generated set is not guaranteed to be finite.

Paper's Contribution. This paper moves one step forward previous work by concentrating only on synchronization of a set of events, and by placing this operation in the context of the computation of the Graver basis of the monoid generated by the linear combination of the set of events to be synchronized. A Graver basis is a union of Hilbert basis of appropriate monoides and the mathematical properties of the Hilbert bases of polyhedral convex cones are exploited to prove finiteness and uniqueness of the set of synchronized events, as well as some properties like associativity and commutativity, as well as the presence of all the smallest synchronized events, which include all the τ interactions.

2 Definitions

In this section we provide the basic definitions for *multisets*, *symbols*, *tags* and *labels* that are needed to define labelled events. Labels associated to events are multisets of "symbols" and "conjugated symbols" (co-symbols for short). We prefer to use the term symbol, instead of action (and co-action) as in CCS or

PBC, since we plan to use in future work the proposed synchronization operator to compose both Petri nets transitions and places.

Multisets. Given a discrete set $A = \{a_1, a_2, \dots, a_k\}$, a *multiset* on A is a function $\mu : A \to \mathbb{N}$ that assigns to each element a of A a non-negative *cardinality* value, denoted with $\mu[\![a]\!]$. Let $\mathcal{U}(A)$ be the set of all multisets of A.

Let $\Sigma = \{a, b, c, \dots\}$ be an alphabet of *symbols*. Given a symbol $a \in \Sigma$, $\hat{a}$ denotes its *co-symbol*, with $\hat{a} \neq a$. As usual we define the *complementary* operation $\hat{\circ}$ as a bijective function that maps symbols with their corresponding co-symbols and vice versa, so that $\hat{\hat{a}} = a$. Given Σ we can define the set of *tags* $\mathcal{V}$ as the set of all symbols in Σ together with their co-symbols. So for $\Sigma = \{a, b, c, \dots\}$ we have $\mathcal{V} = \{a, \hat{a}, b, \hat{b}, c, \hat{c}, \dots\}$. The notion of tag and co-tag is as for symbols.

Labels. Let $\mathcal{U}(\mathcal{V})$ be the set of all natural multisets of tags. Elements of $\mathcal{U}(\mathcal{V})$ are called *labels*. The empty label (empty multiset) is denoted by τ. A label $l \in \mathcal{U}(\mathcal{V})$ is *regular* iff $\forall a \in \mathcal{V} : l[\![a]\!] > 0 \Rightarrow l[\![\hat{a}]\!] = 0$, i.e. a label is regular if it cannot include simultaneously a tag and its conjugated: $a + 2\hat{a} + 2b$ is not a regular label, while $2a + \hat{b} + 2\hat{c}$ is. Let $\mathcal{U}(\mathcal{V})^{\text{reg}}$ be the set of all *regular labels.*

In this paper we shall consistently use plain letters to indicate multisets, and bold letters for their representations as vectors. If we assume some indexing over Σ, then any regular label l can be represented without loss of information by a vector $\mathbf{l}$ in $\mathbb{Z}^{|\Sigma|}$ according to the relation $\mathbf{l}[a] \stackrel{\text{def}}{=} l[\![a]\!] - l[\![\hat{a}]\!]$. In this paper, with a slight abuse of notation, we assume that a tag coincides with its index in the vector, i.e. $\mathbf{l}[a]$ is the value of vector $\mathbf{l}$ for the index of tag a. From now on we shall assume that labels are regular, unless otherwise stated[1].

Let E^{ref} be a set of basic events of reference, and let $lab : E^{\text{ref}} \to \mathcal{U}(\mathcal{V})^{\text{reg}}$ be an *event labeling function* that assigns a regular label to each basic event. We indicate with $\mathcal{E}^{\text{ref}}$ the set of basic labelled events: $\mathcal{E}^{\text{ref}} = \{\langle e, l\rangle : e \in E^{\text{ref}}, \text{ and } l = lab(e)\}$.

Linear combination of events are called *compound labelled event.* We shall use the term event for short hereafter to indicate both a labelled event from the basic set $\mathcal{E}^{\text{ref}}$ or a compound one, if there is no need to distinguish or no risk of confusion. Let $|E^{\text{ref}}| = n$ and $|\Sigma| = c$. Since we assume label regularity, a basic or compound event $\langle e, l\rangle$ can be represented by a vector $[\mathbf{e}|\mathbf{l}]$ in $\mathbb{N}^n \times \mathbb{Z}^c$, with $\mathbf{e}$ being the composed events (or a singleton for basic events), and $\mathbf{l}$ the regular label vector. Consequently, the set of labelled events $\mathcal{E}^{\text{ref}}$ can be represented in matrix form as $[\mathbf{E}^{\text{ref}}|\mathbf{L}^{\text{ref}}]$, where $\mathbf{E}^{\text{ref}} = \mathbf{Id}^{n\times n}$, the identity matrix of size n.

Given a set of labelled events $\mathcal{E}$ we can build on $\mathcal{E}$ the additive *monoid* $\mathcal{M} \subseteq \mathbb{N}^n \times \mathbb{Z}^c$ as the set of all possible linear combinations of the elements of $\mathcal{E}$ with natural coefficients), defined by:

$$\mathcal{M} \stackrel{\text{def}}{=} \{\boldsymbol{\alpha} \cdot \mathcal{E} \mid \boldsymbol{\alpha} \in \mathbb{N}^n\} \tag{1}$$

[1] Non regular labels will be considered only in the context of the comparison with PBC, that allows a transition to be labelled with both a symbol and its co-symbol.

$\mathcal{E}$ is then said to be an $\mathbb{N}$-generator of $\mathcal{M}$. Elements in $\mathcal{M}$ are also vectors of the form $[\mathbf{e}|\mathbf{l}] \in \mathbb{N}^n \times \mathbb{Z}^c$, where $\mathbf{e}$ is the vector representation of a multiset over $\mathcal{E}^{\text{ref}}$, and

$$\mathbf{l} = \mathbf{e} \cdot \mathbf{L}^{\text{ref}} \tag{2}$$

with $\cdot$ the standard vector-matrix multiplication. Note that if $\mathcal{E}^{\text{ref}} \subseteq \mathcal{E}$ then any non-negative integer vector $\mathbf{e}$ is in $\mathcal{M}$, as long as its label $\mathbf{l}$ respects (2).

A consequence of (2) is that, for any pair $[\mathbf{e}_1|\mathbf{l}_1]$, $[\mathbf{e}_2|\mathbf{l}_2] \in \mathcal{M}$: $(\mathbf{e}_1 = \mathbf{e}_2) \Rightarrow (\mathbf{l}_1 = \mathbf{l}_2)$. The vice versa is obviously not true, as the same label can be assigned to different basic events or can be generated by different combinations of basic events.

Elements in $\mathcal{M}$ can also be inductively defined as elementary combinations of basic or compound events.

Definition 1 (Elementary combination). *Given two events in vector form* $[\mathbf{e}_1|\mathbf{l}_1]$ *and* $[\mathbf{e}_2|\mathbf{l}_2]$ *their elementary combination, denoted as* $[\mathbf{e}_1|\mathbf{l}_1] + [\mathbf{e}_2|\mathbf{l}_2]$, *is the event* $[\mathbf{e}_1 + \mathbf{e}_2|\mathbf{l}_1 + \mathbf{l}_2]$.

Obviously, elementary combinations respect Eq. (2).

We shall use $\mathcal{E}$, with $|\mathcal{E}| = m$, to represent a set of (compound) labelled events, and $[\mathbf{E}|\mathbf{L}] \in \mathbb{N}^{m \times n} \times \mathbb{Z}^{m \times c}$, its matrix representation. We can also write $\mathbf{L}$ as $\mathbf{L} = \mathbf{E} \cdot \mathbf{L}^{\text{ref}}$. The matrices in Eq. 3 are an example of the notation used. The basic events (left matrix) are e_0 and e_1, with the labelling function $lab(e_0) = \{\hat{a}\}$ and $lab(e_1) = \{a, b\}$. The matrix on the right has three additional compound events e_2, e_3 and e_4 with coherent labelling (a labelling that respects Eq. (2)). Note the slight abuse of notation as e_i on a columns represents a basic event, while row e_i represents a labelled, possibly compound, event.

$$[\mathbf{E}^{\text{ref}}|\mathbf{L}^{\text{ref}}] = \begin{array}{c} \\ e_0 \\ e_1 \end{array} \begin{array}{c} \begin{array}{cccc} e_0 & e_1 & a & b \end{array} \\ \left[\begin{array}{cc|cc} 1 & 0 & -1 & 0 \\ 0 & 1 & 1 & 1 \end{array}\right] \end{array} \qquad [\mathbf{E}|\mathbf{L}] = \begin{array}{c} \\ e_0 \\ e_1 \\ e_2 \\ e_3 \\ e_4 \end{array} \begin{array}{c} \begin{array}{cccc} e_0 & e_1 & a & b \end{array} \\ \left[\begin{array}{cc|cc} 1 & 0 & -1 & 0 \\ 0 & 1 & 1 & 1 \\ 1 & 1 & 0 & 1 \\ 2 & 1 & -1 & 1 \\ 2 & 0 & -2 & 0 \end{array}\right] \end{array} \tag{3}$$

We are interested in defining *synchronization of events* in terms of compound events of interest.

3 Synchronization of Events

In this section we first define the *synchronization set* $Hc(\mathcal{E})$, the set of events generated from $\mathcal{E}$ by synchronization over the whole set of symbols Σ, to then define $\mathcal{E}\ \mathtt{hc}\ A$, the synchronization of the events in $\mathcal{E}$ over a set of symbols $A \subseteq \Sigma$:

3.1 The Synchronization Set *Hc*

We are interested only in compound events that are the result of a sequence of elementary combination of events with *joinable* labels, according to the following definition.

Definition 2 (Joinable labels). *Two labels* $\mathbf{l}_1$ *and* $\mathbf{l}_2$ *are* joinable, *denoted as* $\mathbf{l}_1 \bowtie \mathbf{l}_2$, *iff* $\exists a \in \Sigma : \mathbf{l}_1[a] \cdot \mathbf{l}_2[a] < 0$.

The definition is trivially extended to *joinable events*, as any pair of events with joinable labels. In the following we shall use the term "combination" to refer to a generic elementary combination of events and "composition" to refer to an elementary combination of joinable events. Note that in the left matrix in Eq. (3), event e_4 is the sum of e_1 with itself, so it is a result of a combination, but it is not the result of a composition. Indeed since we assume that labelling is regular, an events can never be composed with itself.

To define the set of synchronized events of interest, that we shall call $Hc(\mathcal{E})$, we need first to define the subset $\mathcal{J}(\mathcal{E})$ of $\mathcal{M}(\mathcal{E})$ that correspond to *a sequence of events compositions.* We drop the $\mathcal{E}$ specification when obvious from the context. The generation of $\mathcal{J}$ from a set $\mathcal{E}$ of events corresponds to the recursive application of the elementary composition of pairs of joinable events. The recursion may not be finite. To define $\mathcal{J}$ it is convenient to introduce the notion of degree of an element.

Definition 3 (Degree). *The* degree *of* $[\mathbf{e}|\mathbf{l}]$ *is:* $\deg([\mathbf{e}|\mathbf{l}]) \stackrel{\text{def}}{=} \sum_{i=1}^{n} \mathbf{e}[i]$.

Note that since $\mathbf{e} \in \mathbb{N}^n$, it holds that $\deg([\mathbf{e}_1|\mathbf{l}_1] + [\mathbf{e}_2|\mathbf{l}_2]) = \deg([\mathbf{e}_1|\mathbf{l}_1]) + \deg([\mathbf{e}_2|\mathbf{l}_2])$, for any pair $[\mathbf{e}_1|\mathbf{l}_1], [\mathbf{e}_2|\mathbf{l}_2] \in \mathcal{M}$. We can now define the set $\mathcal{J}$ of all possible *joined events*: compound events that result from a sequence of compositions.

Definition 4 (Joined events). *The set of joined events generated from a set* $\mathcal{E}$ *of events is the set* $\mathcal{J} = \bigcup_{i=1}^{\infty} \mathcal{J}^{[i]}$, *where*

$$\begin{aligned} \mathcal{J}^{[i]} = &\{[\mathbf{e}|\mathbf{l}] \mid [\mathbf{e}|\mathbf{l}] \in \mathcal{E} \wedge \deg([\mathbf{e}|\mathbf{l}]) = i\} \cup \\ &\{[\mathbf{e}_1|\mathbf{l}_1] + [\mathbf{e}_2|\mathbf{l}_2] \mid [\mathbf{e}_1|\mathbf{l}_1] \in \mathcal{J}^{[j_1]}, [\mathbf{e}_2|\mathbf{l}_2] \in \mathcal{J}^{[j_2]} \wedge \\ &\quad 1 \le j_1, j_2 < i \;\wedge\; (j_1 + j_2) = i \;\wedge\; \mathbf{l}_1 \bowtie \mathbf{l}_2\} \end{aligned} \tag{4}$$

Obviously $\mathcal{J} \subseteq \mathcal{M}$, and it may not be finite.

To illustrate the construction of $\mathcal{J}$ let us first consider some examples. Figure 1 shows four examples of composition of multilabelled events. Boxes represent labelled events in vector form, identified as $\mathbf{v}_i$ for ease of reference. Vertical alignment indicates elements of equal degree. Arrows allows to identify the pair of events that generate the newly composed ones. For the time being, let's ignore the distinction of dotted/solid boxes and lines.

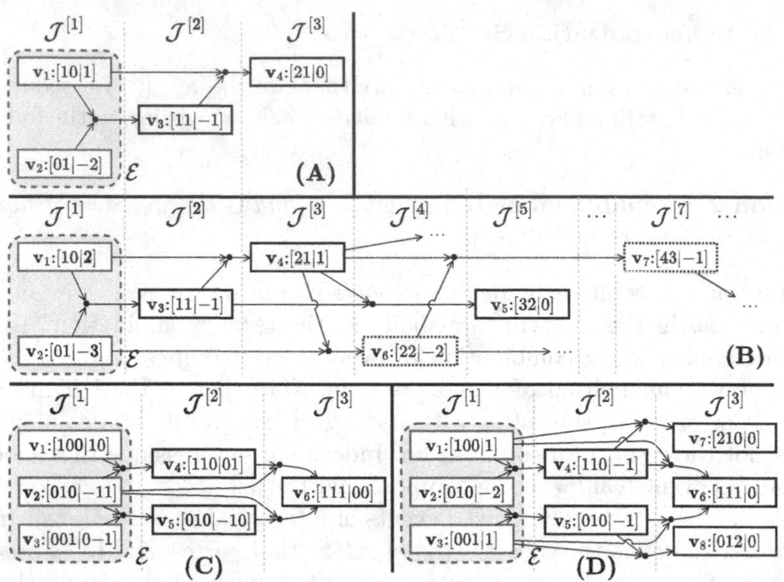

Fig. 1. Simple composition examples.

Example 1. Figure 1(A) is a case of a finite set $\mathcal{J}$ produced from two events e_1, with $lab(e_1) = a$ and e_2, with $lab(e_2) = 2\hat{a}$ (thus labels appears as 1 and -2 in $\mathbf{v}_1$ and $\mathbf{v}_2$, respectively). Composing e_1 with e_2 (both of degree 1) produces the labelled event represented by vector $\mathbf{v}_3$, of degree 2. Composition of $\mathbf{v}_1$ with $\mathbf{v}_3$ produces the τ-labelled event represented by vector $\mathbf{v}_4$, of degree 3. No more compositions are possible, and therefore $\mathcal{J}$ is finite.

Example 2. Figure 1(B) is instead a composition example in which the $\mathcal{J}$ set is infinite. The only difference w.r.t. Fig. 1(A) is the labelling of the two elementary events (the tag cardinality is increased by one). This small difference in the labelling is enough to produce an infinite set $\mathcal{J}$, as indeed vector $\mathbf{v}_7$ has the same labelling as $\mathbf{v}_3$ but it is built on a larger set of events, so the same compositions that leads from $\mathbf{v}_3$ to $\mathbf{v}_7$ can be repeated indefinitely, but each composition produces an event of larger degree, so the $\mathcal{J}$ set is infinite.

Example 3. Figure 1(C) shows an example of a multi-way synchronization realized through two different tags. This is a standard example to show the need for labels defined over sets. Assuming again that the tags are a and b, in the order, and the events are named e_1, e_2, and e_3, with $lab(e_1) = a$, as depicted in vector $\mathbf{v}_1$, $lab(e_2) = \{\hat{a}, b\}$ (vector $\mathbf{v}_2$), and $lab(e_3) = \hat{b}$ (vector $\mathbf{v}_3$). The composition generates the full synchronization of the three events (the τ labelled event of vector $\mathbf{v}_6$), as well as the single synchronizations over a (vector $\mathbf{v}_4$) and over b (vector $\mathbf{v}_5$). No additional compositions are possible, so the $\mathcal{J}$ set is finite.

Example 4. Figure 1(D) is an example of the use of a multiset over a single tag for a three-ways synchronization. This is a standard example to show the need for labels defined over multisets. Using a single tag there are three different ways of obtaining a full synchronization (a τ event), one corresponds to a real three-ways synchronization, as in Example 3 (Fig. 1(C)), while the other two correspond to synchronizing with multiple copies of the same event.

The objective of the composition is to select, among all elements of $\mathcal{J}$, a finite subset Hc which is representative of the whole set. We shall then discuss if the choice is really representative of the whole set, both mathematically and from a modelling point of view.

Note that *Hc* should not only be finite, but should be computable in a finite number of steps: we need to identify an algorithm that generates *Hc* in a finite number of steps, of course without generating the (possibly infinite) set $\mathcal{J}$.

The first step is to define a notion of *reducibility* of an event with respect to an events or a set of events.

Let's recall that the elements of $\mathcal{M}$ have a non-negative $\mathbf{e}$ component. We can then define a notion of reducibility among two elements of $\mathcal{M}$ as:

Definition 5 (Reducibility among elements). *Given two elements $[\mathbf{e}_1|\mathbf{l}_1], [\mathbf{e}_2|\mathbf{l}_2]$ of monoid $\mathcal{M}$, we say that $[\mathbf{e}_1|\mathbf{l}_1]$ reduces $[\mathbf{e}_2|\mathbf{l}_2]$, written as $[\mathbf{e}_1|\mathbf{l}_1] \preceq [\mathbf{e}_2|\mathbf{l}_2]$, if*

$$[\mathbf{e}_1|\mathbf{l}_1] \preceq [\mathbf{e}_2|\mathbf{l}_2] \overset{\text{def}}{=} \mathbf{e}_1 \leq \mathbf{e}_2 \ \wedge\ |\mathbf{l}_1| \leq |\mathbf{l}_2| \ \wedge\ \mathbf{l}_1 \odot \mathbf{l}_2 \geq \mathbf{0}$$

where $\odot$ is the vector Hadamard product (component-wise multiplication), $\mathbf{0}$ is the vector of all zeros, $|\circ|$ is the vector of absolute values, and $\mathbf{a} \leq \mathbf{b}$ is true iff $\mathbf{a}[i] \leq \mathbf{b}[i]$, for all elements i.

The trivial extension to sets is:

Definition 6 (Reducibility against a set). *An element $[\mathbf{e}|\mathbf{l}]$ of monoid $\mathcal{M}$ is reducible by $S \subseteq \mathcal{M}$ if $\exists [\mathbf{e}_1|\mathbf{l}_1] \in S$: $[\mathbf{e}_1|\mathbf{l}_1] \preceq [\mathbf{e}|\mathbf{l}]$.*

Consequently, we say that an element $[\mathbf{e}_1|\mathbf{l}_1]$ is *irreducible* by $[\mathbf{e}_2|\mathbf{l}_2]$, if either $\mathbf{l}_1$ and $\mathbf{l}_2$ have different signs for the same tag, or they have the same sign but the condition $\mathbf{e}_1 \leq \mathbf{e}_2 \ \wedge\ |\mathbf{l}_1| \leq |\mathbf{l}_2|$ does not hold. To identify *Hc* we shall use the following definition of irreducibility against a set. $S \subseteq \mathcal{M}$ as:

Definition 7 (Irreducibility against a set S). *An element $[\mathbf{e}|\mathbf{l}] \in \mathcal{M}$ is irreducible in S, $S \subseteq \mathcal{M}$, if there is no element $[\mathbf{e}_1|\mathbf{l}_1] \in S$ s.t. $[\mathbf{e}_1|\mathbf{l}_1] \preceq [\mathbf{e}|\mathbf{l}]$.*

Obviously, if $[\mathbf{e}, \mathbf{l}]$ is reducible on a set S, then it is surely reducible also on any of its supersets. Vice versa, if $[\mathbf{e}, \mathbf{l}]$ is irreducible against a set S, then it is irreducible against any subset of S.

Definition 8 (Irreducible event). *An event $[\mathbf{e}|\mathbf{l}] \in \mathcal{M}$ is irreducible, if it is irreducible against the whole set $\mathcal{M}$: there is no element $[\mathbf{e}_1|\mathbf{l}_1] \in \mathcal{M}$: $[\mathbf{e}_1|\mathbf{l}_1] \preceq [\mathbf{e}|\mathbf{l}]$.*

We can now define the set Hc of the synchronized events, a subset of the set of joined events $\mathcal{J}$. In the next section we shall indeed prove that Hc is *finite, unique* and *computable in a finite number of steps.*

Definition 9. *The set of synchronized events Hc is the subset of elements of $\mathcal{J}$ that cannot be reduced by any other element of $\mathcal{J}$:*

$$Hc = \{[\mathbf{e}|\mathbf{l}] \in \mathcal{J} \mid \nexists [\mathbf{e}'|\mathbf{l}'] \in \mathcal{J}, [\mathbf{e}'|\mathbf{l}'] \neq [\mathbf{e}|\mathbf{l}] \ \wedge \ [\mathbf{e}'|\mathbf{l}'] \preceq [\mathbf{e}|\mathbf{l}]\} \tag{5}$$

Clearly $Hc \subseteq \mathcal{J} \subseteq \mathcal{M}$. As for the set $\mathcal{J}$, we shall write $Hc(\mathcal{E})$ to make explicit that the set of synchronized events is computed starting from the set $\mathcal{E}$ of events.

Example 5. Let's consider again the example in Fig. 1(B), in which only 7 elements of $\mathcal{J}$ are depicted. Clearly $\mathbf{v}_3$ reduces $\mathbf{v}_7$, since they have the same label and $\mathbf{e}_3 \leq \mathbf{e}_7, \mathbf{e}_3 \neq \mathbf{e}_7$ (so the same labelling is achieved with a smaller number of elementary events). Similarly, $\mathbf{v}_3$ reduces $\mathbf{v}_6$, as $\mathbf{e}_3 \leq \mathbf{e}_6$, labels have the same sign, and $|\mathbf{l}_3[a]| < |\mathbf{l}_6[a]|$. Note that the composition of $\mathbf{v}_7$ with $\mathbf{v}_4$ produces the reducible element $[6\,4|0]$ that can be reduced by $\mathbf{v}_5$. The reader can verify that, in this example, all elements with dotted boxes are reducible and that composition of a reducible event with another event (whether reducible or not) produces another reducible event, so that $Hc = \{\mathbf{v}_1, \mathbf{v}_2, \mathbf{v}_3, \mathbf{v}_4, \mathbf{v}_5\}$, the set of events depicted with solid boxes.

Significance of Hc, Part 1. The objective of the synchronization is, as in CCS and in PBC, to create, when possible, events that are fully synchronized (τ-labelled). As in CCS and PBC, for the sake of future compositions, also partially synchronized events are of interest. The choice of considering the Hc set as a good representative of all possible (infinite) events that are created through successive synchronizations is coherent with the idea of keeping the events that achieve a smaller (or equal) label with a smaller number of event compositions. Indeed being irreducible means that there is no other event that has a smaller label with a fewer number of composition of the basic events. We shall later see an additional modelling interpretation once we have established what are the mathematical properties of the Hc set.

Note that Hc (that we claim is a finite set) is defined by irreducibility w.r.t. $\mathcal{J}$ (often an infinite set). So the above definition is not a constructive one, and the next section takes care of assessing finiteness and uniqueness of Hc to later provide an algorithm that computes Hc directly from the set of events $\mathcal{E}$, without building $\mathcal{J}$.

Synchronization of $\mathcal{E}$ over a Subset of Labels A. Given Definition 9, we can generalize it as a synchronization of the events in $\mathcal{E}$ over the set of labels $A \subseteq \Sigma$, denoted as $\mathcal{E}\,\mathtt{hc}\,A$. This is simply achieved by taking the Hc set using a projection of the label matrix $\mathbf{L}$ over the columns corresponding to the labels A.

Let $\pi_A(\mathbf{L})$ be the projection of $\mathbf{L}$ over the columns of A. Let $\pi_A(\mathcal{E}) = [\mathbf{E}|\pi_A(\mathbf{L})]$ denote the matrix representation of the set $\mathcal{E}$ with labels projected on A. We can then define $\mathcal{E}\,\mathtt{hc}\,A$ as the set of events of $\mathcal{M}$ defined as follows.

Definition 10. *Let* $[\mathbf{E}'|\mathbf{L}'] = Hc(\pi_A(\mathcal{E}))$*, then*

$$\mathcal{E}\, hc\, A \;=\; [\mathbf{E}'|\mathbf{E}' \cdot \mathbf{L}^{ref}] \tag{6}$$

$\mathcal{E}\,\mathrm{hc}\,A$ is therefore defined over all labels, not just A, since the unprojected matrix of labels (with all the labels, not just A) are obtained back using Eq. (2).

Example 6. Let's go back to the example of Fig. 1(C) and consider the effect of the synchronization over the first tag of the label, that we have called a. Considering

$$\mathcal{E} = \{\mathbf{v}_1, \mathbf{v}_2, \mathbf{v}_3\}$$

we get $\mathcal{E}\,\mathrm{hc}\,a = \mathcal{E} \cup \{\mathbf{v}_4\}$, $\mathcal{E}\,\mathrm{hc}\,b = \mathcal{E} \cup \{\mathbf{v}_5\}$, and $(\mathcal{E}\,\mathrm{hc}\,a)\,\mathrm{hc}\,b = \mathcal{E} \cup \{\mathbf{v}_4, \mathbf{v}_5, \mathbf{v}_6\}$. The reader can easily verify that, in this example, synchronization over a followed by synchronization over the second tag of the label, b, is equal to synchronization over b followed by a which it is also equal to the synchronization over the whole set of symbols Σ.

4 Computation of the Synchronization Set

We now connect Definition 9 with its fundamental mathematical structure, which allows us to define its properties. As we shall see, *Hc* is the Graver basis of $\mathcal{M}$.

4.1 Hilbert and Graver Basis

Let's recall that Eq. (1) defines $\mathcal{M}$ as the additive monoid generated by the linear combination with natural coefficients of the elements in $\mathcal{E}$. Let's also recall that, for each $[\mathbf{e}|\mathbf{l}]$ in $\mathcal{M}$, and therefore in $\mathcal{J}$, the $\mathbf{e}$ component is non-negative.

Let then $\boldsymbol{\sigma} \in \{+1, -1\}^c$ be a vector of c signs, and let $\mathcal{M}^{\sigma} \subseteq \mathcal{M}$ be the corresponding set of elements of $\mathcal{M}$ in the orthant induced by $\boldsymbol{\sigma}$ defined as:

$$\mathcal{M}^{\sigma} = \{[\mathbf{e}|\mathbf{l}] \in \mathcal{M} \mid \mathbf{e} \geq 0 \,\wedge\, \mathbf{l} \odot \boldsymbol{\sigma} \geq \mathbf{0}\}$$

Note that $\boldsymbol{\sigma}$ only defines c signs, as, by definition, the first n components of the elements in $\mathcal{M}$ are non-negative. Note also that, if $[\mathbf{e}_1|\mathbf{l}_1] \preceq [\mathbf{e}_2|\mathbf{l}_2]$ there is (at least) a vector of signs $\boldsymbol{\sigma}$ such that $[\mathbf{e}_1|\mathbf{l}_1], [\mathbf{e}_2|\mathbf{l}_2] \in \mathcal{M}^{\sigma}$.

Definition 11. *A subset* $\mathcal{H}^{\sigma} \subseteq \mathcal{M}^{\sigma}$ *is a* Hilbert basis *of* $\mathcal{M}^{\sigma}$ *iff it is a* $\mathbb{N}$*-generating*[2] *set of* $\mathcal{M}^{\sigma}$ *and it is minimal w.r.t. set inclusion [8], i.e., no proper subset of* $\mathcal{H}^{\sigma}$ *is a generating set of* $\mathcal{M}^{\sigma}$.

Theorem 1. *For a given* $\mathcal{M}^{\sigma}$*, the Hilbert basis* $\mathcal{H}^{\sigma}$ *(the set of non-decomposable compound events) is finite and unique.*

[2] A $\mathbb{N}$-generating set generates elements as linear combinations with coefficients in $\mathbb{N}$.

Proof. The proof relies on the well-known *Gordan's Lemma* [13, p. 315] that states that a sufficient criteria for the finiteness/uniqueness of $\mathcal{H}^\sigma$ is that $\mathcal{M}^\sigma$ is contained in a pointed polyhedral convex cone. $\mathcal{M}$ is surely a polyhedral convex cone, since it is $\mathbb{N}$-generated, but in principle it is not pointed. Let $\mathbb{R}^\sigma = \{[\mathbf{x}|\mathbf{y}] \mid \mathbf{x} \in \mathbb{R}^m_{\geq 0},\ \mathbf{y} \in \mathbb{R}^c, \mathbf{y}\cdot\boldsymbol{\sigma} \geq 0\}$ be an orthant of $\mathbb{R}$, which is a pointed convex cone. Observe that $\mathcal{M}^\sigma \subset \mathbb{R}^\sigma$. Now consider the set $\mathcal{Q} = \text{span}([\mathbf{E}^{\text{ref}}|\mathbf{L}^{\text{ref}}]) \cap \mathbb{R}^\sigma$. Such set is a pointed convex cone, since $\mathbf{r}_1, \mathbf{r}_2 \in \mathcal{Q} \Rightarrow \mathbf{r}_1 + \mathbf{r}_2 \in \mathcal{Q}$ and for any $\lambda \geq 0, \mathbf{r} \in \mathcal{Q} \Rightarrow \lambda\mathbf{r} \in \mathcal{Q}$. Since $\mathcal{M}^\sigma$ can be alternatively defined as $\mathcal{Q} \cap \mathbb{Z}^{m+c}$, we conclude that $\mathcal{M}^\sigma$ is contained in a pointed convex cone, and thus it is pointed. Therefore the Hilbert basis $\mathcal{H}^\sigma$ of $\mathcal{M}^\sigma$ is finite and unique. □

When the Hilbert basis is unique, it can be equivalently characterized as the set of all nonzero *irreducible* elements of the monoid [13], a property that we recall in the following.

Property 1 (From [13]). For a given $\mathcal{M}^\sigma$, if the Hilbert basis $\mathcal{H}^\sigma$ is unique, then $\mathcal{H}^\sigma = \{[\mathbf{e}|\mathbf{l}] : [\mathbf{e}|\mathbf{l}] \text{ is irreducible in } \mathcal{M}^\sigma\}$

Using an orthant decomposition [6] of $\mathcal{M}$, we can define the *Graver basis* [6] $\mathcal{G}$ of $\mathcal{M}$ as

$$\mathcal{G} = \bigcup_{\sigma \in \{\pm 1\}^m} \mathcal{H}^\sigma \tag{7}$$

i.e., for every orthant $\mathbb{R}^\sigma$, the set $\mathcal{G} \cap \mathbb{R}^\sigma$ is the unique Hilbert basis for $\mathcal{M}^\sigma$.

Moreover, we can define the Hilbert basis $\mathcal{H}^0$ as the set of all the irreducible elements with $\mathbf{l} = \mathbf{0}$, defined as:

$$\mathcal{H}^0 = \bigcap_{\sigma \in \{\pm 1\}^m} \mathcal{H}^\sigma \tag{8}$$

The set $\mathcal{H}^0$ is therefore the $\mathbb{N}$-generating set for the submonoid of $\mathcal{M}$ consisting of all the elements of $\mathcal{M}$ having $\mathbf{l} = \mathbf{0}$.

4.2 An Algorithm to Compute *Hc* as the Graver Basis of $\mathcal{M}$

Before proceeding to present an algorithm for the computation of *Hc*, let us go a bit more deeply on the relationship between Graver bases and Hilbert bases. A Graver basis $\mathcal{G}$ is a notion associated to a lattice $\mathcal{L}$ which is $\mathbb{Z}$-generated from a set of vectors (as in (1) but with $\boldsymbol{\alpha}$ defined over $\mathbb{Z}^n$). The projection over an orthant of a lattice is not a lattice any longer, but a monoid, and the Graver basis of a lattice is defined as the union of the Hilbert bases of the monoids generated by projection over all orthants. Properties are typically defined for an Hilbert basis (as we shall soon see), but the algorithms in the literature address the Graver basis computation (since the latter is more efficient than computing the Hilbert basis of each orthant separately) from which the Hilbert basis of a specific orthant can be obtained by intersection. The definition in Eq. (7) is therefore slightly more general than the usual one, since we consider a Graver basis of a monoid, and not of a lattice. In our context we call Graver basis of a monoid $\mathcal{M}$ the union of the Hilbert basis of the orthants of $\mathcal{M}$.

Consider Algorithm 1, which is a modified version of the *Pottier Euclidean algorithm in n-dimensions* [12] for the computation of the Graver basis.

Algorithm 1: Computation of $Hc(\mathcal{E})$

---

1 Let $\mathcal{E}$ be a finite set of irreducible elements that are a $\mathbb{N}$-generators of $\mathcal{M}$;
2 Define a set F that initially contains $\mathcal{E}$
3 **for** increasing values of $d > 0$ **do**
4   Consider all pairs $[\mathbf{e}_1|\mathbf{l}_1], [\mathbf{e}_2|\mathbf{l}_2] \in F$ with $\mathbf{l}_1 \bowtie \mathbf{l}_2$ and $\deg([\mathbf{e}_1|\mathbf{l}_1]) + \deg([\mathbf{e}_2|\mathbf{l}_2]) = d$, i.e. have joinable labels and the elementary sum is d.
5   If there is no element $[\mathbf{e}'|\mathbf{l}'] \in F : [\mathbf{e}'|\mathbf{l}'] \preceq [\mathbf{e}_1 + \mathbf{e}_2|\mathbf{l}_1 + \mathbf{l}_2]$, then add $[\mathbf{e}_1 + \mathbf{e}_2|\mathbf{l}_1 + \mathbf{l}_2]$ to F and continue searching pairs.
6   Once all elementary sums of degree d have been tested, increase d by one and continue.
7   The cycle ends when $d > 2 \cdot \max(\deg([\mathbf{e}|\mathbf{l}]) \mid [\mathbf{e}|\mathbf{l}] \in F)$
8 **return** F

Algorithm 1 differs from the Pottier algorithm because it considers a set of $\mathbb{N}$-generators (instead of $\mathbb{Z}$-generators) and considers a graded order [7] for the loop. Therefore, we prove again a set of properties for this algorithm.

Consider a set $\mathcal{E}$ that is assumed to be initially irreducible. Let $F(\mathcal{E})$ be the set returned by Algorithm 1 on input $\mathcal{E}$.

Theorem 2. *$F(\mathcal{E}) \subseteq \mathcal{J}(\mathcal{E})$.*

Proof. Trivial, since $F(\mathcal{E})$ is initialized by $\mathcal{E}$ and any element that is added to F at line 5 has joinable labels, as checked at line 4. □

Theorem 3. *$\mathcal{G} = F(\mathcal{E})$.*

Proof. This proof is an adaptation of the proof (ii) in [12, pag. 41]. We want to prove that any element $[\mathbf{e}|\mathbf{l}] \in \mathcal{M} \setminus \{\mathbf{0}\}$ can be written as a sum of elements

$$[\mathbf{e}|\mathbf{l}] \;=\; [\mathbf{e}_1|\mathbf{l}_1] + \ldots + [\mathbf{e}_k|\mathbf{l}_k], \qquad [\mathbf{e}_1|\mathbf{l}_1], \ldots, [\mathbf{e}_k|\mathbf{l}_k] \in F(\mathcal{E}) \tag{9}$$

such that

$$\forall i \leq k : [\mathbf{e}_i|\mathbf{l}_i] \preceq [\mathbf{e}|\mathbf{l}] \tag{10}$$

which, by Property 1, is equivalent to stating that $\mathcal{G} \subseteq F(\mathcal{E})$. We shall later take care of proving equality.

Since $\mathcal{E}$ is a $\mathbb{N}$-generator of $\mathcal{M}$, and $\mathcal{E} \subseteq F(\mathcal{E})$ by line 2, it is always possible to find a sum that satisfies (9). Consider one such sum.

With $(\mathbf{l})^+$ we denote the vector of non-negative elements of $\mathbf{l}$, and with $(\mathbf{l})^-$ the opposite of the non-positive elements of $\mathbf{l}$. Then $[\mathbf{e}|\mathbf{l}] = [\mathbf{e}|(\mathbf{l})^+ - (\mathbf{l})^-]$. Given a sum (9), if $(\mathbf{l})^+ = (\mathbf{l}_1)^+ + \ldots + (\mathbf{l}_k)^+$ (and consequently $(\mathbf{l})^- = (\mathbf{l}_1)^- + \ldots + (\mathbf{l}_k)^-$), then (10) holds and we are done, since we are assuming that the k terms belong to $F(\mathcal{E})$.

Therefore let's consider the case when there exists two terms, say $[\mathbf{e}_1|\mathbf{l}_1]$ and $[\mathbf{e}_2|\mathbf{l}_2]$, for which:

$$[\mathbf{e}_1 + \mathbf{e}_2|(\mathbf{l}_1 + \mathbf{l}_2)^+] \;<\; [\mathbf{e}_1 + \mathbf{e}_2|(\mathbf{l}_1)^+ + (\mathbf{l}_2)^+] \tag{11}$$

and

$$\deg([\mathbf{e}_1|\mathbf{l}_1]) + \deg([\mathbf{e}_2|\mathbf{l}_2]) \;\leq\; \deg([\mathbf{e}|\mathbf{l}]) \tag{12}$$

(we only write the plus side, since the minus follows). We then have two cases:

- $k > 2$: the inequality of degrees in (12) is strict.
- $k = 2$: the (12) is an equality, but the quantity $(\mathbf{l}_1 + \mathbf{l}_2)^+$ is strictly smaller than $(\mathbf{l}_1)^+ + (\mathbf{l}_2)^+$.

In both cases, we can proceed by induction on $[\mathbf{e}_1 + \mathbf{e}_2|\mathbf{l}_1 + \mathbf{l}_2]$ and rewrite it as a new sum for which (9) and (10) hold. Let

$$[\mathbf{e}_1 + \mathbf{e}_2|\mathbf{l}_1 + \mathbf{l}_2] \;=\; [\mathbf{e}_{k+1}|\mathbf{l}_{k+1}] + \ldots + [\mathbf{e}_h|\mathbf{l}_h] \tag{13}$$

be such rewriting. Since (10) holds, we have

$$(\mathbf{l}_1 + \mathbf{l}_2)^+ = (\mathbf{l}_{k+1})^+ + \ldots + (\mathbf{l}_h)^+ \tag{14}$$

Therefore we have a new rewriting of $[\mathbf{e}|\mathbf{l}]$ that satisfies (9)

$$[\mathbf{e}|\mathbf{l}] \;=\; [\mathbf{e}_3|\mathbf{l}_3] + \ldots + [\mathbf{e}_h|\mathbf{l}_h] \tag{15}$$

and by Eq. (11) and Eq. (14) we conclude that

$$(\mathbf{l})^+ \;\leq\; (\mathbf{l}_3)^+ + \ldots + (\mathbf{l}_h)^+ \;<\; (\mathbf{l}_1)^+ + \ldots + (\mathbf{l}_k)^+$$

This chain of rewritings is therefore finite, and it terminates with a rewriting (15) satisfying also (10). In that case:

1. either $k > 1$, hence $[\mathbf{e}|\mathbf{l}]$ is reducible and can be written in terms of elements of $F(\mathcal{E})$;
2. or $k = 1$, hence $[\mathbf{e}|\mathbf{l}]$ was added to F by Algorithm 1, either because $[\mathbf{e}|\mathbf{l}] \in \mathcal{E}$ (line 2) or because $[\mathbf{e}|\mathbf{l}] = [\mathbf{e}_1|\mathbf{l}_1] + [\mathbf{e}_2|\mathbf{l}_2]$ (line 5).

We now prove that $\mathcal{G} = F(\mathcal{E})$, so there is no element of $F(\mathcal{E})$ that is not in $\mathcal{G}$. We recall that there are no reducible events in the input set $\mathcal{E}$. Assume by contradiction that there is instead a $[\mathbf{e}, \mathbf{l}]$ in $F(\mathcal{E})$, but not in $\mathcal{G}$, which means that $[\mathbf{e}, \mathbf{l}]$ is reducible. Let $[\mathbf{e}', \mathbf{l}']$ be the event that reduces $[\mathbf{e}, \mathbf{l}]$ ($[\mathbf{e}', \mathbf{l}'] \preceq [\mathbf{e}, \mathbf{l}]$), with $[\mathbf{e}', \mathbf{l}'] \neq [\mathbf{e}, \mathbf{l}]$. By definition of $\preceq$, $\deg([\mathbf{e}', \mathbf{l}']) \leq \deg([\mathbf{e}, \mathbf{l}])$. Let's consider first the case $\deg([\mathbf{e}', \mathbf{l}']) < \deg([\mathbf{e}, \mathbf{l}])$. Since Algorithm 1 proceeds in a graded order, and since the operation at line 5 generates new elements with grade d, when

$[\mathbf{e}, \mathbf{l}]$ is added to F (line 5) all irreducible elements of $\mathcal{G}$ with a degree strictly smaller than d are already in F, and therefore a reducible element $[\mathbf{e}, \mathbf{l}]$ would not pass the test at line 5 and it would not be added to F, which contradicts the hypothesis.

We consider instead the case in which $\deg([\mathbf{e}', \mathbf{l}']) = \deg([\mathbf{e}, \mathbf{l}])$, and $[\mathbf{e}, \mathbf{l}]$ is tested at line 5 before $[\mathbf{e}', \mathbf{l}']$ is inserted in F. By definition of $\preceq$: $\mathbf{e}' \leq \mathbf{e}$ and, by hypothesis, their degrees are equal, so $\mathbf{e}' = \mathbf{e}$. But this implies that also $\mathbf{l}' = \mathbf{l}$, so the vectors are equal, which, again, contradicts the hypothesis. □

We can now state the relation between the set of irreducible joined events $Hc(\mathcal{E})$ and the set $F(\mathcal{E})$ computed by Algorithm 1 that forms the Graver basis of $\mathcal{M}$.

Theorem 4. $Hc(\mathcal{E}) = F(\mathcal{E})$.

Proof. By Definition 9, $Hc(\mathcal{E})$ is the set of irreducible events of $\mathcal{J}$ w.r.t. $\mathcal{J}$ itself. By Theorem 3 $F(\mathcal{E})$ is the Graver basis of $\mathcal{M}$, that is to say the set of irreducible elements of $\mathcal{J}$ w.r.t. $\mathcal{M}$. But if an element is irreducible against a set, it is also irreducible against any subset. Therefore, since $F(\mathcal{E}) \subseteq \mathcal{J}$ by Theorem 2, it follows that $Hc(\mathcal{E}) = F(\mathcal{E})$. □

Corollary 1. *$Hc(\mathcal{E}) = \mathcal{G}$. Therefore, by Theorem 1, $Hc(\mathcal{E})$ is finite and unique.*

Corollary 2. *$\mathcal{E} \subseteq Hc(\mathcal{E})$ when $\mathcal{E}$ is irreducible.*

Significance of $Hc(\mathcal{E})$, Part 2. The fact that $Hc(\mathcal{E})$, the set of synchronized events of interest, is a Graver basis for $\mathcal{M}$ leads to another interesting interpretation of this set. By definition (Definition 11) any element of $\mathcal{M}^\sigma$ can be expressed as a sum of elements of $\mathcal{H}^\sigma$, that is to say with labels of the same sign. As a consequence the events in $\mathcal{H}^\sigma$ can be used to "emulate" the synchronization behaviour of any element in $\mathcal{M}^\sigma$. We illustrate this concept through a simple example: let $\mathbf{v}_1 : [100|10]$, $\mathbf{v}_2 : [010|01]$, and $\mathbf{v}_3 : [110|11] = \mathbf{v}_1 + \mathbf{v}_2$. Clearly $\mathbf{v}_1$ and $\mathbf{v}_2$ are irreducible, while $\mathbf{v}_3$ is not since $\mathbf{v}_1, \mathbf{v}_2 \preceq \mathbf{v}_3$. They all belong to the same orthant. Now consider the synchronization of these events with a new event $\mathbf{v}_4 : [001|-1\,-1]$. Clearly $\mathbf{v}_3$ does not add anything to the resulting set, since the same result can be obtained by synchronization (i.e. by the elementary combination) $(\mathbf{v}_1 + \mathbf{v}_4) + \mathbf{v}_2$, or vice versa by $(\mathbf{v}_2 + \mathbf{v}_4) + \mathbf{v}_1$.

4.3 Comparison with CCS and PBC

We shall now compare the hc operator for event composition with the action and co-action synchronization of CCS [11] and of the *Petri Box Calculus* (PBC) [3]. The reason of the choice of CCS and PBC for comparison is that synchronization in CCS and PBC, as with hc, is by pairs (pairs of action and co-action in CCS, pairs of transitions in PBC and pairs of events in hc).

In CCS an agent may evolve through an action a, or its co-action $\hat{a}$, or through the silent action τ. With the terminology of this paper we can say that events in CCS are labelled with a single tag or with τ. The result of composing an agent that can do a with an agent that can do $\hat{a}$ is a new agent that can do either a, $\hat{a}$, or τ.

PBC goes beyond the CCS synchronization by allowing to label transitions with multisets of tags, with the additional freedom that both a symbol and its co-symbol can appear in the label of a single transition, possibly with different cardinalities. Let's recall that, instead, hc assumes that all events are labelled with regular labels.

In PBC events are labelled with multisets of tags, and the composition of an event labelled $2a$ with one labelled $2\hat{a}$ also result in a τ event, with no further composition possible, but the composition of an element labelled $2a$ with one labelled $3\hat{a}$ results in an element labelled $\hat{a}$ that can be further composed with the already generated events. When more tags are present in a single label, the composition may lead to an infinite number of new events. The reader can find in [3, sec. 4.5] the PBC definition of transition synchronization that can be intuitively be described as:

> Repeatedly choose $a, \hat{a}$–pairs of labeled transitions, and <u>each time</u> create a new composed transition from them. Label this new transition with the <u>union of their labels minus a single $a, \hat{a}$–pair.</u> (16)

In PBC, when composing over a symbol a, only the portion of the label that refer to a or $\hat{a}$ is affected. Moreover e_1 and e_2 may actually be the same element, if such an element is labelled with both a and $\hat{a}$ (i.e. if it has a non-regular label). In contrast hc sums up the labels: in $\{[\mathbf{e}_1|\mathbf{l}_1], [\mathbf{e}_2|\mathbf{l}_2]\}\,\texttt{hc}\,a$ is the label a that drives the synchronization: so a new event is created only if $\mathbf{l}_1[a] > 0$ and $\mathbf{l}_2[a] < 0$, or vice versa, but the label of the new event hc is the sum over *all* symbols. Note that when PBC and hc work on labels of the form a (or $\hat{a}$) as in CCS, the result for CCS, PBC and for hc is the same.

The composition of PBC may result in a finite (Example 7) or an infinite set of new transitions. The latter happens every time there is a single transition labeled with both a and $\hat{a}$ (Example 8), or when composing two transitions that are labelled, respectively, with n copies of tag a and m copies of tag $\hat{a}$, with $n, m > 1$ (Example 9). Even when a synchronization over a produces a finite set, it is possible that some of the composed transitions become labelled with a label b and its conjugated, leading to an infinite set of composed transitions if at a later stage another PBC synchronization over b is performed (Example 11). In the examples we shall indicate with $\mathcal{E}_T = \{\langle t_i, l_i \rangle\}$ the set of labelled transition on which the synchronization is applied, and with $\mathcal{E}_T\,\texttt{sy}\,a$ the set of transitions produced by the PBC synchronization over the symbol a.

Example 7. If $\mathcal{E}_T = \{\langle t_1, a \rangle, \langle t_2, \hat{a} \rangle\}$, then $\mathcal{E}_T\,\texttt{sy}\,a = \mathcal{E}_T \cup \{\langle t_1 + t_2, \tau \rangle\}$. The same set is computed by the hc operator.

Example 8. If $\mathcal{E}_T = \{\langle t_1, a + \hat{a} \rangle\}$, the PBC synchronization of transition t_1 composes with itself producing $2t_1$, labelled again with $a + \hat{a}$. The set $\mathcal{E}' = \mathcal{E}_T\,\texttt{sy}\,a$ is therefore the infinite set $\{\langle ct_1, a + \hat{a} \rangle \mid c \geq 1\}$, because any $c_1 t_1$ can compose with any c_2t_1 resulting in $(c_1 + c_2)t_1$, always with label $a + \hat{a}$. This example is not possible for hc because the label of t_1 is not regular.

Example 9. If $\mathcal{E}_T = \{\langle t_1, 2a\rangle, \langle t_2, 2\hat{a}\rangle\}$, then $\mathcal{E}_T\,\mathtt{sy}\,a$ is the *infinite* set $\mathcal{E}_T \cup \{\langle t_1 + t_2, a + \hat{a}\rangle, \langle t_1 + 2t_2, 2\hat{a}\rangle, \langle 2t_1 + t_2, 2a\rangle, \ldots\}$, i.e. containing all elements

$$\langle c_1t_1 + c_2t_2, (c_1 - c_2 + 1)a + (c_2 - c_1 + 1)\hat{a}\rangle, \quad \text{with } c_1, c_2 \in \mathbb{N} \text{ and } |c_1 - c_2| \leq 1$$

The set computed by hc is instead the *finite* set $\mathcal{E}_T \cup \{\langle t_1 + t_2, \tau\rangle$. Note that no τ-labelled transition is ever created by the sy operator

Example 10. We now consider a three-ways synchronization as in Example 3. Let $\mathcal{E}_T = \{\langle t_1, a\rangle, \langle t_2, \hat{a} + b\rangle, \langle t_3, \hat{a}\rangle\}$ then $(\mathcal{E}_T\,\mathtt{sy}\,a)\,\mathtt{sy}\,b = \{\langle t_1, a\rangle, \langle t_2, \hat{a} + b\rangle, \langle t_3, \hat{b}\rangle, \langle t_4, b\rangle, \langle t_5, \hat{a}\rangle, \langle t_6, \tau\rangle\}$. Therefore hc and sy produce the same set of transitions/events. The same is true for the variant in Example 4.

Example 11. If $\mathcal{E}_T = \{\langle t_1, a{+}b\rangle, \langle t_2, \hat{a}{+}\hat{b}\rangle\}$, then $\mathcal{E}_T\,\mathtt{sy}\,a = \mathcal{E}_T \cup \{\langle t_1{+}t_2, b{+}\hat{b}\rangle\}$. This means that a successive composition over b will lead to an infinite net. For $\mathcal{E}_T\,\mathtt{hc}\,a$ we get instead $\mathcal{E}_T \cup \{\langle t_1 + t_2, \tau\rangle\}$. To compare this case with PBC from a practical point of view, let us consider a synchronization over action a followed by a synchronization over b. If we interpret a and b as a single request of two distinct resources, and $\hat{a}$ and $\hat{b}$ as the availability of such resources, then hc produces a single τ transition, while sy of PBC produces first a single transition labelled with both a request for a resource b and its availability, and the successive synchronization over b produces an infinite set. Note that in this example $(\mathcal{E}_T\,\mathtt{hc}\,a) = (\mathcal{E}_T\,\mathtt{hc}\,b)$ since even if it is a that drives the composition, the full label of the two events are summed (the same for b). Moreover $(\mathcal{E}_T\,\mathtt{hc}\,a)\,\mathtt{hc}\,b = (\mathcal{E}_T\,\mathtt{hc}\,b)\,\mathtt{hc}\,a = (\mathcal{E}_T\,\mathtt{hc}\,a)$.

The observation that, in PBC, a single transition with a non-regular label always generates an infinite composition set, is what lead us to the choice of defining hc only over events with regular labels and to ensure that the application of the operator only produces events with regular labels. As we have seen, regularity is a necessary condition for finiteness, but it is not sufficient, and this justifies the notion of irreducibility introduced for the hc definition.

5 A Hilbert Calculus of Events

We now complete the synchronization operator with a parallel composition and a restriction to get a calculus of events, that we name the *Hilbert Calculus of Events* (HCE). Note that the calculus does not have a sequential composition, since the focus of this paper is only on synchronization, a required basis for later defining, in future work, a calculus for Petri nets.

$$\begin{aligned} \mathcal{E} \ ::= \ & \mathcal{E}^{\mathrm{ref}} && \text{set of basic events} \\ & \mid \mathcal{E} \parallel \mathcal{E} && \text{concurrent union} \\ & \mid \mathcal{E}\,\mathtt{hc}\,A && \text{synchronization} \\ & \mid \mathcal{E}\,\mathtt{rs}\,A && \text{restriction} \end{aligned} \tag{17}$$

Basic Events $\mathcal{E}^{\text{ref}}$. A set of basic (non composed) events, expressed in matrix form as $[\mathbf{E}^{\text{ref}}|\mathbf{L}^{\text{ref}}]$ and having $\mathbf{E}^{\text{ref}} = \mathbf{Id}$.

Concurrent Union $\mathcal{E} \parallel \mathcal{E}$. Given $\mathcal{E}_1 = [\mathbf{E}_1|\mathbf{L}_1]$ and $\mathcal{E}_2 = [\mathbf{E}_2|\mathbf{L}_2]$, defined over the same set of symbols Σ and disjoint sets of events. $\mathcal{E}_{12} = \mathcal{E}_1 \parallel \mathcal{E}_2$ is the new set of events defined (in matrix form) as

$$[\mathbf{E}_{12}|\mathbf{L}_{12}] = \left[\begin{array}{cc|c} \mathbf{E}_1 & \mathbf{0} & \mathbf{L}_1 \\ \mathbf{0} & \mathbf{E}_2 & \mathbf{L}_2 \end{array}\right] \quad \text{and} \quad [\mathbf{E}^{\text{ref}}_{12}|\mathbf{L}^{\text{ref}}_{12}] = \left[\begin{array}{cc|c} \mathbf{E}^{\text{ref}}_1 & \mathbf{0} & \mathbf{L}^{\text{ref}}_1 \\ \mathbf{0} & \mathbf{E}^{\text{ref}}_2 & \mathbf{L}^{\text{ref}}_2 \end{array}\right] \tag{18}$$

Synchronization $\mathcal{E}\,\texttt{hc}\,A$. The set of events defined by Eq. (6).

Restriction $\mathcal{E}\,\texttt{rs}\,A$. The restriction operator removes from $\mathcal{E}$ all the events whose labels include either a or $\hat{a}$, for any $a \in A$.

$$\mathcal{E}\,\texttt{rs}\,A = \big\{\langle \mathbf{e}, \mathbf{l}\rangle \;\big|\; \langle \mathbf{e}, \mathbf{l}\rangle \in \mathcal{E} \text{ and } \forall a \in A : \mathbf{l}[a] = 0\big\}$$

In matrix form, restriction eliminates the rows of $[\mathbf{E}|\mathbf{L}]$ having at least a non-zero entry in the columns of the tags in A.

5.1 Basic Properties of the Hilbert Calculus of Events

Concurrent Union Properties

$$\begin{aligned} \mathcal{E}_1 \parallel \mathcal{E}_2 &\equiv \mathcal{E}_2 \parallel \mathcal{E}_1 && \text{[commutative]} \\ \mathcal{E}_1 \parallel (\mathcal{E}_2 \parallel \mathcal{E}_3) &\equiv (\mathcal{E}_1 \parallel \mathcal{E}_2) \parallel \mathcal{E}_3 && \text{[associative]} \end{aligned}$$

Synchronization Properties

$$\begin{aligned} (\mathcal{E}_1\,\texttt{hc}\,A)\,\texttt{hc}\,B &\equiv \mathcal{E}_1\,\texttt{hc}\,(A \cup B) && \text{[associative over tags]} \\ (\mathcal{E}_1\,\texttt{hc}\,A)\,\texttt{hc}\,B &\equiv (\mathcal{E}_1\,\texttt{hc}\,B)\,\texttt{hc}\,A && \text{[commutative]} \\ (\mathcal{E}\,\texttt{hc}\,A)\,\texttt{hc}\,A &\equiv \mathcal{E}\,\texttt{hc}\,A && \text{[idempotent]} \\ \mathcal{E}\,\texttt{hc}\,A &\subseteq \mathcal{E}\,\texttt{hc}(A \cup B) && \text{[synchronization over subsets]} \end{aligned}$$

Restriction Properties

$$\begin{aligned} (\mathcal{E}\,\texttt{rs}\,A)\,\texttt{rs}\,B &\equiv (\mathcal{E}\,\texttt{rs}\,B)\,\texttt{rs}\,A && \text{[commutative]} \\ (\mathcal{E}\,\texttt{rs}\,A)\,\texttt{rs}\,A &\equiv \mathcal{E}\,\texttt{rs}\,A && \text{[idempotent]} \\ (\mathcal{E}_1 \parallel \mathcal{E}_2)\,\texttt{rs}\,A &\equiv (\mathcal{E}_1\,\texttt{rs}\,A) \parallel (\mathcal{E}_2\,\texttt{rs}\,A) && \text{[distributive over } \parallel\text{]} \end{aligned}$$

Composition Properties

$$(\mathcal{E}_1 \parallel \mathcal{E}_2)\,\texttt{hc}\,A \equiv \big((\mathcal{E}_1\,\texttt{hc}\,A) \parallel (\mathcal{E}_2\,\texttt{hc}\,A)\big)\,\texttt{hc}\,A \qquad \text{[propagates through } \parallel\text{]}$$

$$(\mathcal{E}\,\texttt{hc}\,A)\,\texttt{rs}\,B \equiv (\mathcal{E}\,\texttt{rs}\,B)\,\texttt{hc}\,A \quad \text{such that } \forall\,[\mathbf{e}|\mathbf{l}] \in \mathcal{E},$$
$$\big(\exists a \in A : \mathbf{l}[a]{\neq}0 \Rightarrow \forall b \in B : \mathbf{l}[b]{=}0\big) \text{ and } \big(\exists b \in B : \mathbf{l}[b]{\neq}0 \Rightarrow \forall a \in A : \mathbf{l}[a]{=}0\big)$$
$$\text{[non-interfering synchronization and restriction commute]}$$

Fundamental Property of HCE. HCE operators are deterministic and preserve finiteness and irreducibility.

We now proceed to prove the above properties. Some are trivial (like the ones for the concurrent union, which is akin to a set union), but others require more formal proofs.

Theorem 5 (hc is associative w.r.t. tags).
Let $A, B \subseteq \Sigma$. Then $(\mathcal{E}\, hc\, A)\, hc\, B \;=\; \mathcal{E}\, hc(A \cup B)$.

Proof. The proof stems from the idea illustrated by the "project and lift" approach of Hemmecke [7], that the computation of the Hilbert basis of a monoid can be performed "one variable at a time", by working on a sequence of increasingly larger projections of the monoid. Let $\mathcal{M}$ be a monoid on $\mathbb{Z}^{n+c}$, and let $\boldsymbol{\sigma}$ be a vector of $n+c$ signs. Let $\pi_j(\mathcal{M}^{\sigma_j})$ be the projection of $\mathcal{M}^{\sigma}$ over the first j variables of the orthant of sign $\boldsymbol{\sigma}_j$, the first j signs of $\boldsymbol{\sigma}$, and $\mathcal{H}_j$ the Hilbert basis of $\pi_j(\mathcal{M}^{\sigma_j})$.

Lemma 2.2 of [7] ensures that, when j is greater than the rank of the generator matrix of the monoid[3], $G_{j+1} = \bigcup_{[\mathbf{e}|\mathbf{l}_j] \in \mathcal{H}_j} [\mathbf{e}|\mathbf{l}_j, l_{j+1}]$ is a generator set for $\pi_{j+1}(\mathcal{M}^{\sigma_{j+1}})$ if l_{j+1} is chosen such that $[\mathbf{e}|\mathbf{l}_j, l_{j+1}] \in \pi_{j+1}(\mathcal{M}^{\sigma_{j+1}})$. Such choice of l_{j+1} is trivial in our context since it is uniquely determined by $\mathbf{e}$ using (2). We can therefore compute the Hilbert basis of $\pi_{j+1}(\mathcal{M}^{\sigma_{j+1}})$ starting from G_j. By iterating over j, and considering all possible $\boldsymbol{\sigma}$, when $j+1 = n+c$ what we get is an iterative method for computing the Graver basis of $\mathcal{M}$ one variable at a time. Of course in our context we can start with $j = n$, that is to say with the Hilbert bases of $\pi_n(\mathcal{M}^{\sigma_n})$, to compute G_{n+1}. Since n is the upper bound of the rank of the generator matrix (because of Eq. 2), we can apply Lemma 2.2 as reported above.

To show the application of Lemma 2.2 of [7] on our context, for ease of notation, let us now assume that there are only two tags a and b, with $A = \{a\}$ and $B = \{b\}$. As usual n is the number of columns, so we can start with $j = n+1$, and can compute the Hilbert basis of $\pi_{n+1}(\mathcal{M}^{\sigma_{n+1}})$ and obtain from it the set G_{n+2} of generators for $\pi_{n+2}(\mathcal{M}^{\sigma_{n+2}})$ which is exactly $\mathcal{M}^{\sigma}$. Since the Hilbert basis of $\mathcal{M}^{\sigma}$ corresponds to the set of vectors obtained by synchronization over all tags, we get $(\mathcal{E}\, \mathtt{hc}\, a)\, \mathtt{hc}\, b = \mathcal{E}\, \mathtt{hc}(a \cup b)$. The extension over label sets and for all orthants thus unfolds. □

As a consequence of Theorem 5, the $\mathtt{hc}$ operator is commutative w.r.t. the order of application to subsets of symbols:

Corollary 3. *Let $A, B \subseteq \Sigma$. Then $(\mathcal{E}\, \mathtt{hc}\, A)\, \mathtt{hc}\, B \;=\; (\mathcal{E}\, \mathtt{hc}\, B)\, \mathtt{hc}\, A$.*

Proof. Trivial, since $(\mathcal{E}\, \mathtt{hc}\, A)\, \mathtt{hc}\, B \;=\; \mathcal{E}\, \mathtt{hc}(A \cup B)$ and $(\mathcal{E}\, \mathtt{hc}\, B)\, \mathtt{hc}\, A \;=\; \mathcal{E}\, \mathtt{hc}\, (B \cup A)$ □

Corollary 4. *Let $A, B \subseteq \Sigma$. Then $\mathcal{E}\, \mathtt{hc}\, A \;\subseteq\; \mathcal{E}\, \mathtt{hc}\, (A \cup B)$.*

[3] The expression for G_{j+1} is different when j is smaller than the rank of the generator matrix, but this is not reported here since it is not of interest for our proof.

Proof. Immediate consequence of the proof of Theorem 5. □

Theorem 6 (Synchronization propagates through $\|$).
$(\mathcal{E}_1 \parallel \mathcal{E}_2)\,\mathtt{hc}\,A \;=\; \big((\mathcal{E}_1\,\mathtt{hc}\,A) \parallel (\mathcal{E}_2\mathtt{hc}\,A)\big)\,\mathtt{hc}\,A$

Proof. Let $S_L = (\mathcal{E}_1 \parallel \mathcal{E}_2)\,\mathtt{hc}\,A$, and let $S_R = \big((\mathcal{E}_1\,\mathtt{hc}\,A) \parallel (\mathcal{E}_2\,\mathtt{hc}\,A)\big)\,\mathtt{hc}\,A$. Since, by Corollary 2, $\mathcal{E}_1 \subseteq (\mathcal{E}_1\,\mathtt{hc}\,A)$ and $\mathcal{E}_2 \subseteq (\mathcal{E}_2\,\mathtt{hc}\,A)$, it easily follows that $S_L \subseteq S_R$. We now prove that S_R does not have more elements than S_L. Suppose that an element $[\mathbf{e}|\mathbf{l}] \in S_R \setminus S_L$ exists. Surely, such element $[\mathbf{e}|\mathbf{l}]$ is irreducible w.r.t. the columns A and it is not present in $(\mathcal{E}_1 \parallel \mathcal{E}_2)\,\mathtt{hc}\,A$ (i.e. S_L). But Corollary 1 ensures that S_L is a Graver basis with all the irreducible elements w.r.t. the columns A, resulting in a contradiction. □

Theorem 7 (Non-interfering synchronization and restriction commute). *Given that there are no two elements of $\mathcal{E}$ that are labelled with both an element from A and an element from B, formally stated as:* $\forall\,[\mathbf{e}|\mathbf{l}] \in \mathcal{E}$, $\big(\exists a \in A : \mathbf{l}[a]{\neq}0 \Rightarrow \forall b \in B : \mathbf{l}[b]{=}0\big)$ *and* $\big(\exists b \in B : \mathbf{l}[b]{\neq}0 \Rightarrow \forall a \in A : \mathbf{l}[a]{=}0\big)$, *we have that* $(\mathcal{E}\,\mathtt{hc}\,A)\,\mathtt{rs}\,B \;\equiv\; (\mathcal{E}\,\mathtt{rs}\,B)\,\mathtt{hc}\,A$.

Proof. The proof stems from the fact that the set of events over which $\mathtt{hc}\,A$ operates is disjoint from the set of events over which $\mathtt{rs}\,B$ operates, and $\mathtt{hc}\,A$ does not add B labels, as well as $\mathtt{rs}\,B$ does not remove A labels. □

Theorem 8. *HCE operators preserve finiteness and irreducibility.*

Proof. The only operation of the calculus that generates new elements (elements that are not already part of the operands) is the synchronization. But the set produced by $\mathtt{hc}$ is an union of Hilbert bases, which by Theorem 1 is surely finite, unique and made by irreducible events, proving the statement. □

Theorem 9. *HCE is a finite and deterministic calculus of irreducible events.*

Proof. Basic events are irreducible by definition, because $\mathbf{E}^{\mathrm{ref}} = \mathbf{Id}$. Concurrent union treats the set of events as disjoint. Restriction only removes events. As before, the only operation of HCE that generates new elements is the synchronization, and it may only generate irreducible elements (see Corollary 2) and is unique (i.e. deterministic), proving the theorem. □

From Theorem 8 and Theorem 9 follows the fundamental property of HCE.

6 Conclusions

In this paper we propose a novel approach for event synchronization that works by combining pairs of multilabelled events to match symbols and co-symbols. The objective of the synchronization is to produce a finite set of irreducible synchronized events. The rational behind irreducibility is the need to minimize the number of unmatched tags (non τ tags in the label of the synchronized events), through the minimum number of event compositions. We have provided an interpretation of the proposed synchronization from a modelling point of view, also through a number of examples of basic synchronization patterns.

We establish a direct connection between the pairwise synchronization procedure and the theory of Hilbert bases. We prove that the synchronization algorithm that performs such pairwise compositions under the constraint of irreducibility and following a graded order is actually computing a set of Hilbert bases (i.e. a Graver basis) for the monoid representing all possible combinations of events.

This connection allows us to formulate several properties for the composition, and for a novel calculus of events that we named *Hilbert calculus of events*. A brief comparison of HCE with the well known compositional calculus CCS, PBC and the calculus of Anisimov, is also provided.

The Hilbert calculus of events only deals with event synchronization and misses, for instance, sequential compositions of events. Therefore in this form it is useful as a sub-calculus inside other formalisms. As a future work, we plan to propose a calculus of Petri nets where nets are composed over places and transitions following the rules defined for HCE. The well-grounded mathematical structure of the Hilbert bases should ease the task of proving relevant properties for the Petri net calculus.

Acknowledgements. This work has been supported by the PRIN project MEDICA *"Modelling and vErification of alkaptonuria and multiple sclerosis Driven by biomedICAl data"*, Bando 2022, funded by MUR Missione 4 "Istruzione e Ricerca", Componente C2 Investimento 1.1.

Lea Terracini is a member of the INdAM group GNSAGA.

References

1. Amparore, E.G., Donatelli, S.: The ins and outs of Petri net composition. In: Bernardinello, L., Petrucci, L. (eds.) PETRI NETS 2022. LNCS, vol. 13288, pp. 278–299. Springer, Cham (2022). https://doi.org/10.1007/978-3-031-06653-5_15
2. Anisimov, N.A., Golenkov, E.A., Kharitonov, D.I.: Compositional Petri net approach to the development of concurrent and distributed systems. Program. Comput. Softw. **27**(6), 309–319 (2001)
3. Best, E., Devillers, R., Hall, J.G.: The box calculus: a new causal algebra with multi-label communication. In: Rozenberg, G. (ed.) Advances in Petri Nets 1992. LNCS, vol. 609, pp. 21–69. Springer, Heidelberg (1992). https://doi.org/10.1007/3-540-55610-9_167
4. Best, E., Devillers, R., Koutny, M.: Petri Net Algebra. Monographs in Theoretical Computer Science. EATCS. Springer, Heidelberg (2001). https://doi.org/10.1007/978-3-662-04457-5
5. De Cindio, F., et al.: A Petri net model for CSP. In: Proceedings of Convención Informática Latina (CIL 1981), Barcelona, vol. 81, pp. 392–406 (1981)
6. Graver, J.E.: On the foundations of linear and integer linear programming I. Math. Program. **9**(1), 207–226 (1975)
7. Hemmecke, R.: On the computation of Hilbert bases of cones. In: Mathematical software, pp. 307–317. World Scientific (2002). (preprint available at arXiv:math/0203105)

8. Henk, M., Weismantel, R.: On Hilbert bases of polyhedral cones. Technical report SC-96-12, Berlin (1996)
9. Hoare, C.A.R.: Communicating sequential processes. Commun. ACM **21**(8), 666–677 (1978)
10. Kotov, V.E.: An algebra for parallelism based on petri nets. In: Winkowski, J. (ed.) MFCS 1978. LNCS, vol. 64, pp. 39–55. Springer, Heidelberg (1978). https://doi.org/10.1007/3-540-08921-7_55
11. Milner, R.: A Calculus of Communicating Systems. LNCS, vol. 92. Springer, Heidelberg (1980). https://doi.org/10.1007/3-540-10235-3
12. Pottier, L.: The Euclidean algorithm in dimension n. In: Proceedings of the 1996 International Symposium on Symbolic and Algebraic Computation, pp. 40–42 (1996)
13. Schrijver, A.: Theory of Linear and Integer Programming. Wiley-Interscience Series in Discrete Mathematics (1986)

Relational Structures for Interval Order Semantics of Concurrent Systems

Ryszard Janicki[1], Jetty Kleijn[2], Maciej Koutny[3], and Łukasz Mikulski[4(✉)]

[1] Department of Computing and Software, McMaster University, Hamilton, ON L8S 4K1, Canada
janicki@mcmaster.ca

[2] Leiden Institute of Advanced Computer Science, Leiden University, Niels Bohrweg 1, 2333 CA Leiden, The Netherlands
h.c.m.kleijn@liacs.leidenuniv.nl

[3] School of Computing, Newcastle University, 1 Science Square, Newcastle upon Tyne NE4 5TG, UK
maciej.koutny@ncl.ac.uk

[4] Faculty of Mathematics and Computer Science, Nicolaus Copernicus University in Toruń, Chopina 12/18, 87-100 Toruń, Poland
lukasz.mikulski@mat.umk.pl

Abstract. Relational structures like partial orders that are based on acyclic relations capturing a 'before' relationship, can provide versatile frameworks for the modelling and verification of a wide class of concurrent systems behaviour. There are also relational structures with an acyclic 'before' (strong precedence) relationship and a possibly cyclic 'not later than' (weak precedence) relationship, which can be used for more general concurrent behaviours. However, in each of these cases, the execution model is based on sequences or step sequences of executed actions, where actions are assumed to be executed instantaneously. In this paper, we drop this restriction and consider executions modelled by interval orders, where actions are assumed to be executed non-instantaneously. For this execution model, we introduce new relational structures which can capture both strong precedence and weak precedence. This is achieved, in particular, thanks to a novel notion of acyclicity where any mixed cycle of strong and weak precedence is allowed, provided that it contains at least two consecutive weak precedence relationships.

Keywords: interval order · concurrency · relational structure · causality · closure · precedence · weak precedence · acyclicity

1 Introduction

Faithful behaviour capture and effective verification techniques of concurrent systems usually require analyses of relationships, such as causality and independence, between the events (occurrences of actions) involved in a system run. As explained below, there is already ample evidence that such an approach greatly improves the practical design, construction, and verification of complex software and hardware concurrent systems.

L. M. Kristensen and J. M. van der Werf (Eds.): PETRI NETS 2024, LNCS 14628, pp. 153–174, 2024.
https://doi.org/10.1007/978-3-031-61433-0_8

Background. At the heart of the approach to modelling concurrency semantics followed in this paper is the view, that everything an observer can report about a run of a concurrent system is that it comprises a (finite) set of events together with a set of statements about the mutual relationships between these events.

Consider, for example, a system given by a process expression $E = (a.b.c + d)\|e$ describing two concurrent processes, $a.b.c + d$ and e, where the first process is a choice between executing either action d or actions a, b, c sequentially. Assume further that in this case all actions are executed atomically and never at the same time (as in a majority of process algebras and temporal logics). As a result, the only meaningful statement about the relationship between events x and y is that x was observed before y (we denote this by $x \prec y$) or y was observed before x (denoted by $y \prec x$). Moreover, if x is executed before y, then x will be observed before y. Suppose now that a run of E has executed a, b, c, and e. Then, knowing the construction of E, we have that e is executed in parallel with $a.b.c$, while we can expect that any observer of this run will for sure report that $a \prec b$ and $b \prec c$, together with other statements about the (observed) precedence of executed actions. However, no matter what these additional statements are, a report (i.e., an observation made by a single observer) cannot comprise statements which form a *cycle*, such as $a \prec b \prec c \prec a$ or $b \prec e \prec b$, as this would contradict the physical nature of the concurrent run. Moreover, it should be clear that some of the statements can be *derived* from those already made, e.g., , $a \prec c$ follows from $a \prec b$ and $b \prec c$ due to the transitivity of the 'happened before (preceded)' relationship. To summarise, the observer's report would comprise two kinds of precedence relationships: (i) those which result from the construction of the expression (i.e., $a \prec b$ and $b \prec c$); and (ii) additional relationships which do not violate the graph-theoretic acyclicity of the report. Note that, by the expression E, for a run executing d and e, an observer will report $d \prec e$ or $e \prec d$. We would treat these observations separately from the ones above, as we consider this a different run. In other words, observations are not concerned with choice or conflict between events.

The above discussion is an informal justification of the following two key principles of reports which can be supplied for a given run (or execution) of a concurrent system with atomic and non-simultaneous actions:

- each report is an acyclic relation; and
- any acyclic extension of a report is also a valid report.

It is then remarkable that these two rather meagre principles provide the basis for a wide range of methods and techniques. In particular, there are two special kinds of reports with clear semantical meaning:

- *saturated reports*, i.e., those which are complete in the sense that they cannot be extended (no relationship can be added) without violating acyclicity; and
- *closed reports*, i.e., those which comprise all implied relations and cannot be extended without reducing their set of saturated extensions

It follows that in the sequential case, saturated reports are nothing but total orders (often called interleavings, represented as sequences of executed actions), and closed reports are partial orders (often called causal partial orders). For our initial example, there are four observations (viz. $a \prec b \prec c \prec e$, $a \prec b \prec e \prec c$, $a \prec e \prec b \prec c$, and $e \prec a \prec b \prec c$) and a unique causal partial order (with $a \prec b$, $b \prec c$, and $a \prec c$). Moreover, the four reports are all the saturated extensions of the causal partial order, which in turn is the intersection of these reports. Hence, this causal partial order can be used as an unambiguous representation of a *concurrent history* comprising the four interleavings.

Motivating Example. Representing concurrent runs through sequences or step sequences of actions implicitly assumes that these actions are executed instantaneously. Steps are then sets of 'simultaneously' executed actions (i.e., they happen together) and so here simultaneity is a transitive relation. However, when actions take time, runs need to be represented as interval orders.[1] As argued by Wiener in 1914 [33] — and later, more formally, in [14] — any execution that can be observed by a single observer must be an interval order and so the most precise qualitative observational semantics is based on interval orders, where simultaneity is not guaranteed to be transitive.

An interval order representing a system run is a partial order where each element (event) e can be thought of as corresponding to a finite interval $I(e)$ on the real (time) scale. In such a representation, two events, e and f, are ordered iff the end of the first interval $I(e)$ is less than the beginning of the second interval $I(f)$. As a consequence, two events, g and h, are unordered iff the corresponding intervals, $I(g)$ and $I(h)$, overlap (fully or partially). One then interprets the relationship between e and f - with the end of $I(e)$ before the beginning of $I(f)$ - as precedence, and the relationship between g and h - with $I(g)$ and $I(h)$ overlapping - as simultaneity. As an example, in the analysis of concurrent programs, it is quite natural to assume that different threads of control are totally independent unless some synchronisation mechanisms are used. And, due to software and hardware optimisations used in practice, one needs to take into account some interleavings at the level of single thread computation [3].

Let us consider the following concurrent program fragment:

$$\begin{array}{c} init: x = y = 0 \\ a: x = x+1 \; \Big\| \begin{array}{l} b: x = x+2 \\ c: y = x \end{array} \end{array}$$

Is it possible to finish the calculation with $x = 1$ and $y = 2$? If we assume that the executions of the statements a, b, c are instantaneous, or that simultaneity is transitive (like in the case of the sequential or step sequence execution semantics), the answer is no. However, if we assume that the second thread works simultaneously with the first one (i.e., at the same time on different machine/processor/core), this outcome is possible. In such a case, we can resort to the interval order execution semantics. Here, the handling of x is as follows: if b has begun and a begins before b finishes, then a starts with $x = 0$; and if then a finishes after b, it resets x from 2 to 1. And y gets through c the then current value of x when it begins. In terms of the corresponding intervals, $I(b)$ is to the left of $I(c)$, and $I(a)$ overlaps with both $I(b)$ and $I(c)$. Hence b precedes c, while a and b as well as c and a are unordered.

[1] While in small examples one might consider changing the granularity of events, for more involved situations this would lead to an exponential explosion of possibilities obscuring the structure of the run.

The Approach Followed in This Paper. Dealing with the semantics of concurrent systems purely in terms of their individual executions, using e.g., sequences, step sequences, or interval orders, is far from being computationally efficient in terms both of behaviour modelling and of validation. To address this shortcoming, more involved relational structures have been introduced, aiming at a single succinct and faithful representation *rs* of (often exponentially large) sets *RS* of closely related individual executions all having the same executed events. Examples include causal partial orders for sets of sequences, and invariant order structures for sets of step sequences. Succinctness is usually achieved by retaining in *rs* (through intersection) only those relationships which are common to all executions in *RS*. Faithfulness, on the other hand, requires that all potential executions which are extensions of *rs* belong to *RS*. Structures like *rs*, referred to as invariant structures represent concurrent histories and both desired properties (succinctness and faithfulness) follow from generalisations of Szpilrajn's Theorem by which a partial order is the intersection of all its total order extensions. Although an invariant structure *rs* provides a clean theoretical capture of the set *RS*, to turn them into a practical tool (as, e.g., in [24]) one needs to be able to derive them directly from single executions using the relevant structural properties of the concurrent system. This brings into focus relational structures with acyclic relations on events (e.g., dependence graphs introduced in [23] and analysed in detail in [10]), which yield invariant structures after applying a suitable closure operation (e.g., transitive closure).

The approach sketched above has been introduced and investigated in [12, 13] as a generic model that provides general recipes for building analytic frameworks based on acyclicity. It starts from a class $\mathscr{R}$ of relational structures (such as acyclic relations), which are compared w.r.t. the information they convey (expressed by the relation $\trianglelefteq$ with $rs' \trianglelefteq rs$ if rs is obtained from rs' by adding new relationships, i.e., rs is more *concrete* than rs'). Then the *maximal* relational structures (such as total orders) $\mathscr{R}^{max} \subseteq \mathscr{R}$ represent individual executions. Invariant structures (like causal partial orders) are the *closed* relational structures $\mathscr{R}^{clo} \subseteq \mathscr{R}$ representing all executions conforming to a more abstract concurrent history. A crucial aspect of this generic approach is that after specifying the set $\mathscr{R}$ of relational structures for an application specific class of concurrent behaviours (e.g., those based on relevant invariant relationships between events together with a suitable notion of acyclicity), the development of a complete framework is basically automatic. All what remains to be done, is to define a suitable notion of closure for the structures in $\mathscr{R}$ and to provide a convenient representation of the closed structures (e.g., in an axiomatic form).

Related Work. The framework based on causal partial orders has been widely used both in theoretical investigations (e.g., [5, 22, 30]), tool building (e.g., [7, 24]), and applications (e.g., in [1, 31]. The established approach — at first formulated for cases where system runs are represented as sequences (total orders) of executed actions — was extended to situations when reports of executions are step sequences (i.e., sequences of sets of actions that happen together) represented by stratified orders. A comprehensive theory of both cases, i.e., sequences and step sequences, has recently been published in the monograph [13]. The framework resulting from considering executions as step sequences (stratified orders) has already been successfully utilized in computational biology [26], digital graphics [25], and model checking [18]. Still, the (step) sequence frameworks can only be used if event simultaneity is transitive, i.e., all observations of a concurrent system can be represented by stratified orders.

The use of relational structures to model concurrency phenomena originates from [20,21,30], and, in the version used here, from [14]. The initial investigations on the processes of concurrent systems with interval order semantics can be found in [15,16]. Interval orders were also used to investigate communication protocols in [2], in a setting consistent with sending/receiving as intervals boundaries from [19]. In subsequent works, interval semantics (called ST-semantics) was taken into account in the research on Petri nets with read arcs [32] and general discussions on distributability of concurrent systems [9].

Recall that our approach is based on reports that are observations of runs, meaning that choices have been made and thus conflicts resolved. As event structures involve conflicts between events, they are not directly relevant to this paper and we do not compare our approach with Winskel's event structures [34]. We would also like to point at a fundamental difference between the approach pursued in this paper, and Allen's Interval Algebra [4], which employs 13 base relations to capture the possible relationships between two intervals. In essence, the latter approach is semantically close to real-time semantics whereas the former is more abstract. For similar reasons, the interval order semantics used in this paper and the 'interval semantics' or 'interval time semantics' of, e.g., [27,29], are incomparable.

Contribution of This Paper. Although the case of interval order runs has received some attention in the literature, the state-of-the-art is still far from being satisfactory, and this paper aims to address this. More precisely, following the general approach formalised in [12,13], we will capture concurrent histories which are based on two invariant relationships between non-instantaneous events, namely *precedence* and *weak precedence*, intuitively corresponding to the 'earlier than' and 'not later than' positions in individual executions. As a result, the present paper deals with concurrent histories obeying the pattern that whenever a concurrent history comprises two executions with event e preceding event f in one, and f preceding e in the other, then this concurrent history also comprises an execution in which e and f are simultaneous, i.e., happening/existing at the same time (this pattern is referred to as paradigm π_3 in the classification of different kinds of models of concurrency from [13,14]). Inhibitor nets, Boolean networks, reaction systems, and membrane systems are examples of concurrent system models with an execution semantics that adhere to paradigm π_3 [13,17].Thus, while preserving paradigm π_3, in this paper we move from the transitive simultaneity exhibited in the stratified order semantics of step sequences to an interval order semantics where simultaneity is not necessarily transitive.

The main contributions of the paper are:

- a novel (and central to the whole approach) concept of acyclicity tailored for concurrent histories composed of interval orders adhering to paradigm π_3, and the definition of combined interval structures;
- an axiomatisation of interval poset structures, and the proof of their maximality among all combined interval structures;

- the proof of the closedness of invariant combined structures among all combined interval structures;
- a proof of the generalization of Szpilrajn's theorem for invariant combined structures carried out within the general framework ('roadmap') described in [12].

Structure of This Paper. First, after a preliminary section, we introduce a new concept of acyclicity (called *cis-acyclicity*) that defines *combined interval structures* (CI-structures), the relational structures (with precedence and weak precedence relationships) that underpin the proposed framework. In Sect. 4, the maximal CI-structures, called *interval poset structures* (IP-structures) are fully characterised. (Note that IP-structures play the same role as the total or stratified orders for frameworks where actions are instantaneous.) Moreover, the suitability of IP-structures to model interval order behaviours is demonstrated. Next, an axiomatisation of the closed CI-structures, called *invariant combined structures* (IC-structures), and the structure closure operation for CI-structures are provided.

2 Preliminaries

For two binary relations, R and Q, over a set Δ,

- $R \circ Q = \{(x,z) \in \Delta \times \Delta \mid \exists y : xRy \wedge yQz\}$ is the composition of Q and R.
- $R^0 = id_\Delta = \{(x,x) \mid x \in \Delta\}$ is the identity relation over Δ.
- $R^+ = R^1 \cup R^2 \cup R^3 \cup \ldots$, where $R^1 = R$ and $R^i = R^{i-1} \circ R$, for $i > 1$, is the transitive closure of R.
- $R^* = R^0 \cup R^+$ is the reflexive and transitive closure of R.
- R is acyclic if its transitive closure is irreflexive ($R^+ \cap R^0 = \varnothing$).

In this paper, a relational structure is a triple $rs = \langle \Delta, \prec, \sqsubset \rangle$, where Δ is a finite set (the *domain* of rs) and $\prec$ and $\sqsubset$ are two irreflexive[2] binary relations over Δ. Given a relational structure rs, we may use Δ_{rs}, $\prec_{rs}$, and $\sqsubset_{rs}$ to denote its components.

For nonempty sets of relational structures $\mathscr{R}$, $\mathscr{S} \subseteq \mathscr{R}$, and $rs, rs' \in \mathscr{R}$:

- rs' is an extension of rs, denoted $rs \trianglelefteq rs'$, if $\Delta_{rs} = \Delta_{rs'}$, $\prec_{rs} \subseteq \prec_{rs'}$ and $\sqsubset_{rs} \subseteq \sqsubset_{rs'}$.
- $\mathrm{ext}_{\mathscr{S}}(rs) = \{\overline{rs} \in \mathscr{S} \mid rs \trianglelefteq \overline{rs}\}$ are the extensions of rs in $\mathscr{S}$.
- $\mathscr{R}^{max} = \{\overline{rs} \in \mathscr{R} \mid \mathrm{ext}_{\mathscr{R}}(\overline{rs}) = \{\overline{rs}\}\}$ are the maximal structures in $\mathscr{R}$.
- $\max_{\mathscr{R}}(rs) = \mathrm{ext}_{\mathscr{R}^{max}}(rs)$ are the maximal extensions of rs in $\mathscr{R}$.
- $\mathscr{R}^{dom}$ is the set of all nonempty subsets of $\mathscr{R}$ consisting of relational structures with the same domain. If $RS \subseteq \mathscr{R}$ is such that Δ is the domain of all $rs \in RS$, then

$$\bigcap RS = \langle \Delta, \bigcap_{\overline{rs} \in RS} \prec_{\overline{rs}}, \bigcap_{\overline{rs} \in RS} \sqsubset_{\overline{rs}} \rangle$$

is the intersection of RS.
- $\mathscr{R}$ is intersection-closed if $\bigcap RS \in \mathscr{R}$, for every $RS \in \mathscr{R}^{dom}$.

[2] We use irreflexive ('strict') relationships between events.

– If $\mathscr{R}$ is intersection-closed, then

$$\mathscr{R}^{clo} = \{\overline{rs} \in \mathscr{R} \mid \overline{rs} = \bigcap \max_{\mathscr{R}}(\overline{rs})\}$$

are the closed structures in $\mathscr{R}$, and $\mathrm{clo}_{\mathscr{R}} : \mathscr{R} \to \mathscr{R}^{clo}$ where $\mathrm{clo}_{\mathscr{R}}(\overline{rs}) = \bigcap \max_{\mathscr{R}}(\overline{rs})$, for all $\overline{rs} \in \mathscr{R}$, is the structure closure of $\mathscr{R}$.

The following result allows to deal with the closure operation and closed relational structures without referring to intersections of the maximal extensions of structures.

Proposition 1 (Prop.7.8 in [13]**).** *Let $\mathscr{R}$ be an intersection-closed set of relational structures, $\mathscr{S} \subseteq \mathscr{R}$, and $f : \mathscr{R} \to \mathscr{S}$ be a monotonic (i.e., $rs \trianglelefteq rs' \implies f(rs) \trianglelefteq f(rs')$) and non-decreasing (i.e., $rs \trianglelefteq f(rs)$) function. Moreover, for all $rs = \langle \Delta, \prec, \sqsubset \rangle \in \mathscr{S}$ and $x \neq y \in \Delta$, we have:*

- *$f(rs) \trianglelefteq rs$.*
- *If $x \not\prec y$ (or $x \not\sqsubset y$), then there is $\overline{rs} \in \max_{\mathscr{R}}(rs)$ such that $x \not\prec_{\overline{rs}} y$ (resp. $x \not\sqsubset_{\overline{rs}} y$).*

Then f is the structure closure of $\mathscr{R}$, i.e., $\mathscr{S} = \mathscr{R}^{clo}$ and $f(rs) = \mathrm{clo}_{\mathscr{R}}(rs)$, for every $rs \in \mathscr{R}$.

Let $\langle \Delta, \prec \rangle$, where Δ is a set and $\prec$ an irreflexive binary relation over Δ. Then:

- $\langle \Delta, \prec \rangle$ is a partial order if $\prec$ is transitive.
- $\langle \Delta, \prec \rangle$ is a total order if $x \prec y$ or $y \prec x$, for all $x \neq y \in \Delta$.
- $\langle \Delta, \prec \rangle$ is a stratified order if there is a partition $\Delta_1, \ldots, \Delta_n$ of Δ such that $\prec$ is equal to $\bigcup_{1 \leq i < j \leq n} \Delta_i \times \Delta_j$.
- $\langle \Delta, \prec \rangle$ is an interval order if $x \prec y \wedge z \prec w$ implies $x \prec w \vee z \prec y$, for all $x, y, z, w \in \Delta$.

All total orders are stratified, all stratified orders are interval, and all interval orders are partial.

The adjective 'interval' derives from Fishburn's Theorem [8]: *A countable partial order $\langle \Delta, \prec \rangle$ is interval iff there exists a total order $\langle Y, \lhd \rangle$ and two injective mappings $\beta, \varepsilon : \Delta \to Y$ such that, for all $x, y \in \Delta$, $\beta(x) \lhd \varepsilon(x)$ and $x \prec y \iff \varepsilon(x) \lhd \beta(y)$.* Intuitively, the mappings β and ε can then be interpreted as specifying the 'beginnings' and 'endings' of intervals corresponding to the events in Δ.

The relevance of interval orders in concurrency theory follows from an observation, credited to Wiener [33], that any execution of a physical system that can be observed by a single observer is an interval order. Hence the most precise observational semantics should be defined in terms of interval orders or suitable representations thereof (cf. [14]). Note that interval order executions are typically generated after splitting each action in the model of a concurrent system into a begin and end action and, after executing the modified model using sequential semantics, deriving the corresponding interval orders by applying Fishburn's Theorem. However, this can also be done directly [28], i.e., without splitting actions.

Example 1. Consider again the program fragment from the Introduction. Each of its possible computations is modelled by a partial order in Fig. 1. The first three are total orders corresponding to purely sequential observations. In the next two stratified orders,

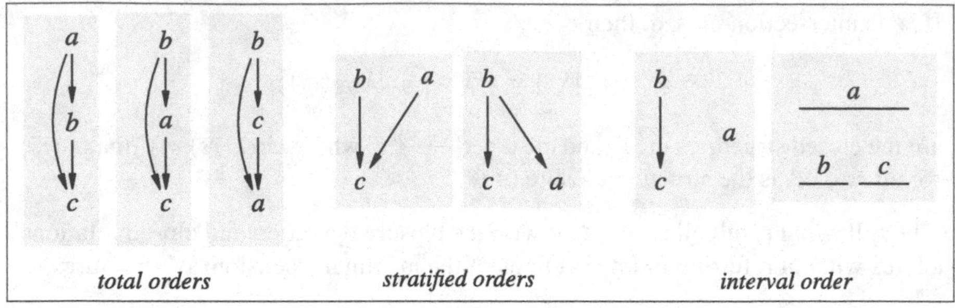

Fig. 1. Possible partial orders modelling a concurrent computation, cf. Example 1.

the execution is given as a sequence of sets of (unordered) actions that may be perceived as executed simultaneously. The last execution is represented by an interval order which is neither total nor stratified. In this case, the execution in the second thread is sequential while a (being independent from both b and c) is observed as starting before b finishes and ending after c starts. This can be seen also in the interval-based representation in the rightmost diagram of Fig. 1. $\diamond$

3 Behavioural Acyclicity in Interval Executions

As mentioned in the Introduction, the first (and crucial) step in the development of a relational structure semantics for a class of concurrent behaviours is the choice of a right notion of 'behavioural acyclicity'. Intuitively, this means that by following any sequence of events while time progresses, one should never reach the first event. Hence, for example, if a relational structure only records execution precedence between events, then behavioural acyclicity becomes the standard notion of acyclicity. In the case of the *combined interval structures (or* CI*-structures)* studied in this paper, the notion of behavioural acyclicity is much more involved. The reason is that the two relations, $\prec$ and $\sqsubset$ in a relational structure $rs = \langle \Delta, \prec, \sqsubset \rangle$ represent here two different notions of precedence between events, namely $e \prec f$ states the standard precedence between the two events (e occurred before f), while $e \sqsubset f$ states the weak precedence between them (e did not occur later than f, or, equivalently, e occurred before or simultaneously with f). Although it is expected that $\prec$ must be acyclic, and $\sqsubset$ can allow cycles (as two or more events can occur simultaneously), it is also necessary to state what kinds of mixed cycles are disallowed and so also what kinds of mixed cycles are allowed.

The next definition provides a graph-based capture of the mixed paths inferred from $\prec$ and $\sqsubset$ which need to be taken into account in order to identify inadmissible cycles in CI-structures as well as to define the closure operation for CI-structures.

Definition 1 (cis-path and cis-cycle). *A* combined interval structure path *(or* cis-path*) of a relational structure* $rs = \langle \Delta, \prec, \sqsubset \rangle$ *is a sequence* $\pi = x_1 k_1 x_2 \ldots x_{n-1} k_{n-1} x_n$ $(n \geq 2)$ *such that the following hold:*

– $x_1, \ldots, x_n \in \Delta$ *and* $k_1, \ldots, k_{n-1} \in \{1,2\}$*;*

– $k_i = 1 \implies x_i \prec x_{i+1}$ *and* $k_i = 2 \implies x_i \sqsubset x_{i+1}$, *for every* $1 \leq i < n$; *and*
– *there is no* $j < n-1$ *satisfying* $k_j = k_{j+1} = 2$.

We denote this by $\pi \in \text{cispaths}_{rs}^{k_1 k_{n-1}}(x_1, x_n)$. *Also,* π *is called a* cis-cycle *if* $x_1 = x_n$.

Remark 1. Intuitively, cis-paths are the only paths in the combined graph of the relations $\prec$ and $\sqsubset$ that can lead to new relationships between events, and so need to be taken into account when constructing the closure of a CI-structure. It may come as a surprise that paths like $x2y2z$ (i.e., $x \sqsubset y \sqsubset z$) are irrelevant. This is because weak precedence is not transitive in the interval setting (unlike in the models based on step sequences from [13]). Indeed, it is possible to assign execution intervals to x, y, and z so that $x \sqsubset y \sqsubset z$ and $x \not\sqsubset z$ as follows: $I(x) = [3,5]$, $I(z) = [0,2]$, and $I(y) = [1,4]$. Then, x starts before y finishes and y starts before z finishes. However, x does not start before z finishes as it starts only after z finishes. ◇

To ease the presentation, we introduce notations for different groups of cis-paths:

$$\begin{aligned}
\text{cispaths}_{rs}^{ij,kl}(x,y) &= \text{cispaths}_{rs}^{ij}(x,y) \cup \text{cispaths}_{rs}^{kl}(x,y)\\
\text{cispaths}_{rs}^{ij,kl,mn}(x,y) &= \text{cispaths}_{rs}^{ij}(x,y) \cup \text{cispaths}_{rs}^{kl}(x,y) \cup \text{cispaths}_{rs}^{mn}(x,y)\\
\text{cispaths}_{rs}(x,y) &= \\
&\text{cispaths}_{rs}^{11}(x,y) \cup \text{cispaths}_{rs}^{12}(x,y) \cup \text{cispaths}_{rs}^{21}(x,y) \cup \text{cispaths}_{rs}^{22}(x,y)\,.
\end{aligned}$$

Two cis-paths, $\pi \in \text{cispaths}_{rs}^{ij}(x,y)$ and $\pi' \in \text{cispaths}_{rs}^{mn}(y,z)$, can be *concatenated* to yield the sequence $\pi \odot \pi' = \pi\pi''$, where π'' is such that $\pi' = y\pi''$. We then have:

Proposition 2. *Let* x, y, z *belong to the domain of a relational structure* rs. *If* $j = 1$ *or* $m = 1$, *then* $\text{cispaths}_{rs}^{ij}(x,y) \odot \text{cispaths}_{rs}^{mn}(y,z) \subseteq \text{cispaths}_{rs}^{in}(x,z)$.

We are now ready to introduce the concept of behavioural acyclicity that will be used to define CI-structures. Real-life behaviours are 'acyclic' due to the underlying time vector which precludes re-visiting the past. Bearing this in mind, the essence of cis-acyclicity can be explained using the following simple example involving three events: $x \sqsubset y \prec z \sqsubset x$. Within the concurrency models discussed in [13], such a cyclic behaviour seems illogical as it suggests that x occurred before itself. However, this view is based on the implicit assumption that events are instantaneous, or that their duration is not taken into account. When we accept that events can take time to complete, the picture changes and the cycle $x \sqsubset y \prec z \sqsubset x$ is not a source of logical or temporal inconsistency. Indeed, similarly as in Remark 1, it is possible to assign execution intervals to x, y, and z so that $x \sqsubset y \prec z \sqsubset x$, e.g., $I(x) = [1,4]$, $I(y) = [0,2]$, and $I(z) = [3,5]$. Then, in the interval semantics, event x is in a weak precedence relationship with both y and z, viz. $z \sqsubset x \sqsubset y$. As a result, it is possible for x to start before y finishes, and for z to start before x finishes. Hence, x can be simultaneous, i.e., overlap in time, with both y and z.

Definition 2 (cis-acyclicity). *A relational structure rs is* cis-acyclic *if*

$$\bigcup_{x \in \Delta_{rs}} \text{cispaths}_{rs}^{11,12,21}(x,x) = \varnothing\,. \tag{1}$$

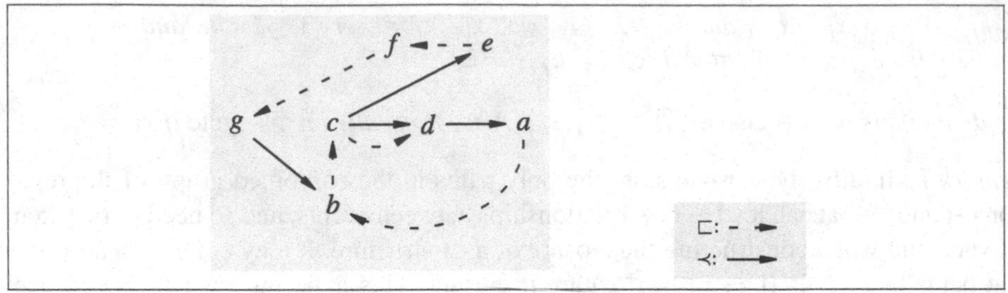

Fig. 2. Relational structure of Example 2.

In other words, cis-acyclicity requires that any cis-cycle includes exactly one pair of consecutive weak precedence relationships. (Note that, by definition, a cis-cycle includes at most one pair of consecutive weak precedence relationships.) Hence, cycles in $\bigcup_{x\in\Delta_{rs}} \text{cispaths}^{22}_{rs}(x,x)$ are allowed as are any cycles outside $\bigcup_{x\in\Delta_{rs}} \text{cispaths}_{rs}(x,x)$.

Example 2. Consider the following relational structure with six events as in Fig. 2:

$$rs = \langle\{a,b,c,d,e,f\},\{\langle c,d\rangle,\langle c,e\rangle,\langle d,a\rangle,\langle g,b\rangle\},\{\langle a,b\rangle,\langle b,c\rangle,\langle c,d\rangle,\langle e,f\rangle,\langle f,g\rangle\rangle\,.$$

It includes $a-b-c-d-a$ (a cis-cycle derived from $b2c1d1a2b \in \text{cispaths}^{22}_{rs}(b,b)$, i.e., a cycle constructed using a path which has its first element equal to its last element) and the cis-cycle $b-c-e-f-g-b$ (derived from $f2g1b2c1e2f \in \text{cispaths}^{22}_{rs}(f,f)$), with a common weak precedence arc $\langle b,c\rangle$. (Note also that $b2c2d1a2b$ is not a cis-path as it involves two consecutive weak precedence relationships.) In both cis-cycles one can find precedence arcs, and in the first one even two in a row. However, both of these cis-cycles show also two consecutive weak precedence arcs: $a \sqsubset b \sqsubset c$ and $e \sqsubset f \sqsubset g$, respectively. This is why *rs* is a cis-acyclic relational structure. ◇

According to the definition of cis-acyclicity of *rs*, by adding a new relationship $\langle x,y\rangle$ to $\prec$ or $\sqsubset$ (and forming in this way an extended relational structure rs') one can push *rs* outside the class of cis-acyclic relational structures only by creating a new cis-path belonging to $\text{cispaths}^{11,12}_{rs'}(x,x)$ or $\text{cispaths}^{21}_{rs'}(x,x)$, respectively (or, equivalently, a new cis-path in $\text{cispaths}^{11,21}_{rs'}(y,y)$ or $\text{cispaths}^{12}_{rs'}(y,y)$). Moreover, if *rs* is a maximal cis-acyclic relational structure, then any new relationship added to $\prec$ or $\sqsubset$ makes the new relational structure non-cis-acyclic. Hence, we have the following, where here and later $\prec^{\langle x,y\rangle}$ and $\sqsubset^{\langle x,y\rangle}$ respectively denote $\prec \cup\{\langle x,y\rangle\}$ and $\sqsubset \cup\{\langle x,y\rangle\}$:

Proposition 3. *Let* $rs = \langle\Delta,\prec,\sqsubset\rangle$ *be a cis-acyclic relational structure, and* $x \neq y \in \Delta$.

- *If* $\langle\Delta,\prec,\sqsubset^{\langle x,y\rangle}\rangle$ *is not cis-acyclic, then* $\text{cispaths}^{11}_{rs}(y,x) \neq \varnothing$.
- *If* $\langle\Delta,\prec^{\langle x,y\rangle},\sqsubset\rangle$ *is not cis-acyclic, then* $\text{cispaths}_{rs}(y,x) \neq \varnothing$.

Moreover, if rs is a maximal cis-acyclic relational structure, then

- *If* $x \not\prec y$, *then* $\langle\Delta,\prec^{\langle x,y\rangle},\sqsubset\rangle$ *is not cis-acyclic.*
- *If* $x \not\sqsubset y$, *then* $\langle\Delta,\prec,\sqsubset^{\langle x,y\rangle}\}\rangle$ *is not cis-acyclic.*

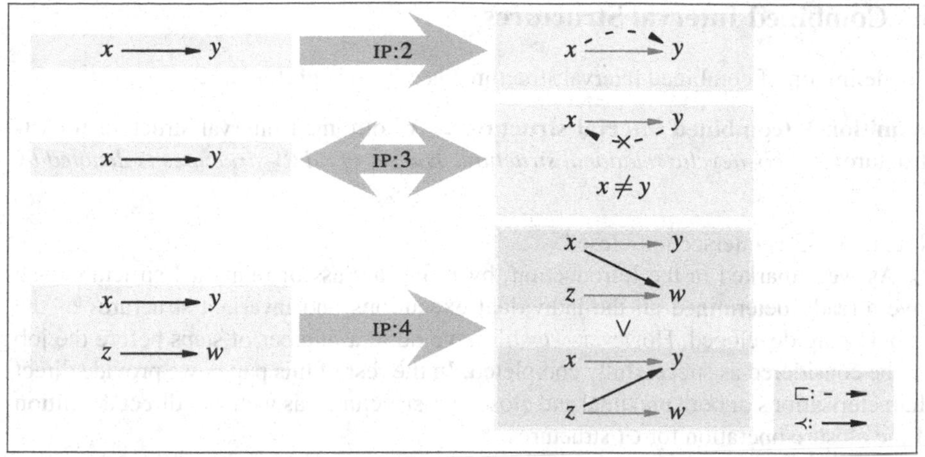

Fig. 3. Axioms IP:2–4. The arc that cannot occur is crossed out.

The above result deals with cases when cis-acyclity is lost after adding new relationships. The next proposition identifies cases when such additions are harmless.

Proposition 4. *Let* $rs = \langle \Delta, \prec, \sqsubset \rangle$ *be a cis-acyclic relational structure and* $x, y, z, w \subseteq \Delta$.

1. $x \prec y$ *implies that* $\langle \Delta, \prec, \sqsubset^{\langle x,y \rangle} \rangle$ *is cis-acyclic.*
2. $x \prec y \prec z$ *implies that* $\langle \Delta, \prec^{\langle x,z \rangle}, \sqsubset \rangle$ *is cis-acyclic.*
3. $x \prec y \sqsubset z \vee x \sqsubset y \prec z$ *implies that* $\langle \Delta, \prec, \sqsubset^{\langle x,z \rangle} \rangle$ *is cis-acyclic.*
4. $x \prec y \sqsubset z \prec w$ *implies that* $\langle \Delta, \prec^{\langle x,w \rangle}, \sqsubset \rangle$ *is cis-acyclic.*
5. $x \sqsubset y \prec z \sqsubset w \neq x$ *implies that* $\langle \Delta, \prec, \sqsubset^{\langle x,w \rangle} \rangle$ *is cis-acyclic.*

Proof. In each case below, we apply Proposition 3 to infer the existence of some cis-path π, and then obtain a contradiction with the cis-acyclicity of rs.

(1) Otherwise, there is $\pi \in \text{cispaths}_{rs}^{11}(y,x)$, and so $\pi 1 y \in \text{cispaths}_{rs}^{11}(y,y)$.

(2) Otherwise, there is $\pi \in \text{cispaths}_{rs}(z,x)$, and so $\pi 1 y 1 z \in \text{cispaths}_{rs}^{11,21}(z,z)$.

(3) Otherwise, there is $\pi \in \text{cispaths}_{rs}^{11}(z,x)$, and so $y 2 \pi 1 y \in \text{cispaths}_{rs}^{21}(y,y)$ or $y 1 \pi 2 y \in \text{cispaths}_{rs}^{12}(y,y)$.

(4) Otherwise, there is $\pi \in \text{cispaths}_{rs}(w,x)$, and so $\pi 1 y 2 z 1 w \in \text{cispaths}_{rs}^{11,21}(w,w)$.

(5) Otherwise, there is $\pi \in \text{cispaths}_{rs}^{11}(w,x)$, and so $\pi 2 y 1 z 2 w \in \text{cispaths}_{rs}^{12}(w,w)$. □

We end this section with an immediate result which also introduces two new notations, $\text{pre}(\prec, \sqsubset)$ and $\text{wpre}(\prec, \sqsubset)$, in preparation for a characterisation of the structure closure mapping introduced later.

Proposition 5. *Let* $rs = \langle \Delta, \prec, \sqsubset \rangle$ *be a relational structure and* $x, y \in \Delta$.

1. $\text{cispaths}_{rs}^{11}(x,y) \neq \varnothing$ *iff* $\langle x,y \rangle \in \text{pre}(\prec, \sqsubset) = \prec \circ (\prec \cup (\sqsubset \circ \prec))^*$.
2. $\text{cispaths}_{rs}(x,y) \neq \varnothing$ *iff* $\langle x,y \rangle \in \text{wpre}(\prec, \sqsubset) = (\prec \cup (\sqsubset \circ \prec))^+ \cup (\prec \cup (\sqsubset \circ \prec))^* \circ \sqsubset$.

4 Combined Interval Structures

The definition of combined interval structures is now straightforward.

Definition 3 (combined interval structure). *A* combined interval structure (or CI-structure) *is a cis-acyclic relational structure. The set of all* CI*-structures is denoted by* CIS.

Note that CIS is intersection-closed.

As we remarked in the Introduction, by fixing a class of relational structures, we have already determined all the individual executions and invariant structures of the model being developed. However, we still have quite a number of steps before the job can be considered as successfully completed. In the rest of this paper, we provide direct characterisations of both maximal and closed CI-structures, as well as a direct definition of the closure operation for CI-structures.

4.1 Maximal Combined Interval Structures

In this subsection, our goal is two-fold. First, we want to characterise the maximal CI-structures through a suitable set of axioms given below. The second is to demonstrate that such structures — intended to represent individual interval executions — are in fact interval orders in disguise.

Definition 4 (interval poset structure). Interval poset structures (or IP-structures) IPS *are triples ips* $= \langle \Delta, \prec, \sqsubset \rangle$ *such that* $\prec$ *and* $\sqsubset$ *are finite binary relations on a finite set* Δ. *Moreover, for all* $x, y, z, w \in \Delta$, *it is required that:*

$$x \not\sqsubset x \quad : \ \text{IP:1} \qquad\qquad x \prec y \iff y \not\sqsubset x \neq y \quad : \ \text{IP:3}$$
$$x \prec y \implies x \sqsubset y \quad : \ \text{IP:2} \qquad\qquad x \prec y \wedge z \prec w \implies x \prec w \vee z \prec y \quad : \ \text{IP:4}$$

Axioms IP:2–4 are illustrated in Fig. 3. Note that the last axiom is the defining property of interval orders. Moreover, as shown next, no axiom in IP:1–4 is redundant.

Proposition 6. *The set of axioms in Definition 4 is minimal.*

Proof. Let $rs_1, \ldots, rs_4$ be the following relational structures with x, y, z, w all distinct (see Fig. 4):

$$rs_1 = \langle \{x\}, \varnothing, \{\langle x,x\rangle\}\rangle \quad rs_3 = \langle \{x,y\}, \{\langle x,y\rangle\}, \{\langle x,y\rangle, \langle y,x\rangle\}\rangle$$
$$rs_2 = \langle \{x,y\}, \{\langle x,y\rangle\}, \varnothing\rangle \quad rs_4 = \langle \{x,y,z,w\}, \{\langle x,y\rangle, \langle z,w\rangle\}, \{\langle x,y\rangle, \langle z,w\rangle\}\rangle\,.$$

None of these relational structures is an IP-structure. Moreover, each rs_i satisfies all the axioms in Definition 4 except for IP:i. Hence, dropping any of the axioms IP:1-IP:4 would lead to a strictly larger set of structures satisfying the remaining axioms. □

IP-structures are not only maximal CI-structures, as we prove in Theorem 1, but also are closely related to interval orders, which provides a justification for some of the terminology used.

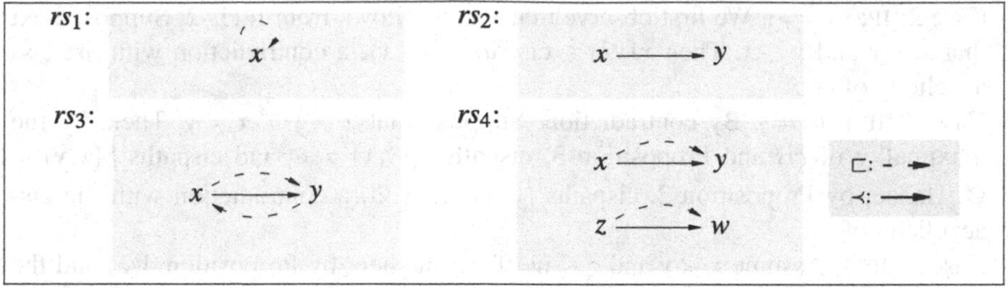

Fig. 4. Relational structures in Proposition 6.

Example 3. Consider again the relational structure *rs* from Example 2. One can extend it to an IP-structure rs' by adding five precedence arcs, $\langle g,a\rangle, \langle g,c\rangle, \langle g,d\rangle, \langle g,e\rangle$ and $\langle c,a\rangle$, and filling the gaps of missing weak precedence arcs according to axioms IP:2 and IP:3. The resulting structure (without weak precedence arcs and with weak precedence arcs from $\sqsubset \setminus \prec$ drawn for clarity in gray) is depicted on the left of Fig. 5. The right side shows the relationship between rs' and the corresponding interval order illustrated as a set of time intervals. ◇

Apart from verifying the correctness of the axiomatisation of the maximal CI-structures, the next result justifies the claimed suitability of CI-structures to model interval order behaviours. The latter follows after extending each interval order *ipo* to a relational structure $\langle \Delta_{ipo}, \prec_{ipo}, \sqsubset_{ipo}\rangle$, where

$$\sqsubset_{ipo} = \{\langle x,y\rangle \mid x \neq y \in \Delta_{ipo} \wedge y \not\prec_{ipo} x\}$$

is a derived relation capturing the 'not later than' relationship at the level of interval orders. Intuitively, $\langle \Delta_{ipo}, \prec_{ipo}, \sqsubset_{ipo}\rangle$ is an '*ipo* in disguise'.

Theorem 1. $\mathsf{CIS}^{max} = \mathsf{IPS} = \{\langle \Delta_{ipo}, \prec_{ipo}, \sqsubset_{ipo}\rangle \mid$ *ipo a finite interval order*$\}$.

Proof. We observe that the second equality follows directly from the definitions.
($\mathsf{IPS} \supseteq \mathsf{CIS}^{max}$) Let $cis = \langle \Delta, \prec, \sqsubset\rangle \in \mathsf{CIS}^{max}$. We show that the axioms hold.

– Case 1: IP:1 and IP:2. The first axiom clearly holds, and the second one holds by Proposition 4(1), as *cis* is maximal.

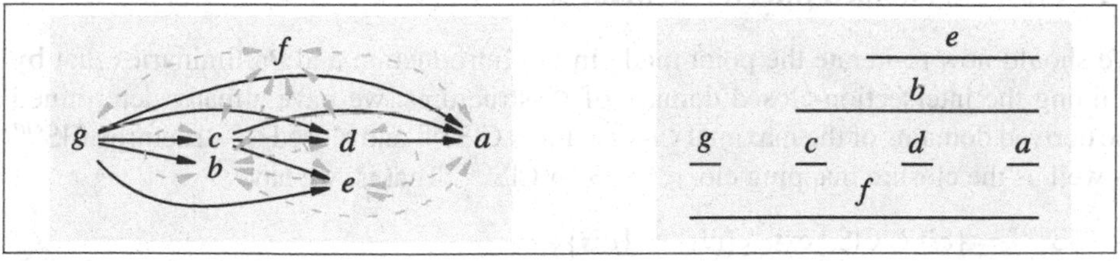

Fig. 5. CI-structure (left) and its interval representation (right) in Example 3.

- Case 2: IP:3 ($\Longrightarrow$). We first observe that $x \neq y$ follows from IP:1–2. Suppose next that $x \prec y$ and $y \sqsubset x$. Then $x1y2x \in \text{cispaths}^{12}_{cis}(x,x)$, a contradiction with the cis-acyclicity of *cis*.
- Case 3: IP:3 ($\Longleftarrow$). By contradiction. Suppose that $x \not\prec y \not\sqsubset x \neq y$. Then, by the maximality of *cis* and Proposition 3, $\text{cispaths}_{cis}(y,x) \neq \varnothing$ and $\text{cispaths}^{11}_{cis}(x,y) \neq \varnothing$. Hence, by Proposition 2, $\text{cispaths}^{11,12}_{cis}(x,x) \neq \varnothing$, a contradiction with the cis-acyclicity of *cis*.
- Case 4: IP:4. Assume $x \prec y$ and $z \prec w$. If $x = w$ then, by Proposition 4(2) and the maximality of *cis*, $z \prec y$. Similarly, if $z = y$, then $x \prec w$. If $x \not\prec w \neq x$ and $z \not\prec y \neq z$, then, by the maximality of *cis* and Proposition 3, $\text{cispaths}_{cis}(w,x) \neq \varnothing$ and $\text{cispaths}_{cis}(y,z) \neq \varnothing$. Hence, by applying Proposition 2 three times, we obtain that $\text{cispaths}^{11,12}_{cis}(x,x) \neq \varnothing$, contradicting the cis-acyclicity of *cis*.

($\mathsf{IPS} \subseteq \mathsf{CIS}^{max}$) Let $ips = \langle \Delta, \prec, \sqsubset \rangle \in \mathsf{IPS}$.

Suppose that $ips \notin \mathsf{CIS}$. Then there is $x \in \Delta$ and a cis-cycle $\pi = x_1 k_1 \ldots k_{n-1} x_n \in \text{cispaths}^{11,12,21}_{ips}(x,x)$ (note that $x = x_1 = x_n$). Moreover, we can choose x and π in such a way that n has the smallest possible value. Suppose, without loss of generality, that $k_1 = 1$ and consider three cases.

- Case 1: $n = 2$. Then $\pi = x1x$, contradicting IP:1–2.
- Case 2: $n = 3$ and $k_2 = 2$. Then we obtain a contradiction with IP:3.
- Case 3: $n = 3$ and $k_2 = 1$, or $n > 3$. Then there is $1 < j < n$ such that $k_j = 1$, and so, by IP:4, $x_1 \prec x_{j+1}$ or $x_j \prec x_2$. If $x_1 \prec x_{j+1}$, then $x_1 1 x_{j+1} \ldots k_{n-1} x_1 \in \text{cispaths}^{11,12}_{ips}(x,x)$. Moreover, if $x_j \prec x_2$, then $x_2 k_2 \ldots x_j 1 x_2 \in \text{cispaths}^{11,21}_{ips}(x_2,x_2)$. In either case, we obtain a contradiction with the choice made.

Hence $ips \in \mathsf{CIS}$. To show $ips \in \mathsf{CIS}^{max}$, suppose that $x \neq y \in \Delta$ and consider two cases.

- Case 1: $x \not\prec y$ and $ips' = \langle \Delta, \prec^{\langle x,y \rangle}, \sqsubset \rangle \in \mathsf{IPS} \subseteq \mathsf{CIS}$. Then, by the cis-acyclicity of ips', $y \not\sqsubset x$. Hence, by IP:3, $x \prec y$, a contradiction.
- Case 2: $x \not\sqsubset y$ and $ips'' = \langle \Delta, \prec, \sqsubset^{\langle x,y \rangle} \} \rangle \in \mathsf{IPS} \subseteq \mathsf{CIS}$. Then, by IP:3, $y \prec x$, contradicting the cis-acyclicity of ips''. □

In order to single out IP-structures extending CI-structures, we will use the function $\text{cis2IPS} : \mathsf{CIS} \to 2^{\mathsf{IPS}}$ such that $\text{cis2IPS}(cis) = \text{ext}_{\mathsf{IPS}}(cis)$, the extensions of *cis* in IPS, for every $cis \in \mathsf{CIS}$.

4.2 Closed Combined Interval Structures

We should now re-iterate the point made in the Introduction and Preliminaries that by defining the intersection-closed domain of CI-structures, we have already determined the derived domains of the maximal CI-structures CIS^{max} and closed CI-structures CIS^{clo} as well as the closure mapping $\text{clo}_{\mathsf{CIS}} : \mathsf{CIS} \to \mathsf{CIS}^{clo}$. That is, we have:

- $\mathsf{CIS}^{max} = \{cis \in \mathsf{CIS} \mid \text{ext}_{\mathsf{CIS}}(cis) = \{cis\}\}$.
- $\mathsf{CIS}^{clo} = \{ics \in \mathsf{CIS} \mid ics = \bigcap \text{max}_{\mathsf{CIS}}(ics)\}$.
- $\text{clo}_{\mathsf{CIS}}(cis) = \bigcap \text{max}_{\mathsf{CIS}}(cis)$, for every $cis \in \mathsf{CIS}$.

However, the above set-theoretic definitions would reveal a fundamental drawback if someone wanted to apply them, e.g., to gain insights into the subtle relationships between events involved in concurrent behaviours. We have therefore already provided an order-theoretic characterisation of the maximal CI-structures in Theorem 1, and in the rest of this paper we provide order-theoretic characterisations for the other two notions.

We start by adapting axioms from [14] (some of which one can find also in [21]) that will subsequently provide an axiomatisation of the closed CI-structures. Axioms IC:2–6 are illustrated in Fig. 6.

Definition 5 (invariant combined structure). Invariant combined structures (or IC-structures) ICS *are triples* $ics = \langle \Delta, \prec, \sqsubset \rangle$ *such that* $\prec$ *and* $\sqsubset$ *are finite binary relations on a finite set* Δ. *Moreover, for all* $x, y, z \in \Delta$, *it is required that:*

$$
\begin{array}{rcll}
 & & x \not\sqsubset x & : \text{IC:1} \\
x \prec y & \Longrightarrow & x \sqsubset y \not\sqsubset x & : \text{IC:2} \\
x \prec y \prec z & \Longrightarrow & x \prec z & : \text{IC:3}
\end{array}
\qquad
\begin{array}{rcll}
x \sqsubset y \prec z \lor x \prec y \sqsubset z & \Longrightarrow & x \sqsubset z & : \text{IC:4} \\
x \prec y \sqsubset z \prec w & \Longrightarrow & x \prec w & : \text{IC:5} \\
x \sqsubset y \prec z \sqsubset w \neq x & \Longrightarrow & x \sqsubset w & : \text{IC:6}
\end{array}
$$

Proposition 7. *The set of axioms in Definition 5 is minimal.*

Proof. Let $rs_1, \ldots, rs_6$ be the following relational structures with x, y, z, w all distinct (see Fig. 7): $rs_1 = \langle \{x\}, \varnothing, \{\langle x, x \rangle\} \rangle$, $rs_2 = \langle \{x, y\}, \{\langle x, y \rangle\}, \varnothing \rangle$,

$$
\begin{aligned}
rs_3 &= \langle \{x,y,z\}, \{\langle x,y\rangle, \langle y,z\rangle\}, \{\langle x,y\rangle, \langle y,z\rangle, \langle x,z\rangle\} \rangle \\
rs_4 &= \langle \{x,y,z\}, \{\langle x,y\rangle\}, \{\langle x,y\rangle, \langle y,z\rangle\} \rangle \\
rs_5 &= \langle \{x,y,z,w\}, \{\langle x,y\rangle, \langle z,w\rangle\}, \{\langle x,y\rangle, \langle y,z\rangle, \langle x,z\rangle, \langle y,w\rangle, \langle z,w\rangle, \langle x,w\rangle\} \rangle \\
rs_6 &= \langle \{x,y,z,w\}, \{\langle y,z\rangle\}, \{\langle x,y\rangle, \langle y,z\rangle, \langle x,z\rangle, \langle y,w\rangle, \langle z,w\rangle\} \rangle
\end{aligned}
$$

None of these relational structures is an IC-structure. On the other hand, each rs_i satisfies all the axioms in Definition 5 except for IC:i. □

IP-structures are also IC-structures (i.e., IPS $\subseteq$ ICS). This is seen by observing that $ips \in$ IPS satisfies all the axioms IC:1-6, namely: IC:1–2 are weaker versions of IP:1–3; IC:3&5 are special cases of IP:4; and IC:4&6 follow from the fact that, respectively, $z \not\prec x$ and $w \not\prec x$, and so we can use IP:3.

The next result shows that the IC-structures are nothing but the closed CI-structures. Moreover, the closure of CI-structures can be expressed directly using the two formulas generating precedence and weak precedence relationships introduced in Proposition 5.

Theorem 2. *The mapping* cis2ics : CIS $\rightarrow$ ICS, *for every* $cis \in$ CIS *given by:*

$$\text{cis2ics}(cis) = \langle \Delta_{cis}, \text{pre}(\prec_{cis}, \sqsubset_{cis}), \text{wpre}(\prec_{cis}, \sqsubset_{cis}) \setminus id_{\Delta_{cis}} \rangle$$

is the structure closure of CIS. *Note: This means that* ICS $=$ CIS^{clo} *and* cis2ics $=$ clo$_{\text{CIS}}$.

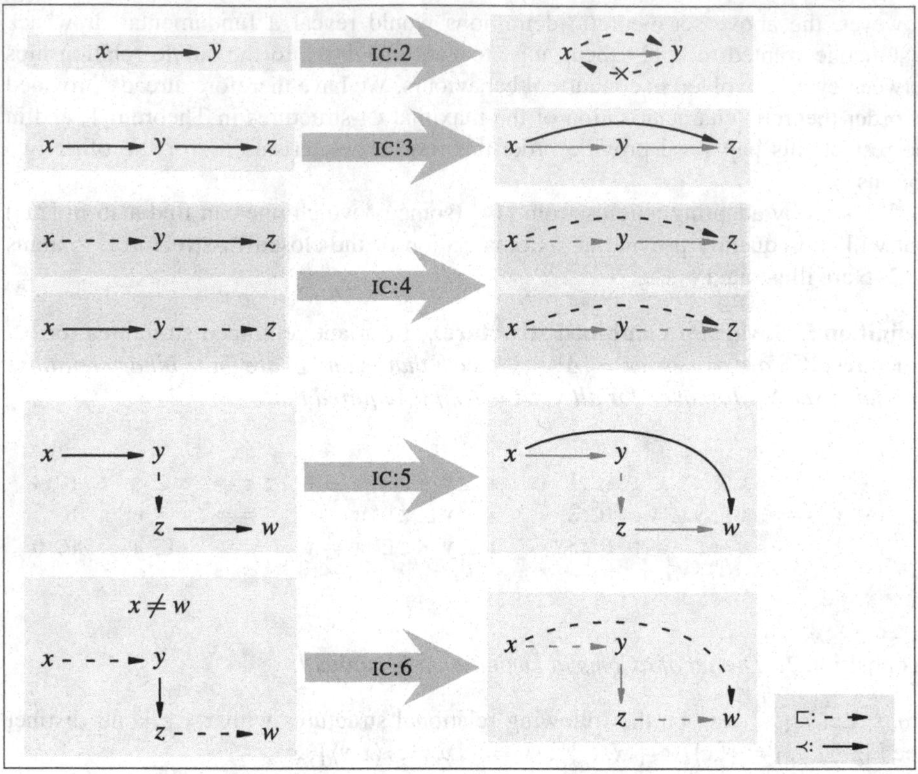

Fig. 6. Axioms IC:2–6. The arc that cannot occur is crossed out.

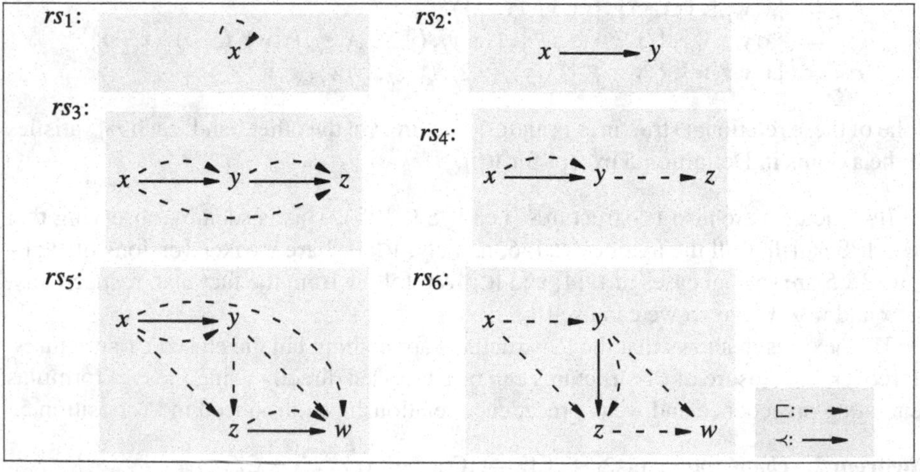

Fig. 7. Relational structures in Proposition 7.

Proof.

Lemma 1. *Let* $ics = \langle \Delta, \prec, \sqsubset \rangle \in \mathsf{ICS}$ *and* $x \neq y \in \Delta$.

1. $\text{cispaths}^{11}_{ics}(x,y) \neq \varnothing$ *implies* $x \prec y$. *(i.e., by Proposition 5,* $\text{pre}(\prec, \sqsubset) \subseteq \prec$*)*
2. $\text{cispaths}_{ics}(x,y) \neq \varnothing$ *implies* $x \sqsubset y$. *(i.e., by Proposition 5,* $\text{wpre}(\prec, \sqsubset) \subseteq \sqsubset$*)*

Proof. (1) Let π be a shortest cis-path in $\text{cispaths}^{11}_{ics}(x,y)$. Then, by IC:3 and the definition of cis-path, if $\pi = \dots ktm \dots$, then $k \neq m$. We have therefore the following two cases.

- Case 1: $\pi = x1y$. Then $x \prec y$.
- Case 2: $\pi = x1z2w1t \dots y$ (possibly $t = y$). Then, by IC:5, $\pi = x1t \dots y$ is a shorter cis-path in $\text{cispaths}^{11}_{ics}(x,y)$, yielding a contradiction.

(2) Let π be a shortest cis-path in $\text{cispaths}_{ics}(x,y)$. Then, by IC:3 and the definition of cis-path, if $\pi = \dots ktm \dots$, then $k \neq m$. We have therefore the following five cases.

- Case 1: $\pi = x2y$. Then $x \sqsubset y$.
- Case 2: $\pi = x1y$. Then, by IC:2, $x \sqsubset y$.
- Case 3: $\pi = x1z2y$ or $\pi = x2z1y$. Then, by IC:4, $x \sqsubset y$.
- Case 4: $\pi = x1z2w1t \dots y$ (possibly $t = y$). Then, by IC:5, $\pi = x1t \dots y$ is a shorter cis-path in $\text{cispaths}^{11}_{ics}(x,y)$, yielding a contradiction.
- Case 5: $\pi = x2z1w2t \dots y$ (possibly $t = y$). Then, by IC:6, $\pi = x2t \dots y$ is a shorter cis-path in $\text{cispaths}^{11}_{ics}(x,y)$, yielding a contradiction. □

Lemma 2. *Let* $rs = \langle \Delta, \prec, \sqsubset \rangle \in \mathsf{ICS}$ *and* $x \neq y \in \Delta$.

1. $x \not\prec y$ *implies* $\langle \Delta, \prec, \sqsubset^{\langle y,x \rangle} \rangle \in \mathsf{CIS}$.
2. $x \not\sqsubset y$ *implies* $\langle \Delta, \prec^{\langle y,x \rangle}, \sqsubset \rangle \in \mathsf{CIS}$.
3. $x \not\sqsubset y \not\sqsubset x$ *implies* $\langle \Delta, \prec, \sqsubset \cup \{\langle x,y \rangle, \langle y,x \rangle\} \rangle \in \mathsf{CIS}$.

Proof. (1) Otherwise, by Proposition 3, $\text{cispaths}^{11}_{rs}(x,y) \neq \varnothing$. Hence, by Lemma 1(1), $x \prec y$. As a result, we obtained a contradiction.

(2) Otherwise, by Proposition 3, $\text{cispaths}_{rs}(x,y) \neq \varnothing$. Hence, by Lemma 1(2), $x \sqsubset y$. As a result, we obtained a contradiction.

(3) Let $rs' = \langle \Delta, \prec, \sqsubset^{\langle x,y \rangle} \} \rangle$ and $rs'' = \langle \Delta, \prec, \sqsubset^{\langle x,y \rangle} \cup \langle y,x \rangle \} \rangle$.
If $rs' \notin \mathsf{CIS}$, then, by Proposition 3, $\text{cispaths}^{11}_{rs}(y,x) \neq \varnothing$. Hence, by Lemma 1(1), $y \prec x$, and so (by IC:2) $y \sqsubset x$, yielding a contradiction. Hence $rs' \in \mathsf{CIS}$.
If $rs'' \notin \mathsf{CIS}$, then, by Proposition 3, there is $\pi \in \text{cispaths}^{11}_{rs'}(x,y)$. If $\pi \in \text{cispaths}^{11}_{rs}(x,y)$, then, by Lemma 1(1), $x \prec y$ and so (by IC:2) $x \sqsubset y$, yielding a contradiction. If $\pi \notin \text{cispaths}^{11}_{rs}(x,y)$, then $\pi = \dots x2y \dots$. Hence $\text{cispaths}^{11,12}_{rs}(x,x) \neq \varnothing$, and so $rs \notin \mathsf{CIS}$, yielding a contradiction. □

We can now proceed with the proof proper. We first observe that cis2ics is well-defined, since the axioms IC:1–6 hold for $ics = \text{cis2ics}(cis)$, where $cis \in \mathsf{CIS}$, which follows from a straightforward application of Propositions 2 and 5 together with the cis-acyclicty of *cis*. We then observe that Proposition 1 can be applied. Clearly, CIS is intersection-closed and cis2ics is both monotonic and non-decreasing. Suppose now that $ics = \langle \Delta, \prec, \sqsubset \rangle \in \mathsf{ICS}$. Then $\text{cis2ics}(ics) \trianglelefteq ics$, which follows from Lemma 1. Suppose next that $x \neq y \in \Delta$ and consider two cases.

– Case 1: $x \not\prec y$. Then, by Lemma 2(1), $rs' = \langle \Delta, \prec, \sqsubset^{\langle y,x \rangle} \rangle \in \mathsf{CIS}$. Hence, there is $\overline{rs} \in \max_{\mathscr{R}}(rs') \subseteq \max_{\mathscr{R}}(rs)$ such that $x \not\prec_{\overline{rs}} y$.
– Case 2: $x \not\sqsubset y$. Then, by Lemma 2(2), $rs' = \langle \Delta, \prec^{\langle y,x \rangle}, \sqsubset \rangle \in \mathsf{CIS}$. Hence, there is $\overline{rs} \in \max_{\mathscr{R}}(rs') \subseteq \max_{\mathscr{R}}(rs)$ such that $x \not\sqsubset_{\overline{rs}} y$. □

Example 4. Consider again the relational structure *rs* from Example 2. It is easy to see that *rs* is not closed. However, since *rs* is cis-acyclic, one can apply the mapping cis2ics. The result is depicted in Fig. 8. To improve clarity, weak precedence arcs induced by strong precedence arcs according to axiom IC:2, are omitted while pairs of opposite weak dependence arcs are depicted as double-headed arcs.
In contrast to the relational structure rs' discussed in Example 3, there are only eight precedence arcs - all four of them that are not present in *rs* are implied by axiom IC:3 (e.g., $\langle c,a \rangle$) or by the axiom IC:5 (e.g., $\langle g,e \rangle$). The remaining arc $\langle g,c \rangle$ is here weak and obtained using axiom IC:4. Finally, an example of a weak precedence arc implied by axiom IC:6 is $\langle b,f \rangle$ (since $b \sqsubset c \prec e \sqsubset f$). ◇

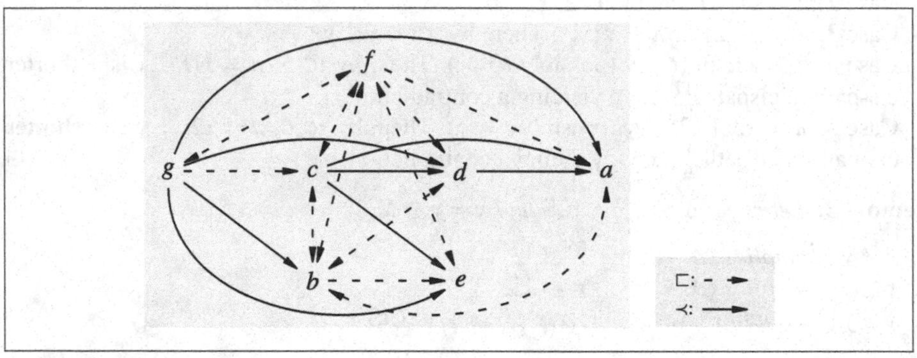

Fig. 8. Relational structure of Example 4.

We have therefore succeeded in providing order-theoretic characterisations of the maximal CI-structures, the closed CI-structures, and the closure for CI-structures. We will now re-state some of the results developed for the generic setup developed in [13].

Theorem 3. *Let cis* $\in \mathsf{CIS}$.

1. $\text{cis2IPS}(cis) = \text{cis2IPS}(\text{cis2ics}(cis)) \neq \varnothing$.
2. $\text{cis2ics}(cis) = \bigcap \text{cis2IPS}(cis)$.
3. $cis \in \mathsf{ICS}$ *iff* $\max_{\mathsf{CIS}}(cis) = \max_{\mathsf{CIS}}(rs)$ *implies* $rs = cis$*, for every* $rs \in \text{ext}_{\mathsf{CIS}}(cis)$.

Proof. Given Theorems 1 and 2, all the parts follow from Prop. 7.1 and 7.5 in [13]. □

Example 5. Figure 9 shows four maximal extensions of *rs* from Example 3, which are also extensions of cis2ics(*rs*) from Example 4.
If we consider events *e* and *d*, one can find two maximal extensions where they have

opposite precedence relationships. Hence, they cannot be precedence-related in *rs* nor in cis2ics(rs). On the other hand, in cis2ics(rs), $f \sqsubset a$. Hence, there is a maximal extension in which f weakly precedes a.

We can treat the maximal extensions of *rs* depicted in Fig. 9 as different schedules of a fragment of a concurrent program, and one can compute their 'widths' and 'lengths'. The lengths of these executions are four or five, while the widths are three and four. The second interval order has the minimal width and maximal length, while the fourth one has the maximal width and minimal length. The most interesting is the third case, where both parameters are minimal. ◇

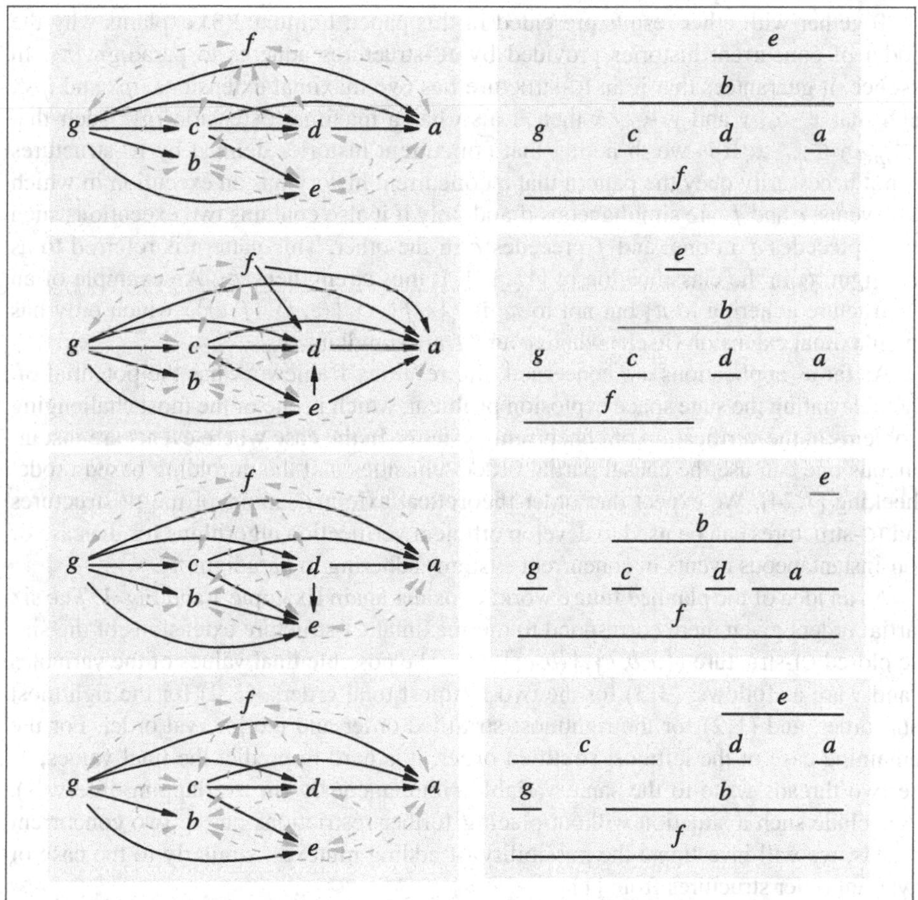

Fig. 9. CI-structures (left) and their interval representations (right) in Example 5.

5 Concluding Remarks

In this paper, we extended the general approach from [12, 13] to deal with the semantics of concurrent systems to execution models where actions are not instantaneously executed. Our aim was to capture concurrent histories based on two invariant relationships between events, namely *precedence* and *weak precedence*, intuitively corresponding to the 'earlier than' and 'not later than' positions in individual executions. To provide a solution, we introduced a novel concept of *cis-acyclicity* and three new classes of relational structures, viz. the *combined interval structures* (CI-structures), *interval poset structures* (IP-structures), and *invariant combined structures* (IC-structures), as well as a structure closure operation for the CI-structures.

Together with other results presented in this paper, Lemma 2(3) explains why the model of concurrent histories provided by IC-structures adheres to paradigm π_3. In essence, it guarantees that if an IC-structure has two maximal extensions, *ips* and *ips'*, such that $x \prec_{ips} y$ and $y \prec_{ips'} x$ then it also has a maximal extension ips'' such that $x \sqsubset_{ips''} y \sqsubset_{ips''} x$. It is worth noting that concurrent histories defined by IC-structures do not necessarily obey the pattern that a concurrent history has an execution in which two events, e and f, are simultaneous if and only if it also contains two executions such that e precedes f in one, and f precedes e in the other. This pattern is referred to as paradigm π_8 in the classification of [13, 14]. It thus strengthens π_3. An example of an IC-structure adhering to π_3 but not to π_8 is $\langle \{e,f\}, \varnothing, \{\langle e,f\rangle, \langle f,e\rangle\}\rangle$ which only has one maximal extension (itself) where e and f are simultaneous.

As far as applications are concerned, the resulting framework has the potential of, e.g., alleviating the state space explosion problem, which is one of the most challenging problems in the verification of concurrent systems. In the case where events are instantaneous one can use the causal partial order semantics and the unfolding based model checking [7, 24]. We expect that order-theoretical axiomatisations of the IP-structures and IC-structures can be used to develop efficient verification algorithms for the case of non-instantaneous events in concurrent systems adhering to paradigm π_3.

As an idea of the planned future work, consider again Example 1 and Fig. 1. The six partial orders given there correspond to the maximal CI-structure extensions of the single closed CI-structure $\langle \{a,b,c\}, \{\langle b,c\rangle\}, \varnothing\rangle$. The possible final values of the variables x and y are as follows: $(3,3)$ for the two leftmost total orders, $(3,2)$ for the rightmost total order, and $(1,2)$ for the rightmost stratified order and the interval order. For the remaining case of the leftmost stratified order, it is hard to predict the final values, as the two threads write to the same variable simultaneously (in overlapping intervals). To exclude such a situation without placing further restrictions on the two concurrent threads, we will investigate the possibility of adding mutexes, similarly to the case of invariant order structures from [11].

Another idea of the planned work is related to the theory of traces [6]. Traces that can be interpreted as sets of interval orders and represent concurrent histories — called *interval traces* — were proposed and discussed in [16]. Though [16] analysed their relationship with interval relational structures from [14, 15], the relationship with the structures proposed in this paper is yet to be developed. Such a development will necessarily require a suitable notion of *dependence* CI-structures corresponding to the dependence graphs in the theory of traces [10]. The existing treatment of paradigm π_3 in [13, 14] does not provide means of defining such structures. However, a class of CI-structures with elements labelled by action names can provide the required device, providing a further justification of the framework proposed in this paper.

Acknowledgement. Partial support by the Discovery NSERC of Canada grant No. 6466-15, and the Leverhulme Trust grant RPG-2022-025 is acknowledged. The authors are grateful to the three anonymous referees, whose comments significantly contributed to the final version of this paper.

Disclosure of Interests. The authors have no competing interests to declare that are relevant to the content of this article.

References

1. van der Aalst, W.M.P.: Process Mining - Discovery, Conformance and Enhancement of Business Processes. Springer, Heidelberg (2011)
2. Abraham, U., Ben-David, S., Magidor, M.: On global-time and inter-process communication. In: Kwiatkowska, M., Shields, M.W., Thomas, R.M. (eds.) Semantics for Concurrency, Workshops in Computing, pp. 311–323. Springer, London (1990). https://doi.org/10.1007/978-1-4471-3860-0_19
3. Alglave, J., Maranget, L., Tautschnig, M.: Herding cats: modelling, simulation, testing, and data mining for weak memory. ACM Trans. Program. Lang. Syst. **36**(2), 1–74 (2014)
4. Allen, J.F.: Maintaining knowledge about temporal intervals. Commun. ACM **26**(11), 832–843 (1983)
5. Best, E., Devillers, R., Koutny, M.: Petri Net Algebra. EATCS Monographs on Theoretical Computer Science, Springer, Heidelberg (2001). https://doi.org/10.1007/978-3-662-04457-5
6. Diekert, V., Rozenberg, G. (eds.): The Book of Traces. World Scientific (1995)
7. Esparza, J., Heljanko, K.: Unfoldings - A Partial-Order Approach to Model Checking. Monographs in Theoretical Computer Science. An EATCS Series, Springer, Heidelberg (2008)
8. Fishburn, P.C.: Intransitive indifference with unequal indifference intervals. J. Math. Psychol. **7**, 144–149 (1970)
9. Glabbeek, R.J.V., Goltz, U., Schicke-Uffmann, J.W.: On characterising distributability. Logical Methods Comput. Sci. **9**(3), 1–58 (2013)
10. Hoogeboom, H.J., Rozenberg, G.: Dependence graphs. In: Diekert, V., Rozenberg, G. (eds.) The Book of Traces, pp. 43–67. World Scientific (1995)
11. Janicki, R., Kleijn, J., Koutny, M., Mikulski, Ł: Invariant structures and dependence relations. Fund. Inform. **155**(1–2), 1–29 (2017)
12. Janicki, R., Kleijn, J., Koutny, M., Mikulski, Ł: Relational structures for concurrent behaviours. Theort. Comput. Sci. **862**, 174–192 (2021). [Open Access]
13. Janicki, R., Kleijn, J., Koutny, M., Mikulski, Ł: Paradigms of Concurrency - Observations, Behaviours, and Systems - a Petri Net View, Studies in Computational Intelligence, vol. 1020. Springer, Heidelberg (2022)
14. Janicki, R., Koutny, M.: Structure of concurrency. Theoret. Comput. Sci. **112**(1), 5–52 (1993)
15. Janicki, R., Koutny, M.: Fundamentals of modelling concurrency using discrete relational structures. Acta Informatica **34**(5), 367–388 (1997)
16. Janicki, R., Yin, X.: Modeling concurrency with interval traces. Inf. Comput. **253**, 78–108 (2017)
17. Kleijn, J., Koutny, M.: Synchrony and asynchrony in membrane systems. In: Hoogeboom, H.J., Păun, G., Rozenberg, G., Salomaa, A. (eds.) WMC 2006. LNCS, vol. 4361, pp. 66–85. Springer, Heidelberg (2006). https://doi.org/10.1007/11963516_5

18. Laarman, A.: Stubborn transaction reduction. In: Dutle, A., Muñoz, C., Narkawicz, A. (eds.) NFM 2018. LNCS, vol. 10811, pp. 280–298. Springer, Cham (2018). https://doi.org/10.1007/978-3-319-77935-5_20
19. Lamport, L.: Time, clocks, and the ordering of events in a distributed system. Commun. ACM **21**(7), 558–565 (1978)
20. Lamport, L.: The mutual exclusion problem: part I - a theory of interprocess communication. J. ACM **33**(2), 313–326 (1986)
21. Lamport, L.: On interprocess communication: part i: basic formalism. Distrib. Comput. **1**, 77–85 (1986)
22. Mazurkiewicz, A.: Concurrent program schemes and their interpretations. DAIMI Rep. PB 78, Aarhus University (1977)
23. Mazurkiewicz, A.: Trace theory. In: Brauer, W., Reisig, W., Rozenberg, G. (eds.) ACPN 1986. LNCS, vol. 255, pp. 278–324. Springer, Heidelberg (1987). https://doi.org/10.1007/3-540-17906-2_30
24. McMillan, K.L.: Using unfoldings to avoid the state explosion problem in the verification of asynchronous circuits. In: von Bochmann, G., Probst, D.K. (eds.) CAV 1992. LNCS, vol. 663, pp. 164–177. Springer, Heidelberg (1993). https://doi.org/10.1007/3-540-56496-9_14
25. Nagy, B., Akkeleş, A.: Trajectories and traces on non-traditional regular tessellations of the plane. In: Brimkov, V.E., Barneva, R.P. (eds.) IWCIA 2017. LNCS, vol. 10256, pp. 16–29. Springer, Cham (2017). https://doi.org/10.1007/978-3-319-59108-7_2
26. Paulevé, L.: Goal-oriented reduction of automata networks. In: Bartocci, E., Lio, P., Paoletti, N. (eds.) CMSB 2016. LNCS, vol. 9859, pp. 252–272. Springer, Cham (2016). https://doi.org/10.1007/978-3-319-45177-0_16
27. Pelz, E.: Full axiomatisation of timed processes of interval-timed Petri nets. Fundam. Informaticae **157**(4), 427–442 (2018)
28. Pietkiewicz-Koutny, M., Koutny, M.: Synthesising elementary net systems with interval order semantics. In: Gomes, L., Leitão, P., Lorenz, R., van der Werf, J.M.E.M., van Zelst, S.J. (eds.) Joint Proceedings of the Workshop on Algorithms & Theories for the Analysis of Event Data and the International Workshop on Petri Nets for Twin Transition co-located with the 44th International Conference on Application and Theory of Petri Nets and Concurrency (Petri Nets 2023), Caparica, Portugal, June 25-30, 2023. CEUR Workshop Proceedings, vol. 3424. CEUR-WS.org (2023)
29. Popova-Zeugmann, L., Pelz, E.: Algebraical characterisation of interval-timed Petri nets with discrete delays. Fundam. Informaticae **120**(3–4), 341–357 (2012)
30. Pratt, V.R.: Modeling concurrency with partial orders. Int. J. Parallel Prog. **15**(1), 33–71 (1986)
31. Sokolov, D., Khomenko, V., Mokhov, A., Dubikhin, V., Lloyd, D., Yakovlev, A.: Automating the design of asynchronous logic control for AMS electronics. IEEE Trans. Comput. Aided Des. Integr. Circuits Syst. **39**(5), 952–965 (2020)
32. Vogler, W.: Partial order semantics and read arcs. Theoret. Comput. Sci. **286**(1), 33–63 (2002)
33. Wiener, N.: A contribution to the theory of relative position. In: Proceedings of the Cambridge Philosophical Society, vol. 33, no. 2, pp. 313–326 (1914)
34. Winskel, G.: An introduction to event structures. In: de Bakker, J.W., de Roever, W.-P., Rozenberg, G. (eds.) REX 1988. LNCS, vol. 354, pp. 364–397. Springer, Heidelberg (1989). https://doi.org/10.1007/BFb0013026

Token Trail Semantics II - Petri Nets And Their Net Language

Jakub Kovář[1(✉)] and Robin Bergenthum[2]

[1] Lehrgebiet Programmiersysteme, FernUniversität in Hagen, Hagen, Germany
jakub.kovar@fernuni-hagen.de

[2] Fakultät für Mathematik und Informatik, FernUniversität in Hagen, Hagen, Germany

Abstract. There are various semantics for Petri nets. Some semantics can express concurrency well, others are good at modelling conflicts. Yet, every semantics has its drawbacks. State graphs explode in size when there is concurrency. Sequential and partial languages explode in size if there is conflict. In our previous paper on token trail semantics, we introduced the concept of the net language of a marked Petri net. The net language is a set of labelled nets so that we can specify both conflict and concurrency very naturally. We proved that the token trail semantics faithfully covers state graphs, sequential languages, and partial languages. In this paper, we show token trail semantics covers synchronous net morphisms and prove the net language of a Petri net includes all its finite unfoldings. Furthermore, we show that a Petri net simulates the state-transition behaviour of all labelled nets of its net language and prove the step language of a Petri net is the union of the step languages of all labelled nets of its net language. Finally, we present an algorithm and an implementation deciding the net language inclusion problem.

Keywords: Token trail semantics · Net language · Petri net · Labelled net · Net morphism · Unfolding · Simulation

1 Introduction

Petri nets (see for example [6,15,30,34]) have a formal semantics, an intuitive graphical representation, and can express both conflict and concurrency between the occurrences of actions. Petri nets model actions with transitions, local states with places, and the dependencies between actions and local states with weighted directed arcs. Petri nets model a state by marking each place with tokens. Firing a transition models the execution of an action and changes the state of the Petri net. A transition can fire only if its preconditions are marked with tokens. If it fires, it consumes the tokens from its preconditions and it marks its postconditions with new tokens. This firing rule is very intuitive and easy to formalize. Based on this firing rule we find different semantics of Petri nets in the literature. Every semantics has its own advantages and disadvantages in different applications.

L. M. Kristensen and J. M. van der Werf (Eds.): PETRI NETS 2024, LNCS 14628, pp. 175–196, 2024.
https://doi.org/10.1007/978-3-031-61433-0_9

Repeatedly processing the firing rule creates *firing sequences*. The set of all firing sequences is called the language of a Petri net. This language is very easy to work with, but it cannot express concurrency of actions. Moreover, a firing sequence cannot directly specify conflict. The *reachability graph* of a Petri net is its set of reachable states with all possible transitions. In contrast to firing sequences, the state graph can very conveniently express conflict, merging, and looping of sequences of actions. But again, a state graph cannot express concurrency. If there is concurrency, there is the so-called state space explosion where the number of global states grows exponentially in the number of local states. To express concurrency, we can extend firing sequences and state graphs to *steps*, i.e. multisets of transitions firing concurrently. Using steps we can specify concurrent sets of actions, but still, the number of global states explodes just like in a regular state graph. Furthermore, sequences of steps are rather technical, thus it is neither easy nor intuitive to specify behaviour using combinations of sequences of steps. *Partially ordered runs* and *process nets* [10,22,33,39] can directly express concurrency. In both semantics a run is still a firing sequence, but the sequence is a partial, not a total order. These runs can easily model concurrent behaviour, but just like firing sequences, they cannot directly specify conflict. The partial language of a Petri net contains a run for every combination of conflicting options. *Branching processes* [39] extend runs by an additional conflict relation. This semantics can specify concurrency, but it can also merge identical prefixes of runs. Unlike state graphs, once split, local states cannot merge. Therefore, these structures fan-out and it is hard to keep track of the relations between the different conflict-free sets of partially ordered nodes. Moreover, since branching processes cannot merge states, they are not able to directly specify looping behaviour.

Neither of the above-mentioned semantics provide a nice graphical representation of the behaviour of a Petri net if it contains both conflict and concurrency. Thus, in our previous work [7] we followed Wolfgang Reisigs claim *"the semantics of a net is a net"* [35] and defined the net language of a marked Petri net as a set of marked labelled nets. We showed that this semantics is a true meta-semantics because it covers the previously mentioned semantics. In [7] we proved that if a labelled net is a sequence, a state graph or a partial order, this labelled net is in the net language of a marked Petri net if and only if there is a related matching firing sequence, reachability graph or enabled partially ordered run.

However, the definition of the net language is not restricted to labelled nets modelling firing sequences, state graphs or runs. If there is a valid token trail, any labelled net can be part of a net language. In this paper, we build upon the results of our previous work, and further investigate the formal relation between a Petri net and its net language. We show that token trail semantics covers synchronous net morphisms. Thus, the net language contains all finite unfoldings of a marked Petri net. Furthermore, we show that a Petri net simulates the state-transition behaviour of all labelled nets of its net language. This proves, that the labelled step language of any labelled net of the net language is in the step language of the Petri net. Additionally, we prove that the step language of a Petri net is the union of all labelled step languages of its net language. Moreover, we show that net language inclusion is distinct from step language inclusion. Finally, we introduce an algorithm and its implementation which decides the net language inclusion problem.

2 Preliminaries

We denote the non-negative integers as $\mathbb{N}$. Let f be a function and B be a subset of the domain of f. We write $f|_B$ to denote the restriction of f to B. Let b be an element of B. We write $f|_b$ to denote $f|_{\{b\}}$. Let A be a set, we denote the cardinality of A by $|A|$. We call $m : A \to \mathbb{N}$ a multiset and write $m = \sum_{a\in A} m(a) \cdot a$ to denote multiplicities of elements in m. Let $m' : A \to \mathbb{N}$ be another multiset. We write $m \leq m'$ iff $\forall a \in A : m(a) \leq m'(a)$ holds. We write $\mathbb{N}^A$ to denote the set of all multisets of A. Let B be a set, we call $\alpha : A \times B \to \mathbb{N}$ a multirelation from A to B. Let $a \in A$, $b \in B$, we denote $\alpha(a, b)$ as α_{ab}. The application of α to $m : A \to \mathbb{N}$ is the multiset $\alpha(m) : B \to \mathbb{N}$ so that $\forall b \in B : \alpha(m)|_b = \sum_{a\in A} \alpha_{ab} \cdot m(a)$ holds. We can depict a multirelation α as a $|A| \times |B|$ matrix of non-negative integers. Thus, the application of α to m is the multiplication of m, depicted as a row vector of size $|A|$, and the α-matrix. We model distributed systems by Petri nets with arc weights [6,15,16,30,34]. These nets are also called Place/Transition nets in the literature.

Definition 1 (Petri Net). *A Petri net is a tuple (P, T, W) where P is a finite set of places, T is a finite set of transitions so that $P \cap T = \emptyset$ holds, and $W : (P \times T) \cup (T \times P) \to \mathbb{N}$ is a multiset of directed arcs. A marking of (P, T, W) is a multiset $m : P \to \mathbb{N}$. Let m_0 be a marking. $N = (P, T, W, m_0)$ is a marked Petri net and we call m_0 the initial marking of N.*

Petri nets have a simple firing rule. Let $N = (P, T, W, m)$ be a marked Petri net and $t \in T$ be a transition. We denote $\circ t = \sum_{p\in P} W(p, t) \cdot p$ the weighted pre-set of t. We denote $t\circ = \sum_{p\in P} W(t, p) \cdot p$ the weighted post-set of t. We call a multiset of transitions $u : T \to \mathbb{N}$ a step. We extend the notation of weighted pre- and post-sets of transitions to their multisets as $\circ u = \sum_{t\in u} u(t) \cdot \circ t$ and $u\circ = \sum_{t\in u} u(t) \cdot t\circ$. A step u is enabled in marking m if $m \geq \circ u$ holds. All transitions of an enabled step can fire concurrently. Firing a step changes the marking m to $m' = m - \circ u + u\circ$. If u is enabled in m and firing u changes m to m', we write $m \xrightarrow{u} m'$. We call a sequence of steps $u_1 \dots u_n$ enabled in marking m if there is a sequence of markings $m_1 \dots m_n$ so that $m \xrightarrow{u_1} m_1 \xrightarrow{u_2} \dots \xrightarrow{u_n} m_n$ holds. The set of all enabled step sequences in the initial marking of N is the step language $\mathcal{L}(N)$ of N. This step semantics is equivalent to various partial order semantics [9,23,26,36].

Figure 1 depicts a marked Petri net. Transitions are rectangles, places are circles, the multiset of arcs is depicted as weighted arcs, and the initial marking is depicted by black dots called tokens. In the initial marking, only transition A is enabled. When it fires, it consumes the token in place p_1 and produces one token in p_2, two tokens in p_3 and one token in p_4. Figure 2 depicts this new

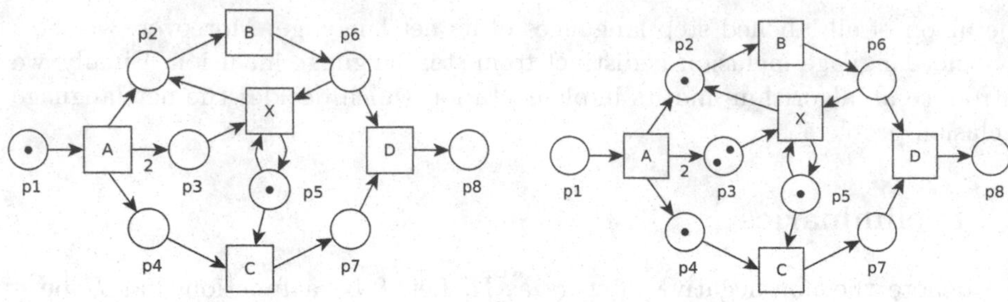

Fig. 1. A marked Petri net N.

Fig. 2. N after firing A.

marking. In the new marking the step $B + C$ is enabled. Firing $B + C$ consumes one token from p_2, p_4, and p_5 and produces one token in p_6 and in p_7.

In this paper, we model the behaviour of a distributed system by labelled nets. This is kind of similar to applications using occurrence nets [19,31] or labelled workflow nets [1,2] to model behaviour. However, we use general labelled nets.

Definition 2 (Labelled Net). *A labelled net is a tuple (C, E, F, A, λ) where (C, E, F) is a Petri net, the set of places C is sometimes called conditions, the set of transitions E is sometimes called events, A is a finite set of actions, and $\lambda : E \longrightarrow A$ is a total labeling function. Let m_0 be a marking of (C, E, F), we call $(C, E, F, A, \lambda, m_0)$ a marked labelled net.*

Obviously, we can turn every Petri net (P, T, W) into a labelled net by defining the set of actions as T and using the identity as the labelling function. Whenever we say a Petri net is a labelled net, we use this interpretation.

Let $L = (C, E, F, A, \lambda, m_0)$ be a marked labelled net. We call a sequence of multisets of actions $v_1 \dots v_n$ enabled in marking m_0 iff there is a sequence of steps $u_1 \dots u_n$ enabled in (C, E, F, m_0) with $\forall i : v_i = \lambda(u_i)$. Remark, here we extend the labelling function to steps. The set of all sequences of multisets of actions enabled in (C, E, F, m_0) is the (labelled) step language $\mathcal{L}(L)$ of L.

To define the net language of a Petri net, we introduced the rise and the level of a transition in [7].

Definition 3 (Rise of a Transition). *Let $N = (P, T, W, m)$ be a marked Petri net and $t \in T$ be a transition. The level of t in marking m, denoted as $t^{\blacktriangle}(m)$, is the weighted sum of tokens $t^{\blacktriangle}(m) := \sum_{(p,t) \in W} W(p, t) \cdot m(p)$ in the pre-set of t. Similarly, we define $t^{\blacktriangle}(m) := \sum_{(t,p) \in W} W(t, p) \cdot m(p)$ the sum of tokens in the post-set of t in m. Using these notions, we define the rise $t^{\triangle}(m)$ of a transition t in marking m as $t^{\triangle}(m) := t^{\blacktriangle}(m) - t^{\blacktriangle}(m)$.*

As an example, we compute the rise of all transitions of Fig. 2 using the depicted marking m. $X^{\blacktriangle}(m) = 3$ because there are 3 tokens in the places p_3, p_5, and p_6. $X^{\blacktriangle}(m) = 2$ because there are 2 tokens in p_2 and p_5. Thus, $X^{\triangle}(m) = -1$. The level $B^{\blacktriangle}(m) = 1$ and $B^{\blacktriangle}(m) = 0$. Therefore, the rise $B^{\triangle}(m) = -1$. The rise $C^{\Delta}(m) = -2$. The rise of transition D is 0. Due to the arc weights, the rise of transition A is $A^{\Delta}(m) = (1 + 2 \cdot 2 + 1) - 0 = 6$.

We use level and rise to define a token trail marking of a labelled net and the net language of a marked Petri net [7].

Definition 4 (Token Trail, Net Language). *Let $N = (P, T, W, m_0)$ be a marked Petri net, $L = (C, E, F, T, \lambda, m_i)$ be a marked labelled net, and $p \in P$ be a place of N. A marking m of L is a token trail for p iff the following three conditions hold.*

(I) $\forall e \in E : e^{\blacktriangle}(m) \geq W(p, \lambda(e))$,
(II) $\forall e \in E : e^{\Delta}(m) = W(\lambda(e), p) - W(p, \lambda(e))$, *and*
(III) $\sum_{c \in C} m_i(c) \cdot m(c) = m_0(p)$.

We call L enabled in N iff for every $p \in P$ there is a token trail m_p in L. We call the set of all enabled labelled nets the net language $\mathcal{N}(N)$ of N.

Figure 3 depicts the marked Petri net of Fig. 1 without the places p_3 and p_5. In the initial marking, we can only fire transition A. After firing A, we can fire transitions C concurrently to the loop of B and X. At the end, transition D synchronises the two concurrent parts.

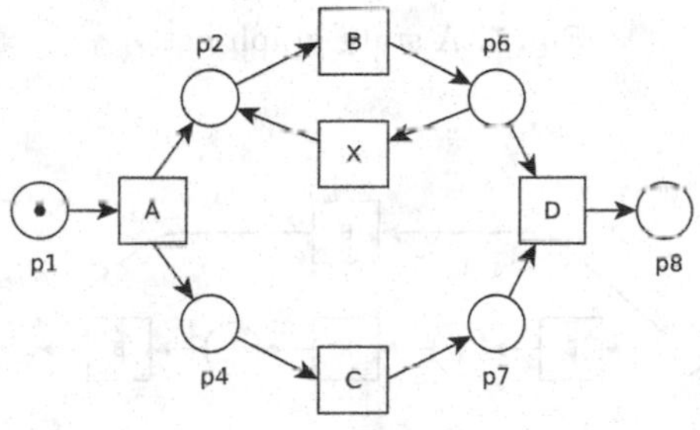

Fig. 3. The Petri net from Fig. 1 without places p_3 and p_5.

Figures 4, 5, 6, and 7 depict four different marked labelled nets of the net language of the marked Petri net from Fig. 3. Figure 4 depicts an acyclic, conflict-free labelled net modelling a partial order. In [7] we proved, that a labelled net modelling a partial order of the partial language of the marked Petri net is in its net language, if and only if, the modelled partial order is in the partial language of the marked Petri net. Figure 5 depicts a concurrency-free labelled net modelling a state graph. In [7] we proved, that a labelled net modelling a state graph is in the net language of a marked Petri net, if and only if, the modelled state graph is a sub-graph of the reachability graph of the Petri net. Petri nets depicted in Fig. 6 and F. 7 model neither a partial order, nor a state graph. Figure 6 depicts a labelled net with both conflict and concurrency. The Petri net of Fig. 7 even contains a loop.

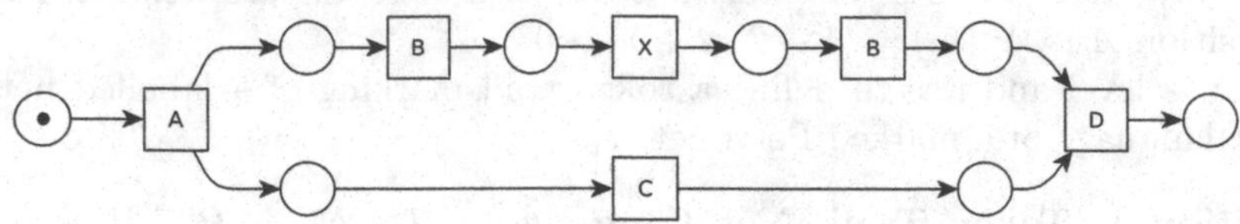

Fig. 4. A partial order net.

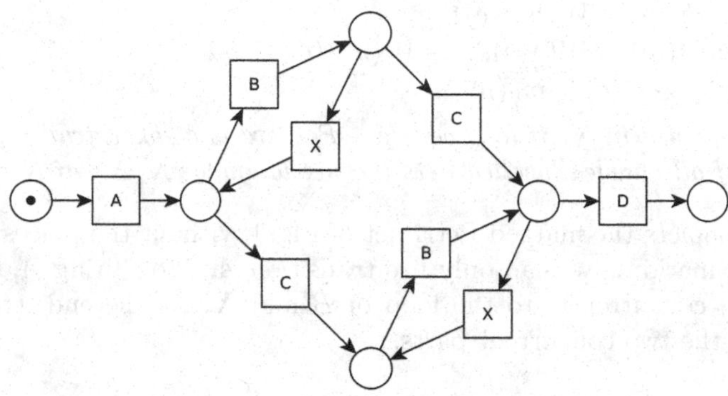

Fig. 5. A state graph net.

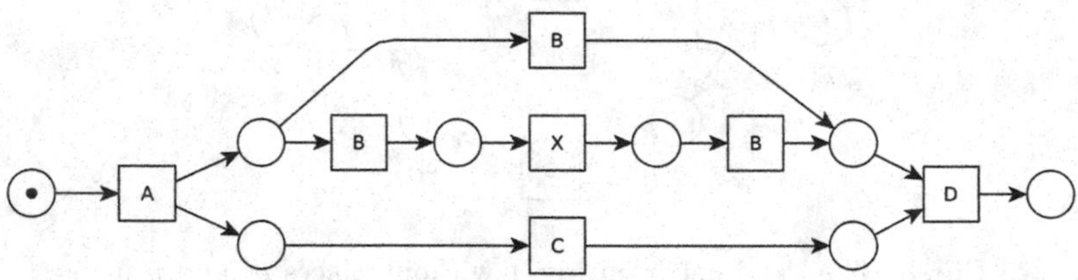

Fig. 6. A labelled net containing both conflict and concurrency.

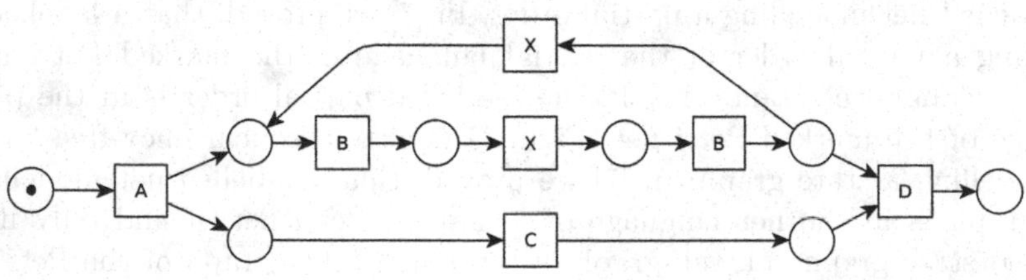

Fig. 7. A labelled net containing a loop.

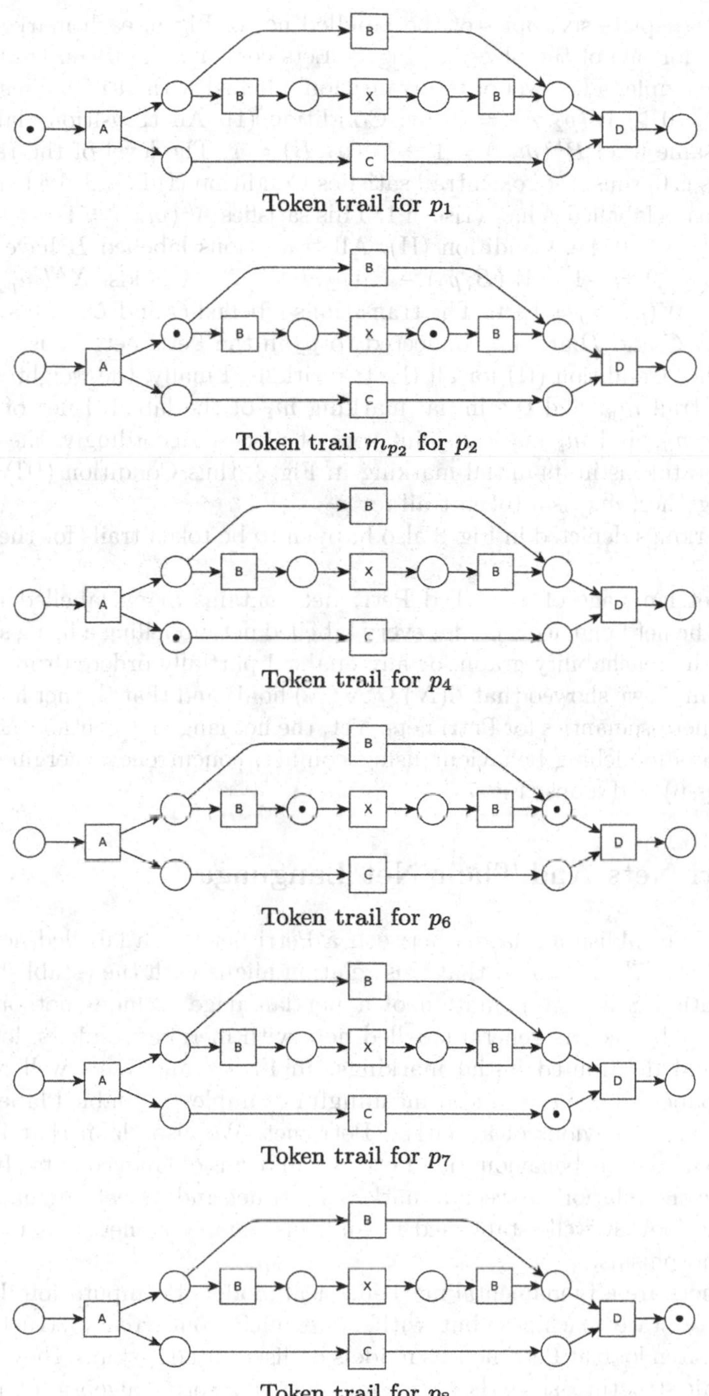

Fig. 8. Token trails of Fig. 6 for all places of the Petri net from Fig. 3.

Figure 8 depicts six copies of the labelled net of Fig. 6, each marked with a token trail for one of the places of Fig. 3. Lets consider the token trail m_{p_2} for p_2 as an example. The level of the transition labelled with A is 0. This satisfies $A^{\blacktriangle}(m_{p_2}) = 0 \geq W(p_2, A) = 0$, i.e. Condition (I). All transitions labelled B have the same level $B^{\blacktriangle}(m_{p_2}) = 1 \geq W(p_2, B) = 1$. The level of the remaining transitions is 0, thus, the token trail satisfies Condition (I) for all the transitions. The transition labelled A has a rise of 1. This satisfies $A^{\vartriangle}(m_{p_2}) = 1 = W(A, p_2) - W(p_2, A) = 1 - 0$, i.e. Condition (II). All transitions labelled B have rise -1. Thus, $B^{\vartriangle}(m_{p_2}) = -1 = W(B, p_2) - W(p_2, B) = 0 - 1$ holds. $X^{\vartriangle}(m_{p_2}) = 1 = W(X, p_2) - W(p_2, X) = 1 - 0$. The transitions labelled C and D have a rise of 0. Transitions C and D are not connected to p_2 in the Petri net. Thus, the token trail satisfies Condition (II) for all the transitions. Finally, the weighted sum of the token trail m_{p_2} and the initial marking m_i of the labelled net of Fig. 6 is 0 because m_{p_2} and m_i mark disjoint sets of places. Accordingly, the place p_2 contains no tokens in its initial marking in Fig. 3, thus Condition (III) holds as well. Altogether, m_{p_2} is a token trail for p_2.

All markings depicted in Fig. 8 also happen to be token trails for the labelled net of Fig 7.

The net language of a marked Petri net contains many labelled nets. For example, the net language includes every labelled net modelling a firing sequence, a part of the reachability graph, or any enabled partially ordered run. Roughly speaking, in [7] we showed that $\mathcal{L}(N) \subseteq \mathcal{N}(N)$ holds and that the net language is kind of a meta-semantics for Petri nets. Yet, the net language contains additional labelled nets modelling behaviour using conflict, concurrency, merging of local states (Fig. 6) and loops (Fig. 7).

3 Petri Nets And Their Net Language

Token trails establish a relation between a Petri net and a labelled net. In our previous work [7], we proved that this relation aligns with the established Petri net semantics. Still, the definition of a net language includes not only these special cases but covers general labelled nets with merging conflicts, loops, arc-weights, and distributed initial markings. In Figs. 6 and 7, as well as in our previous paper [7], we introduced meaningful examples of enabled labelled nets modelling the behaviour of a marked Petri net. We also claim that it is very natural to model the behaviour of a Petri net in terms of labelled nets. To further investigate the relation between a marked Petri net and its net language, we, in this paper, look at well-established formalisms namely at net homomorphisms and net morphisms.

Petri nets are a fundamental mathematical model of computation, like finite and infinite state machines, but with an explicit concurrency structure [38]. While we often look at Petri nets as models of distributed systems, they also have an algebraic structure, since they are equivalent to 2-sorted algebras on multisets [38]. In [38], Glynn Winskel derives various combinators from the mathematical structure of Petri nets, to compose their behaviour. At the time, other formalisms were used to compose concurrent systems [11,21,28,38]. Due to his belief about the fundamental nature of Petri nets, the aim of Winskel was to derive these operations not from some other calculus but from the mathematical structure of Petri nets themselves. To achieve this, he replaced an older definition of net morphisms [12] with a new one. We first discuss net morphisms informally and then give the formal definition.

A net morphism is a special case of a net homomorphism. A net homomorphism is a pair of multirelations (η, β) between two Petri nets. The application of the multirelation β to markings (multisets of places) of one Petri net maps them to markings of the other Petri net, thus it effectively maps states of one Petri net to states of the other Petri net. Similarly, the application of the multirelation η to steps (multisets of transitions) in one Petri net maps them to steps of transitions in the other Petri net. A pair of multirelations (η, β) is a net homomorphism only if the mappings satisfy two conditions, (i) β maps the initial marking of one Petri net to the initial marking of the other Petri net and (ii) both η and β preserve pre- and post-sets of elements. A net morphism is simply a homomorphism, where the multirelation η is a total or a partial function, i.e. it maps each transition of one Petri net to one or no transition of the other Petri net. Unlike in the Definition 1 of this paper, Winskel allows for Petri nets where the sets of places and transitions are infinite [38].

Definition 5 (Net Homomorphism, Net Morphism).
Let $N = (P, T, W, m_0)$ and $N' = (P', T', W', m'_0)$ be marked Petri nets. A net homomorphism from N' to N is a pair of multirelations (η, β) with $\eta : T' \times T \to \mathbb{N}$ and $\beta : P' \times P \to \mathbb{N}$ which satisfies (i) $\beta(m'_0) = m_0$ and (ii) $\forall u \in \mathbb{N}^{T'} : {\circ}\eta(u) = \beta({\circ}u) \wedge \eta(u){\circ} = \beta(u{\circ})$. A net homomorphism is called finitary if for every transition $t' \in T'$, the multiset $\eta|_{t'}$ is finite.

A net morphism from N' to N is a net homomorphism (η, β) where η is a partial function. A net morphism is called synchronous if η is a total function.

Petri nets with homomorphisms form a category [38], where Petri nets with morphisms or synchronous morphisms form subcategories [38]. Winskel defines compositions as operations in these categories and shows that finitary homomorphisms and morphisms preserve the initial marking. Additionally, if $m \xrightarrow{u} m'$ in N', then $\beta(m) \xrightarrow{\eta(u)} \beta(m')$ in N, thus, finitary homomorphisms preserve reachable markings [38].

In the new context of net morphisms, we reexamine our Definition 4 and identify two multirelations. The labelling function λ maps steps in the labelled net to steps in the Petri net. The second multirelation is determined by the token trail markings and its structure hides in Condition (III). We call this multirelation S and if m_p is a token trail marking of the labelled net, with a set of places C, for a place $p \in P$, then the values of the multirelation S are $\forall c \in C, \forall p \in P : S_{cp} = m_p(c)$. As a first result of this paper, we prove by counterexample that while the pair of these two multirelations (λ, S) relates a marked labelled net to a marked Petri net, they form neither a net morphism nor a net homomorphism.

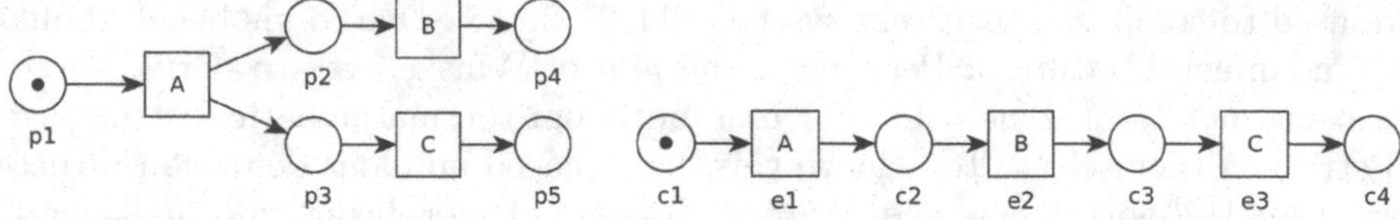

Fig. 9. A marked Petri net containing a parallel split.

Fig. 10. A labelled net modelling the sequence of actions $A\ B\ C$.

Figure 9 depicts a marked Petri net. After firing A, the remaining two transitions B and C can fire concurrently. Figure 10 depicts a labelled net which models the sequential firing of actions A, B and C. Because the firing sequence $A\ B\ C$ is obviously in the sequential language of the Petri net from Fig. 9, the labelled net is in the net language of the Petri net. With an arbitrary but fixed ordering of the nodes of both nets, we can write the multirelations S and λ as matrices. Figure 11 depicts these matrices. The matrix depicted on the left side represents the multirelation S. The values $S_{c_i p_j}$ in the matrix (multirelation) are determined by the values of the token trail marking $m_{p_j}(c_i)$ for place p_j of the Petri net of Fig. 9 and the place c_i of the labelled net of Fig. 10. Thus, the j-th column of matrix S (a marking of the labelled net) is a token trail for place p_j of the Petri net. The second matrix represents the labelling function λ. Each row relates to a transition of the labelled net and each column to a transition of the Petri net. In this simple example, this function is the identity.

$$S = \begin{array}{c} \\ c_1 \\ c_2 \\ c_3 \\ c_4 \end{array} \begin{array}{c} p_1\ p_2\ p_3\ p_4\ p_5 \\ \begin{pmatrix} 1 & 0 & 0 & 0 & 0 \\ 0 & 1 & 1 & 0 & 0 \\ 0 & 0 & 1 & 1 & 0 \\ 0 & 0 & 0 & 1 & 1 \end{pmatrix} \end{array} \qquad \lambda = \begin{array}{c} \\ e_1 \\ e_2 \\ e_3 \end{array} \begin{array}{c} A\ B\ C \\ \begin{pmatrix} 1 & 0 & 0 \\ 0 & 1 & 0 \\ 0 & 0 & 1 \end{pmatrix} \end{array}$$

Fig. 11. Matrix representation of the multirelations S and λ.

λ is a total function, so if these mutlirelations form a homomorphism, it would be a synchronous morphism as well. It is easy to see that the multirelation S satisfies condition (i) of Definition 5 because when we apply it to the initial marking c_1 of the labelled net it maps it to the initial marking $S(c_1) = p_1$ of the Petri net. Now, we examine condition (ii). It requires that for all steps u of the labelled net, $\circ\lambda(u) = S(\circ u)$ and $\lambda(u)\circ = S(u\circ)$ holds. We take $u = e_2$. First, we evaluate $S(\circ u)$ and $S(u\circ)$. For the pre-set, we get $\circ e_2 = c_2$. We apply the multirelation S to c_2 and get the multiset of places $S(c_2) = p_2 + p_3$ of the Petri net. This is intuitive because the place c_2 of the labelled net of Fig. 10 models the state of the Petri net of Fig. 9 after transition A fires. For the post-set, we get $e_2\circ = c_3$. Again, we apply the multirelation S to c_3 and get the multiset of places $S(c_3) = p_3 + p_4$ of the Petri net. This matches our expectation because the condition c_3 of Fig. 10 models the state of the Petri net of Fig. 9 after both A and B fired. Now, we check if the multirelation λ preserves the pre- and post-set. The application of the multirelation λ maps the step e_2 in the labelled net of Fig. 10 to the step $\lambda(e_2) = B$ in the Petri net of Fig. 9. The weighted pre-set of this step is $\circ B = p_2$, the weighted post-set of this step is $B\circ = p_4$. Clearly, condition (ii) of Definition 5 does not hold, since $S(\circ u) = p_2 + p_3 \neq p_2 = \circ\lambda(u)$ and $S(u\circ) = p_3 + p_4 \neq p_4 = \lambda(u)\circ$. Therefore, the pair of multirelations is neither a net homomorphism nor a net morphism. This example shows the fundamental difference between net morphisms and token trails. In a net morphism pre- and post-sets must always match. In a token trail, a single place can model a distributed state.

In the following theorem, we prove the converse direction. Namely, every synchronous net morphism defines a token trail.

Theorem 1. *Let $N = (P, T, W, m_0)$ and $N' = (C, E, F, m_i)$ be marked Petri nets. If (λ, S) is a synchronous morphism from N' to N, the labelled net $L = (C, E, F, T, \lambda, m_i)$ is in the net language of N and $\forall p \in P$ the marking m_p of L with $\forall c \in C : m_p(c) = S_{cp}$ is a token trail for p.*

Proof. First we show, for all $p \in P$, m_p is a token trail for p.

(λ, S) is a homomorphism, thus $\forall u \in \mathbb{N}^E : S(\circ u) = \circ\lambda(u)$ holds. Obviously, $\forall e \in E : S(\circ e) = \circ\lambda(e)$ holds as well. Fix arbitrary $p \in P$, $e \in E$, we get

$$\begin{aligned} e^{\blacktriangle}(m_p) &= \textstyle\sum_{(c,e)\in F} F(c,e) \cdot m_p(c) && \text{by Definition 3} \\ &= \textstyle\sum_{c\in\circ e} F(c,e) \cdot m_p(c) && \text{by application of } S \text{ to } \circ e \\ &= S(\circ e)|_p && \text{by homomorphism (ii)} \\ &= \circ\lambda(e)|_p \\ &= W(p, \lambda(e)). \end{aligned}$$

Thus, for all p, m_p satisfies $\forall e \in E : e^{\blacktriangle}(m_p) \geq W(p, \lambda(e))$, i.e. Condition (I) of Definition 4.

Again, (λ, S) is a homomorphism, thus $\forall u \in \mathbb{N}^E : S(\circ u) = \circ\lambda(u) \wedge S(u\circ) = \lambda(u)\circ$ holds. Obviously, $\forall e \in E : S(\circ e) = \circ\lambda(e) \wedge S(e\circ) = \lambda(e)\circ$ holds as well. Fix arbitrary $p \in P$, $e \in E$, we get

$$\begin{aligned} e^{\triangle}(m_p) &= e^{\blacktriangle}(m_p) - e^{\blacktriangle}(m_p) && \text{by Definition 3} \\ &= \textstyle\sum_{(e,c)\in F} F(e,c) \cdot m_p(c) - \sum_{(c,e)\in F} F(c,e) \cdot m_p(c) \\ &= \textstyle\sum_{c\in e\circ} F(e,c) \cdot m_p(c) - \sum_{c\in\circ e} F(c,e) \cdot m_p(c) && \text{by applications of } S \\ &= S(e\circ)|_p - S(\circ e)|_p && \text{by homomorphism (ii)} \\ &= \lambda(e)\circ|_p - \circ\lambda(e)|_p \\ &= W(\lambda(e), p) - W(p, \lambda(e)). \end{aligned}$$

Thus, for all p, m_p satisfies $\forall e \in E : e^{\triangle}(m_p) = W(\lambda(e), p) - W(p, \lambda(e))$, i.e. Condition (II) of Definition 4.

(λ, S) is a homomorphism, thus $S(m_i) = m_0$ holds. We get $\forall p \in P : \sum_{c\in C} m_i(c) \cdot m_p(c) = S(m_i)|_p = m_0(p)$, i.e. Condition (III) of Definition 4.

There is a token trail m_p for every place p of N and $L \in \mathcal{N}(N)$ holds. □

All synchronous net morphisms respect the token trail constrains. Whenever there is a synchronous net morphism (λ, S) from a Petri net N' to a Petri net N, we simply use λ as a labelling function, the set of transitions of N as the set of actions, and consider N' to be a labelled net. Theorem 1 states, this labelled net N' is in the net language of the Petri net N.

In the following, we deduce Corollary 1 directly from Theorem 1 to show, that token trails cover unfoldings. An unfolding is a Petri net and a folding morphism representing executions of a Petri net. A folding morphism is a net morphism with additional constrains [32,38,39]. The folding morphism relates the semantics of the unfolding to the original Petri net [32]. Using unfoldings, we can easily model concurrency, but conflict introduces branching alternatives to model all possible non-sequential executions of a net [32]. Thus, branching processes [39] are prohibitively large, or infinite in size [32]. Therefore, other, more sophisticated, unfolding techniques exist which fold local states to get a more compact representation [32]. Examples of such techniques are unravel nets [13,14,32], merged processes [25,32], and trellis processes [18,32]. Most recently, spread nets were introduced as a higher-level abstraction, that unifies these techniques in a single theoretical unfolding framework [32]. Anyways, all the above-mentioned semantics, like branching processes, unravel nets, merged processes, trellis processes, and spread nets all use folding morphisms and therefore respect the token trails semantics.

Corollary 1. *Any finite unfolding of a Petri net N is in the net language of N.*

Proof. An unfolding comes with a folding morphism (λ, S). The multirelation λ is the labelling function relating transitions of the unfolding to the transitions of the Petri net N. Theorem 1 applies. □

Remark, Corollary 1 covers finite unfoldings. In his work, Winskel guarantees that for safe Petri nets of infinite size net morphisms still have a meaningful definition, which does not break because of infinite sums [38]. Since our motivation for the token trails semantics is to model and specify behaviour of Petri nets with labelled nets, which are created by people, or algorithms, our definition currently does not account for Petri nets of infinite size. We think, our theory could be generalised, but leave this open for future work.

Originally, Winskel introduced his new definition of homomorphisms, because, unlike the previous definition [12], it preserves the behaviour of Petri nets [38]. Since the token trails semantics covers net morphisms, and not the other way around, we do not inherit this property for free. In Theorem 2 we prove, that we can derive the same property from the, apparently weaker, constrains enforced by the net language definition. In automata theory the relation we are going to prove is called simulation [29].

Theorem 2. *Let $L = (C, E, F, T, \lambda, m_i)$ be a marked labelled net and $N = (P, T, W, m_0)$ be a marked Petri net. If L is in the net language of N (a token trail m_p exists for every $p \in P$) and we define a multirelation $S : C \times P \to \mathbb{N}$ as $\forall p \in P, \forall c \in C : S_{cp} = m_p(c)$, which maps a marking m of L to a marking $S(m)$*

of N with $S(m) = \sum_{p \in P} \sum_{c \in C} m(c) \cdot m_p(c) \cdot p$, then for all markings $m \in \mathbb{N}^C$ and all steps $u \in \mathbb{N}^E$ the following holds.

(A) If a step u is enabled in marking m, then the step $\lambda(u)$ is enabled in marking $S(m)$.
(B) If firing an enabled step u in marking m produces the marking m', then firing the step $\lambda(u)$ in $S(m)$ produces the marking $S(m')$.
(C) The initial marking m_0 of N is $S(m_i)$.

Proof. We prove (A). Step u is enabled in m, therefore $m \geq \sum_{e \in u} u(e) \cdot {\circ}e$ holds. We fix a place $p \in P$. For the image $S(m)$ we get the following.

$$\begin{aligned}
S(m)|_p &= \textstyle\sum_{c \in C} m(c) \cdot m_p(c) && \text{by implication premise}\\
&\geq \textstyle\sum_{e \in u} u(e) \sum_{c \in {\circ}e} {\circ}e(c) \cdot m_p(c) && \text{see proof Theorem 1}\\
&= \textstyle\sum_{e \in u} u(e) \cdot e^{\blacktriangle}(m_p) && \text{by Definition 4 (I)}\\
&\geq \textstyle\sum_{e \in u} u(e) \cdot W(p, \lambda(e))\\
&= \textstyle\sum_{e \in u} u(e) \cdot {\circ}\lambda(e)|_p.
\end{aligned}$$

Thus, for all steps u and all places p, $S(m)$ satisfies the firing condition $S(m) \geq \sum_{e \in u} u(e) \cdot {\circ}\lambda(e)$ and therefore the step $\lambda(u)$ is enabled in $S(m)$.

We prove (B). $m \xrightarrow{u} m'$, therefore $m' = m - {\circ}u + u{\circ}$ holds. For the image $S(m')$ we get the following.

$$\begin{aligned}
S(m') &= S(m - {\circ}u + u{\circ})\\
&= \textstyle\sum_{p \in P} \sum_{c \in C} (m(c) - {\circ}u(c) + u{\circ}(c)) \cdot m_p(c) \cdot p\\
&= \textstyle\sum_{p \in P} \sum_{c \in C} m(c) \cdot m_p(c) \cdot p\\
&\quad - \textstyle\sum_{p \in P} \sum_{c \in C} {\circ}u(c) \cdot m_p(c) \cdot p\\
&\quad + \textstyle\sum_{p \in P} \sum_{c \in C} u{\circ}(c) \cdot m_p(c) \cdot p\\
&= S(m)\\
&\quad - \textstyle\sum_{e \in u} u(e) \cdot \sum_{p \in P} \sum_{c \in C} {\circ}e(c) \cdot m_p(c) \cdot p\\
&\quad + \textstyle\sum_{e \in u} u(e) \cdot \sum_{p \in P} \sum_{c \in C} e{\circ}(c) \cdot m_p(c) \cdot p && \text{see proof Theorem 1}\\
&= S(m) + \textstyle\sum_{e \in u} u(e) \cdot \left(- \sum_{p \in P} e^{\blacktriangle}(m_p) \cdot p + \sum_{p \in P} e^{\blacktriangle}(m_p) \cdot p\right)\\
&= S(m) + \textstyle\sum_{e \in u} u(e) \cdot \sum_{p \in P} e^{\Delta}(m_p) \cdot p && \text{by Definition 4 (II)}\\
&= S(m) + \textstyle\sum_{e \in u} u(e) \cdot \sum_{p \in P} (W(\lambda(e), p) - W(p, \lambda(e))) \cdot p\\
&= S(m)\\
&\quad - \textstyle\sum_{e \in u} u(e) \cdot \sum_{p \in P} W(p, \lambda(e)) \cdot p\\
&\quad + \textstyle\sum_{e \in u} u(e) \cdot \sum_{p \in P} W(\lambda(e), p) \cdot p\\
&= S(m) - \textstyle\sum_{e \in u} u(e) \cdot {\circ}\lambda(e) + \sum_{c \in u} u(e) \cdot \lambda(e){\circ}\\
&= S(m) - {\circ}\lambda(u) + \lambda(u){\circ}.
\end{aligned}$$

Thus, $S(m') = S(m) - {\circ}\lambda(u) + \lambda(u){\circ}$ and $S(m) \xrightarrow{\lambda(u)} S(m')$ holds.

Finally, Proposition (C) follows immediately from Definition 4 (III) and the definition of S. □

A Petri net can simulate the states and state-transitions of all the labelled nets in its net language. Therefore, token trails, just like homomorphisms, preserve the behaviour of nets.

Obviously, if a marked Petri net can simulate all the steps of a marked labelled net, then the step language of the Petri net is a super set of the labelled step language.

Corollary 2. *Let a marked labelled net L be in the net language of a marked Petri net N, then $\mathcal{L}(L) \subseteq \mathcal{L}(N)$ holds.*

Proof. We prove by induction on the reachability graph of L. Due to (C), the initial marking of L maps to the initial marking of N. Due to (A), for every marking m of L the set of enabled labelled steps is a subset of the set of enabled steps in $S(m)$ of N. Due to (B) this property is preserved when any step fires. □

The converse of Corollary 2 does not hold. We demonstrate this with a counter example. Figure 12 depicts a labelled net similar to the labelled net of Fig. 6. The step language of this net consists of two maximal step sequences $ABCD$ and $ABX(B+C)D$. This step language is obviously included in the step language of the Petri net from Fig. 1. Figure 12 is not in the net language of Fig. 1, because there is no token trail for the place p_3 of Fig. 1. We will show how to decide if there is a token trail for some place in the next section. Token trail semantics is different from labelled step language inclusion.

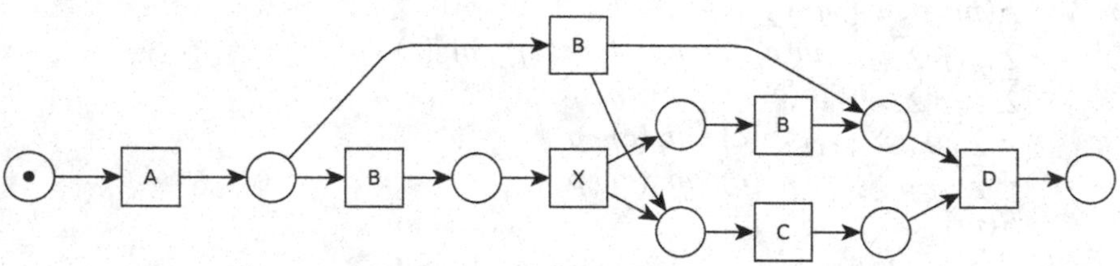

Fig. 12. Counter example to the converse of Corollary 2.

Finally, in our third theorem, we now prove that the union of the step languages of all labelled nets in a net language is the step language of its Petri net.

Theorem 3. *Let N be a marked Petri net. $\mathcal{L}(N) = \bigcup_{L \in \mathcal{N}(N)} \mathcal{L}(L)$ holds.*

Proof. We construct a synchronous net morphism from N to N. We simply map places and transitions using identity functions. We get $N \in \mathcal{N}(N)$ because of Theorem 1. Thus, $\mathcal{L}(N) \subseteq \bigcup_{L \in \mathcal{N}(N)} \mathcal{L}(L)$ holds.

Assume $\mathcal{L}(N) \subsetneq \bigcup_{L \in \mathcal{N}(N)} \mathcal{L}(L)$ holds. There is a labelled net $L \in \mathcal{N}(N)$ with $\mathcal{L}(L) \setminus \mathcal{L}(N) \neq \emptyset$. This contradicts Corollary 2. □

Summing up, a net language contains all the labelled nets, which can be simulated by its marked Petri net. Moreover, we now understand, that the Petri net model is an upper bound on the behaviour of its net language, i.e. any labelled net modelling more behaviour, cannot be included in the net language. This is an important difference between the token trail semantics and unfoldings. Unfoldings always unfold the behaviour completely with respect to some property [32]. Labelled nets of a net language can also model parts of the complete behaviour of a Petri net. This makes modelling and specification of behaviour with labelled nets easier, because we can use multiple smaller nets and example runs instead of one complex structure. Finally, we understand that the concept of net language inclusion is fundamentally different from the concept of step language inclusion. A Petri net must not only be able to reproduce the enabled step sequences of its net language but must also reflect the state space structure of the labelled nets of its net language. Additionally, we showed that the token trail semantics covers, but is not equal to the notion of unfoldings and the notion of net homomorphisms as defined by Winskel.

4 Solving Net Language Inclusion

We introduced the net language in [7] and discussed its properties in the last section. We claim it is very natural to model the behaviour of a Petri net as a set of labelled nets [7,8]. For example, in the domain of business process modelling, modelling languages like workflow nets [1,2] and BPMN [17,37] allow labelling of activities. If we want to specify behaviour of a system using labels, we need to decide if a labelled net is in the net language of a Petri net model. We will tackle this net language inclusion problem in this section.

To decide net language inclusion, we need to decide the token trail problem. Let $N = (P, T, W, m_0)$ be a marked Petri net, $p \in P$ be a place of N, $L = (C, E, F, T, \lambda, m_i)$ be a marked labelled net. Is there a marking m of L so that m is a token trail for p? Obviously, if we can decide the token trail problem for all places of a marked Petri net, we can decide net language inclusion.

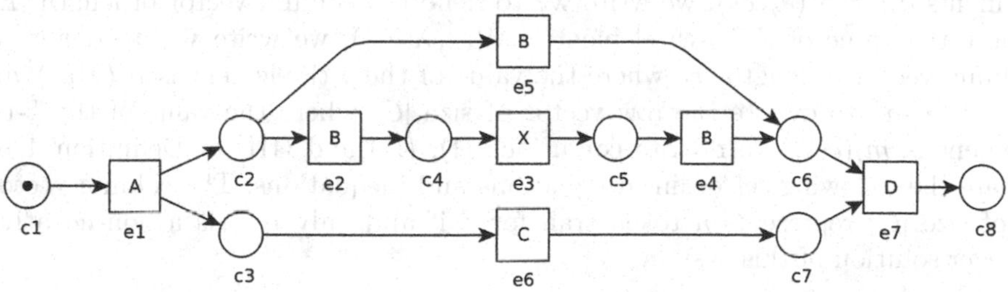

Fig. 13. The labelled net from Fig. 6 with identifiers.

As an example, we consider the labelled net from Fig. 13 as a specification of behaviour. The Figure depicts the labelled net from Fig. 6, but includes an identifier for every node. We already know, this net is in the net language of the Petri net from Fig. 3, because the six token trails for each of the places of the Petri net are depicted in Fig. 8. To show that the labelled net from Fig. 6 is not in the net language of the more complex Petri net from Fig. 1 we look at the place p_3 and formalise the conditions of Definition 4 as a set of equations. Transition X from Fig. 1 has one in-going arc from place p_3. The only transition in the labelled net depicted in Fig. 13 labelled with X is e_3. From condition (II) we get that a token trail marking m for p_3 must satisfy $e_3^{\Delta}(m) = W(X, p_3) - W(p_3, X) = 0 - 1 = -1$. Rewriting $e_3^{\Delta}(m)$ using the rise definition, we get $m(c_5) - m(c_4) = -1$. This matches our intuition, because transition e_3 has a rise of -1 in the token trail marking m if there is one less token in c_5 than in c_4. Transition B is not connected to the place p_3, therefore all transitions labelled B must have a rise of 0. We apply condition (II) to all three transitions to get $m(c_2) = m(c_4) = m(c_5) = m(c_6)$. Again, e_3 labelled X implies $m(c_4) = m(c_5) + 1$. Obviously, both cannot be true at the same time and there is no token trail for p_3. The labelled net from Fig. 13 is not in the net language of the Petri net from Fig. 1.

This result also makes sense on a semantics level. The labelled net from Fig. 13 specifies, that both, skipping the loop and executing the loop, reaches the same local state c_6. This is obviously not the case for the Petri net from Fig. 1, because the place p_3 tracks the number of executed loops.

We use the same argument to show that the labelled net from Fig. 12 is not in the net language of Fig. 1. We get the conflicting equations for a token trail for p_3, by analysing the constrains imposed by transitions B and X of the labelled net. By the same argument with local states, we get that in the labelled net from Fig. 12, the execution of the firing sequences AB, and $ABXB$ lands in the same state, which is not the case for the Petri net from Fig. 1.

After these two examples, we show how to decide the token trail problem in general. We rewrite Definition 4 using vectors and matrices and form a system of linear equations and inequations. Let $N = (P, T, W, m_0)$ be a marked Petri net, $p \in P$ be a place, $L = (C, E, F, T, \lambda, m_i)$ be a marked labelled net. We fix an arbitrary ordering of C and E. We write $\mathbf{I}$ to denote the $|E| \times |C|$ matrix with elements $i_{jk} = F(c_k, e_j)$, we write $\mathbf{O}$ to denote the $|E| \times |C|$ matrix with elements $o_{jk} = F(e_j, c_k)$, we write $\mathbf{w}_I$ to denote a column vector of length $|E|$ where the value of the j-th element is $W(p, \lambda(e_j))$, we write $\mathbf{w}_O$ to denote a column vector of length $|E|$ where the value of the j-th element is $W(\lambda(e_j), p)$, we write $\mathbf{m}_i^{\intercal}$ to denote the row vector of size $|C|$ where the value of the k-th element is $m_i(c_k)$. We rewrite conditions (I), (II) and (III) of Definition 4 to create the following set of linear equations and inequations. The column vector $\mathbf{x}$ of size $|C|$ represents a token trail for p if and only if it is a non-negative integer solution of this system.

$$
\begin{array}{lll}
(1) & \mathbf{I} \cdot \mathbf{x} - \mathbf{w}_I & \geq \mathbf{0} \\
(2) & (\mathbf{O} - \mathbf{I}) \cdot \mathbf{x} - \mathbf{w}_O + \mathbf{w}_I & = \mathbf{0} \\
(3) & \mathbf{m}_i^{\intercal} \cdot \mathbf{x} - m_0(p) & = 0
\end{array}
$$

The value of the k-th element of a non-negative integer solution vector $\mathbf{x}$ of the system is the number of tokens in place c_k of a token trail marking for p. Notice, that the values of the two $\mathbf{w}$ vectors depend on the place p.

$$\mathbf{I} = \begin{array}{c} \\ e_1 \\ e_2 \\ e_3 \\ e_4 \\ e_5 \\ e_6 \\ e_7 \end{array} \begin{array}{c} \begin{array}{cccccccc} c_1 & c_2 & c_3 & c_4 & c_5 & c_6 & c_7 & c_8 \end{array} \\ \begin{pmatrix} 1 & 0 & 0 & 0 & 0 & 0 & 0 & 0 \\ 0 & 1 & 0 & 0 & 0 & 0 & 0 & 0 \\ 0 & 0 & 0 & 1 & 0 & 0 & 0 & 0 \\ 0 & 0 & 0 & 0 & 1 & 0 & 0 & 0 \\ 0 & 1 & 0 & 0 & 0 & 0 & 0 & 0 \\ 0 & 0 & 1 & 0 & 0 & 0 & 0 & 0 \\ 0 & 0 & 0 & 0 & 0 & 1 & 1 & 0 \end{pmatrix} \end{array} \qquad \mathbf{O} = \begin{array}{c} \\ e_1 \\ e_2 \\ e_3 \\ e_4 \\ e_5 \\ e_6 \\ e_7 \end{array} \begin{array}{c} \begin{array}{cccccccc} c_1 & c_2 & c_3 & c_4 & c_5 & c_6 & c_7 & c_8 \end{array} \\ \begin{pmatrix} 0 & 1 & 1 & 0 & 0 & 0 & 0 & 0 \\ 0 & 0 & 0 & 1 & 0 & 0 & 0 & 0 \\ 0 & 0 & 0 & 0 & 1 & 0 & 0 & 0 \\ 0 & 0 & 0 & 0 & 0 & 1 & 0 & 0 \\ 0 & 0 & 0 & 0 & 0 & 1 & 0 & 0 \\ 0 & 0 & 0 & 0 & 0 & 0 & 1 & 0 \\ 0 & 0 & 0 & 0 & 0 & 0 & 0 & 1 \end{pmatrix} \end{array}$$

$$\mathbf{w}_I = \begin{pmatrix} W(p_3, \lambda(e_1)) \\ W(p_3, \lambda(e_2)) \\ W(p_3, \lambda(e_3)) \\ W(p_3, \lambda(e_4)) \\ W(p_3, \lambda(e_5)) \\ W(p_3, \lambda(e_6)) \\ W(p_3, \lambda(e_7)) \end{pmatrix} = \begin{pmatrix} 0 \\ 0 \\ 1 \\ 0 \\ 0 \\ 0 \\ 0 \end{pmatrix} \qquad \mathbf{w}_O = \begin{pmatrix} W(\lambda(e_1), p_3) \\ W(\lambda(e_2), p_3) \\ W(\lambda(e_3), p_3) \\ W(\lambda(e_4), p_3) \\ W(\lambda(e_5), p_3) \\ W(\lambda(e_6), p_3) \\ W(\lambda(e_7), p_3) \end{pmatrix} = \begin{pmatrix} 2 \\ 0 \\ 0 \\ 0 \\ 0 \\ 0 \\ 0 \end{pmatrix} \qquad \mathbf{m}_i = \begin{pmatrix} 1 \\ 0 \\ 0 \\ 0 \\ 0 \\ 0 \\ 0 \\ 0 \end{pmatrix}$$

Fig. 14. The matrices **I** and **O** of the labelled net of Fig. 13, the vectors $\mathbf{w}_I$ and $\mathbf{w}_O$ for the place p_3 of the Petri net of Fig. 1, and the vector $\mathbf{m}_i$ in its column form.

As an example, we show these matrices and vectors, for the place p_3 of the Petri net from Fig. 1 and the labelled net from Fig. 13. We use the numbering of the places and transitions of Fig. 13 to determine the ordering for the matrices and vectors depicted in Fig. 14. The first two matrices model the flow relation F of the labelled net from Fig. 13. The j-th row represents the transition e_j, and the k-th column represents the place c_k of the labelled net. The entries in the **I** matrix represent weights of arcs that connect places to transitions, the entries in the **O** matrix represent weights of arcs that connect transitions to places. The first two vectors depicted in Fig. 14 relate the transitions of the labelled net to the transitions of the Petri net from Fig. 1. The j-th entry in each vector is the arc weight that connects the fixed place p_3 with the transition $\lambda(e_j)$ of the Petri net that shares a label with the transition e_j of the labelled net. The vector $\mathbf{w}_I$ contains the weight of the arc which leads from p_3 to the transition $\lambda(e_j)$, the vector $\mathbf{w}_O$ contains the weight of the arc in the opposite direction. Notice, that in this case, the 2nd, 4th, and 5th entries of these vectors must always have the same value, regardless of the fixed place p, since the transitions e_2, e_4 and e_5 share the same label. The final vector depicted in Fig. 14 represents the initial marking m_i of the labelled net of Fig. 13. The k-th entry contains the number of tokens in the initial marking of the place c_k. If we run a solver and try to compute a non-negative integer solution for the system of Fig. 14, we will get no solution. This is true, because of our arguments from the beginning of this section.

To decide the net language inclusion problem, we simply iterate over all places of the marked Petri net N and solve the token trail problem for each place using our system of inequations. Notice, that for each place only the two vectors $\mathbf{w}_I$ and $\mathbf{w}_O$ change. If there is a token trail for every place, the labelled net is in the net language of the marked Petri net. If there is no solution for one of the places, we can decide net language inclusion early because the labelled net will not be part of the net language. Alternatively, we can always iterate over all the places and return the sets of all valid and all invalid places. Additionally, if the answer to the inclusion problem is positive, we can return the multirelation S that maps the states of the labelled net to the states of the Petri net. We can construct it by combining the solution vectors $\mathbf{x}$ from every iteration into a single matrix.

Solving a general integer linear program is a well-known NP-hard problem [24]. However, Petri nets in practical applications produce sparse matrices, furthermore the separation of the constrains (1), (2) and (3) into three distinct sets creates a natural block-structuredness of our problem. Integer programming with these properties can be solved in fixed polynomial and fixed-parameter tractable time [27]. We hope to show a better upper bound for the token trail problem in future research. If we restrict token trails to partial orders, they are equivalent to compact token flows [3,7,8]. In fact, token trails are kind of the generalised version of compact tokenflows on arbitrary labelled nets. We can solve the compact tokenflow problem for partial orders in polynomial time [3,5]. In the future, we hope to get similar results for token trails.

5 A Tool For Calculating Token Trails

As a proof of concept, we implemented the algorithm as part of our previous work on token trails [7]. The tool is part of the *I ♥ Petri nets* web tool kit and its updated version is available at www.fernuni-hagen.de/ilovepetrinets/fox. By clicking the symbol a user can select files containing a Petri net and a labelled net. Files can also be drag-and-dropped onto the symbol either from the client computer, or from the *examples* section of the tool website. The tool supports files in the standard PNML file format, as well as, a JSON format specific to the *I ♥ Petri nets* website. As soon as, we load a marked Petri net and a labelled net, the tool runs the net language inclusion algorithm and constructs the related systems of linear inequations for each place of the Petri net. We use an integer linear programming solver, to find a non-negative integer solution of this system. The tool marks all valid places green and all invalid places red. If all places of the model are green, then the labelled net is in the net language of the marked Petri net. Additionally, if we click on a green place, it is highlighted in the Petri net, and the related token trail marking is displayed in the labelled net.

Figure 15 depicts a screenshot of our tool. The Petri net displayed in the top part is the net from Fig. 1. The labelled net displayed in the bottom part is the net from Fig. 13. We can clearly see that six of the places are valid and therefore coloured green, the remaining two places are invalid. As discussed in the previous section, the labelled net is not in the net language of the Petri net. Finally, we see that the place p_2 of the Petri net has a slightly darker colour. This means it is currently selected and the marking depicted in the labelled net is a token trail for p_2. The depicted token trail marking is the same as in Fig. 8.

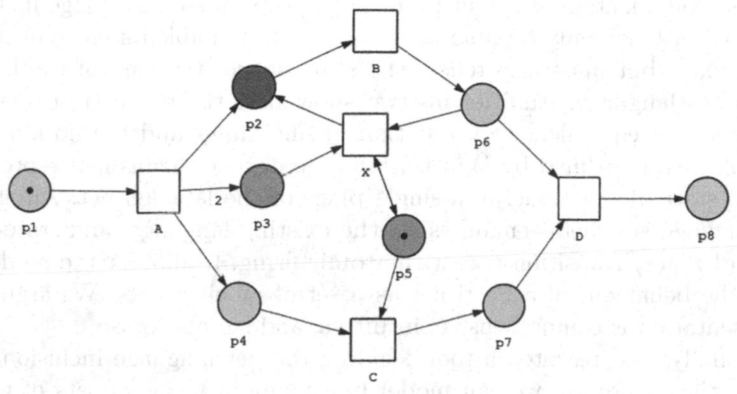

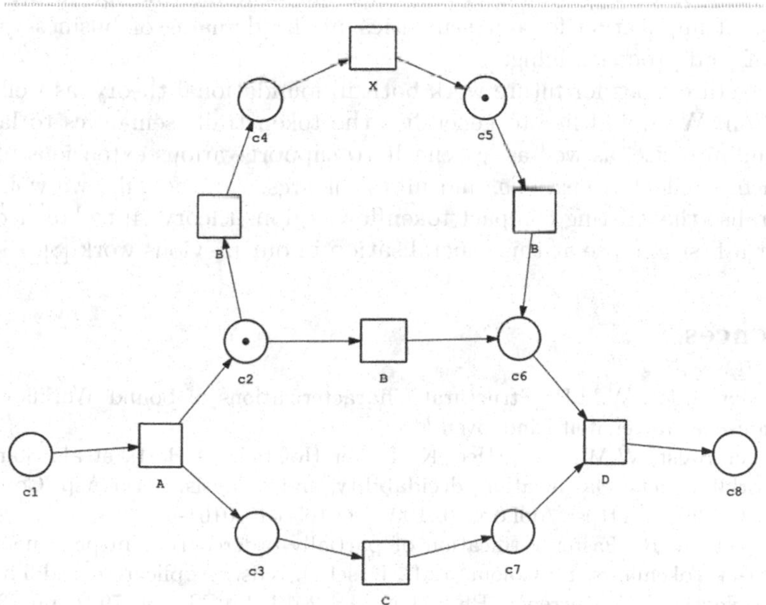

Fig. 15. Screenshot of the tool

6 Conclusion

In [7], we introduced the token trail semantics for Petri nets and discussed its relation to step semantics [20], branching processes [39], tokenflows [23], and compact tokenflows [3]. In this paper, we added the discussion on net morphisms [38] and unfoldings [32]. We extend the results of [7] and show that, the net language contains all labelled nets, which can be simulated by the marked Petri net. We now understand that the Petri net model is an upper bound of the behaviour of the net language and any labelled net modelling more behaviour, is not included in the net language. We show that the concept of net language inclusion is fundamentally different from the concept of step language inclusion. A Petri net must not only be able to reproduce the enabled step sequences of its net language but must also reflect the state space structure of the labelled nets of its net language. Additionally, we show that the token trail semantics covers, but is not equivalent to the notion of unfoldings and the notion of net homomorphisms as defined by Winskel [38]. Here, token trails can represent a distributed state of the model in a single place of the labelled net. Altogether, the net language is a meta-semantics for the existing language- and state-based semantics of Place/Transition nets with atomic firing. It allows us to model and represent the behaviour of a Petri net as a set of labelled nets. We argue that this representation is comprehensive, intuitive, and simple to use.

Additionally, we presented a tool deciding the net language inclusion problem. Using this approach, we can model behaviour in terms of sets of general labelled nets and see if a marked Petri net model fulfils this specification. We see a lot of application for our semantics in the domains of business process modelling and process mining.

We see directions for future work both in foundational theory, as well as, in application. We would like to generalise the token trails semantics to labelled nets of infinite size, as well as, extend it to support various extensions of Petri nets, such as silent transitions, and inhibitor arcs. Additionally, we would like to generalise the existing compact tokenflow regions theory [4] to labelled nets. We offer a first glimpse at this generalisation in our previous work [8].

References

1. Van der Aalst, W.M.P.: Structural Characterizations of Sound Workflow Nets. Technische Universiteit Eindhoven (1996)
2. van der Aalst, W.M.P., van Hee, K.M., ter Hofstede, A.H.M., et al.: Soundness of workflow nets: classification, decidability, and analysis. Form Asp. Comp. **23**, 333–363 (2011). https://doi.org/10.1007/s00165-010-0161-4
3. Bergenthum, R.: Faster verification of partially ordered runs in petri nets using compact Tokenflows. In: Colom, J.M., Desel, J. (eds.) Application and Theory of Petri Nets and Concurrency. PETRI NETS 2013. LNCS, vol. 7927, pp 330–348. Springer, Berlin (2013). https://doi.org/10.1007/978-3-642-38697-8_18
4. Bergenthum, R.: Synthesizing petri nets from hasse diagrams. In: Carmona, J., Engels, G., Kumar, A. (eds.) Business Process Management. BPM 2017. LNCS, vol. 10445, pp. 22–39. Springer, Cham (2017). https://doi.org/10.1007/978-3-319-65000-5_2
5. Bergenthum, R.: Firing partial orders in a petri net. In: Buchs, D., Carmona, J. (eds.) Application and Theory of Petri Nets and Concurrency. PETRI NETS 2021.

LNCS, vol. 12734, pp. 399–419. Springer, Cham (2021). https://doi.org/10.1007/978-3-030-76983-3_20
6. Bergenthum, R.: Prozessmanagement und Process-Mining, chap. Grundlagen der Formalen Prozessanalyse. De Gruyter, Petrinetze (2021)
7. Bergenthum, R., Folz-Weinstein, S., Kovář, J.: Token trail semantics - modeling behavior of petri nets with labeled petri nets. In: Gomes, L., Lorenz, R. (eds.) Application and Theory of Petri Nets and Concurrency. PETRI NETS 2023. LNCS, vol. 13929, pp. 286–306. Springer, Cham (2023). https://doi.org/10.1007/978-3-031-33620-1_16
8. Bergenthum, R., Kovář, J.: A first glimpse at petri net regions. In: Proceedings of ATAED 2022, pp. 60–68. CEUR Workshop Proceedings 3167 (2022)
9. Bergenthum, R., Lorenz, R.: Verification of scenarios in petri nets using compact Tokenflows. Fundamenta Informaticae **137**, 117–142 (2015)
10. Best, E., Devillers, R.: Sequential and concurrent behaviour in petri net theory. Theor. Comput. Sci. **55**, 87–136 (1987)
11. Boudol, G., Roucairol, G., de Simone, R.: Petri Nets and Algebraic Calculi of Processes. Springer, Cham (1986)
12. Brauer, W. (ed.): Net theory and applications. In: Proceedings of the Advanced Course on General Net Theory of Processes and Systems,, LNCS, vol. 84. Springer, Cham (1980)
13. Casu, G., Pinna, G.M.: Merging relations: a way to compact petri nets behaviors uniformly. In: Drewes, F., Martín-Vide, C., Truthe, B. (eds.) Language and Automata Theory and Applications. LATA 2017. LNCS, vol. 10168, pp. 325–337. Springer, Cham (2017). https://doi.org/10.1007/978-3-319-53733-7_24
14. Casu, G., Pinna, G.M.: Petri nets and dynamic causality for service-oriented computations. In: Proceedings of the Symposium on Applied Computing, pp. 1326–1333 (2017)
15. Desel, J., Juhás, G.: What is a petri net?. Unifying Petri Nets, pp. 1–25 (2001)
16. Desel, J., Reisig, W.: Lectures on Petri Nets I: Basic Models. In: Advances in Petri Nets, chap. Place/transition Petri Nets, pp. 122–173. Springer, Berlin (1998). https://doi.org/10.1007/3-540-65306-6
17. Dumas, M., La Rosa, M., Mendling, J., Reijers, H.A., et al.: Fundamentals of Business Process Management. Springer, Cham (2018). https://doi.org/10.1007/978-3-662-56509-4
18. Fabre, E.: Trellis processes: a compact representation for runs of concurrent systems. Discrete Event Dyn. Syst. **17**, 267–306 (2007)
19. Goltz, U., Reisig, W.: The non-sequential behaviour of petri nets. Inf. Control **57**, 125–147 (1983)
20. Grabowski, J.: On partial languages. Fundamenta Informaticae **42**, 427–498 (1981)
21. Györy, G., Knuth, E., Romai, L.: Grammatical projections. In: Working paper of Computer and Autom. Institute. Hungarian Academy of Sciences (1979)
22. Janicki, R., Koutny, M.: Structure of concurrency. Theor. Comput. Sci. **112**, 5–52 (1993)
23. Juhás, G., Lorenz, R., Desel, J.: Can i execute my scenario in your net?. In: Ciardo, G., Darondeau, P. (eds.) Applications and Theory of Petri Nets 2005. ICATPN 2005. LNCS, vol. 3536, pp. 289–308 Springer, Berlin (2005). https://doi.org/10.1007/11494744_17
24. Karp, R.M.: Reducibility among combinatorial problems, pp. 85–103. Springer, US (1972)

25. Khomenko, V., Kondratyev, A., Koutny, M., Vogler, W.: Merged processes: a new condensed representation of petri net behaviour. Acta Informatica **43**, 307–330 (2006)
26. Kiehn, A.: On the interrelation between synchronized and behaviour non-synchronized of petri nets. J. Inf. Proc. Cybern. **24**, 3–18 (1988)
27. Koutecký, M., Levin, A., Onn, S.: A parameterized strongly polynomial algorithm for block structured integer programs. In: Proceedings of ICALP, pp. 85:1–85:14. Schloss Dagstuhl – Leibniz-Zentrum für Informatik (2018)
28. Mazurkiewicz, A.: Concurrent Program Schemes and Their Interpretations. DAIMI Report Series (1977)
29. Park, D.: Concurrency and automata on infinite sequences. In: Deussen, P. (eds.) Theoretical Computer Science. LNCS, vol. 104, pp. 167–183. Springer, Berlin (1981). https://doi.org/10.1007/BFb0017309
30. Peterson, J.L.: Petri Net Theory and the Modeling of Systems. Prentice Hall PTR, Hoboken (1981)
31. Petri, C.A.: Non-Sequential Processes. GMD-ISF Report (1977)
32. Pinna, G.M., Fabre, E.: Spreading nets: a uniform approach to unfoldings. J. Logical Algebraic Methods Program. **112**, 1–33 (2020)
33. Pratt, V.: Modeling concurrency with partial orders. Int. J. Parall. Program. **15**, 33–71 (1986)
34. Reisig, W.: Understanding Petri Nets: Modeling Techniques, Analysis Methods. Springer, Case Studies (2013)
35. Smith, E., Reisig, W.: The semantics of a net is a net. In: Voss, K., Genrich, H.J., Rozenberg, G. (eds.) Concurrency and Nets. Springer, Berlin (1987). https://doi.org/10.1007/978-3-642-72822-8_29
36. Vogler, W. (ed) Modular Construction and Partial Order Semantics of Petri Nets. Springer, Berlin (1992). https://doi.org/10.1007/3-540-55767-9_4
37. Weske, M., et al.: Concepts, Languages. Architectures, Business Process Management (2007)
38. Winskel, G.: Petri nets, morphisms and compositionality. In: Rozenberg, G. (ed.) Advances in Petri Nets 1985. APN 1985. LNCS, vol. 222, pp 453–477. Springer, Berlin (1986). https://doi.org/10.1007/BFb0016226
39. Winskel, G.: Event Structures. In: Petri Nets: Applications and Relationships to Other Models of Concurrency, pp. 325–392. Springer, Berlin (1986)

Languages of Higher-Dimensional Timed Automata

Amazigh Amrane[1], Hugo Bazille[1(✉)], Emily Clement[2], and Uli Fahrenberg[1]

[1] EPITA Research Laboratory (LRE), Le Kremlin-Bicêtre, France
hugo@lrde.epita.fr
[2] Université Paris Cité, CNRS, IRIF, Paris, France

Abstract. We present a new language semantics for real-time concurrency. Its operational models are higher-dimensional timed automata (HDTAs), a generalization of both higher-dimensional automata and timed automata. We define languages of HDTAs as sets of interval-timed pomsets with interfaces. As an application, we show that language inclusion of HDTAs is undecidable. On the other hand, using a region construction we can show that untimings of HDTA languages have enough regularity so that untimed language inclusion is decidable.

Keywords: higher-dimensional timed automaton · real-time concurrency · timed automaton · higher-dimensional automaton

1 Introduction

In order to model non-interleaving concurrency, models such as Petri nets [46], event structures [45], configuration structures [51,52], or higher-dimensional automata (HDAs) [26,47,48] allow several events to happen simultaneously. The interest of such models, compared to other models such as automata or transition systems, is the possibility to distinguish concurrent and interleaving executions; using CCS notation [44], parallel compositions $a \parallel b$ are not the same as choices $a.b + b.a$.

Semantically, concurrency in non-interleaving models is represented by the fact that their languages do not consist of words but rather of partially ordered multisets (*pomsets*). As an example, Fig. 1 shows Petri net and HDA models which execute the parallel composition of $a.c$ and b; their language is generated

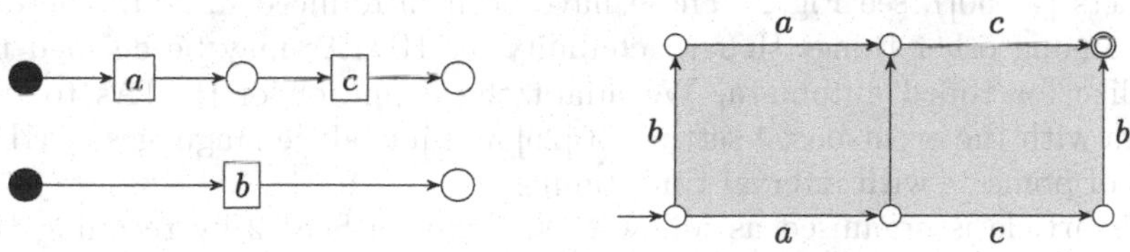

Fig. 1. Petri net and HDA models for $a.c \parallel b$.

L. M. Kristensen and J. M. van der Werf (Eds.): PETRI NETS 2024, LNCS 14628, pp. 197–219, 2024.
https://doi.org/10.1007/978-3-031-61433-0_10

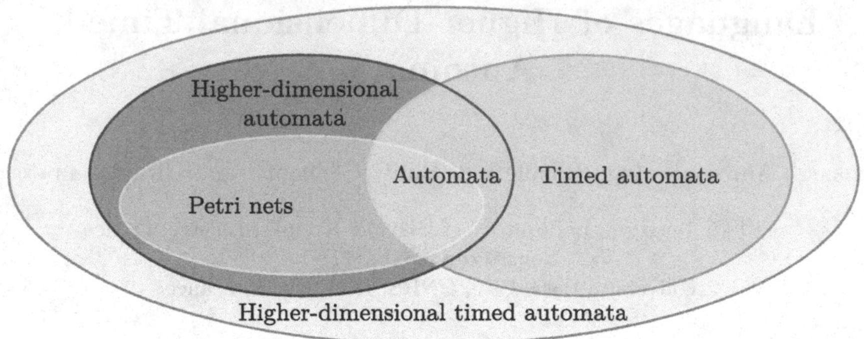

Fig. 2. Taxonomy of some models for time and concurrency

by the pomset $\left[\begin{smallmatrix} a\to c\\ b\end{smallmatrix}\right]$ in which there is no order relation between a and b nor between b and c. However, these models and pomsets use logical time and make no statements about the precise durations or timings of events.

When we consider models for real-time systems, such as for example timed automata [3] which can model precise durations and timings of events, the distinction between concurrency and interleaving is usually left behind. Their languages are sets of *timed words*, that is, sequences of symbols each of which is associated with a timestamp that records when the associated event took place.

In this article, our goal is to propose a language-based semantics for concurrent real-time systems. Our aim is to combine the two semantics above, timed words for interleaving real time and pomsets for non-interleaving logical time.

Another such proposal was developed in [12], where, going back to [11], languages of time Petri nets [43] are given as sets of pomsets with timestamps on events, see also [18–20,34]. Nevertheless, this creates problems of causality, as explained in [20] which notes that *"[t]ime and causality [do] not necessarily blend well in [...] Petri nets"*.

We put forward a different language-based semantics for real-time concurrency, inspired by recent work on interval-order semantics of higher-dimensional automata [5,26–28,32]. We use pomsets with *interval timestamps* on events, that is, every event has a start time and an end time, and the partial order respects these timestamps.

Our operational models for real-time concurrent systems are higher-dimensional timed automata (HDTAs), a simultaneous extension of timed automata [2,3] and higher-dimensional automata [26,47,48] (which in turn generalize (safe) Petri nets [49,50]), see Fig. 2. These have been introduced in [25], where it is shown among other things that reachability for HDTAs may be decided using zones like for timed automata. We adapt the definition of HDTAs to better conform with the event-based setting of [26] and introduce languages of HDTAs as sets of pomsets with interval timestamps.

This article is organised as follows. We begin in Sect. 2 by recalling timed automata and expressing their language semantics using two perspectives: delay words and timed words. In Sect. 3, we revisit higher-dimensional automata and their languages, again focusing on two complementary perspectives, of step sequences and pomsets with interfaces. In Sect. 4 we recall the definition of higher-dimensional timed automata and give examples.

The following sections present our proper contributions. In Sect. 5, we present two formalisms for languages for real-time concurrency: interval delay words and timed pomsets with interfaces, generalizing the dual view on languages of timed automata and of HDAs and showing their equivalence. Then in Sect. 6, we define languages of higher-dimensional timed automata using the formalisms previously introduced, and prove two main results: language inclusion is undecidable for higher-dimensional timed automata, but untimed language inclusion is decidable. We refer to the long version [4] for proofs of our results.

2 Timed Automata and Their Languages

Timed automata extend finite automata with clock variables and invariants which permit the modeling of real-time properties. For a set C (of *clocks*), $\Phi(C)$ denotes the set of *clock constraints* defined as

$$\Phi(C) \ni \phi_1, \phi_2 ::= c \bowtie k \mid \phi_1 \wedge \phi_2 \qquad (c \in C, k \in \mathbb{N}, \bowtie \in \{<, \leq, \geq, >\}).$$

Hence a clock constraint is a conjunction of comparisons of clocks to integers.

A *clock valuation* is a mapping $v : C \to \mathbb{R}_{\geq 0}$, where $\mathbb{R}_{\geq 0}$ denotes the set of non-negative real numbers. The *initial* clock valuation is $v^0 : C \to \mathbb{R}_{\geq 0}$ given by $v^0(c) = 0$ for all $c \in C$. For $v \in \mathbb{R}_{\geq 0}^C$, $d \in \mathbb{R}_{\geq 0}$, and $R \subseteq C$, the clock valuations $v + d$ and $v[R \leftarrow 0]$ are defined by

$$(v + d)(c) = v(c) + d \qquad v[R \leftarrow 0](c) = \begin{cases} 0 & \text{if } c \in R, \\ v(c) & \text{if } c \notin R. \end{cases}$$

For $v \in \mathbb{R}_{\geq 0}^C$ and $\phi \in \Phi(C)$, we write $v \models \phi$ if v satisfies ϕ.

A *timed automaton* is a structure $(\Sigma, C, Q, \bot, \top, I, E)$, where Σ is a finite set (alphabet), C is a finite set of clocks, Q is a finite set of locations with initial and accepting locations $\bot, \top \subseteq Q$, $I : Q \to \Phi(C)$ assigns invariants to states, and $E \subseteq Q \times \Phi(C) \times \Sigma \times 2^C \times Q$ is a set of guarded transitions. We will often take the liberty to omit Σ and C from the signature of timed automata.

Timed automata have a long and successful history in the modeling and verification of real-time computing systems. Several tools exist such as Uppaal[1] [13,14,41], TChecker[2], IMITATOR[3] [6,7], and Romeo[4] [35,42], some of which are routinely applied in industry. The interested reader is referred to [1,16,40].

The *operational semantics* of a timed automaton $A = (Q, \bot, \top, I, E)$ is the (usually uncountably infinite) transition system $[\![A]\!] = (S, S^\bot, S^\top, \leadsto)$, with $\leadsto \subseteq S \times (\Sigma \cup \mathbb{R}_{\geq 0}) \times S$, given as follows:

[1] https://uppaal.org/.
[2] https://www.labri.fr/perso/herbrete/tchecker/.
[3] https://www.imitator.fr/.
[4] https://romeo.ls2n.fr/.

$$
\begin{aligned}
S &= \{(q, v) \in Q \times \mathbb{R}_{\geq 0}^{C} \mid v \models I(q)\} \\
S^{\perp} &= \{(q, v^{0}) \mid q \in \perp\} \qquad S^{\top} = S \cap \top \times \mathbb{R}_{\geq 0}^{C} \\
\rightsquigarrow &= \{((q, v), d, (q, v + d)) \mid \forall 0 \leq d' \leq d : v + d' \models I(q)\} \\
&\cup \{((q, v), a, (q', v')) \mid \exists (q, \phi, a, R, q') \in E : v \models \phi, v' = v[R \leftarrow 0]\}
\end{aligned}
$$

Tuples in $\rightsquigarrow$ of the first type are called *delay moves* and denoted $\rightsquigarrow^{d}$, tuples of the second kind are called *action moves* and denoted $\overset{a}{\rightsquigarrow}$.

The definition of $\rightsquigarrow$ ensures that actions are immediate: for any $(q, \phi, a, R, q') \in E$, then A passes from (q, v) to (q', v') without any delay. Time progresses only during delays $(q, v) \rightsquigarrow (q, v + d)$ in locations. A path π in $[\![A]\!]$ is a finite sequence of consecutive moves of $\rightsquigarrow$:

$$
\pi = (l_0, v_0) \rightsquigarrow (l_1, v_1) \rightsquigarrow \cdots \rightsquigarrow (l_{n-1}, v_{n-1}) \rightsquigarrow (l_n, v_n) \tag{1}
$$

It is accepting if $(l_0, v_0) \in S^{\perp}$ and $(l_n, v_n) \in S^{\top}$.

The *language semantics* of timed automata is defined in terms of timed words. There are two versions of these in the literature, and we will use them both. The first, which we call *delay words* here, is defined as follows. The label of a delay move $\delta = (q, v) \rightsquigarrow (q, v + d)$ is $\mathsf{ev}(\delta) = d$. That of an action move $\sigma = (l, v) \overset{a}{\rightsquigarrow} (l', v')$ is $\mathsf{ev}(\sigma) = a$. Finally, the label $\mathsf{ev}(\pi)$ of π as in (1) is

$$
\mathsf{ev}((l_0, v_0) \rightsquigarrow (l_1, v_1)) \cdots \mathsf{ev}((l_{n-1}, v_{n-1}) \rightsquigarrow (l_n, v_n)).
$$

Delay words are elements of the quotient of the free monoid on $\Sigma \cup \mathbb{R}_{\geq 0}$ by the equivalence relation $\sim$ which allows to add up subsequent delays and to remove zero delays. Formally, $\sim$ is the congruence on $(\Sigma \cup \mathbb{R}_{\geq 0})^*$ generated by the relations

$$
dd' \sim d + d', \qquad 0 \sim \epsilon. \tag{2}
$$

The *delay language* $\mathcal{L}(A)$ of the timed automaton A is the set of delay words labeling accepting paths in $[\![A]\!]$:

$$
\mathcal{L}(A) = \{\mathsf{ev}(\pi) \mid \pi \text{ accepting path in } [\![A]\!]\} \subseteq (\Sigma \cup \mathbb{R}_{\geq 0})^*{}_{/\sim}
$$

Any equivalence class of delay words has a unique representative of the form $d_0 a_0 d_1 a_1 \ldots a_n d_{n+1}$ in which delays $d_i \in \mathbb{R}_{\geq 0}$ and symbols $a_i \in \Sigma$ alternate.

The second language semantics of timed automata is given using words with timestamps, which we will call *timed words* here. In the literature [1,16,40] these are usually defined as elements of the free monoid on $\Sigma \times \mathbb{R}_{\geq 0}$ in which the real components form an increasing sequence. Formally, this is the subset $\mathsf{TW}' \subseteq (\Sigma \times \mathbb{R}_{\geq 0})^*$ given as

$$
\mathsf{TW}' = \{w = (a_0, t_0) \ldots (a_n, t_n) \mid \forall i = 0, \ldots, n - 1 : t_i \leq t_{i+1}\}.
$$

The notions of delay words and timed words do not match completely, as delay words allow for a delay *at the end* of a run while timed words terminate with the last timestamped symbol. In order for the language semantics to better match the operational semantics, we prefer to allow for these extra delays. Let $\mathsf{TW} \subseteq (\Sigma \times \mathbb{R}_{\geq 0})^* \mathbb{R}_{\geq 0}$ be the subset

$$\mathsf{TW} = \{w = (a_0, t_0) \dots (a_n, t_n)\, t_{n+1} \mid \forall i = 0, \dots, n : t_i \leq t_{i+1}\}.$$

Concatenation in TW is defined by shifting timestamps. For $w = (a_0, t_0) \dots (a_n, t_n)\, t_{n+1}$, $w' = (a'_0, t'_0) \dots (a'_n, t'_n)\, t'_{n+1} \in \mathsf{TW}$:

$$ww' = (a_0, t_0) \dots (a_n, t_n)(a'_0, t_{n+1} + t'_0) \dots (a'_n, t_{n+1} + t'_n)(t_{n+1} + t'_{n+1}).$$

The monoids $(\Sigma \cup \mathbb{R}_{\geq 0})^*{}_{/\sim}$ and TW are then isomorphic via the mapping

$$d_0 a_0 d_1 a_1 \dots a_n d_{n+1} \mapsto \\ (a_0, d_0)\,(a_1, d_0 + d_1) \dots (a_{n-1}, d_0 + \dots + d_n)\,(d_0 + \dots + d_{n+1}),$$

and the *timed language* of a timed automaton A is the image of its delay language $\mathcal{L}(A)$ under this isomorphism.

3 Higher-Dimensional Automata and Their Languages

Higher-dimensional automata (HDAs) extend finite automata with extra structure which permits to specify independence or concurrency of events. We focus in this section on the *languages* of HDAs and refer to [26,32] for more details.

3.1 Higher-Dimensional Automata

An HDA is a set X of *cells* which are connected by *face maps*. Each cell has a list of events which are active, and face maps permit to pass from a cell to another in which some events have not yet started or are terminated.

We make this precise. A *conclist* (*concurrency list*) over a finite alphabet Σ is a tuple $U = (U, \dashrightarrow, \lambda)$, consisting of a finite set U (of events), a strict total order $\dashrightarrow \subseteq U \times U$ (the event order),[5] and a labeling $\lambda : U \to \Sigma$. Let $\square$ denote the set of conclists over Σ.

A *precubical set* on a finite alphabet Σ,

$$X = (X, \mathsf{ev}, \{\delta^0_{A,U}, \delta^1_{A,U} \mid U \in \square, A \subseteq U\}),$$

consists of a set of cells X together with a function $\mathsf{ev} : X \to \square$. For $U \in \square$ we write $X[U] = \{x \in X \mid \mathsf{ev}(x) = U\}$. Further, for every $U \in \square$ and $A \subseteq U$ there are face maps $\delta^0_{A,U}, \delta^1_{A,U} : X[U] \to X[U \setminus A]$ which satisfy

$$\delta^\nu_{A,U} \delta^\mu_{B,U \setminus A} = \delta^\mu_{B,U} \delta^\nu_{A,U \setminus B} \tag{3}$$

for $A \cap B = \emptyset$ and $\nu, \mu \in \{0, 1\}$.[6] The upper face maps δ^1_A transform a cell x into one in which the events in A have terminated; the lower face maps δ^0_A transform

[5] A strict *partial* order is a relation which is irreflexive and transitive; a strict *total* order is a relation which is irreflexive, transitive, and total. We may omit the "strict".

[6] We will omit the extra subscript "U" in $\delta^\nu_{A,U}$ from here on.

x into a cell where the events in A have not yet started. (3) expresses the fact that these transformations commute for disjoint sets of events.

A *higher-dimensional automaton* (*HDA*) $A = (\Sigma, X, \bot, \top)$ consists of a finite alphabet Σ, a finite precubical set X on Σ, and subsets $\bot, \top \subseteq X$ of initial and accepting cells. The *dimension* of A is $\dim(A) = \max\{|\mathsf{ev}(x)| \mid x \in X\}$.

3.2 Pomsets with Interfaces

The *language semantics* of HDAs is defined in terms of ipomsets which we define now; see again [26,32] for more details. First, a *partially ordered multiset*, or *pomset*, over a finite alphabet Σ is a structure $P = (P, <, \dashrightarrow, \lambda)$ consisting of a finite set P, two strict partial orders $<, \dashrightarrow \subseteq P \times P$ (precedence and event order), and a labeling $\lambda : P \to \Sigma$, such that for each $x \neq y$ in P, at least one of $x < y$, $y < x$, $x \dashrightarrow y$, or $y \dashrightarrow x$ holds.[7]

A *pomset with interfaces*, or *ipomset*, over a finite alphabet Σ is a tuple $(P, <, \dashrightarrow, S, T, \lambda)$ consisting of a pomset $(P, <, \dashrightarrow, \lambda)$ and subsets $S, T \subseteq P$ of *source* and *target interfaces* such that the elements of S are $<$-minimal and those of T are $<$-maximal. Note that different events of ipomsets may carry the same label; in particular we do *not* exclude autoconcurrency. Source and target events are marked by "$\bullet$" at the left or right side, and if the event order is not shown, we assume that it goes downwards.

An ipomset P is *interval* if its precedence order $<_P$ is an interval order [33], that is, if it admits an interval representation given by functions $\sigma^-, \sigma^+ : P \to \mathbb{R}$ such that $\sigma^-(x) \leq \sigma^+(x)$ for all $x \in P$ and $x <_P y$ iff $\sigma^+(x) < \sigma^-(y)$ for all $x, y \in P$. We will only use interval ipomsets here and hence omit the qualification "interval". The set of (interval) ipomsets over Σ is denoted iPoms.

For ipomsets P and Q we say that Q *subsumes* P and write $P \sqsubseteq Q$ if there is a bijection $f : P \to Q$ for which

(1) $f(S_P) = S_Q$, $f(T_P) = T_Q$, and $\lambda_Q \circ f = \lambda_P$;
(2) $f(x) <_Q f(y)$ implies $x <_P y$;
(3) $x \not<_P y$, $y \not<_P x$ and $x \dashrightarrow_P y$ imply $f(x) \dashrightarrow_Q f(y)$.

That is, f respects interfaces and labels, reflects precedence, and preserves essential event order. (Event order is essential for concurrent events, but by transitivity, it also appears between non-concurrent events. Subsumptions ignore such non-essential event order.)

Isomorphisms of ipomsets are invertible subsumptions, *i.e.*, bijections f for which items (2) and (3) above are strengthened to

(2′) $f(x) <_Q f(y)$ iff $x <_P y$;
(3′) $x \not<_P y$ and $y \not<_P x$ imply that $x \dashrightarrow_P y$ iff $f(x) \dashrightarrow_Q f(y)$.

Due to the requirement that all elements are ordered by $<$ or $\dashrightarrow$, there is at most one isomorphism between any two ipomsets. Hence we may switch freely between

[7] The event order is needed to identify concurrent events, see [26,32] for details.

ipomsets and their isomorphism classes. We will also call these equivalence classes ipomsets and often conflate equality and isomorphism.

Serial composition of pomsets [37] generalises to a *gluing* composition for ipomsets which continues interface events across compositions and is defined as follows. Let P and Q be ipomsets such that $T_P = S_Q$, $x \dashrightarrow_P y$ iff $x \dashrightarrow_Q y$ for all $x, y \in T_P = S_Q$, and the restrictions $\lambda_{P|T_P} = \lambda_{Q|S_Q}$, then $P * Q = (P \cup Q, <, \dashrightarrow, S_P, T_Q, \lambda)$, where

- $x < y$ if $x <_P y$, $x <_Q y$, or $x \in P \setminus T_P$ and $y \in Q \setminus S_Q$;
- $\dashrightarrow$ is the transitive closure of $\dashrightarrow_P \cup \dashrightarrow_Q$;
- $\lambda(x) = \lambda_P(x)$ if $x \in P$ and $\lambda(x) = \lambda_Q(x)$ if $x \in Q$.

Gluing is, thus, only defined if the targets of P are equal to the sources of Q *as conclists.*

An ipomset P is a *word* (with interfaces) if $<_P$ is total. Conversely, P is *discrete* if $<_P$ is empty (hence $\dashrightarrow_P$ is total). Conclists are discrete ipomsets without interfaces. A *starter* is a discrete ipomset U with $T_U = U$, a *terminator* one with $S_U = U$. The intuition is that a starter does nothing but start the events in $A = U - S_U$, and a terminator terminates the events in $B = U - T_U$. These will be so important later that we introduce special notation, writing ${}_A{\uparrow}U$ and $U{\downarrow}_B$ for the above. Discrete ipomsets U with $S_U = T_U = U$ are identities for the gluing composition and written id_U. Note that $\mathsf{id}_U = {}_\emptyset{\uparrow}U = U{\downarrow}_\emptyset$. The empty ipomset is $\mathsf{id}_\emptyset$.

3.3 Step Sequences

Any ipomset can be decomposed as a gluing of starters and terminators [28,39]. Such a presentation is called a *step decomposition.* If starters and terminators are alternating, the step decomposition is called *sparse.* [32, Prop. 4] shows that every ipomset has a unique sparse step decomposition.

We develop an algebra of step decompositions. Let $\mathsf{St}, \mathsf{Te}, \mathsf{Id} \subseteq \mathsf{iPoms}$ denote the sets of starters, terminators, and identities over Σ, then $\mathsf{Id} = \mathsf{St} \cap \mathsf{Te}$. Let $\mathsf{St}_+ = \mathsf{St} \setminus \mathsf{Id}$ and $\mathsf{Te}_+ = \mathsf{Te} \setminus \mathsf{Id}$. The following notion was introduced in [5].

Definition 1. *A word $P_1 \dots P_n \in (\mathsf{St} \cup \mathsf{Te})^*$ is* coherent *if the gluing $P_1 * \dots * P_n$ is defined.*

Let $\mathsf{Coh} \subseteq (\mathsf{St} \cup \mathsf{Te})^*$ denote the subset of coherent words and $\sim$ the congruence on Coh generated by the relations

$$I \sim \epsilon \quad (I \in \mathsf{Id}),$$
$$S_1 S_2 \sim S_1 * S_2 \quad (S_1, S_2 \in \mathsf{St}), \qquad T_1 T_2 \sim T_1 * T_2 \quad (T_1, T_2 \in \mathsf{Te}). \tag{4}$$

Here, ϵ denotes the empty *word* in $(\mathsf{St} \cup \mathsf{Te})^*$, not the empty ipomset $\mathsf{id}_\emptyset \in \mathsf{Id}$.

Definition 2. *A* step sequence *is an element of the set* $\mathsf{SSeq} = \mathsf{Id}.\mathsf{Coh}_{/\sim}.\mathsf{Id}$.

The identities in the beginning and end of step sequences are used to "fix" events in the source and target interfaces, *i.e.*, which are already running in the beginning or continue beyond the end. For example, $\bullet a \bullet . \bullet a \bullet$ denotes an event labeled a which is neither started nor terminated. Technically we only need these identities when the inner part (in $(\mathsf{St} \cup \mathsf{Te})^*/_\sim$) is empty (and even then we would only need one of them); but we prefer a more verbose notation that will be of interest when we introduce time, see Definition 12.

Lemma 3. *Every element of* SSeq *has a unique representative* $I_0 P_1 \dots P_n I_{n+1}$, *for* $n \geq 0$, *with the property that* $(P_i, P_{i+1}) \in \mathsf{St}_+ \times \mathsf{Te}_+ \cup \mathsf{Te}_+ \times \mathsf{St}_+$ *for all* $1 \leq i \leq n-1$. *Such a representative is called* sparse.

Proof. Directly from [32, Prop. 4].

Concatenation of step sequences is inherited from the monoid $(\mathsf{St} \cup \mathsf{Te})^*$ where $I_0.w.I_n \cdot I'_0.w'.I'_m$ is defined iff $I_n = I'_0$. Concatenations of sparse step sequences may not be sparse.

Remark 4. SSeq forms a *local partial monoid* [29] with left and right units $\mathrm{id}_U \mathrm{id}_U$, for $U \in \square$, and $(UV)W = U(VW)$ when the concatenation is defined. Local partial monoids may admit many units and are equivalent to categories: here, the objects are the conclists in $\square$ and the morphisms from $U \in \square$ to $V \in \square$ are the step sequences $\mathrm{id}_U w\, \mathrm{id}_V$. We refer to [29] for more details.

For a coherent word $P_1 \dots P_n \in \mathsf{Coh} \subseteq (\mathsf{St} \cup \mathsf{Te})^*$ we define $\mathsf{Glue}(P_1 \dots P_n) = P_1 * \cdots * P_n$. It is clear that for $w_1, w_2 \neq \epsilon$, $w_1 \sim w_2$ implies $\mathsf{Glue}(w_1) = \mathsf{Glue}(w_2)$. That is, Glue induces a mapping Glue : SSeq $\to$ iPoms.

Lemma 5 (**[32, Prop. 4]**)**.** Glue : SSeq $\to$ iPoms *is a bijection.*

See below for examples of step sequences and ipomsets.

3.4 Languages of HDAs

Paths in an HDA X are sequences $\alpha = (x_0, \phi_1, x_1, \dots, x_{n-1}, \phi_n, x_n)$ consisting of cells x_i of X and symbols ϕ_i which indicate face map types: for every $i \in \{1, \dots, n\}$, (x_{i-1}, ϕ_i, x_i) is either

- $(\delta^0_A(x_i), \nearrow^A, x_i)$ for $A \subseteq \mathsf{ev}(x_i)$ (an *upstep*)
- or $(x_{i-1}, \searrow_A, \delta^1_A(x_{i-1}))$ for $A \subseteq \mathsf{ev}(x_{i-1})$ (a *downstep*).

Downsteps terminate events, following upper face maps, whereas upsteps start events by following inverses of lower face maps.

The *source* and *target* of α as above are $\mathsf{src}(\alpha) = x_0$ and $\mathsf{tgt}(\alpha) = x_n$. A path α is *accepting* if $\mathsf{src}(\alpha) \in \bot_X$ and $\mathsf{tgt}(\alpha) \in \top_X$. Paths α and β may be concatenated if $\mathsf{tgt}(\alpha) = \mathsf{src}(\beta)$. Their concatenation is written $\alpha * \beta$ or simply $\alpha\beta$.

The observable content or *event ipomset* $\mathsf{ev}(\alpha)$ of a path α is defined recursively as follows:

- if $\alpha = (x)$, then $\mathsf{ev}(\alpha) = \mathrm{id}_{\mathsf{ev}(x)}$;
- if $\alpha = (y \nearrow^A x)$, then $\mathsf{ev}(\alpha) = {}_A{\uparrow}\mathsf{ev}(x)$;

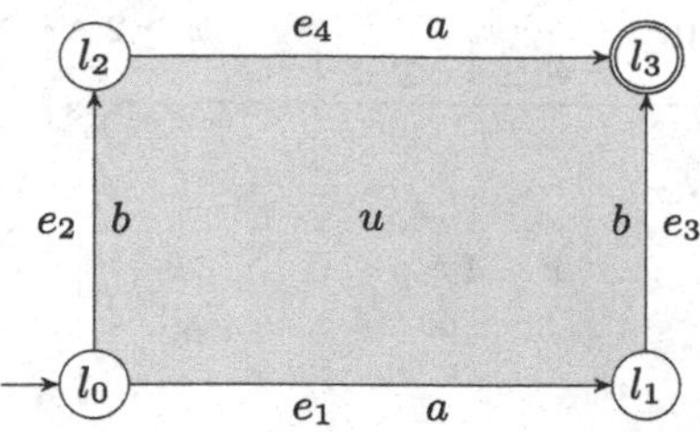

Fig. 3. HDA of Example 7. The grayed area indicates that a and b may occur concurrently, *i.e.*, there is a two-dimensional cell, u in this instance

- if $\alpha = (x \searrow_A y)$, then $\mathsf{ev}(\alpha) = \mathsf{ev}(x)\downarrow_A$;
- if $\alpha = \alpha_1 * \cdots * \alpha_n$ is a concatenation, then $\mathsf{ev}(\alpha) = \mathsf{ev}(\alpha_1) * \cdots * \mathsf{ev}(\alpha_n)$.

Note that upsteps in α correspond to starters in $\mathsf{ev}(\alpha)$ and downsteps correspond to terminators.

For $A \subseteq \mathsf{iPoms}$, $A\downarrow = \{P \in \mathsf{iPoms} \mid \exists Q \in A \colon P \sqsubseteq Q\}$ denotes its subsumption closure. The language of an HDA X is $\mathcal{L}(X) = \{\mathsf{ev}(\alpha) \mid \alpha \text{ accepting path in } X\}$. A language is *regular* if it is the language of a finite HDA. It is *rational* if it is constructed from $\emptyset$, $\{\mathrm{id}_\emptyset\}$ and discrete ipomsets using $\cup$, $*$ and $^+$ (Kleene plus) [27]. Languages of HDAs are closed under subsumption, that is, if L is regular, then $L\downarrow = L$ [26,27]. The rational operations above have to take this closure into account.

Theorem 6 ([27]). *A language is regular if and only if it is rational.*

Example 7. The HDA of Fig. 3 is two-dimensional and consists of nine cells: the corner cells $X_0 = \{l_0, l_1, l_2, l_3\}$ in which no event is active (for all $z \in X_0$, $\mathsf{ev}(z) = \emptyset$), the transition cells $X_1 = \{e_1, e_2, e_3, e_4\}$ in which one event is active ($\mathsf{ev}(e_1) = \mathsf{ev}(e_4) = a$ and $\mathsf{ev}(e_2) = \mathsf{ev}(e_3) = b$), and the square cell u where a and b are active: $\mathsf{ev}(u) = \left[\begin{smallmatrix} a \\ b \end{smallmatrix}\right]$. When we have two concurrent events a and b with $a \dashrightarrow b$, we will draw a horizontally and b vertically. Concerning face maps, we have for example $\delta^1_{ab}(u) = l_3$ and $\delta^0_{ab}(u) = l_0$.

This HDA admits several accepting paths, for example

$$\alpha_1 = l_0 \nearrow^{ab} u \searrow_{ab} l_3, \qquad \alpha_2 = l_0 \nearrow^{a} e_1 \nearrow^{b} u \searrow_{b} e_4 \searrow_{a} l_3,$$
$$\alpha_3 = l_0 \nearrow^{a} e_1 \searrow_{a} l_1 \nearrow^{b} e_3 \searrow_{b} l_3, \qquad \alpha_4 = l_0 \nearrow^{b} e_2 \searrow_{b} l_2 \nearrow^{a} e_4 \searrow_{b} l_3,$$

where $\mathsf{ev}(\alpha_1) = \left[\begin{smallmatrix} a\bullet \\ b\bullet \end{smallmatrix}\right] * \left[\begin{smallmatrix} \bullet a \\ \bullet b \end{smallmatrix}\right] = \mathsf{ev}(\alpha_2) = a\bullet * \left[\begin{smallmatrix} \bullet a\bullet \\ b\bullet \end{smallmatrix}\right] * \left[\begin{smallmatrix} \bullet a\bullet \\ \bullet b \end{smallmatrix}\right] * \bullet a = \left[\begin{smallmatrix} a \\ b \end{smallmatrix}\right]$, $\mathsf{ev}(\alpha_3) = a\bullet * \bullet a * b\bullet * \bullet b = ab$, and $\mathsf{ev}(\alpha_4) = b\bullet * \bullet b * a\bullet * \bullet a = ba$. Its language is $\{\left[\begin{smallmatrix} a \\ b \end{smallmatrix}\right]\}\downarrow = \{\left[\begin{smallmatrix} a \\ b \end{smallmatrix}\right], ab, ba\}$.

Observe that α_1 and α_2 induce the coherent words $w_1 = \left[\begin{smallmatrix} a\bullet \\ b\bullet \end{smallmatrix}\right]\left[\begin{smallmatrix} \bullet a \\ \bullet b \end{smallmatrix}\right]$ and $w_2 = a\bullet \left[\begin{smallmatrix} \bullet a\bullet \\ b\bullet \end{smallmatrix}\right]\left[\begin{smallmatrix} \bullet a \\ \bullet b\bullet \end{smallmatrix}\right] \bullet b$ such that $w_1 \sim w_2$ and $s = \mathrm{id}_\emptyset w_1 \mathrm{id}_\emptyset$ is their corresponding sparse step sequence with $\mathsf{Glue}(s) = \left[\begin{smallmatrix} a \\ b \end{smallmatrix}\right]$.

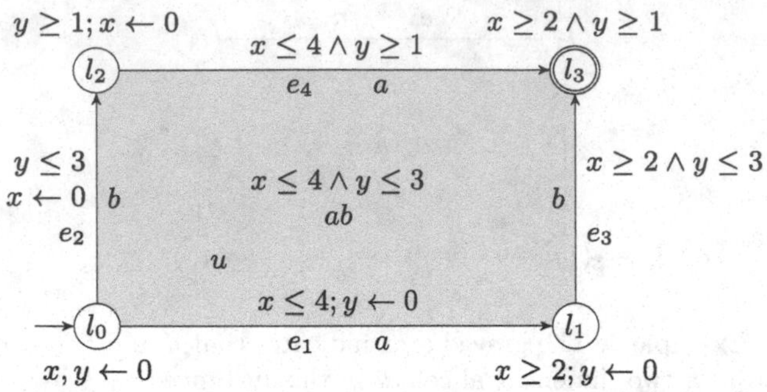

Fig. 4. HDTA of Example 9.

4 Higher-Dimensional Timed Automata

Unlike timed automata, higher-dimensional automata make no formal distinction between states (0-cells), transitions (1-cells), and higher-dimensional cells. We transfer this intuition to higher-dimensional timed automata, so that each cell has an invariant which specifies when it is enabled and an exit condition giving the clocks to be reset when leaving. Semantically, this implies that time delays can occur in any n-cell, not only in states as in timed automata; hence actions are no longer instantaneous.

Definition 8. *A* higher-dimensional timed automaton (HDTA) *is a structure* $(\Sigma, C, Q, \bot, \top, \mathsf{inv}, \mathsf{exit})$, *where* $(\Sigma, Q, \bot, \top)$ *is a finite higher-dimensional automaton and* $\mathsf{inv} : Q \to \Phi(C)$, $\mathsf{exit} : Q \to 2^C$ *assign* invariant *and* exit *conditions to each cell of* Q.

As before, we will often omit Σ and C from the signature.

The *operational semantics* of an HDTA $A = (Q, \bot, \top, \mathsf{inv}, \mathsf{exit})$ is a (usually uncountably infinite) transition system $[\![A]\!] = (S, S^\bot, S^\top, \leadsto)$, with the set of transitions (moves) $\leadsto \subseteq S \times (\mathsf{St} \cup \mathsf{Te} \cup \mathbb{R}_{\geq 0}) \times S$ given as follows:

$$S = \{(q, v) \in Q \times \mathbb{R}_{\geq 0}^C \mid v \models \mathsf{inv}(q)\}$$
$$S^\bot = \{(q, v^0) \mid q \in \bot\} \qquad S^\top = S \cap \top \times \mathbb{R}_{\geq 0}^C$$
$$\leadsto = \{((q, v), d, (q, v + d)) \mid \forall 0 \leq d' \leq d : v + d' \models \mathsf{inv}(q)\}$$
$$\cup \{((\delta_A^0(q), v), {}_A{\uparrow}\mathsf{ev}(q), (q, v')) \mid A \subseteq \mathsf{ev}(q), v' = v[\mathsf{exit}(\delta_A^0(q)) \leftarrow 0] \models \mathsf{inv}(q)\}$$
$$\cup \{((q, v), \mathsf{ev}(q){\downarrow}_A, (\delta_A^1(q), v')) \mid A \subseteq \mathsf{ev}(q), v' = v[\mathsf{exit}(q) \leftarrow 0] \models \mathsf{inv}(\delta_A^1(q))\}$$

Note that in the first line of the definition of $\leadsto$ above, we allow time to evolve in any cell of Q. As before, these are called *delay moves* and denoted $\leadsto^d$ for some delay $d \in \mathbb{R}_{\geq 0}$. The second line in the definition of $\leadsto$ defines the start of concurrent events A (denoted $\leadsto^A$) and the third line describes what happens when finishing a set A of concurrent events (denoted $\leadsto_A$). These are again called *action moves*. Exit conditions specify which clocks to reset when leaving a cell.

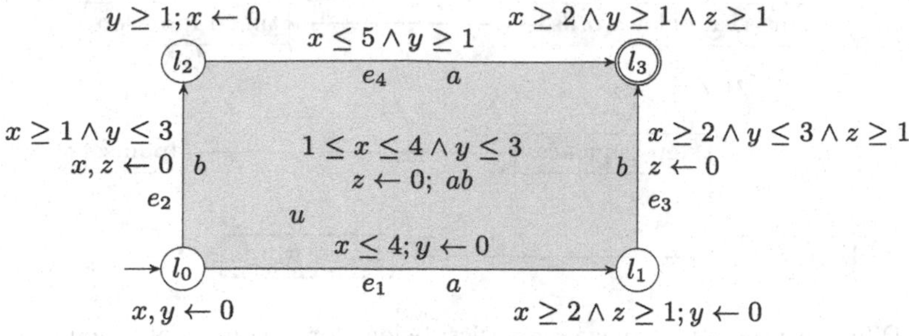

Fig. 5. HDTA of Example 10

Example 9. We give a few examples of two-dimensional timed automata. The first, in Fig. 4, is the HDA of Fig. 3 with time constraints. It models two actions, a and b, which can be performed concurrently. This HDTA models that performing a takes between two and four time units, whereas performing b takes between one and three time units. To this end, we use two clocks x and y which are reset when the respective actions are started and then keep track of how long they are running.

Hence $\mathsf{exit}(l_0) = \{x, y\}$, and the invariants $x \leq 4$ at the a-labeled transitions e_1, e_4 and at the square u ensure that a takes at most four time units. The invariants $x \geq 2$ at l_1, e_3 and l_3 take care that a cannot finish before two time units have passed. Note that x is also reset when exiting e_2 and l_2, ensuring that regardless when a is started, whether before b, while b is running, or after b is terminated, it must take between two and four time units.

Example 10. The HDTA in Fig. 5 models the following additional constraints:

- b may only start after a has been running for one time unit;
- once b has terminated, a may run one time unit longer;
- and b must finish one time unit before a.

To this end, an invariant $x \geq 1$ has been added to the two b-labeled transitions and to the ab-square (at the right-most b-transition $x \geq 1$ is already implied), and the condition on x at the top a-transition has been changed to $x \leq 5$. To enforce the last condition, an extra clock z is introduced which is reset when b terminates and must be at least 1 when a is terminating.

Note that the left edge is now unreachable: when entering it, x is reset to zero, but its edge invariant is $x \geq 1$. This is as expected, as b should not be able to start before a. Further, the right b-labeled edge is deadlocked: when leaving it, z is reset to zero but needs to be at least one when entering the accepting state. Again, this is expected, as a should not terminate before b. As both vertical edges are now permanently disabled, the accepting state can only be reached through the square.

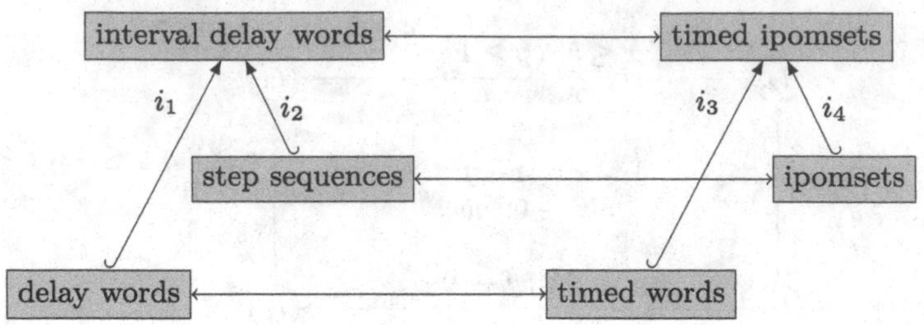

Fig. 6. Different types of language semantics: below, for languages of timed automata; middle, for languages of HDAs; top, for languages of HDTAs. Vertical arrows denote injections, horizontal arrows bijections.

5 Concurrent Timed Languages

We introduce two formalisms for concurrent timed words: interval delay words which generalize delay words and step sequences, and timed ipomsets which generalize timed words and ipomsets. Figure 6 shows the relations between the different language semantics used and introduced in this paper.

5.1 Interval Delay Words

Intuitively, an interval delay word is a step sequence interspersed with delays. These delays indicate how much time passes between starts and terminations of different events.

Definition 11. *A word $x_1 \ldots x_n \in (\mathsf{St} \cup \mathsf{Te} \cup \mathbb{R}_{\geq 0})^*$ is* coherent *if, for all $i < k$ such that $x_i, x_k \in \mathsf{St} \cup \mathsf{Te}$ and $\forall i < j < k : x_j \in \mathbb{R}_{\geq 0}$, the gluing $x_i * x_k$ is defined.*

Let $\mathsf{tCoh} \subseteq (\mathsf{St} \cup \mathsf{Te} \cup \mathbb{R}_{\geq 0})^*$ denote the subset of coherent words. Let $\sim$ be the congruence on tCoh generated by the relations

$$dd' \sim d + d', \qquad 0 \sim \epsilon, \qquad I \sim \epsilon \quad (I \in \mathsf{Id}),$$
$$S_1 S_2 \sim S_1 * S_2 \quad (S_1, S_2 \in \mathsf{St}), \qquad T_1 T_2 \sim T_1 * T_2 \quad (T_1, T_2 \in \mathsf{Te}).$$

Again, ϵ denotes the empty word in $(\mathsf{St} \cup \mathsf{Te} \cup \mathbb{R}_{\geq 0})^*$ above. That is, successive delays may be added up and zero delays removed, as may identities, and successive starters or terminators may be composed. Note how this combines the identifications in (2) for delay words and the ones for step sequences in (4).

Definition 12. *An* interval delay word (idword) *is an element of the set*

$$\mathsf{IDW} = \mathsf{Id}.\mathsf{tCoh}_{/\sim}.\mathsf{Id}.$$

Lemma 13. *Every element of IDW has a unique representative $I_0 d_0 P_1 d_1 \ldots P_n d_n I_{n+1}$, for $n \geq 0$, with the property that for all $1 \leq i \leq n$, $P_i \notin \mathsf{Id}$ and for all $1 \leq i \leq n-1$, if $d_i = 0$, then $(P_i, P_{i+1}) \in \mathsf{St}_+ \times \mathsf{Te}_+ \cup \mathsf{Te}_+ \times \mathsf{St}_+$. Such a representative is called* sparse.

Fig. 7. Tipomsets T_1 (left) and T_2 (right) of Example 15

This is analogous to Lemma 3, except that here, we must admit successive starters or terminators if they are separated by non-zero delays (see Example 21 below for an example).

With concatenation of idwords inherited from the monoid $(\mathsf{St} \cup \mathsf{Te} \cup \mathbb{R}_{\geq 0})^*$, IDW forms a partial monoid (successive identities are composed using $\sim$). Concatenations of sparse idwords are not generally sparse. The identities for concatenation are the words $\mathrm{id}_U \mathrm{id}_U \sim \mathrm{id}_U\, 0\, \mathrm{id}_U$ for $U \in \square$.

5.2 Timed Ipomsets

Timed ipomsets are ipomsets with timestamps which mark beginnings and ends of events:

Definition 14. *Let P be a set, $\sigma^-, \sigma^+ : P \to \mathbb{R}_{\geq 0}$, $\sigma = (\sigma^-, \sigma^+)$, and $d \in \mathbb{R}_{\geq 0}$. Then $P = (P, <_P, \dashrightarrow, S, T, \lambda, \sigma, d)$ is a* timed ipomset (tipomset) *if*

- *$(P, <_P, \dashrightarrow, S, T, \lambda)$ is an ipomset,*
- *for all $x \in P$, $0 \leq \sigma^-(x) \leq \sigma^+(x) \leq d$,*
- *for all $x \in S$, $\sigma^-(x) = 0$,*
- *for all $x \in T$, $\sigma^+(x) = d$, and*
- *for all $x, y \in P$, $\sigma^+(x) < \sigma^-(y) \implies x <_P y \implies \sigma^+(x) \leq \sigma^-(y)$.*

The *activity interval* of event $x \in P$ is $\sigma(x) = [\sigma^-(x), \sigma^+(x)]$; we will always write $\sigma(x)$ using square brackets because of this. The *untiming* of P is its underlying ipomset, *i.e.*, $\mathrm{unt}(P) = (P, <_P, \dashrightarrow, S, T, \lambda)$. We will often write tipomsets as (P, σ, d) or just P.

Example 15. Figure 7 depicts the following tipomsets:

- $T_1 = (\{x_1, x_2, x_3\}, <_1, \dashrightarrow, \{x_1, x_3\}, \{x_1\}, \lambda_1, \sigma_1, 3)$ with
  - $<_1 = \{(x_3, x_2)\}$, $\dashrightarrow = \{(x_1, x_2), (x_1, x_3)\}$,
  - $\lambda_1(x_1) = a$, $\lambda_1(x_2) = d$, $\lambda_1(x_3) = c$, and
  - $\sigma_1(x_1) = [0, 3], \sigma_1(x_2) = [1.5, 3], \sigma_1(x_3) = [0, 1.5]$
- $T_2 = (\{x_4, x_5, x_6\}, <_2, x_4 \dashrightarrow x_5 \dashrightarrow x_6, \{x_4\}, \emptyset, \lambda_2, \sigma_2, 4)$ with
  - $<_2 = \emptyset$, $\lambda_2(x_4) = a$, $\lambda_2(x_5) = b$, $\lambda_2(x_6) = c$, and
  - $\sigma_2(x_4) = [0, 2], \sigma_2(x_5) = [0.5, 3.5], \sigma_2(x_6) = [1, 3]$.

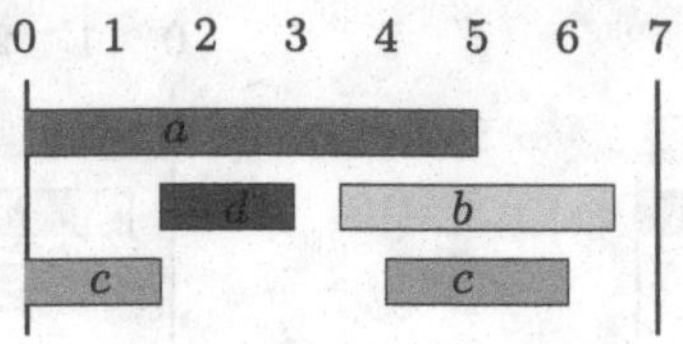

Fig. 8. Gluing $T_1 * T_2$, see Example 17

Note that in T_1, the d-labeled event x_2 is not in the terminating interface as it ends exactly at time 3. Further, the precedence order is *not* induced by the timestamps in T_1: we have $\sigma_1^+(x_3) = \sigma_1^-(x_2)$ but $x_3 <_1 x_2$; setting $<_1 = \emptyset$ instead would *also* be consistent with the timestamps. For the underlying ipomsets,

$$\mathrm{unt}(T_1) = \begin{bmatrix} \bullet a \bullet \\ \bullet c \to d \end{bmatrix}, \qquad \mathrm{unt}(T_2) = \begin{bmatrix} \bullet a \\ b \\ c \end{bmatrix}.$$

We generalize the gluing composition of ipomsets to tipomsets.

Definition 16. *Given two tipomsets (P, σ_P, d_P) and (Q, σ_Q, d_Q), the gluing composition $P * Q$ is defined if $\mathrm{unt}(P) * \mathrm{unt}(Q)$ is. Then, $P * Q = (U, \sigma_U, d_U)$, where*

- *$U = P * Q$ and $d_U = d_P + d_Q$,*
- *$\sigma_U^-(x) = \sigma_P^-(x)$ if $x \in P$ and $\sigma_U^-(x) = \sigma_Q^-(x) + d_P$ else,*
- *$\sigma_U^+(x) = \sigma_Q^+(x) + d_P$ if $x \in Q$ and $\sigma_U(x) = \sigma_P^+(x)$ else.*

The above definition is consistent for events $x \in T_P = S_Q$: here, $\sigma_U^-(x) = \sigma_P^-(x)$ and $\sigma_U^+(x) = \sigma_Q^+(x) + d_P$.

Example 17. Continuing Example 15, Fig. 8 depicts the gluing of T_1 and T_2, which is the tipomset $T = (\{x_1, x_2, x_3, x_5, x_6\}, <, \dashrightarrow, \{x_1, x_3\}, \emptyset, \lambda, \sigma, 7)$ with

- $< = \{(x_3, x_2), (x_2, x_5), (x_3, x_5), (x_2, x_6), (x_3, x_6)\}$,
- $\dashrightarrow = \{(x_1, x_2), (x_1, x_3), (x_1, x_5), (x_5, x_6), (x_1, x_6)\}$,
- $\lambda(x_1) = a$, $\lambda(x_2) = d$, $\lambda(x_3) = c$, $\lambda(x_5) = b$, $\lambda(x_6) = c$,
- $\sigma(x_1) = [0, 5]$, $\sigma(x_2) = [1.5, 3]$, $\sigma(x_3) = [0, 1.5]$, $\sigma(x_5) = [3.5, 6.5]$, and $\sigma(x_6) = [4, 6]$.

In this example, events x_1 and x_4 have been glued together. Thus

$$\mathrm{unt}(T) = \mathrm{unt}(T_1) * \mathrm{unt}(T_2) = \begin{bmatrix} \bullet a & & \\ \bullet c & \longrightarrow d \longrightarrow & b \\ & & c \end{bmatrix}.$$

Here, dashed arrows indicate event order, and full arrows indicate precedence order.

The next lemma, whose proof is trivial, shows that untiming respects gluing composition.

Lemma 18. *For all tipomsets P and Q, $P * Q$ is defined iff* $\mathrm{unt}(P) * \mathrm{unt}(Q)$ *is, and in that case,* $\mathrm{unt}(P) * \mathrm{unt}(Q) = \mathrm{unt}(P * Q)$. □

Definition 19. *An* isomorphism *of tipomsets (P, σ_P, d_P) and (Q, σ_Q, d_Q) is an ipomset isomorphism $f : P \to Q$ for which $\sigma_P = \sigma_Q \circ f$ and $d_P = d_Q$.*

In other words, two tipomsets are isomorphic if they share the same activity intervals, durations, precedence order, interfaces, and essential event order. As for (untimed) ipomsets, isomorphisms between tipomsets are unique, hence we may switch freely between tipomsets and their isomorphism classes.

Remark 20. Analogously to ipomsets, one could define a notion of subsumption for tipomsets such that isomorphisms would be invertible subsumptions. We refrain from doing this here, mostly because we have not seen any need for it. Note that as per Example 27 below, untimings of HDTA languages are not closed under subsumption.

5.3 Translations

We now provide the translations shown in Fig. 6. First, the bijection between idwords and tipomsets. Let $I_0 d_0 P_1 d_1 \dots P_n d_n I_{n+1}$ be an interval delay word in sparse normal form. Define the ipomset $P = P_1 * \dots * P_n$ and let $d_P = \sum_{i=0}^{n} d_i$. In order to define the activity intervals, let $x \in P$ and denote

$$\mathsf{first}(x) = \min\{i \mid x \in P_i \lor x \in I_i\}, \qquad \mathsf{last}(x) = \max\{i \mid x \in P_i \lor x \in I_i\}.$$

If $\mathsf{first}(x) = 0$, then let $\sigma^-(x) = 0$, otherwise, $\sigma^-(x) = \sum_{i=0}^{\mathsf{first}(x)-1} d_i$. Similarly, if $\mathsf{last}(x) = n+1$, then let $\sigma^+(x) = d_P$; otherwise, $\sigma^+(x) = \sum_{i=0}^{\mathsf{last}(x)-1} d_i$. Using Lemma 13, this defines a mapping tGlue from idwords to tipomsets.

Example 21. Tipomset T of Example 17 is the translation of the following sparse interval delay word:

$$\left[\begin{smallmatrix} \bullet a \bullet \\ \bullet c \bullet \end{smallmatrix}\right] 1.5 \left[\begin{smallmatrix} \bullet a \bullet \\ \bullet c \end{smallmatrix}\right] 0 \left[\begin{smallmatrix} \bullet a \bullet \\ d \bullet \end{smallmatrix}\right] 1.5 \left[\begin{smallmatrix} \bullet a \bullet \\ \bullet d \end{smallmatrix}\right] 0.5 \left[\begin{smallmatrix} \bullet a \bullet \\ b \bullet \end{smallmatrix}\right] 0.5 \left[\begin{smallmatrix} \bullet a \bullet \\ \bullet b \bullet \\ c \bullet \end{smallmatrix}\right] 1 \left[\begin{smallmatrix} \bullet a \\ \bullet b \bullet \\ \bullet c \bullet \end{smallmatrix}\right] 1 \left[\begin{smallmatrix} \bullet b \bullet \\ \bullet c \end{smallmatrix}\right] 0.5 \bullet b \, 0.5 \, \mathrm{id}_\emptyset$$

Lemma 22. *The mapping* tGlue *is a bijection between idwords and tipomsets.*

Lemma 23. *The vertical mappings $i_1, \dots, i_4$ of Fig. 6 are injective and commute with the horizontal bijections.*

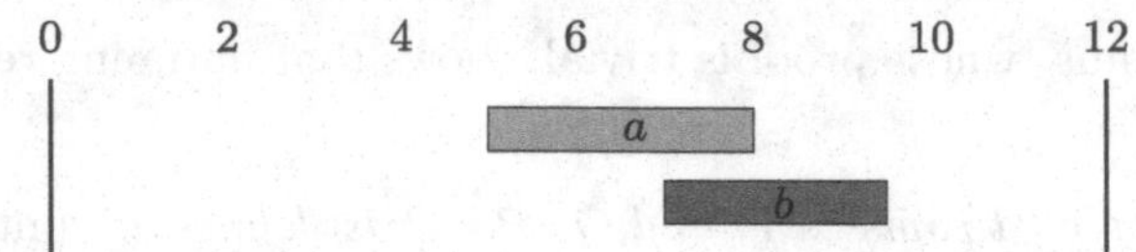

Fig. 9. Tipomset of accepting path in HDTA of Example 27

6 Languages of HDTAs

We are now ready to introduce languages of HDTAs as sets of timed ipomsets. Let $A = (\Sigma, C, L, \bot, \top, \mathsf{inv}, \mathsf{exit})$ be an HDTA and $[\![A]\!] = (S, S^\bot, S^\top, \leadsto)$. A *path* π in $[\![A]\!]$ is a finite sequence of consecutive moves $s_1 \leadsto s_2 \leadsto \cdots \leadsto s_n$, where each $s_i \leadsto s_{i+1}$ is either $s_i \leadsto^{d_i} s_{i+1}$, $s_i \leadsto^U s_{i+1}$ or $s_i \leadsto_U s_{i+1}$ for $d_i \in \mathbb{R}_{\geq 0}$ and $U \in \square$. As usual, π is *accepting* if $s_1 \in S^\bot$ and $s_n \in S^\top$.

Definition 24. *The observable content* $\mathsf{ev}(\pi)$ *of a path* π *in* $[\![A]\!]$ *is the tipomset* $(P, <_P, \dashrightarrow_P, S_P, T_P, \lambda_P, \sigma_P, d_P)$ *defined recursively as follows:*

- *if* $\pi = (l, v)$, *then* $(P, <_P, \dashrightarrow_P, S_P, T_P, \lambda_P) = \mathrm{id}_{\mathsf{ev}(l)}$, $\sigma_P(x) = [0, 0]$ *for all* $x \in P$, *and* $d_P = 0$;
- *if* $\pi = (l, v) \leadsto^d (l, v + d)$, *then* $(P, <_P, \dashrightarrow_P, S_P, T_P, \lambda_P) = \mathrm{id}_{\mathsf{ev}(l)}$, $\sigma_P(x) = [0, d]$ *for all* $x \in P$, *and* $d_P = d$;
- *if* $\pi = (l_1, v_1) \leadsto^U (l_2, v_2)$, *then* $(P, <_P, \dashrightarrow_P, S_P, T_P, \lambda_P) = {}_U{\uparrow}\mathsf{ev}(l_2)$, $\sigma_P(x) = [0, 0]$ *for all* $x \in P$, *and* $d_P = 0$;
- *if* $\pi = (l_1, v_1) \leadsto_U (l_2, v_2)$, *then* $(P, <_P, \dashrightarrow_P, S_P, T_P, \lambda_P) = \mathsf{ev}(l_1){\downarrow}_U$, $\sigma_P(x) = [0, 0]$ *for all* $x \in P$, *and* $d_P = 0$;
- *if* $\pi = \pi_1 \pi_2$, *then* $\mathsf{ev}(\pi) = \mathsf{ev}(\pi_1) * \mathsf{ev}(\pi_2)$.

Observe that by definition of $[\![A]\!]$, the first item above is a special case of the second one with $d = 0$.

Definition 25. *The* language *of an HDTA* A *is*

$$\mathcal{L}(A) = \{\mathsf{ev}(\pi) \mid \pi \text{ accepting path of } A\}.$$

Remark 26. With a few simple changes to Definition 24 above, we can define the observable content of an HDTA path as an idword instead of a tipomset. (By Lemma 22 this is equivalent.) If we define $\mathsf{ev}((l, v) \leadsto^d (l, v+d)) = d$ in the second case above and use concatenation of idwords instead of gluing composition in the last case, then $\mathsf{ev}(\pi) \in \mathsf{IDW}$. Thus, the language of an HDTA can be seen as a set of tipomsets or as a set of idwords.

Example 27. We compute the language of the HDTA A of Fig. 5. As both vertical transitions are disabled, any accepting path must proceed along the location sequence (l_0, e_1, u, e_4, l_3). The general form of accepting paths is thus

$$\begin{aligned}\pi = (l_0, v^0) &\leadsto^{d_1} (l_0, v^0 + d_1) \leadsto^a (e_1, v_2) \leadsto^{d_2} (e_1, v_2 + d_2)\\ &\leadsto^b (u, v_3) \leadsto^{d_3} (u, v_3 + d_3) \leadsto_b (e_4, v_4)\\ &\leadsto^{d_4} (e_4, v_4 + d_4) \leadsto_a (l_3, v_5) \leadsto^{d_5} (l_3, v_5 + d_5).\end{aligned}$$

Fig. 10. Two HDTAs pertaining to Remark 28

There are no conditions on d_1, as both clocks x and y are reset when leaving l_0. The conditions on x at the other four locations force $1 \leq d_2 \leq 4$, $1 \leq d_2 + d_3 \leq 4$, and $2 \leq d_2 + d_3 + d_4 \leq 5$. As y is reset when leaving e_1, we must have $1 \leq d_3 \leq 3$ and $1 \leq d_3 + d_4$, and the condition on z at l_3 forces $1 \leq d_4$. As there are no upper bounds on clocks in l_3, there are no constraints on d_5.

To sum up, $\mathcal{L}(A)$ is the set of tipomsets

$$(\{x_1, x_2\}, \emptyset, x_1 \dashrightarrow x_2, \emptyset, \emptyset, \lambda, \sigma, d_1 + \cdots + d_5)$$

with $\lambda(x_1) = a$, $\lambda(x_2) = b$, $\sigma(x_1) = [d_1, d_1 + \cdots + d_4]$ and $\sigma(x_2) = [d_1 + d_2, d_1 + d_2 + d_3]$, or equivalently the set of idwords

$$\mathrm{id}_\emptyset \, d_1 \, a{\bullet} \, d_2 \left[\begin{smallmatrix} \bullet a \bullet \\ b \bullet \end{smallmatrix}\right] d_3 \left[\begin{smallmatrix} \bullet a \\ \bullet b \bullet \end{smallmatrix}\right] d_4 \, {\bullet}b \, d_5 \, \mathrm{id}_\emptyset,$$

in which the delays satisfy the conditions above. As an example,

$$\begin{aligned} \pi = (l_0, (0,0,0)) &\leadsto^5 (l_0, (5,5,5)) \leadsto^a (e_1, (0,0,5)) \leadsto^2 (e_1, (2,2,7)) \\ &\leadsto^b (u, (2,0,7)) \leadsto^1 (u, (3,1,7)) \leadsto_b (e_4, (3,1,0)) \\ &\leadsto^{1.5} (e_4, (4.5, 2.5, 1.5)) \leadsto_a (l_3, (4.5, 2.5, 1.5)) \leadsto^{2.5} (l_3, (7,5,4)) \end{aligned}$$

is an accepting path whose associated tipomset is depicted in Fig. 9. Its interval delay word is

$$\mathrm{id}_\emptyset \, 5 \, a{\bullet} \, 2 \left[\begin{smallmatrix} \bullet a \bullet \\ b \bullet \end{smallmatrix}\right] 1 \left[\begin{smallmatrix} \bullet a \\ \bullet b \bullet \end{smallmatrix}\right] 1.5 \, {\bullet}b \, 4.5 \, \mathrm{id}_\emptyset$$

Note that $\mathrm{unt}(\mathcal{L}(A)) = \{\left[\begin{smallmatrix} a \\ b \end{smallmatrix}\right]\}$ which is not closed under subsumption.

Remark 28. We can now show why the precedence order of a tipomset *cannot* generally be induced from the timestamps. Figure 10 shows two HDTAs in which events labeled a and b happen instantly. On the left, a precedes b, and the language consists of the tipomset ab with duration 0 and $\sigma(a) = \sigma(b) = [0, 0]$. On the right, a and b are concurrent, and the language contains the tipomset $\left[\begin{smallmatrix} a \\ b \end{smallmatrix}\right]$ with the same duration and timestamps.

6.1 Language Inclusion is Undecidable

[25] introduces a translation from timed automata to HDTAs which we review below. We show that the translation preserves languages. It is not simply an

embedding of timed automata as one-dimensional HDTAs, as transitions in HDTAs are not instantaneous. We use an extra clock to force immediacy of transitions and write $\mathsf{idw}(w)$ for the idword induced by a delay word w below.

Let $A = (\Sigma, C, Q, \bot, \top, I, E)$ be a timed automaton and $C' = C \uplus \{c_T\}$, the disjoint union. In the following, we denote the components of a transition $e = (q_e, \phi_e, \ell_e, R_e, q'_e) \in E$. We define the HDTA $H(A) = (L, \bot, \top, \mathsf{inv}, \mathsf{exit})$ by $L = Q \uplus E$ and, for $q \in Q$ and $e \in E$,

$$\mathsf{ev}(q) = \emptyset, \qquad \mathsf{ev}(e) = \{\ell_e\}, \qquad \delta^0_{\ell_e}(e) = q_e\,, \qquad \delta^1_{\ell_e}(e) = q'_e\,,$$
$$\mathsf{inv}(q) = I(q), \qquad \mathsf{exit}(e) = R_e, \qquad \mathsf{inv}(e) = \phi_e \wedge c_T \leq 0, \qquad \mathsf{exit}(q) = \{c_T\}.$$

Example 29. The HDTA on the left of Fig. 10 is isomorphic to the translation of the timed automaton with the same depiction. (Because of the constraint $x \leq 0$ in the accepting location, the extra clock c_T may be removed.)

Lemma 30. *For any $q_1, q_2 \in Q$ and $v_1, v_2 : C \to \mathbb{R}_{\geq 0}$, $(q_1, v_1) \overset{a}{\leadsto} (q_2, v_2)$ is an action move of $[\![A]\!]$ if and only if $(q_1, v'_1) \leadsto^a (e, v')$ and $(e, v') \leadsto_a (q_2, v'_2)$ are moves of $[\![H(A)]\!]$ such that*

- *for all $c \in C$, $v'_1(c) = v'(c) = v_1(c)$ and $v'_2(c) = v_2(c)$;*
- *$v'_1(c_T) \in \mathbb{R}_{\geq 0}$ and $v'(c_T) = v'_2(c_T) = 0$.*

In addition, $\mathsf{idw}(\mathsf{ev}((q_1, v_1) \overset{a}{\leadsto} (q_2, v_2))) = \mathsf{ev}((q_1, v'_1) \leadsto^a (e, v') \leadsto_a (q_2, v'_2))$.

Lemma 31. *For any $a \in \Sigma$ and $d, d' \in \mathbb{R}_{\geq 0}$, $d\,a\,d'$ is the label of some path in $[\![A]\!]$ if and only if $d\,a\!\bullet 0 \bullet\! a\,d'$ is the label of some path in $[\![H(A)]\!]$.*

Theorem 32. *For any timed automaton A, $\mathcal{L}(H(A)) = \{\mathsf{idw}(w) \mid w \in \mathcal{L}(A)\}$.*

By the above theorem, we can reduce deciding inclusion of languages of timed automata to deciding inclusion of HDTA languages. It follows that inclusion of HDTA languages is undecidable:

Corollary 33. *For HDTAs A_1, A_2, it is undecidable whether $\mathcal{L}(A_1) \subseteq \mathcal{L}(A_2)$.*

6.2 Untimings of HDTA Languages are (Almost) Regular

We revisit the notions of region equivalence and region automaton from [25] in order to study untimings of languages of HDTAs. For $d \in \mathbb{R}_{\geq 0}$ we write $\lfloor d \rfloor$ and $\langle d \rangle$ for the integral, respectively fractional, parts of d, so that $d = \lfloor d \rfloor + \langle d \rangle$.

Let $A = (\Sigma, C, L, \bot_L, \top_L, \mathsf{inv}, \mathsf{exit})$ be an HDTA. Denote by M the maximal constant which appears in the invariants of A and let $\cong$ denote the region equivalence on $\mathbb{R}^C_{\geq 0}$ induced by A. That is, valuations $v, v' : C \to \mathbb{R}_{\geq 0}$ are *region equivalent* if

- $\lfloor v(x) \rfloor = \lfloor v'(x) \rfloor$ or $v(x), v'(x) > M$, for all $x \in C$, and
- $\langle v(x) \rangle = 0$ iff $\langle v'(x) \rangle = 0$, for all $x \in C$, and
- $\langle v(x) \rangle \leq \langle v(y) \rangle$ iff $\langle v'(x) \rangle \leq \langle v'(y) \rangle$ for all $x, y \in C$.

[25] introduces a notion of untimed bisimulation for HDTA and shows that $\cong$ is an untimed bisimulation. An immediate consequence is the following.

Lemma 34. *Let $t = (l_1, v_1) \leadsto (l_2, v_2)$ be a transition in $[\![A]\!]$. For all $v_1' \cong v_1$ there exists a transition $t' = (l_1, v_1') \leadsto (l_2, v_2')$ such that $v_2' \cong v_2$.* □

As usual, a *region* is an equivalence class of $\mathbb{R}^C_{\geq 0}$ under $\cong$. Let $R = \mathbb{R}^C_{\geq 0/\cong}$ denote the set of regions, then R is finite [3].

Definition 35. *The* region automaton *of A is the transition system $R(A) = (S, S^\perp, S^\top, \twoheadrightarrow)$ given as follows:*

$$
\begin{aligned}
&S = \{(l, r) \in L \times R \mid r \subseteq [\![\mathsf{inv}(l)]\!]\} \cup \{(l^0_\perp, \{v^0\}) \mid l^0 \in \bot_L\} \\
&S^\perp = \{(l^0_\perp, \{v^0\}) \mid l^0 \in \bot_L\} \qquad S^\top = S \cap \top_L \times R \\
&\twoheadrightarrow = \{((l^0_\perp, \{v^0\}), \mathrm{id}_{\mathsf{ev}(l^0)}, (l^0, \{v^0\})) \mid l^0 \in \bot_L\} \\
&\quad \cup \{((l, r), \mathrm{id}_{\mathsf{ev}(l)}, (l, r')) \mid \exists v \in r, v' \in r', d \in \mathbb{R}_{\geq 0} : (l, v) \leadsto^d (l, v'), v' = v + d\} \\
&\quad \cup \{((l, r), {}_U{\uparrow}\mathsf{ev}(l'), (l', r')) \mid \exists v \in r, v' \in r' : (l, v) \leadsto^U (l', v')\} \\
&\quad \cup \{((l, r), \mathsf{ev}(l){\downarrow}_U, (l', r')) \mid \exists v \in r, v' \in r' : (l, v) \leadsto_U (l', v')\}
\end{aligned}
$$

Extra copies of start locations are added in order to avoid paths on the empty word ϵ. This construction is similar to the construction of an ST-automaton from an HDA [5]. The region automaton of A is a standard finite automaton whose transitions are labeled by elements of $\mathsf{St} \cup \mathsf{Te}$. Its language is a set of coherent words of $(\mathsf{St} \cup \mathsf{Te})^*$ [5].

Lemma 36. *A path $(l^0, v_0) \leadsto \cdots \leadsto (l_p, v_p)$ is accepting in $[\![A]\!]$ if and only if $(l^0_\perp, r_0) \twoheadrightarrow (l^0, r_0) \twoheadrightarrow \cdots \twoheadrightarrow (l_p, r_p)$ with $v_i \in r_i$ is accepting in $R(A)$.*

Theorem 37. *For any HDTA A, $\mathcal{L}(R(A)) = \mathrm{unt}(\mathcal{L}(A))$.*

By the Kleene theorem for finite automata, $\mathcal{L}(R(A))$ is represented by a regular expression over $\mathsf{St} \cup \mathsf{Te}$. Since $P_0 * P_1 * \cdots * P_n$ is accepted by A if and only if the coherent word $\mathrm{unt}(P_0)\,\mathrm{unt}(P_1) \ldots \mathrm{unt}(P_n)$ is accepted by $R(A)$, Theorems 6 and 37 now imply the following.

Corollary 38. *For any HDTA A, $\mathrm{unt}(\mathcal{L}(A)){\downarrow}$ is a regular ipomset language.*

In [5] it is shown that inclusion of regular ipomset languages is decidable. Now untimings of HDTA languages are not regular because they are not closed under subsumption, but the proof in [5], using ST-automata, immediately extends to a proof of the fact that also inclusion of untimings of HDTA languages is decidable:

Corollary 39. *For HDTAs A_1 and A_2, it is decidable whether $\mathrm{unt}(\mathcal{L}(A_1)) \subseteq \mathrm{unt}(\mathcal{L}(A_2))$.*

7 Conclusion and Perspectives

We have introduced a new language-based semantics for real-time concurrency, informed by recent work on higher-dimensional timed automata (HDTAs) and on languages of higher-dimensional automata. On one side we have combined the delay words of timed automata with the step sequences of higher-dimensional automata into interval delay words. On the other side we have generalized the timed words of timed automata and the ipomsets (*i.e.*, pomsets with interfaces) of higher-dimensional automata into timed ipomsets. We have further shown that both approaches are equivalent.

Higher-dimensional timed automata model concurrency with higher-dimensional cells and real time with clock constraints. Analogously, timed ipomsets express concurrency by partial orders and real time by interval timestamps on events. Compared to related work on languages of time Petri nets, what is new here are the interfaces and the fact that each event has two timestamps (instead of only one), the first marking its beginning and the second its termination. This permits to introduce a gluing operation for timed ipomsets which generalizes serial composition for pomsets. It further allows us to generalize step decompositions of ipomsets into a notion of interval delay words which resemble the delay words of timed automata.

As an application, we have shown that language inclusion of HDTAs is undecidable, but that the untimings of their languages have enough regularity to imply decidability of untimed language inclusion.

Perspectives. We have seen that unlike languages of higher-dimensional automata, untimings of HDTA languages are not closed under subsumption. This relates HDTAs to partial higher-dimensional automata [23,31] and calls for the introduction of a proper language theory of these models.

Secondly, higher-dimensional automata admit Kleene and Myhill-Nerode theorems [27,32], but for timed automata this is more difficult [8]. We are wondering how such properties will play out for HDTAs.

Timed automata are very useful in real-time model checking, and our language-based semantics opens up first venues for real-time concurrent model checking using HDTAs and some linear-time logic akin to LTL. What would be needed now are notions of simulation and bisimulation— we conjecture that as for timed automata, these should be decidable for HDTAs— and a relation with CTL-type logics. One advantage of HDTAs is that they admit a partial-order semantics, so partial-order reduction (which is difficult for timed automata [15,36]) should not be necessary.

Finally, a note on robustness. Adding information about durations and timings of events to HDTAs raises questions similar to those already existing in timed automata. Indeed, the model of timed automata supports unrealistic assumptions about clock precision and zero-delay actions, and adding concurrency makes the need for robustness in HDTAs even more crucial. It is thus pertinent to study the robustness of HDTAs and their languages under delay perturbations, similarly for example to the work done in [17,21,22].

Robustness may be formalized using notions of distances and topology, see for example [9,10,24,30,38]. Distances between timed words need to take permutations of symbols into account [9], and it seems promising to use partial orders and timed ipomsets to formalize this.

References

1. Aceto, L., Ingólfsdóttir, A., Larsen, K.G., Srba, J.: Reactive Systems. Cambridge University Press, Cambridge (2007)
2. Alur, R., Dill, D.: Automata for modeling real-time systems. In: Paterson, M.S. (ed.) Automata, Languages and Programming. Lecture Notes in Computer Science, vol. 443, pp. 322–335. Springer, Berlin (1990). https://doi.org/10.1007/bfb0032042
3. Alur, R., Dill, D.L.: A theory of timed automata. Theoret. Comput. Sci. **126**(2), 183–235 (1994)
4. Amrane, A., Bazille, H., Clement, E., Fahrenberg, U.: Languages of higher-dimensional timed automata. CoRR, abs/2401.17444 (2024)
5. Amrane, A., Bazille, H., Fahrenberg, U., Ziemianski, K.: Closure and decision properties for higher-dimensional automata. In: Abraham, E., Dubslaff, C., Tarifa, S.L.T. (eds.) Theoretical Aspects of Computing - ICTAC 2023. Lecture Notes in Computer Science, vol. 14446, pp. 295–312. Springer, Cham (2023). https://doi.org/10.1007/978-3-031-47963-2_18
6. André, É.: IMITATOR: a tool for synthesizing constraints on timing bounds of timed automata. In: Leucker, M., Morgan, C. (eds.) Theoretical Aspects of Computing - ICTAC 2009. Lecture Notes in Computer Science, vol. 5684, pp. 336–342. Springer, Berlin (2009). https://doi.org/10.1007/978-3-642-03466-4_22
7. André, É.: IMITATOR 3: synthesis of timing parameters beyond decidability. In: Silva, A., Leino, K.R.M. (eds.) Computer Aided Verification. Lecture Notes in Computer Science(), vol. 12759, pp. 552–565. Springer, Cham (2021). https://doi.org/10.1007/978-3-030-81685-8_26
8. Asarin, E.: Challenges in timed languages: from applied theory to basic theory. Bull. EATCS **83**, 106–120 (2004)
9. Asarin, E., Basset, N., Degorre, A.: Entropy of regular timed languages. Inf. Comput. **241**, 142–176 (2015)
10. Asarin, E., Basset, N., Degorre, A.: Distance on timed words and applications. In: Jansen, D., Prabhakar, P. (eds.) Formal Modeling and Analysis of Timed Systems. Lecture Notes in Computer Science(), vol. 11022, pp. 199–214. Springer, Cham (2018). https://doi.org/10.1007/978-3-030-00151-3_12
11. Aura, T., Lilius, J.: A causal semantics for time Petri nets. Theoret. Comput. Sci. **243**(1–2), 409–447 (2000)
12. Balaguer, S., Chatain, T., Haar, S.: A concurrency-preserving translation from time Petri nets to networks of timed automata. Formal Methods Syst. Des. **40**(3), 330–355 (2012)
13. Behrmann, G., David, A., Larsen, K.G.: A tutorial on Uppaal. In: Bernardo, M., Corradini, F. (eds.) Formal Methods for the Design of Real-Time Systems. Lecture Notes in Computer Science, vol. 3185, pp. 200–236. Springer, Berlin (2004). https://doi.org/10.1007/978-3-540-30080-9_7
14. Behrmann, G., et al.: Uppaal 4.0. In: QEST, pp. 125–126. IEEE Computer Society (2006)

15. Bønneland, F.M., Jensen, P.G., Larsen, K.G., Muñiz, M., Srba, J.: Start pruning when time gets urgent: partial order reduction for timed systems. In: Chockler, H., Weissenbacher, G. (eds.) Computer Aided Verification. Lecture Notes in Computer Science(), vol. 10981, pp. 527–546. Springer, Cham (2018). https://doi.org/10.1007/978-3-319-96145-3_28
16. Bouyer, P., Fahrenberg, U., Larsen, K.G., Markey, N., Ouaknine, J., Worrell, J.: Model checking real-time systems. In: Clarke, E., Henzinger, T., Veith, H., Bloem, R. (eds.) Handbook of Model Checking, pp. 1001–1046. Springer, Cham (2018). https://doi.org/10.1007/978-3-319-10575-8_29
17. Bouyer, P., Fang, E., Markey, N.: Permissive strategies in timed automata and games. Electr. Commun. EASST **72** (2015)
18. Chatain, T.: Concurrency in real-time distributed systems, from unfoldings to implementability. PhD thesis, École normale supérieure de Cachan (2013)
19. Chatain, T., Jard, C.: Complete finite prefixes of symbolic unfoldings of safe time petri nets. In: Donatelli, S., Thiagarajan, P.S. (eds.) Petri Nets and Other Models of Concurrency - ICATPN 2006. Lecture Notes in Computer Science, vol. 4024, pp. 125–145. Springer, Berlin (2006). https://doi.org/10.1007/11767589_8
20. Chatain, T., Jard, C.: Back in time Petri nets. In: Braberman, V., Fribourg, L. (eds.) Formal Modeling and Analysis of Timed Systems. Lecture Notes in Computer Science, vol. 8053, pp. 91–105. Springer, Berlin (2013). https://doi.org/10.1007/978-3-642-40229-6_7
21. Clement, E.: Robustness of timed automata: computing the maximally-permissive strategies. PhD thesis, University of Rennes 1, France (2022)
22. Clement, E., Jéron, T., Markey, N., Mentré, D.: Computing maximally-permissive strategies in acyclic timed automata. In: Bertrand, N., Jansen, N. (eds.) Formal Modeling and Analysis of Timed Systems. Lecture Notes in Computer Science(), vol. 12288, pp. 111–126. Springer, Cham (2020). https://doi.org/10.1007/978-3-030-57628-8_7
23. Dubut, J.: Trees in partial higher dimensional automata. In: Bojanczyk, M., Simpson, A. (eds.) Foundations of Software Science and Computation Structures. Lecture Notes in Computer Science(), vol. 11425, pp. 224–241. Springer, Cham (2019). https://doi.org/10.1007/978-3-030-17127-8_13
24. Fahrenberg, U.: A generic approach to quantitative verification. Habilitation thesis, Paris-Saclay University (2022)
25. Fahrenberg, U.: Higher-dimensional timed and hybrid automata. Leibniz Trans. Embed. Syst. **8**(2), 1–16 (2022)
26. Fahrenberg, U., Johansen, C., Struth, G., Ziemianski, K.: Languages of higher-dimensional automata. Math. Struct. Comput. Sci. **31**, 1–39 (2021)
27. Fahrenberg, U., Johansen, C., Struth, G., Ziemiański, K.: A Kleene theorem for higher-dimensional automata. In: Klin, B., Lasota, S., Muscholl, A. (eds.) CONCUR, volume 243 of LIPIcs, pp. 1– 18. Schloss Dagstuhl - Leibniz-Zentrum für Informatik (2022)
28. Fahrenberg, U., Johansen, C., Struth, G., Ziemiański, K.: Posets with interfaces as a model for concurrency. Inf. Comput. **285**(B), 104914 (2022)
29. Fahrenberg, U., Johansen, C., Struth, G., Ziemiański, K.: Catoids and modal convolution algebras. Algebra Univers. **84**(10) (2023)
30. Fahrenberg, U., Legay, A.: The quantitative linear-time-branching-time spectrum. Theoret. Comput. Sci. **538**, 54–69 (2014)
31. Fahrenberg, U., Legay, A.: Partial higher-dimensional automata. In: Moss, L.S., Sobocinski, P. (eds.) CALCO, volume 35 of LIPIcs, pp. 101–115 (2015)

32. Fahrenberg, U., Ziemiański, K.: A Myhill-Nerode theorem for higher-dimensional automata. In: Gomes, L., Lorenz, R. (eds.) Application and Theory of Petri Nets and Concurrency. Lecture Notes in Computer Science, vol. 13929, pp. 167–188. Springer, Cham (2023). https://doi.org/10.1007/978-3-031-33620-1_9
33. Fishburn, P.C.: Interval Orders and Interval Graphs: A Study of Partially Ordered Sets. Wiley, Hoboken (1985)
34. Fleischhack, H., Stehno, C.: Computing a finite prefix of a time Petri net. In: Esparza, J., Lakos, C. (eds.) Application and Theory of Petri Nets 2002. Lecture Notes in Computer Science, vol. 2360, pp. 163–181. Springer, Berlin (2002). https://doi.org/10.1007/3-540-48068-4_11
35. Gardey, G., Lime, D., Magnin, M., Roux, O.H.: Romeo: a tool for analyzing time Petri nets. In: Etessami, K., Rajamani, S.K. (eds.) Computer Aided Verification. Lecture Notes in Computer Science, vol. 3576, pp. 418–423. Springer, Berlin (2005). https://doi.org/10.1007/11513988_41
36. Govind, R., Herbreteau, F., Srivathsan, B., Walukiewicz, I.: Abstractions for the local-time semantics of timed automata: a foundation for partial-order methods. In: Baier, C., Fisman, D. (eds.) LICS, pp. 1–14. ACM (2022)
37. Grabowski, J.: On partial languages. Fund. Inform. **4**(2), 427 (1981)
38. Gupta, V., Henzinger, T.A., Jagadeesan, R.: Robust timed automata. In: Maler, O. (ed.) Hybrid and Real-Time Systems. Lecture Notes in Computer Science, vol. 1201, pp. 331–345. Springer, Berlin (1997). https://doi.org/10.1007/BFb0014736
39. Janicki, R., Koutny, M.: Operational semantics, interval orders and sequences of antichains. Fund. Inform. **169**(1–2), 31–55 (2019)
40. Larsen, K.G., Fahrenberg, U., Legay, A.: From timed automata to stochastic hybrid games. In: Dependable Software Systems Engineering, pp. 60–103. IOS Press (2017)
41. Larsen, K.G., Pettersson, P., Yi, W.: Uppaal in a nutshell. Int. J. Softw. Tools Technol. Transfer **1**(1-2), 134–152 (1997)
42. Lime, D., Roux, O.H., Seidner, C., Traonouez, L.M.: Romeo: a parametric model-checker for Petri nets with stopwatches. In: Kowalewski, S., Philippou, A. (eds.) Tools and Algorithms for the Construction and Analysis of Systems. Lecture Notes in Computer Science, vol. 5505, pp. 54–57. Springer, Berlin (2009). https://doi.org/10.1007/978-3-642-00768-2_6
43. Merlin, P.M., Farber, D.J.: Recoverability of communication protocols - implications of a theoretical study. IEEE Trans. Commun. **24**(9), 1036–1043 (1976)
44. Milner, R.: Communication and Concurrency. Prentice Hall, Hoboken (1989)
45. Nielsen, M., Plotkin, G.D., Winskel, G.: Petri nets, event structures and domains, part i. Theoret. Comput. Sci. **13**, 85–108 (1981)
46. Petri, C.A.: Kommunikation mit Automaten. Institut für Instrumentelle Mathematik, Bonn. Schriften des IIM Nr. 2 (1962)
47. Pratt, V.R.: Modeling concurrency with geometry. In: Wise, D.S. (ed.) POPL, pp. 311–322. ACM Press (1991)
48. van Glabbeek, R.J.: Bisimulations for higher dimensional automata. Email message (1991). http://theory.stanford.edu/~rvg/hda
49. van Glabbeek, R.J.: On the expressiveness of higher dimensional automata. Theor. Comput. Sci. **356**(3), 265–290 (2006). See also [50]
50. van Glabbeek, R.J.: Erratum to "on the expressiveness of higher dimensional automata". Theor. Comput. Sci. **368**(1–2), 168–194 (2006)
51. van Glabbeek, R.J., Plotkin, G.D.: Configuration structures. In: LICS, pp. 199–209. IEEE Computer Society (1995)
52. van Glabbeek, R.J., Plotkin, G.D.: Configuration structures, event structures and Petri nets. Theoret. Comput. Sci. **410**(41), 4111–4159 (2009)

Reachability

Petri Net Synthesis from a Reachability Set

Eike Best[1] and Raymond Devillers[2(✉)]

[1] Department of Computing Science, Carl von Ossietzky Universität Oldenburg, 26111 Oldenburg, Germany
eike.best@informatik.uni-oldenburg.de

[2] Département d'Informatique, Université Libre de Bruxelles, 1050 Brussels, Belgium
raymond.devillers@ulb.be

Abstract. Classical Petri net synthesis is a method to generate a Petri net from a labelled transition system. In this paper, by contrast, it is assumed that a (finite) set of markings is given, and an algorithm is described which generates a Petri net of some class having exactly this set as its reachability set. A notion of T-monotonicity simplifying the problem is introduced, but it is also shown that for some non-T-monotonic classes, the synthesis may nevertheless be solved algorithmically.

Keywords: Petri Nets · Synthesis · Reachability Set · Net Classes · T-monotonicity

1 Introduction: Place and Transition Synthesis

Those last years, several works have been dedicated to the synthesis of various classes of place/transition Petri nets from a given labelled transition system [1–5]. The idea is to determine if a net of some class exists whose reachability graph is isomorphic to the given LTS, to build one if the answer is positive, and otherwise, to explain why if the answer is negative. This means that the set of transition names is known and some places have to be found (and connected to the transitions) to progressively restrict the reachability graph until the target has been reached (the names of the places are irrelevant).

A symmetric kind of problem may be considered, where a reachability set is given, meaning that the number of places is known (their names are irrelevant) and it is asked if some net of a fixed class may be found which generates a corresponding set of reachable marking. This means here that we have to find transitions connected to the places in such a way that progressively the correct reachability set of markings is obtained (here the names of the transitions are irrelevant: only the way they are connected to the places is interesting). Again, if we face an impossibility, we want to know why. If the construction works, we have thus built a system of the adequate kind visiting all the configurations of the given predefined set, and only them.

L. M. Kristensen and J. M. van der Werf (Eds.): PETRI NETS 2024, LNCS 14628, pp. 223–243, 2024.
https://doi.org/10.1007/978-3-031-61433-0_11

The structure of the paper is as follows: after recalling the context, Sect. 3 presents a simple solution when no special constraint is imposed. The next section considers the cases where a target class is searched for, which satisfies a T-monotonicity property, and several kinds of such net or system classes are illustrated in Sect. 5. Then, several classes are considered that show how it is possible to get rid of the T-monotonicity constraint. Some complexity issues are discussed in Sect. 7, and a somewhat concrete example is described in the penultimate section. The last section, as usual, concludes and suggests possible future developments.

2 The Context

A labelled transition system with initial state, *LTS* for short, is a quadruple $TS = (S, \rightarrow, T, \imath)$ where S is the set of states, T is the set of labels, $\rightarrow \subseteq (S \times T \times S)$ is the transition relation, and $\imath \in S$ is the initial state.
A label t is *enabled* at $s \in S$, written $s[t\rangle$, if $\exists s' \in S \colon (s, t, s') \in \rightarrow$, in which case s' is said to be *reachable* from s by the firing of t, and we write $s[t\rangle s'$. Generalising to any sequence $\sigma \in T^*$, $s[\varepsilon\rangle$ and $s[\varepsilon\rangle s$ are always true, with ε being the empty sequence; and $s[t\sigma\rangle s'$, i.e., $t\sigma$ is *enabled* from state s and leads to s', if there is some state s'' with $s[t\rangle s''$ and $s''[\sigma\rangle s'$.
A state s' is *reachable* from state s if $\exists \sigma \in T^* : s[\sigma\rangle s'$. The set of states reachable from s is denoted by $[s\rangle$.

A (finite, place-transition, arc-weighted) Petri net is a triple $PN = (P, T, F)$ such that P is a finite set of places, T is a finite set of transitions, with $P \cap T = \emptyset$, $F \colon ((P \times T) \cup (T \times P)) \rightarrow \mathbb{N}$ is the flow function ($F(p, t)$ is the weight of the arc from p to t, and $F(t, p)$ is the weigth of the arc from t to p in the net PN).
The incidence matrix C of a Petri net is the integer place-transition matrix with components $C(p, t) = F(t, p) - F(p, t)$, where p is a place and t is a transition. For any transition $t \in T$, we shall denote by $F(., t)$ the P-vector such that for any $p \in P$, $F(., t)(p) = F(p, t)$, and similarly for $C(., t)$ and $F(t, .)$. For any place $p \in P$ and transition $t \in T$, if $F(p, t) > 0 < F(t, p)$, there is a side-loop (or side condition) between p and t, $\min(F(p, t), F(t, p))$ characterising the level of the side-loop (level 0 meaning no side-loop).

A marking is a mapping $M \colon P \rightarrow \mathbb{N}$, indicating the number of (black) tokens in each place. Markings are provided with a partial order ($M \leq M'$ iff $M(p) \leq M'(p)$ for each $p \in P$). The latter may be promoted into a strict partial order $M \lneqq M'$ by requesting $M \leq M'$ and $M \neq M'$; it is then said that M is dominated by M'. Markings may also be added, subtracted and multiplied by an integer. A Petri net system is a net provided with an initial marking (P, T, F, M_0).

A transition $t \in T$ is enabled at a marking M, denoted by $M \xrightarrow{t}$, if $M \geq F(., t)$. If t is enabled at M, then t can occur (or fire) in M, leading to the marking M' defined by $M' = M + C(., t)$ and denoted by $M \xrightarrow{t} M'$. The reachability set of PN is the set $[M_0\rangle$ of all markings reachable from M_0; the reachability graph RG of PN is the labelled transition system whose initial state is M_0, whose vertices are the reachable markings (the reachability set), and whose edges are

$\{(M,t,M') \mid M \xrightarrow{t} M'\}$. Many subclasses of Petri nets or Petri net systems may be defined, based on the structure of the net and/or of its initial marking and/or of its reachability graph. Some of them will be considered below.

An elementary property of Petri nets is the state equation which expresses that, if $M[\sigma\rangle M'$, then $M' = M + C \cdot \Psi(\sigma)$, where $\Psi(\sigma)$ is the Parikh vector of the firing sequence σ, which counts the number of occurrences of each transition in the sequence.

A labelled transition system is PN-solvable if it is isomorphic to the reachability graph of a Petri net system (called the solution); it is $\mathcal{C}$-solvable if there is a solution in class $\mathcal{C}$. The problem of finding a Petri net having a given reachability graph can be qualified as a place synthesis, since it corresponds to find, if possible, places with some initial marking (respecting the needed constraints), connected to the transitions labelling the given transition system, generating an adequate reachability graph.

Let $\mathcal{M}$ be a set of (non-negative) vectors with index set P (so that here we know the place set; usually, we shall assume $P = \{p_1, p_2, \ldots, p_{|P|}\}$) and $M_0 \in \mathcal{M}$ be a selected initial marking. Here, we shall assume that $\mathcal{M}$ is finite, say $\mathcal{M} = \{M_0, M_1, \ldots, M_{|\mathcal{M}|-1}\}$, with M_0 coming first, and that M_0 is not dominated by any marking in $\mathcal{M}$ (otherwise, the reachability set may not be finite and cannot correspond to $\mathcal{M}$, since $(M_0 \xrightarrow{\sigma} M' \wedge M_0 \lneqq M') \Rightarrow M_0 \xrightarrow{\sigma^k} M_0 + k \cdot (M' - M_0)$ for any $k \in \mathbb{N}$). The problem is to find a (finite) set of transitions connected to the places so that the corresponding Petri net system belongs to some chosen class $\mathcal{C}$ and has $\mathcal{M}$ as reachability set. This will be called a transition-($\mathcal{C}-$)synthesis from $\mathcal{M}$, since places are known from the beginning and adequate transitions have to be found, if possible. It is also possible to let M_0 unfixed, and we may try to find one (among the non-dominated ones) leading to an adequate solution, but we shall not consider this variant here.

3 The Unconstrained Case

Since $\mathcal{M}$ is finite and M_0 is not dominated in it, there is always a transition-synthesis that may be obtained easily: for $M \in \mathcal{M} \setminus \{M_0\}$, we shall introduce a transition t_M such that $F(., t_M) = M_0$ and $F(t, .) = M$. Then, we can see that $M_0 \xrightarrow{t_M} M$ and no $t_{M'}$ may fire from M, for any $M' \in \mathcal{M} \setminus \{M_0\}$ (including M), due to the assumption that M_0 is not dominated in $\mathcal{M}$.

This is illustrated by Fig. 1 where there is a single place p_1, the desired reachability set is $\mathcal{M}_1 = \{(5), (0), (1), (2), (3)\}$, and the initial configuration is $M_0 = (5)$. The net built by means of the above procedure then has 4 transitions, and its reachability graph has 4 arcs (with distinct labels).

Of course, this will in general not work if the considered problem targets special net classes, i.e., if additional constraints are added, or if some optimisation is searched for (in general, this may be reduced to satisfy some constraints; for instance, if we want to optimise the number of transitions, we may search if some solution exists with k transitions, but not with $k-1$ transitions).

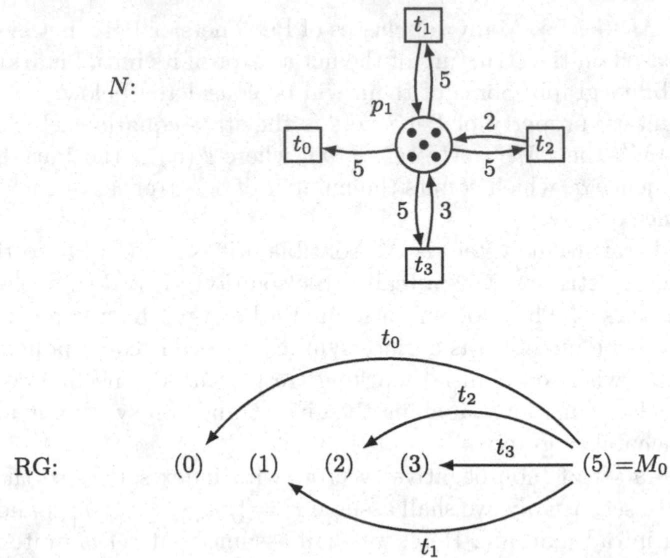

Fig. 1. An unconstrained solution, starting from $\mathcal{M}_1 = \{(5), (0), (1), (2), (3)\}$.

4 A General T-Monotonic Constrained Schema

A subclass $\mathcal{C}$ of Petri net systems will be called T-monotonic if, whenever a system (P, T, F, M_0) belongs to it, so does any system obtained by dropping one or more transitions from it. In particular, this means that $(P, \emptyset, \emptyset, M_0)$ is still in $\mathcal{C}$, and that if (P, T, F, M_0) does not belong to $\mathcal{C}$, neither does any extension of it obtained by adding some transitions. We shall here (in this section) assume we consider a T-monotonic class[1] $\mathcal{C}$ of systems.

Typically, a T-monotonic class will be defined by a (conjunction of) constraints of the following kind:

1.
$$\sum_{t \in T} f(t, P) \leq k \tag{1}$$
$f(t, P)$ being a non-negative weight function characterising the way transition t is connected to the places in P, the total weight being bounded from above by k. Clearly, if one or more transitions are dropped, the constraint is relaxed, hence its T-monotonicity.

2.
$$\sum_{M \in \mathcal{M}} g(M, T) \leq k \tag{2}$$
$g(M, T)$ being a non-negative non-decreasing (in terms of T) weight function characterising the number of the arcs labelled by transitions from T around

[1] Of course, we shall also assume that the membership to this class is decidable.

marking M, the total weight being bounded from above by k. Clearly, if one or more transitions are dropped, the reachability set may only be reduced (or stay the same) and the constraint is relaxed, hence the T-monotonicity.

3.

$$\bigcap_{t \in T} \alpha(t, P) \tag{3}$$

$\alpha(t, P)$ being a predicate restricting the way transition t is connected to the places in P. Again, if one or more transitions are dropped, the constraint is relaxed, hence the T-monotonicity.

4.

$$\bigcap_{M \in [M_0\rangle} \bigcap_{t_1, t_2 \in T} \beta(t_1, t_2, M) \tag{4}$$

$\beta(t_1, t_2, M)$ being a predicate restricting the way the pair of transitions $t_1 - t_2$ behave at each reachable marking in $\mathcal{M}$. If one or more transitions are dropped, the reachability set may only be reduced (or stay the same) and the constraint on the remaining pairs of transitions is preserved, hence the T-monotonicity.

5.

$$\bigcap_{p \in P} \gamma(p, T) \tag{5}$$

$\gamma(p, T)$ being a predicate restricting the way a place p is connected to the transitions in T, anti-monotonic in T (i.e., $T_1 \subset T_2 \Rightarrow (\neg\gamma(p, T_1) \Rightarrow \neg\gamma(p, T_2))$). If one or more transitions are dropped, since P is unchanged, the constraint is preserved, hence the T-monotonicity.

We assume to start from a non-connected (unless $|\mathcal{M}| = 1$) transition system TS_0, composed of the node set $\mathcal{M}$, without any transition and thus without any (labelled) arc, so that $[M_0\rangle = \{M_0\}$. We then execute the algorithm in Fig. 2.

The basic mechanism of this algorithm is shown schematically in Fig. 3.

Since this is only the skeleton of an algorithm, some comments need to be added.

Ill-formed problems may, for example, include the following ones:

- some constraints are unknown;
- the place set is assumed to be bounded and P is too large;
- the reachability set is assumed to be bounded and $\mathcal{M}$ is too large;
- some marking(s) of $\mathcal{M}$ are not built on P;
- the net is assumed to be safe (no place may receive more than one token) and some $M \in \mathcal{M}$ is not a $\{0, 1\}$-vector;
- more generally, the net is assumed to be k-bounded (no place may receive more than k tokens), and $M(p) > k$ for some $M \in \mathcal{M}$ and $p \in P$.

Updating the transition set means adding transition t (together with $F(., t)$ and $F(t, .)$) to the Petri net under construction, and a problem must be launched if this leads out of the target class $\mathcal{C}$. Updating the arc set consists, for each marking $M \in \mathcal{M}$ such that $M \geq F(., t)$, to add an arc $M \xrightarrow{t} M + F(t, .) - F(., t)$.

```
input the constraints, P, 𝓜 and M_0
if a problem occurred then output "ill-formed problem" and stop
end if
while [M_0⟩ ≠ 𝓜 do
    choose a triple (M_1, t, M_2) such that
        • M_1 ∈ [M_0⟩
        • M_2 ∈ 𝓜 \ [M_0⟩
        • M_1 --t--> M_2
            plus some additional constraints depending on the target class
    update the sets of transitions and arcs in the transition system
    if some problem occurs then
        wipe out the last updates and choose another triple (backtracking)
    end if
end while
output "solution found", the transition set and, if desired, the arc set
```

Fig. 2. A general schema for synthesising a T-monotonic system.

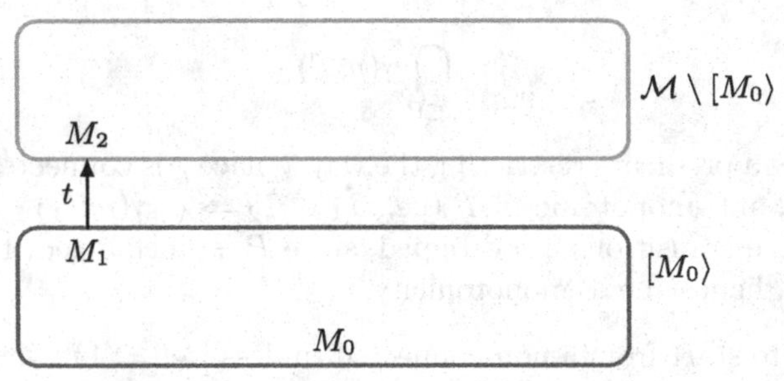

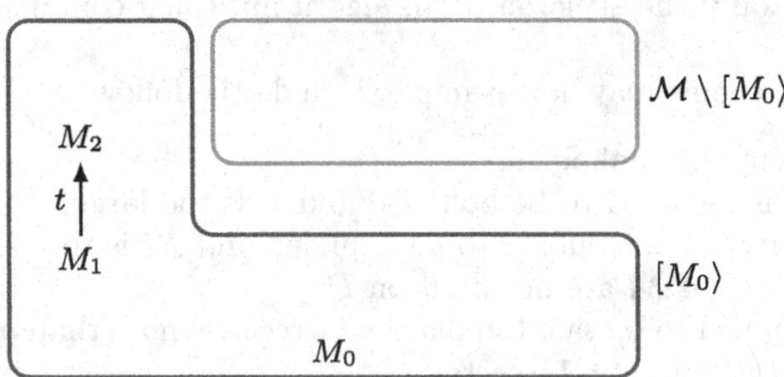

Fig. 3. The general scheme. Updating means adding a transition from a previously reachable marking M_1 (before updating) to a newly reachable marking M_2 (after updating). If for some reason this is not possible (i.e., some problem occurs), backtracking will reject the triple $M_1 \xrightarrow{t} M_2$ and choose another one.

Of course, a problem must be launched if $(M + F(t, .) - F(., t)) \notin \mathcal{M}$,[2] or if some added arcs do not fulfill the constraints of the target class, leading to wipe out the last updates and to choose another triple.

Another kind of problem occurs if one has to choose another triple but no new one is available (all the possibilities to extend the transition set have been exhausted). Then wiping out the last updates occurs before adding new transitions and arcs, so that wiping and choosing a new triple will occur at the previous level of the while loop (and this can occur several times in a row). If this leads to go before the first level of the while loop, this means that all possibilities to build a solution have been exhausted; the algorithm will thus output "no solution has been found" and stop.

If $M_2 > M_1$ (we may not have $M_2 = M_1$ since they belong to disjoint subsets of markings), transition t may be executed as many times as we want and, since $\mathcal{M}$ is finite, it is certain that during the arc updates we will reach a situation where $(M + F(t, .) - F(., t)) \notin \mathcal{M}$. It is thus a good idea to browse M_1 following a non-increasing topological sort of $[M_0\rangle$, and to browse M_2 following an increasing topological sort of $\mathcal{M} \setminus [M_0\rangle$, rejecting it if $M_1 \lneq M_2$.

Even if we already know the marking $M_1 \in [M_0\rangle$ which is reachable and the marking $M_2 \in \mathcal{M} \setminus [M_0\rangle$ which we want to reach next, the choice of a transition t effecting $M_1 \xrightarrow{t} M_2$ is not unique in general. For instance, Fig. 4 shows three transitions t_2, t_2', t_2'', all of which lead from the state (5) to the state (2) in the previous example (Fig. 1).

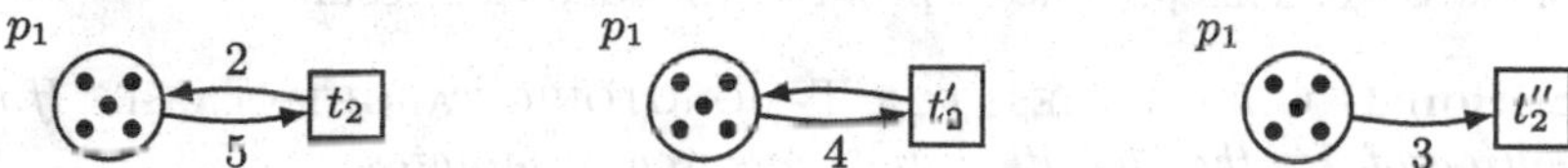

Fig. 4. Three transitions t_2, t_2', and t_2'', leading from $M_1 = (5)$ to $M_2 = (2)$. They correspond, respectively, to $X = (0)$, $X = (1)$ and $X = (2)$.

In general, we are free to choose a place-based integer vector X such that

$$F(., t) = M_1 - X \geq 0 \text{ and } F(t, .) = M_2 - X \geq 0 \tag{6}$$

because the state equation is satisfied with all of them. This choice allows different arc sets between t and the places of the net, as exemplified in Fig. 4. For example, with $X = (1)$, Equation (6) yields $F(., t) = 4$ and $F(t, .) = 1$, that is, the transition t_2' shown in the middle of the figure. This choice is not completely arbitrary, because X is bounded as follows:

$$0 \leq X \leq \min(M_1, M_2) \tag{7}$$

where the min operation is understood componentwise.[3] This is because $0 \nleq X$ creates a transition t which cannot be executed, while $X \nleq \min(M_1, M_2)$ leads

[2] This is a kind of incoherence of t with respect to $\mathcal{M}$.
[3] Such that, for example, $\min((1, 2), (2, 1)) = (1, 1)$.

to negative arc weights, which are disallowed in place/transition Petri nets. The vector X controls the number and levels of the side-loops around t; if X is null, we have the maximum number and levels of side-loops, while they all disappear if $X = \min(M_1, M_2)$.

As a rule of thumb, the greater X is, the more permissive is t. For $X =$ null, a choice which was made implicitly in Sect. 3 (the unconstrained case), it is certain that t cannot be iterated and does not lead out of the domain of target markings $\mathcal{M}$. In general, all X vectors (subject to (7)) have to be scanned in the algorithm, in order to reach a given target. Nevertheless, some general optimisation is possible: if, during an arc update with some choice of X, it has been detected that t leads out of $\mathcal{M}$, it is pointless to choose a larger X with the same M_1 and M_2, since this will generate the same problem (choosing a larger X means to choose a smaller $F(.,t)$, so that all arcs previously allowed will still be allowed, including the ones leading out of $\mathcal{M}$).

Since we considered a T-monotonic subclass, it may be observed that we never need to continue to add transitions when $[M_0\rangle = \mathcal{M}$. In other words, if we have a solution and dropping a transition keeps the stopping condition $[M_0\rangle = \mathcal{M}$ valid, the latter system is still a solution. Also, if there is a solution, it is possible to get it by adding transitions one by one, as done by the algorithm. And if adding a transition launches a problem, we need to backtrack since adding more transitions will never lead to a solution of the adequate class. Finally, at each round of the while loop, if we do not have to backtrack, $[M_0\rangle$ strictly increases, so that it is sure the algorithm stops, either with a solution or with an error message. This provides a proof of the following result.

Proposition 1. CORRECTNESS FOR T-MONOTONIC TARGET CLASSES *If there is a solution of $\mathcal{M}$, the algorithm finds one and terminates.*
Otherwise a failure message is produced before stopping.

5 Some Constraints, and Target Classes

We shall now consider several constraints that may be imposed on the result of a synthesis, and their impact on the algorithm of the previous section. The constraints considered in this section always lead to T-monotonic classes, and they may be combined (together with their impacts). Some constraints concern the net itself, while some other ones concern the reachability graph of the system obtained by adding the initial marking specified by the problem.

5.1 Limited Net Size

We may search for synthesised nets with limited characteristics, leading to constraints of type (1).

For instance we may limit the number of transitions. Searching for a net with at most k transitions is quite simple: when looping for completing the reachability set, if $[M_0\rangle \neq \mathcal{M}$ and the present number of transitions is k, we must go back, i.e., wipe out the last updates and search for a new triple. This also allows to search for an optimal solution in terms of the number of transitions: one simply has to search for the minimal number of transitions allowing synthesis by dichotomy, between 0 and $k_{\max}$; the maximum number $k_{\max}$ of transitions to be considered may be estimated a priori: $k_{\max} = |\mathcal{M}| \cdot (|\mathcal{M}| - 1)$ since the latter is the number of pairs of different markings to examine to extend the reachability set from M_0, but in general it will be far too large since, when searching for a transition $M_1 \xrightarrow{t} M_2$, M_1 and M_2 are chosen in two disjoint subsets of markings, and we should not consider a marking M_2 dominating M_1. Hence it is usually preferable to proceed differently: we first search a solution without imposing a limit; if there is no solution we may not have an "optimal" one; otherwise we may take the number of transitions in the found solution as $k_{\max}$ and start the dichotomy.

We may also limit the number of connections from places to transitions and/or from transitions to places by some integer bounds: when choosing a triple (M_1, t, M_2), we may count the number of non-null weights $F(p,t)$ and $F(t,p)$ and add them to the previous sums: if this exceeds one of the bounds, we must backtrack and search for another t (and possibly also for other markings M_1 and M_2). A refined version consists in adding the weights $F(p,t)$ and/or $F(t,p)$, and again if a chosen limit is exceeded we have to go back and search for another solution. Again also a dichotomic search allows to optimise the number of connections or the sum of weights; the maximum possible values may be estimated a priori by examining the reachability set and Equation (7) for each pair of marking, but this may be less than fully efficient, so that we may again first search a solution without limitation and use its characteristics to start the dichotomic search.

Figure 5 illustrates a solution for the same reachability set $\mathcal{M}_1$ as before, with at most 2 transitions (as it turns out, 2 is the optimal case) instead of 4. The algorithm detects this solution by trial and error. Starting with a transition such as t_0 or t_1 in Fig. 1 will never lead to a solution with at most 2 transitions. However, if the algorithm (clairvoyantly, so to speak) starts with $M_1 = (5)$ and $M_2 = (3)$, and a transition $t = t_3$ using $X = (3)$, it turns out that not only can (3) be reached from (5), but also (1) from (3), so that both (3) and (1) can be added to the set of reachable states. After that, a next (again, clairvoyant) iteration can add $t = t_2$ (with $X = (2)$), by which (2) and (0) can be reached from (5) and (3), respectively. In practice, of course, much backtracking can be expected to take place before this solution is found. In this particular case, state (0) can also be reached from (2) by the previously created transition t_3, and this is not problematic; but in general, when creating a new transition, it must be checked whether or not this leads out of the set of markings allowed by $\mathcal{M}$.

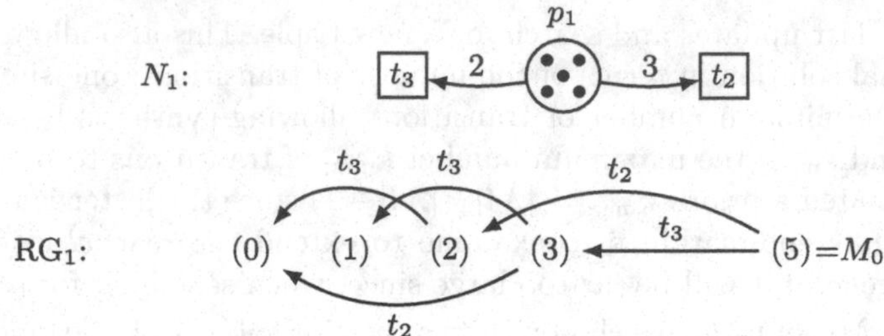

Fig. 5. A solution of $\mathcal{M}_1 = \{(5), (0), (1), (2), (3)\}$ with at most (in fact, exactly) 2 transitions.

5.2 Limited Reachability Graph

Instead of limiting the net size, we may want to limit the reachability graph size, leading to constraints of type (2). Since the number of reachable states is fixed ($|\mathcal{M}|$), we simply have to bound the number of arcs (of the reachability graph). This may be done when new arcs are added to the system, and if the limit is exceeded, we need to go back. Optimal solutions may be obtained by dichotomy, as before.

It may be observed that the solution for $\mathcal{M}_1$ in Fig. 5 is minimal in terms of the size of the net, but not in terms of the reachability graph. Indeed, the solution in Fig. 1 has less arcs in RG (and it is minimal in that respect since it has 4 arcs and each marking except the initial one must have an arc leading to it).

5.3 Pureness

If we want that the sought net is pure, i.e., there is no side-loops, this correspond to a constraint of type (3) (as well as type (5). Then, when searching a transition linking M_1 to M_2, we simply have to choose, for each place p:

- $F(p,t) = M_1(p) - M_2(p)$ and $F(t,p) = 0$ if $M_1(p) > M_2(p)$,
- $F(p,t) = 0$ and $F(t,p) = M_2(p) - M_1(p)$ if $M_1(p) < M_2(p)$,
- $F(p,t) = 0 = F(t,p)$ if $M_1(p) = M_2(p)$.

This amounts to choose a maximal vector X.

For instance, for the reachability set $\mathcal{M}_1$, the solution of Fig. 5 is pure (but the one in Fig. 1 is not).

5.4 Plainness

If we want the sought net to be plain (sometimes also called ordinary [6]), i.e., such that for each transition t and place p, $F(p,t)$ and $F(t,p)$ are not larger than 1, this also corresponds to a constraint of type (3) (as well as type (5). Then, when searching to extend the reachability set from M_0, we simply have to choose (if possible, otherwise backtracking is necessary) M_1 and M_2 so that, for each place p, $M_1(p) - M_2(p) \in \{-1, 0, 1\}$. For the new transition t, we thus have:

- $F(p,t) = 1$ and $F(t,p) = 0$ if $M_1(p) > M_2(p)$,
- $F(p,t) = 0$ and $F(t,p) = 1$ if $M_1(p) < M_2(p)$,
- $F(p,t) = 0 = F(t,p)$ or $F(p,t) = 1 = F(t,p)$ if $M_1(p) = M_2(p)$.

The reachability set $\mathcal{M}_1$ (Fig. 5) has no plain solution, since M_0 is at least 2 above any other required marking. This changes if we consider the reachability set $\mathcal{M}_2 = \mathcal{M}_1 \cup \{(4)\}$. Then we have the solution represented in Fig. 6; it is plain as requested, but also pure and minimal.

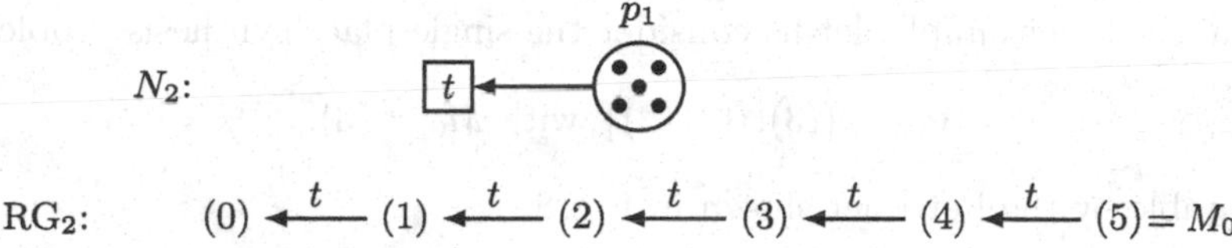

Fig. 6. A plain solution of $\mathcal{M}_2 = \{(5), (0), (1), (2), (3), (4)\} = \mathcal{M}_1 \cup \{(4)\}$.

5.5 Persistence

Let us now assume that we search for a persistent net [7], i.e., such that, for each $M \in [M_0\rangle$ and $t \neq u \in T$, if $M \xrightarrow{t}$ and $M \xrightarrow{u}$, then $M \xrightarrow{tu}$ and $M \xrightarrow{ut}$ (closing what is usually called a "diamond"). Such a constraint is of type (4). This may easily be taken into account in the algorithm: when an arc $M \xrightarrow{t} M'$ is built, if there is a previous arc $M \xrightarrow{u} M''$, then we must have $M' \geq F(.,u)$ and $M'' \geq F(.,t)$ (otherwise a problem must be launched). We may observe that, if $M' \geq F(.,u)$, when u was introduced, it was checked that $M' \xrightarrow{u} \widetilde{M} \in \mathcal{M}$, so that also (from the state equation) if $M'' \geq F(.,t)$, we have $M'' \xrightarrow{t} \widetilde{M} \in \mathcal{M}$.

Note that, strictly speaking, in order to stay in the persistent class of nets, we should only check the markings reachable after the introduction of the new transition; however we should then check all the pairs of (different) transitions. Since we know from the beginning which markings should become reachable at some point if the problem is solvable, it is preferable to check all the markings in $\mathcal{M}$ in order to avoid to recheck again and again the pairs that were already in the system. Then it is only necessary to check the pairs containing the newly introduced transition t.

The example shown in Fig. 6 is trivially persistent since there is a single transition. The nets in Figs. 11 and 12 (depicted later) are also persistent since there is no diamond to close. A more interesting case corresponds to the reachability set $\mathcal{M}_3 = \{(1,1), (1,0), (0,1), (0,0)\}$ with $M_0 = (1,1)$. This leads to the (only) solution N_3 and the (diamond-shaped) persistent reachability graph in Fig. 7.

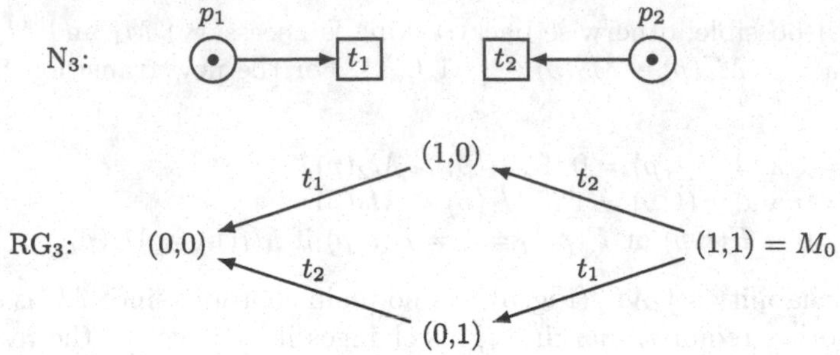

Fig. 7. A persistent solution from $\mathcal{M}_3 = \{(1,1),(1,0),(0,1),(0,0)\}$.

As a negative example, let us consider the single-place synthesis problem

$$\mathcal{M}_4 = \{(3),(0),(2)\}, \text{with } M_0 = (3).$$

It is solvable by the Petri net shown in Fig. 8.

Fig. 8. A nonpersistent solution of $\mathcal{M}_4 = \{(3),(0),(2)\}$.

However, there is no persistent solution. To see this, note first that no transition can be inserted from (0) to (2), because that would create infinitely many reachable markings. Moreover, no transition can be inserted from (2) to (0), because that would also allow marking (1) to be reached from (3). Hence the states (0) and (3) cannot be reached in any other way than as already shown in Fig. 8, and the half-diamond shown in that figure cannot be closed to a full diamond.

5.6 Choice-Freeness

Let us now assume that we search for a choice-free net [8], i.e., such that for each $p \in P$ and $t \neq u \in T$, if $F(p,t) > 0$ then $F(p,u) = 0$ (meaning that only one transition may take tokens from any place, which implies persistence). This corresponds to a constraint of type (5), and may again easily be taken into account in the algorithm: when a triple (M_1, t, M_2) is chosen, we must check that, for each place $p \in P$, if $F(p,t) > 0$, then no previous transition u is such that $F(p,u) > 0$ (otherwise, we have to search for another triple).

The systems in Figs. 6, 7, 11, and 12 are choice-free. From them, it could be suspected that all persistent systems are choice-free. This is not true. Let us for instance consider the reachability set $\mathcal{M}_5 = \{(1,1,1),(1,0,1),(0,1,1),(0,0,1)\}$ with $M_0 = (1,1,1)$, with three places. Depending on the way we consider the triples (M_1, t, M_2), we might obtain the (plain) system in Fig. 9, which is not choice-free. But from the same reachability set, we may obtain (by chance or by backtracking) a choice-free solution, where p_3 is isolated.

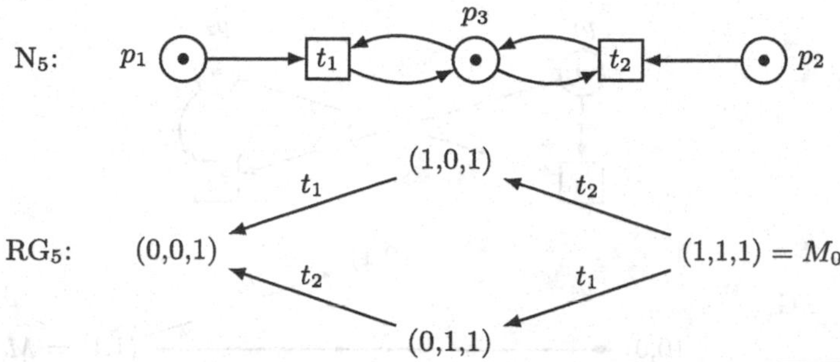

Fig. 9. A persistent but not choice-free solution from $\mathcal{M}_5$.

5.7 Equal Conflict

Let us now assume that we search for an equal conflict net [9], i.e., such that for any $p \in P$ and $t \neq u \in T$, if $F(p,t) > 0 < F(p,u)$ then $F(.,u) = F(.,t)$ (meaning that whenever t is enabled, so is u; notice that choice-free nets are trivially equal-conflict too). This again may easily be taken into account in the algorithm: when a triple (M_1, t, M_2) is chosen, we must check that if, for some place $p \in P$, $F(p,t) > 0$ and for some previous transition u, $F(p,u) > 0$, then $F(.,t) = F(.,u)$ (otherwise, we have to search for another triple).

It is also possible to consider a restricted form of equal conflictness, by requesting that if for some $p \in P$ and $t \neq u \in T$, $F(p,t) > 0 < F(p,u)$ then $F(p,u) = F(p,t)$, and for any other place p' we have $F(p',t) = 0 = F(p',u)$. This again may easily be taken into account in the algorithm: when a triple (M_1, t, M_2) is chosen, we need to check that, if for some place $p \in P$, $F(p,t) > 0$ and for some previous transition u, $F(p,u) > 0$, then $F(p.t) = F(p,u)$ and for any other place p' we have $F(p',t) = 0 = F(p',u)$ (otherwise, we have to backtrack).

For instance, the system in Fig. 1 is equal-conflict (and so are the systems in Figs. 6, 7, 11, and 12, since they are choice-free). By contrast, the systems N_1 in Fig. 5 and N_5 in Fig. 9 are not equal-conflict.

5.8 Free-Choiceness

A net is (extended) free-choice [10] iff it is both plain and equal-conflict. Its handling is thus obtained by combining the corresponding modifications of the general algorithm.

Similarly, it is (restricted) free-choice iff it is plain and restricted equal-conflict, and again its handling may be obtained by combining the corresponding modifications of the general algorithm.

Note that plain choice-free nets are also (restricted) free-choice. In particular, the systems in Figs. 6, 7, 11, and 12 are restricted free-choice. Figure 10 exhibits a different example, for the reachability set $\mathcal{M}_6 = \{(1,1),(0,1),(0,0)\}$ with $M_0 = (1,1)$.

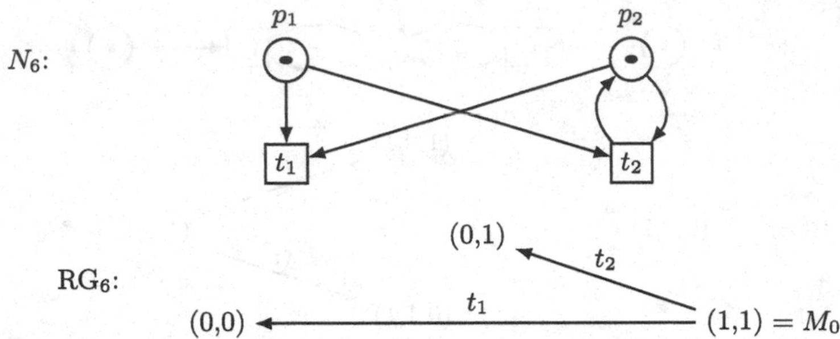

Fig. 10. An (extended) free-choice solution from $\mathcal{M}_6 = \{(1,1),(0,1),(0,0)\}$.

6 Beyond T-Monotonicity

We shall now consider some non-monotonic target classes, and see that they may nevertheless be coped with by some modifications of the algorithm in Fig. 2.

6.1 Reversible Nets

A net (P, T, F) is sometimes called reversible [11,12] if, for each transition u, there is a reverse transition $\widehat{u}$ such that $\forall p \in P\colon F(p,\widehat{u}) = F(u,p) \wedge F(\widehat{u},p) = F(p,u)$. This class of nets is not T-monotonic, because if we drop transition u, the companion $\widehat{u}$ remains lonely. However, it is not too far from being T-monotonic, since if we drop, not individual transitions, but pairs of companions $\{u,\widehat{u}\}$, then the problem disappears. The other way round, in the algorithm of Fig. 2, when a triple (M_1, t, M_2) is selected, the idea is to add both t and its related transition $\widehat{t}$ to the transition set (and check both for coherence with $\mathcal{M}$). Note that in this case, all the members of $\mathcal{M}$ must be incomparable, i.e., $\forall M, M' \in \mathcal{M} : M \leq M' \wedge M \leq M' \Rightarrow M = M'$; indeed, if for instance $M \lneqq M'$, in any tentative solution, since there are evolutions $M_0 \xrightarrow{\sigma} M$ and $M_0 \xrightarrow{\sigma'} M'$, we also have an evolution $M \xrightarrow{\widehat{\sigma}} M_0$ (where $\widehat{\sigma}$ is obtained by reversing σ and replacing each transition by its reverse one), hence an evolution $M \xrightarrow{\widehat{\sigma}\sigma'} M'$, and the firing sequence $\widehat{\sigma}\sigma'$ may be repeated as many times we want, leading to infinitely many different (increasing) reachable markings, while $\mathcal{M}$ is assumed to be finite. This check may be included in a "pre-synthesis" phase checking quickly whether there is any hope that a solution exists (and explaining why it is impossible in the negative case).

For instance, with two places p_1 and p_2 and the reachability set $\mathcal{M}_7 = \{(1,0),(0,1)\}$, we get the (plain and pure) system represented in Fig. 11. Note that if reversibility is not enforced, $\widehat{t}$ will not be included since, with t alone, all the states in $\mathcal{M}_7$ are already visited.

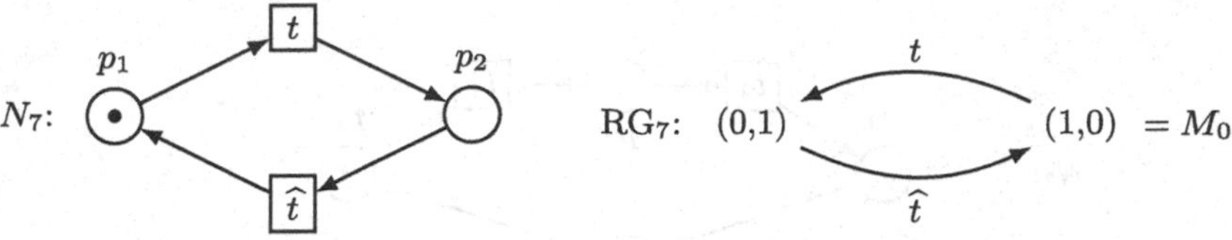

Fig. 11. A reversible solution from $\mathcal{M}_7 = \{(1,0),(0,1)\}$.

6.2 Reversible Graphs

Reversibility may also be defined at the level of a reachability graph [13] (hence at the level of a Petri net system (P,T,F,M_0), and not only at the level of an uninitialised Petri net): a reachability graph is reversible if $\forall M \in [M_0\rangle \colon M_0 \in [M\rangle$. In other works, this property is called cyclicity [14]. Of course, the reachability graph of an initialised reversible net is reversible, but it may happen that an initialised non-reversible net has nevertheless a reversible reachability graph. Like for reversible nets, and for a similar reason, all the states visited in a reversible reachability graph must be incomparable. In a labelled transition system, similarly to $[M\rangle$, we shall define $\langle M]$ as the set of states in $[M\rangle$ from which M may be reached. A reachability graph is thus reversible while reaching $\mathcal{M}$ if $[M_0\rangle = \mathcal{M} = \langle M_0]$.

This class of systems is not T-monotonic, since if we drop one or more transitions from a Petri net system with a reversible reachability graph, it may well happen that reversibility is lost. To mend the algorithm of Fig. 2 to cope with this target class of systems, the idea is to apply the following modifications:
- the while loop becomes: While($[M_0\rangle \neq \mathcal{M}$ or $\langle M_0] \neq \mathcal{M}$);
- the choice of a triple (M_1, t, M_2) is based either on the conditions

$$M_1 \in [M_0\rangle \wedge M_2 \in \mathcal{M} \setminus [M_0\rangle \wedge M_1 \xrightarrow{t} M_2$$

or on the conditions

$$M_2 \in \langle M_0] \wedge M_1 \in \mathcal{M} \setminus \langle M_0] \wedge M_1 \xrightarrow{t} M_2.$$

In both cases the coherence of the newly introduced transition with $\mathcal{M}$ is checked as usual (if $M \xrightarrow{t} M'$ with $M \in \mathcal{M}$, then we must have $M' \in \mathcal{M}$).

Figure 12 illustrates a possible result of this modified algorithm for the reachability set $\mathcal{M}_8 = \{(1,0,0),(0,1,0),(0,0,1)\}$. It may be noticed that the net N_8 will not be produced by the unmodified algorithm, since, for instance, the latter will stop before introducing t_3.

6.3 Liveness

A Petri net system (P,T,F,M_0) is called live [15] if $\forall t \in T \forall M \in [M_0\rangle \exists M' \in [M\rangle : M' \xrightarrow{t}$. We may observe that reversible nets and systems are always live (provided each transition occurs at least once in the reachability graph, which is the case for all the systems constructed in this paper), but the reverse is not true, as illustrated by the system in Fig. 13 (no arc allows to go back to M_0).

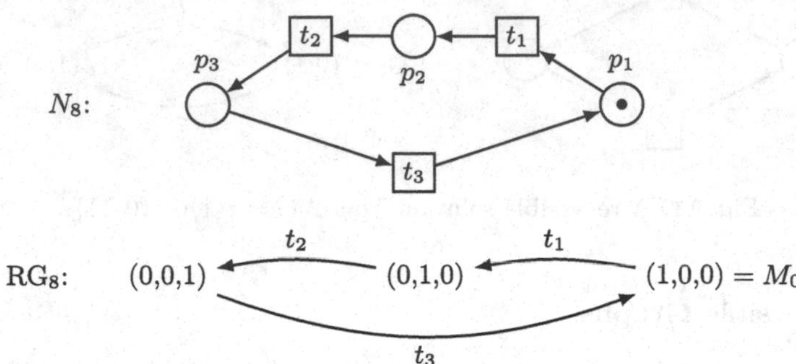

Fig. 12. A reversible reachability graph from $\mathcal{M}_8 = \{(1,0,0),(0,1,0),(0,0,1)\}$.

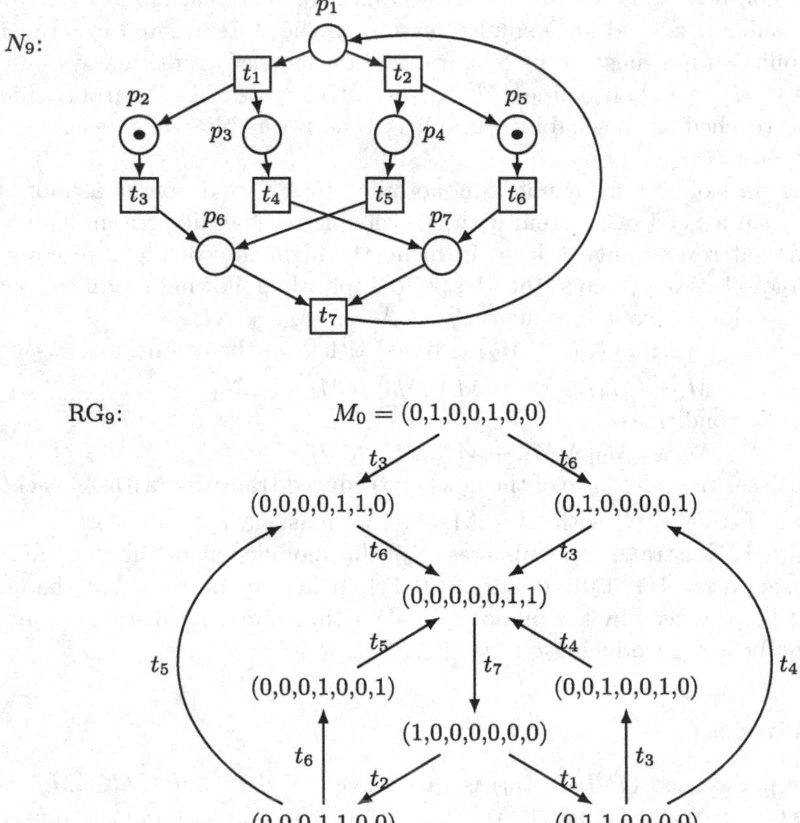

Fig. 13. A live (strict free-choice) net which is not reversible.

T-monotonicity is not fulfilled for this class, since dropping one or more transitions may make M' in the previous formula unreachable. By contrast, dropping a transition t with no effect (i.e., if $M \xrightarrow{t} M$ for some $M \in [M_0\rangle$, hence for any reachable marking enabling t), or a duplicate transition t (i.e., if there is some transition $u \neq t$ such that $F(.,t) = F(.,u)$ and $F(t,.) = F(u,.)$) is harmless; and useless transitions (i.e., which are never enabled in $[M_0\rangle$) should never occur. From the constraints (6) and (7), and the finiteness of $\mathcal{M}$, this means that we only have to consider finitely many transitions in our transition syntheses.

In the algorithm, we need to reach a situation where $[M_0\rangle = \mathcal{M}$, but it may happen that we need to continue to add transitions (as long as they do not lead out of the initially specified set of markings). This is the case, for instance, if we consider two places and the set of markings $\{(0,1),(1,0)\}$ with $M_0 = (0,1)$. Reachability can be solved by adding a single transition from $(0,1)$ to $(1,0)$, but the resulting net is not live. Adding another transition from $(1,0)$ to $(0,1)$ yields a (reversible, hence) live net solving this synthesis problem.

Once we have built the reachability graph of a bounded system, it is not complicated to check if it is live or not. Indeed, with classical algorithms such as Tarjan's [16] (see also https://en.wikipedia.org/wiki/Tarjan's_strongly_connected_components_algorithm), the search for the strongly connected components of the reachability graph,[4] and especially the terminal ones[5] is efficient (linear in the size of nodes and arcs). We may then see that a bounded Petri net system is live if and only if each transition is enabled at some marking in each terminal strongly connected component of its reachability graph.

When $[M_0\rangle = \mathcal{M}$, the idea is thus, while the system is not live, to (try to) add a transition $M_1 \xrightarrow{t} M_2$ where M_1 belongs to a terminal strongly connected component missing to enable some transition constructed up to now, and M_2 is out of this component (of course, backtracking may be necessary during the extension, and possibly to go before the completion of $[M_0\rangle$).

If we consider for instance the system in Fig. 13, the first part of the algorithm (making $\mathcal{M}$ reachable) may add successively transitions t_3, t_6, t_7, t_1, t_2, then stops since all the needed markings are reached (and only them). The resulting system is not live since all the strongly connected component of the reachability graph are singletons, and no transitions are allowed from the terminal ones $\{(0,0,1,0,0,1,0)\}$ and $\{(0,0,0,1,0,0,1)\}$. If we add now t_4, some markings are gathered in a (non-terminal) strongly connected component, but $(0,0,0,1,0,0,1)$ still has no output arc. Finally, if we add t_5, all the markings but M_0 are gathered in a terminal strongly connected component and all transitions are allowed somewhere in it.

---

[4] i.e., the maximal subsets of nodes $\mathcal{S}$ such that $\forall M_1, M_2 \in \mathcal{S} \exists \sigma \in T^* : M_1 \xrightarrow{\sigma} M_2$.

[5] i.e., such that $M_1 \in \mathcal{S} \wedge M_1 \xrightarrow{\sigma} M_2 \Rightarrow M_2 \in \mathcal{S}$.

7 Some Complexity Considerations

Every one of the problems described in the previous sections gives rise to a decision problem:

- **Given:** A finite set of n places, a finite set of markings $\mathcal{M}$, a designated marking $M_0 \in \mathcal{M}$ and a set $\mathcal{P}$ of properties.
- **To decide:** Does there exist a finite place/transition net N with properties $\mathcal{P}$ such that the reachability set of (N, M_0) is exactly $\mathcal{M}$?

where $\mathcal{P}$ means any of the above constraints, or a combination thereof (no constraint / having at most k transitions / plain / pure / ... / live).

Every problem of this kind is in NP. This can be seen by considering a straightforward nondeterministic guess-and-check algorithm. The size of a marked Petri net (N, M_0) solving $\mathcal{M}$ can be bounded from above as follows. From the analyses above, it occurs that, in each case considered in the present paper,[6] if there is a solution, then there is one with at most $|\mathcal{M}| \cdot (|\mathcal{M}| - 1)$ transitions and, from the bounds (6) and (7), their weights to/from each place p are bounded by $\max_{M \in \mathcal{M}} M(p)$. Hence we may assume that the net to be checked satisfies those characteristics, so that its size is polynomial in the size of the problem. Still polynomially in the size of the problem, we may check that this net generates exactly $\mathcal{M}$ as reachability set (by constructing the reachability graph),[7] and satisfies property $\mathcal{P}$.[8]

In the case "no constraint" the algorithm presented in Sect. 3 exhibits a solution which is in fact linear in the size of the problem, but we suspect that if constraints are added, the problem becomes more complex.

8 A Slightly Larger Case Study

In an art gallery, two groups of visitors are to be led around from the entry point (the start) to the exit point (end). There is also a breakpoint in between where snacks and prosecco are served. The groups may take different routes, but they should also meet at break time, and no group is allowed to proceed to the end point before the other group has arrived for a break. Thus, we have six places: p_1, p_2, p_3 for the first group (p_1 for start, p_2 for break, p_3 for end), and similarly,

---

[6] This would not be true if, for instance the constraints on the target class request that the number of transitions is $2^{|\mathcal{M}|}$.

[7] For each marking $M \in \mathcal{M}$ and each transition t, we may check if t is enabled by M and leads to a marking in $\mathcal{M}$, and check that all markings in $\mathcal{M}$ are reachable from M_0.

[8] This is even true for liveness since, as seen above, we only have to construct the terminal strongly connected components of the reachability graph and to check whether all of them contain all transitions.

p_4, p_5, p_6 for the second group. The allowed markings are

$$\mathcal{M} = \left\{ \begin{pmatrix}1\\0\\0\\1\\0\\0\end{pmatrix}, \begin{pmatrix}0\\1\\0\\0\\1\\0\end{pmatrix}, \begin{pmatrix}0\\0\\1\\0\\0\\1\end{pmatrix}, \begin{pmatrix}0\\1\\0\\1\\0\\0\end{pmatrix}, \begin{pmatrix}1\\0\\0\\0\\1\\0\end{pmatrix}, \begin{pmatrix}0\\0\\1\\0\\1\\0\end{pmatrix}, \begin{pmatrix}0\\1\\0\\0\\0\\1\end{pmatrix} \right\} \text{ with } M_0 = \begin{pmatrix}1\\0\\0\\1\\0\\0\end{pmatrix}$$

In addition, we want the solution to be graph-reversible, because the next day, and the days after that, more groups will visit the museum. Since an unconstrained solution already has 6 transitions and is not live, we aim for around 7 (or maybe some more) transitions. Figure 14 shows a solution with 7 transitions, as it can be obtained by the algorithm described in this paper.

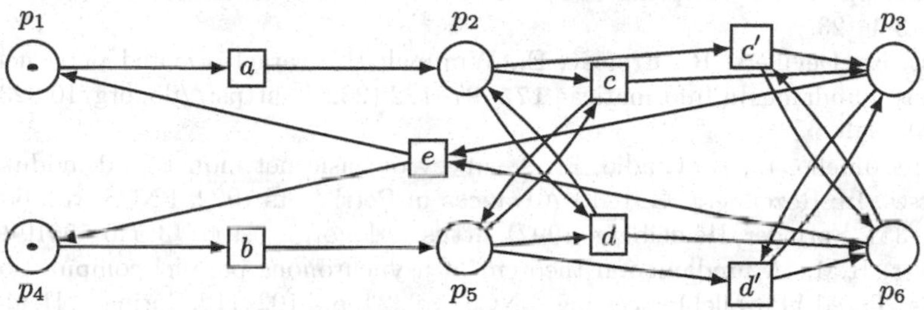

Fig. 14. A solution for the art gallery.

9 Concluding Remarks and Perspectives

We showed how to build a place-transition Petri net of some subclass with a given reachability set. We specifically considered T-monotonic classes, but we also adapted the strategy to some non-T-monotonic targets.

Our approach, which seems to be novel, opens up a variety of avenues for possible future research. For instance, many other constraints could be considered (such as acyclic nets and acyclic reachability graphs, T-restricted nets, asymmetric-choice nets). The techniques we developed are rather brute-force and it should be possible to search for more effective, dedicated algorithms, depending on the desired constraints (presently, most of the time, the algorithms are exponential). In addition, more pre-synthesis analyses would be welcome. Another possible and desirable extension of the problem proposed in this paper is to take into account infinite reachability sets, for instance ones specified finitely in the form of a linear or a semi-linear set. Also, it could be possible to search for approximate solutions if no exact ones are available, and explore what could be said if the reachability set is only partly known.

Acknowledgements. The encouragements, remarks and suggestions of anonymous referees were appreciated.

References

1. Badouel, E., Bernardinello, L., Darondeau, P.: Polynomial algorithms for the synthesis of bounded nets. In: Mosses, P.D., Nielsen, M., Schwartzbach, M.I. (eds.) CAAP 1995. LNCS, vol. 915, pp. 364–378. Springer, Heidelberg (1995). https://doi.org/10.1007/3-540-59293-8_207
2. Desel, J., Reisig, W.: The synthesis problem of petri nets. Acta Inf. **33**(4), 297–315 (1996)
3. Badouel, E., Bernardinello, L., Darondeau, P.: Petri Net Synthesis. TTCSAES, Springer, Heidelberg (2015). https://doi.org/10.1007/978-3-662-47967-4
4. Schlachter, U.: Over-approximative petri net synthesis for restricted subclasses of nets. In: Klein, S.T., Martín-Vide, C., Shapira, D. (eds.) LATA 2018. LNCS, vol. 10792, pp. 296–307. Springer, Cham (2018). https://doi.org/10.1007/978-3-319-77313-1_23
5. Best, E., Devillers, R., Erofeev, E., Wimmel, H.: Target-oriented petri net synthesis. Fundamenta Informaticae **175**, 97–122 (2020). https://doi.org/10.3233/FI-2020-1949
6. Bernardinello, L., De Cindio, F.: A survey of basic net models and modular net classes. In: Rozenberg, G. (ed.) Advances in Petri Nets 1992. LNCS, vol. 609, pp. 304–351. Springer, Heidelberg (1992). https://doi.org/10.1007/3-540-55610-9_177
7. Keller, R.M.: A fundamental theorem of asynchronous parallel computation. In: Feng, T. (ed.) Parallel Processing. LNCS, vol. 24, pp. 102–112. Springer, Heidelberg (1975). https://doi.org/10.1007/3-540-07135-0_113
8. Teruel, E., Colom, J.M., Silva, M.: Choice-free petri nets: a model for deterministic concurrent systems with bulk services and arrivals. IEEE Trans. Syst. Man Cybernet. Part A **27**(1), 3–83 (1997). https://doi.org/10.1109/3468.553226
9. Teruel, E., Silva, M.: Structure theory of Equal Conflict systems. Theor. Comput. Sci. **153**(1&2), 271–300 (1996). https://doi.org/10.1016/0304-3975(95)00124-7
10. Desel, J., Esparza, J.: Free Choice Petri Nets. vol. 40 of Cambridge Tracts in Theoretical Computer Science. Cambridge University Press, New York, USA (1995)
11. Barylska, K., Koutny, M., Mikulski, Ł, Piątkowski, M.: Reversible computation vs. reversibility in petri nets. In: Devitt, S., Lanese, I. (eds.) RC 2016. LNCS, vol. 9720, pp. 105–118. Springer, Cham (2016). https://doi.org/10.1007/978-3-319-40578-0_7
12. Barylska, K., Mikulski, L., Piatkowski, M., Koutny, M., Erofeev, E.: Reversing transitions in bounded petri nets. In: Proceedings of the 25th International Workshop on Concurrency, Specification and Programming, Rostock, Germany, 28–30 September 2016, pp. 74-85 (2016). https://ceur-ws.org/Vol-1698/CS&P2016_08_Barylska&Mikulski&Piatkowski&Koutny&Erofeev_Reversing-Transitions-in-Bounded-Petri-Nets.pdf
13. Hujsa, T., Delosme, J., Kordon, A.M.: On liveness and reversibility of equal-conflict petri nets. Fundam. Inform. **146**(1), 83–119 (2016). https://doi.org/10.3233/FI-2016-1376
14. Bouziane, Z., Finkel, A.: Cyclic petri net reachability sets are semi-linear effectively constructible. In: Moller, F. (ed.) Second International Workshop on Verification of Infinite State Systems, Infinity 1997, Bologna, Italy, 11–12 July 1997, vol. 9

of Electronic Notes in Theoretical Computer Science. Elsevier, pp. 15–24 (1997). https://doi.org/10.1016/S1571-0661(05)80423-2
15. Murata, T.: Petri nets: properties, analysis and applications. Proc. IEEE **77**(4), 541–80 (1989)
16. Tarjan, R.E.: Depth-first search and linear graph algorithms. SIAM J. Comput. **1**(2), 146–160 (1972). https://doi.org/10.1137/0201010

Symbolic Domains and Reachability for Nets with Trajectories

Loïc Hélouët[1(✉)] and Prerak Contractor[2]

[1] Univ. Rennes, IRISA, CNRS & INRIA, Rennes, France
loic.helouet@inria.fr
[2] IIT Bombay, Mumbai, India

Abstract. This paper considers verification of timed models handling additional quantities progressing linearly such as distance of moving objects to a target. We introduce a variant of Petri nets called *trajectory nets* where some places are standard control places containing tokens, and other places contain a trajectory of an object. We give a semantics for this model, and propose an abstraction of sets of equivalent trajectories into symbolic domains. These domains cannot be represented by Difference Bound Matrices, but one can compute in polynomial time a symbolic representation of successor configurations. Furthermore domains are closed under this successor relation, and the set of domains of a trajectory net is finite. A consequence is that, when the control part of a trajectory net is bounded, reachability, coverability and verification of safety properties involving distances are PSPACE-Complete.

1 Introduction

Some properties of cyber-physical systems such as transport networks call for the verification of quantitative properties addressing time, but also continuous values such as distances. A typical example is safety of metro networks, where one wants to guarantee safety headways, or bound the number of trains in tunnels to guarantee safe evacuation of passengers in case of power failure. Models such as timed automata [3] or time Petri nets [20] only address time, and cannot be used to handle such problems. Models that can address both time and continuous values such as distances rapidly have the expressive power of hybrid automata [2], for which most problems become undecidable.

This paper introduces *trajectory nets*, a model tailored for the analysis of safety properties of systems involving both time and distances such as metro networks. Trajectory nets are a variant of time Petri nets, where some places are dedicated to control, and other places depict object movements with simplified representations called *trajectories*. Configurations assign an integral number of token to control places, and a trajectory, representing the remaining time and distance to the end of a trip to a subset of trajectory places. Dealing with trajectories allows to define properties that address both distances and time. Verification of a safety property of the form "At each instant, less than K trains

L. M. Kristensen and J. M. van der Werf (Eds.): PETRI NETS 2024, LNCS 14628, pp. 244–265, 2024.
https://doi.org/10.1007/978-3-031-61433-0_12

are in a tunnel" amounts to a reachability question for sets of configurations depicting forbidden positions of objects.

As a first contribution of this paper, we define trajectory nets and give their semantics in terms of configurations, discrete events (end of progress of a trajectory, creation of a new one) and timed moves. As in many continuous models, the set of possible configurations is infinite. This comes on one hand from the unboundedness of the discrete contents of control places, and on the other hand from the continuous representation of trajectories. We show that in their full generality, trajectory nets can simulate a two-counters machine, and are hence Turing Powerful. As a consequence, safety properties relying on coverability or reachability of a configuration are undecidable.

As a second contribution of the paper, we show that the continuous part of configurations can be represented symbolically by sets of linear inequalities called *domains*. Abstracting time with regions and zones in timed automata [3] or domains in variants of Petri nets [5,14,18] is a standard approach. However, for trajectory nets, domains have to abstract away two types of continuous values: time and distance. They define sets of solutions that cannot be represented with the usual zones, and cannot be encoded by Difference Bound Matrices. Nevertheless, we show that we can compute in polynomial time a successor relation on domains, that domains are closed under this relation, and that the set of reachable domains of a given trajectory net is finite. A consequence is that, for trajectory nets with bounded control places, one can compute a sound and complete symbolic abstraction of the timed behaviour called a *state class graph*, and use it to verify properties of the original model. We then show that checking coverability, reachability and safety properties involving distances are PSPACE-Complete problems for bounded trajectory nets. For space reasons, some proofs are only sketched, but can be found in a long version of this work [15].

2 Preliminaries

In the rest of the paper, we will denote respectively by $\mathbb{R}, \mathbb{Q}, \mathbb{N}$ the sets of reals, rationals, and non-negative integers. We will denote by $\mathbb{R}^{\geq 0}, \mathbb{Q}^{\geq 0}$ the sets of positive reals and rationals, and by $\mathcal{I}_{\mathbb{Q}}$ the set of intervals of the form $[a, b]$ or $[a, \infty)$ where $a, b \in \mathbb{Q}$. Let $X = \{x_1, \cdots, x_n\}$ be a set of variables. A *linear constraint* over X with rational coefficients (or simply constraint for short) is an expression of the form $a_1 \cdot x_1 + a_2 \cdot x_2 + \cdots a_n \cdot x_n \leq b$, where b is a rational value and $a_i's$ are rational coefficients (which can have value 0). A constraint is *two-dimensional* if it has at most two variables with non-zero coefficients.

A *valuation* for a set of variables X is a map $\mu : X \rightarrow \mathbb{R}$. We will say that a valuation μ *satisfies* a linear constraint $C(X) ::= \sum a_i \cdot x_i \leq b$ iff replacing every x_i by its valuation $\mu(x_i)$ in $C(X)$ yields a tautology.

A *system* of linear constraints over a set of variables X is a set of linear constraints. It is two-dimensional iff all its constraints are two-dimensional, i.e. all its linear inequalities are of the form $a_i \cdot x_i \leq b_i$, or $a_i \cdot x_i - b_j \cdot x_j \leq c_{i,j}$. A valuation satisfies a system S iff it satisfies all linear constraints in S (i.e.

systems are conjunctions of constraints over X). A valuation that satisfies S is called *a solution* for S. We will denote by $[\![S]\!]$ the set of solutions for S and say that S is *satisfiable* iff $[\![S]\!] \neq \emptyset$. Slightly abusing our definition, we will sometimes adopt a compact notation and write $a \leq \mathit{expr} \leq b$ instead of a conjunction of constraints of the form $-\mathit{expr} \leq -a$ and $\mathit{expr} \leq b$.

A two-dimensional system S involving only inequalities of the form $a_i \leq x_i, x_i \leq b_i$, or $x_i - x_j \leq c_{i,j}$ is called a *zone*. It can be encoded by a Difference Bound Matrix (DBM for short) [10], that is a matrix DB_S indexed by $x_\perp, x_1, \cdots, x_n$, where variable $x_\perp$ is a dummy variable representing value 0. In a DBM DB_S, a cell $DB_S[x_i, x_j]$ holding value $b_{i,j}$ encodes inequality $x_i - x_j \leq b_{i,j}$ and a constraint of the form $x_i \leq b_i$ is represented by an entry $DB_S[x_i, x_\perp] = b_i$. Alternatively, zones and DBMs can be represented with *constraint graphs*, i.e. weighted graphs whose vertices are variables $x_\perp, x_1, \cdots, x_n$, and whose edges $(x_i, w_{i,j}, x_j)$ are weighted by $w_{i,j} = DB[x_i, x_j]$. DBMs and their graph representations allow for efficient polynomial algorithms to check satisfiability, compute canonical forms, intersections... (see [6] for a survey). For instance, checking non-emptiness of $[\![S]\!]$ amounts to verifying that the constraint graph for S does not contain negative cycles.

3 Trajectory Nets

This section describes a new model called *trajectory nets* that can represent movements of objects in a one-dimensional space. This model can describe conveyors, metro networks,... and refers to positions of objects. We show in Sect. 6 how to address safety properties that depend on simultaneous positions of objects. As explained in introduction, such properties are useful for metro operators to guarantee that passengers can be safely evacuated in a short amount of time in case of power failure. This means setting a limit on the number of trains that are positioned on a dangerous zone (tunnel, bridge...) at any instant, and verifying that this limit is never exceeded. Granting such properties means that one is able to refer to positions of trains.

Petri nets are a rather straightforward choice to model transport networks. Stations and track segments between stations can be represented with places, arrivals and departures of vehicles with transitions, and the structure of the network itself is depicted by the flow relation of the net. Addressing trips and dwells durations can be done via a time(d) extension of Petri nets. For instance, one can use time Petri nets [20], that assign a static interval $[\alpha_t^s, \beta_t^s]$ to every transition to represent possible durations of dwells in stations and of trips from one station to the next one. A standard semantics of Time Petri nets is to keep track at each instant of the time remaining before firing of each enabled transition. However, time(d) variants of Petri nets cannot handle real valued variables beyond timing information related to transitions or tokens, and are hence not expressive enough to address properties referring to positions of vehicles.

Consider, for instance, the pictures in Fig. 1. These diagrams are standard representations of metro trajectories called *space-time diagrams*. They represent

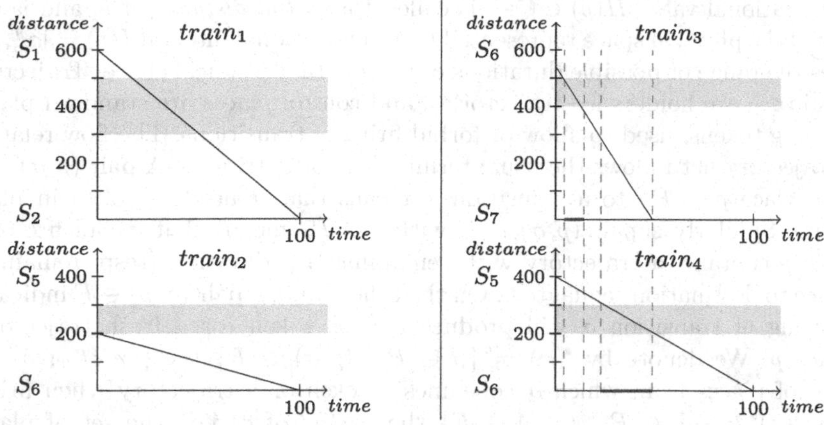

Fig. 1. Space time diagrams: trains positions are needed to address safety issues

trips with functions mapping time elapsed with the remaining distance to the next station. The two diagrams on the left represent movements of two trains $train_1, train_2$ traveling respectively on a track from a station S_1 to a station S_2, and from a station S_5 to a station S_6. The track segments represented on these diagrams contain dangerous zones, symbolized by red areas. The remaining trip duration for $train_1$ and $train_2$ is 100 in both diagrams, but one can easily notice that $train_1$ moves faster than $train_2$. From this situation, both trains cannot be simultaneously in their danger zone in the future, because $train_2$ already exited this area. So, knowing durations of trips does not suffice to distinguish two trajectories or detect dangers. Now consider the two space-time diagrams on the right of Fig. 1. The two trains represented will enter and leave their danger zone. However, as $train_3$ is fast, it will leave its danger zone before $train_4$ enters it. These two examples illustrate the fact that addressing safety of moving objects requires to consider more information than a remaining trip duration for each object. A model such as Time Petri nets [20] is hence not precise enough. A solution is to work with hybrid models, that can handle variables representing evolution of objects positions. It is however well known that most problems are undecidable for hybrid automata [4]. In the rest of this section, we introduce a new model that can address time and distance, without reaching the expressive power of hybrid automata. This model contains transitions that represent objects arrivals or departures, standard places holding tokens, and a new type of place containing space-time diagrams representing forecast trajectories of objects.

Definition 1. *A* Trajectory net *is a tuple* $\mathcal{N} = (P, T, F, I, H)$ *where* $P = P_T \uplus P_C$ *is a set of places. We distinguish a set* P_T *of trajectory places, and a set* P_C *of control places.* $T = \{\sigma_1, \dots \sigma_{|T|}\}$ *is a set of transitions,* $F \subseteq P \times T \cup T \times P$ *is a flow relation. The function* $I : P_T \to \mathcal{I}_{\mathbb{Q}}$ *associates a rational interval* $[\alpha_p^s, \beta_p^s]$ *to every trajectory place* $p \in P_T$*, and the function* $H : P_T \to \mathbb{Q}$ *associates a rational distance to every trajectory place.*

The rational value $H(p) \in \mathbb{Q}^{\geq 0}$ is called the *initial distance* of p, and is the length of the physical space represented by p. The rational interval $I(p) = [\alpha_p^s, \beta_p^s]$ defines the range of possible durations of trips in that physical space. Trajectory places in P_T are holders for trajectories, and control places are standard places containing tokens, used to allow or forbid firing of transitions. The flow relation of a trajectory net follows the usual terminology of Petri nets. A pair $(p, \sigma) \in F$ from a place $p \in P_C$ to a transition σ means that σ needs a token in place p to fire. Similarly a pair $(p, \sigma) \in F$ with $p \in P_T$ means that σ can fire only if place p contains a trajectory with remaining trip duration (resp. remaining distance to destination) equal to 0. On the other hand, a pair $(\sigma, p) \in F$ indicates that firing of transition σ will produce a fresh token (or a fresh trajectory) in place p. We denote by ${}^\bullet(\sigma) = \{p \in P \mid (p, \sigma) \in F\}$ the *preset* of σ, i.e. the set of places from which σ consumes a token or a trajectory when firing, and by $(\sigma)^\bullet = \{p \in P \mid (\sigma, p) \in F\}$ the *postset* of σ, i.e., the set of places where tokens or trajectories are added when firing σ. In the rest of the paper, to simplify semantics, we consider trajectory nets where $|{}^\bullet(\sigma) \cap P_T| \leq 1$ and $|(\sigma)^\bullet \cap P_T| \leq 1$. Slightly abusing notations, we will hence write $p = {}^\bullet(\sigma) \cap P_T$ instead of $\{p\} = {}^\bullet(\sigma) \cap P_T$, and similarly for $(\sigma)^\bullet \cap P_T$.

Figure 2 shows the basic elements of trajectory net: two trajectory places p_1, p_2 represented by large circles, two control places p_3, p_4, represented by small circles, a transition σ represented by a rectangle. The flow relation is represented as usually in Petri nets with arrows connecting places and transitions. On this example, we have ${}^\bullet(\sigma) = \{p_1, p_3\}$ and $(\sigma)^\bullet = \{p_2, p_4\}$. Place p_3 contains a token, and place p_1 a trajectory.

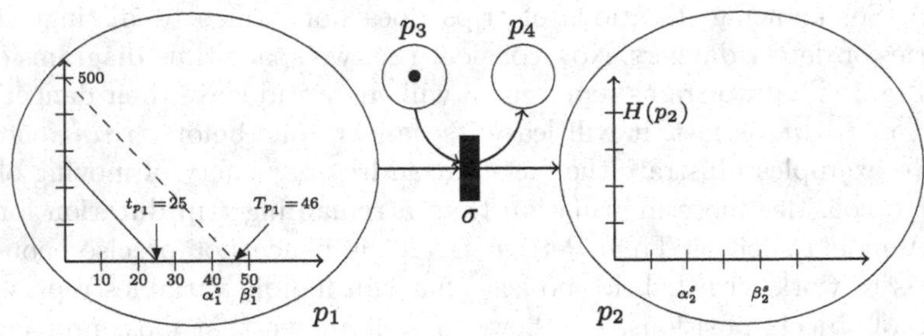

Fig. 2. Basic elements of a trajectory net: places, transitions, trajectories.

Definition 2. *Let $p \in P_T$ be a trajectory place, and $I(p) = [\alpha_p^s, \beta_p^s]$ be the interval depicting the possible duration of a movement in place p. A* trajectory *in p is a pair of real numbers $tr_p = (T_p, t_p)$, where $T_p \in [\alpha_p^s, \beta_p^s]$ denotes the initial duration of a movement in the physical space represented by place p and $t_p \leq T_p$ is the current remaining trip time in that place. A trajectory tr_p is* progressing *if $t_p > 0$ and is* blocked *otherwise.*

As the initial distance $H(p)$ is fixed for every trajectory place $p \in P_T$, choosing non deterministically an initial duration T_p of a trajectory in interval $[\alpha^s, \beta^s]$

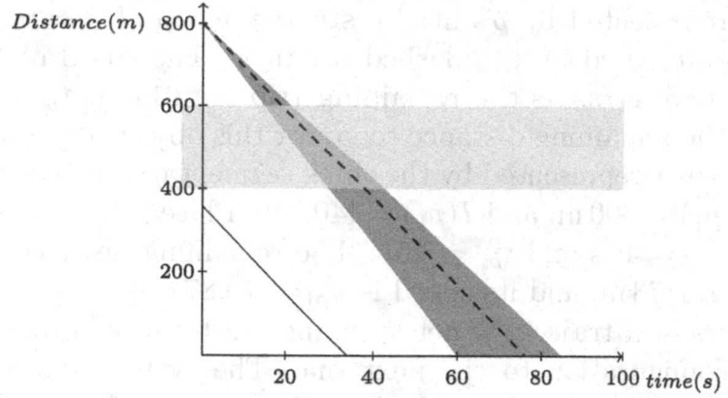

Fig. 3. A Trajectory (T_p, t_p) in a place p with $H(p) = 800m$ and $I(p) = [60, 85]$. We have $T_p = 76$ and $t_p = 34$. The blue cone represents all initial trajectories in p for possible initial values for $T_p \in I(p)$. The red area represents a danger zone (e.g. a tunnel) between distance 400 to 600, that needs to be considered for safety. (Color figure online)

amounts to choosing a speed for an object. A consequence is that T_p and t_p provide information on the speed v_p and remaining distance d_p of an object following trajectory (T_p, t_p). We have $v_p = \frac{H(p)}{T_p}$, and $d_p = t_p \cdot \frac{H(p)}{T_p}$. Given a trajectory place $p \in P_T$, we denote by $\mathcal{T}\nabla(p)$ the set of all trajectories that may appear in p, that is the set of all pairs $\mathcal{T}\nabla(p) = \{(T_p, t_p) \mid T_p \in [\alpha_p^s, \beta_p^s] \wedge 0 \leq t_p \leq T_p\}$. We assume that represented objects have a constant speed $v = H(p)/T_p$ during their trip, which allows us to represent their trajectories as segments. The semantics of a trajectory net is defined in terms of configurations, that assign an integral number of tokens to control places in P_C, and a trajectory to a subset of places in P_T. We explicitly differentiate blocked and progressing trajectories.

More formally, a *configuration* is a triple $C = (M, B, \mathcal{T})$, where $M : P_C \to \mathbb{N}$ is a marking of control places, $B \subseteq P_T$ is a subset of trajectory places containing blocked trajectories, and $\mathcal{T} : P_T \to \bigcup_{p \in P_T} \mathcal{T}\nabla(p)$ associates a progressing trajectory to a subset of places in P_T. For consistency, we require that $\mathcal{T}(p) \in \mathcal{T}\nabla(p)$. Given a marking, we will write $M \geq {}^\bullet(\sigma)$ iff $M(p) \geq 1$ for every $p \in {}^\bullet(\sigma) \cap P_C$. We will denote by $M - {}^\bullet(\sigma)$ the marking M' such that $M'(p) = M(p)$ for every place $p \notin {}^\bullet(\sigma) \cap P_C$ and $M'(p) = M(p) - 1$ for every place $p \in {}^\bullet(\sigma) \cap P_C$. We will denote by $M + (\sigma)^\bullet$ the marking M' such that $M'(p) = M(p)$ for every place $p \notin (\sigma)^\bullet \cap P_C$ and $M'(p) = M(p) + 1$ for every place $p \in (\sigma)^\bullet \cap P_C$.

We represent a configuration $C = (M, B, \mathcal{T})$, with the following graphical convention: we draw $M(p)$ tokens (black dots) in each control place $p \in P_C$, and for a trajectory place $p \in P_T$ we represent trajectory $\mathcal{T}(p) = (T_p, t_p)$ by a dashed segment with coordinates $[(0, H(p)); (T_p, 0)]$ representing the initial trajectory of an object, and a thick segment with coordinates $[(0, d_p); (t_p, 0)]$ representing the remaining displacement. Figure 3, shows a possible trajectory in a place $p \in P_T$ with $H(p) = 800m$, and $I(p) = [60, 85]$. The trajectory represented is (T_p, t_p) with $T_p = 76s$ and $t_p = 34s$. The object moving in the physical space represented by p initially started a trip of duration $76s$ and of length $800m$, represented by the dashed segment. The speed of this object is $v_p = 800/76 = 10.52\,\mathrm{m/s}$. As the remaining trip duration is $t_p = 34\,\mathrm{s}$, one can easily compute the remaining distance to go for this object, namely $d_p = 357$ m. The remaining trip is represented by the thick segment in the diagram. Similarly, in Fig. 2, let $H(p_1) = 500\,\mathrm{m}$, and $I(p_1) = [40, 50]$. Place p_1 contains a trajectory (T_{p_1}, t_{p_1}) with $T_{p_1} = 46\,\mathrm{s}$ and $t_{p_1} = 25\,\mathrm{s}$. The remaining distance to go for this object is $d_{p_1} = 271, 73\,\mathrm{m}$, and its speed is $v_{p_1} = 10.87\,\mathrm{m/s}$.

The semantics of a trajectory net is defined in terms of timed and discrete moves from a configuration to the next one. The system starts in an initial configuration $C_0 = (M_0, B_0, \mathcal{T}_0)$ such that $B_0 = \emptyset$, and for every p such that $\mathcal{T}_0$ is defined, $\mathcal{T}_0(p) = (T_p^0, t_p^0)$ with $T_p^0 = t_p^0 \in [\alpha_p^s, \beta_p^s]$. The main idea of the semantics is that a transition σ can fire if the control places in the preset ${}^\bullet(\sigma)$ allow firing of σ and the objects in the trajectory places of ${}^\bullet(\sigma)$ have reached their final destination. In other terms, all trajectories in the preset of a fired transition must be blocked.

Upon firing of a transition σ, control tokens in ${}^\bullet(\sigma) \cap P_C$ are consumed, blocked trajectories in ${}^\bullet(\sigma) \cap P_T$ are deleted, and new trajectories in $(\sigma)^\bullet \cap P_T$ and new tokens in control places of $(\sigma)^\bullet \cap P_C$ are created. We adopt an exclusive semantics w.r.t. trajectory places, i.e. a transition σ can fire only if the trajectory places in $(\sigma)^\bullet$ are empty. When modeling metro networks, this semantics is appropriate to represent a fixed block policy, where tracks are divided into exclusive blocks that can contain at most one train at any instant[1]. Upon firing, for each place $p \in (\sigma)^\bullet \cap P_T$ new trajectories are sampled: their initial duration T_p is a value chosen non deterministically in $[\alpha_p^s, \beta_p^s]$ and we set $\mathcal{T}(p) = (T_p, T_p)$. On the other hand, elapsing time allows for the progress of existing trajectories. However, we consider an urgent semantics, that allows elapsing $\delta > 0$ time units in a configuration C only if no discrete move can occur in C.

Discrete Moves: Discrete moves are either the blocking of a trajectory or the firing of a transition. Blocking a trajectory $tr_p = (T_p, t_p)$ in place p is possible only if $t_p = 0$, and consists in deleting tr_p, and adding p to the list of places containing a blocked trajectory. Formally, it is defined by the following operational semantics rule:

$$\frac{\begin{array}{c} p \in P_T \\ \mathcal{T}(p) = (x, 0) \text{ for some } x \in [\alpha_p^s, \beta_p^s] \\ \mathcal{T}'(p_i) = \begin{cases} \mathcal{T}'(p_i) \text{ if } p_i \neq p \\ \text{is undefined otherwise} \end{cases} \\ B' = B \cup \{p\} \end{array}}{C = (M, B, \mathcal{T}) \xrightarrow{block\ p} C' = (M, B', \mathcal{T}')}$$

We will say that a transition σ is *firable* in a configuration $C = (M, B, \mathcal{T})$ iff $M \geq {}^\bullet(\sigma)$, the trajectory in ${}^\bullet(\sigma) \cap P_T$ is blocked, and the place depicting

[1] Though this fixed block semantics may seem very constrained, many metro networks in the world are operated this way.

the physical space needed to perform action σ is free, that is, $\forall p \in (\sigma)^\bullet \cap P_T$, $p \notin B$ and $\mathcal{T}(p)$ is undefined. The effect of a firing σ is the consumption of all tokens in ${}^\bullet(\sigma) \cap P_C$, the production of a new token in each place of $(\sigma)^\bullet \cap P_C$, the deletion of the blocked trajectory in ${}^\bullet(\sigma) \cap B$, and the creation of a new trajectory $\mathcal{T}(p') = (T_{p'}, T_{p'})$ in place $p' = (\sigma)^\bullet \cap P_T$ with $T_{p'} \in [\alpha^s_{p'}, \beta^s_{p'}]$. We will write $M[\sigma\rangle M'$ when M' is the marking obtained after firing of σ from M, i.e. when $M' = M - {}^\bullet(\sigma) + (\sigma)^\bullet$. Then, the firing of a transition σ is formally defined by the following operational semantics rule:

$$\frac{\begin{array}{c} M \geq {}^\bullet(\sigma) \wedge M[\sigma\rangle M' \\ \forall p \in (\sigma)^\bullet \cap P_T, p \notin B \wedge \mathcal{T}(p) \text{ is undefined} \\ {}^\bullet(\sigma) \cap P_T \subseteq B \ \wedge \ B' = B \setminus {}^\bullet(\sigma) \\ \mathcal{T}'(p) = \begin{cases} (T_p, T_p), \text{ with} T_p \in [\alpha^s_p, \beta^s_p] \text{ if } p = (\sigma)^\bullet \cap P_T \\ \mathcal{T}(p) \text{ otherwise} \end{cases} \end{array}}{C = (M, B, \mathcal{T}) \xrightarrow{\sigma} C' = (M', B', \mathcal{T}')}$$

When a new trajectory is added in a trajectory place, we necessarily have $d_p = H(p)$. As $H(p)$ is constant, choosing T_p is equivalent to choosing a speed v_p for a vehicle, and memorizing this initial choice. Then, knowing t_p allows to compute d_p after several timed moves.

Timed Moves: The effect of time elapsing is to reduce the remaining trip time of progressing transitions. Elapsing δ time units is allowed if this duration does not exceed remaining trip time of any progressing trajectory. As in Time Petri nets [20], we adopt an *urgent semantics*, that is we forbid time progress if a discrete event can occur. Time progress of δ is hence forbidden if some transition is firable, or if a trajectory gets blocked less than δ time units after the current date. For a given description of trajectories $\mathcal{T}$, we denote by $\mathcal{T} + \delta$ the function that associates the pair $(\mathcal{T} + \delta)(p) = (T_p, t_p - \delta)$ with place p if $\mathcal{T}(p) = (T_p, t_p)$.

$$\frac{\begin{array}{c} 0 < \delta \leq \min\{t_p \mid \exists T_p, (T_p, t_p) \in \mathcal{T}(P_T)\} \\ \forall \sigma \in T, \sigma \text{ is not firable} \end{array}}{C = (M, B, \mathcal{T}) \xrightarrow{\delta} C' = (M, B, \mathcal{T} + \delta)}$$

Obviously, our semantics enjoys time additivity, i.e. if $C_1 \xrightarrow{\delta_1} C_2$ and $C_2 \xrightarrow{\delta_2} C_3$, then $C_1 \xrightarrow{\delta_1 + \delta_2} C_3$. Notice also that timed and discrete moves are exclusive. It is hence natural to describe runs of a trajectory net as an alternation of timed and discrete moves. A *run* of a trajectory net from a configuration $C_0 = (M_0, B_0, \mathcal{T}_0)$ is a sequence $\rho = (M_0, B_0, \mathcal{T}_0) \xrightarrow{\delta_0} (M_0, B_0, \mathcal{T}_0 + \delta_0) \xrightarrow{e_1} (M_1, B_1, \mathcal{T}_1) \cdots$ of timed and discrete moves, where each move $(M_i, B_i, \mathcal{T}_i) \xrightarrow{\delta_i} (M_i, B_i, \mathcal{T}_i + \delta_i)$ is a legal timed move, and each $(M_i, B_i, \mathcal{T}_i) \xrightarrow{e_i} (M_{i+1}, B_{i+1}, \mathcal{T}_{i+1})$ is a legal discrete move, that is a blocking of a trajectory, $(M_i, B_i, \mathcal{T}_i) \xrightarrow{block\ p} (M_{i+1}, B_{i+1}, \mathcal{T}_{i+1})$ or a firing of a transition $(M_i, B_i, \mathcal{T}_i) \xrightarrow{\sigma_i} (M_{i+1}, B_{i+1}, \mathcal{T}_{i+1})$.

We will write $(M, B, \mathcal{T}) \xrightarrow{*} (M', B', \mathcal{T}')$ if there exists a sequence of discrete and timed moves leading from $(M, B, \mathcal{T})$ to $(M', B', \mathcal{T}')$. Without loss of generality, we assume that a net starts in an initial configuration $C_0 = (M_0, B_0, \mathcal{T}_0)$

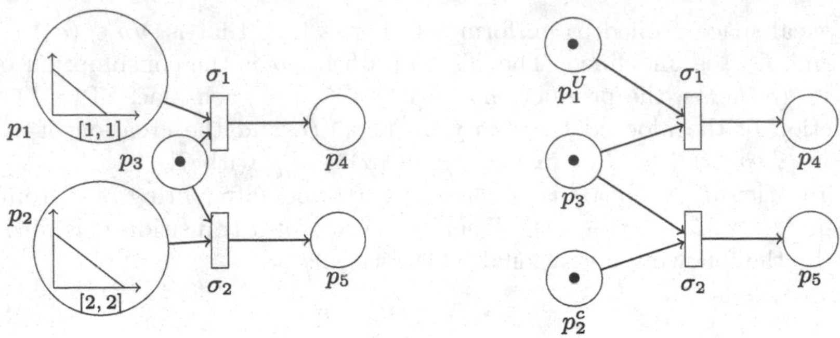

Fig. 4. A trajectory net and its untimed abstraction into a Petri net.

without blocked trajectories, i.e. such that $B_0 = \emptyset$. We also assume that the trajectories of all trains are at their beginning, i.e. for every $p \in P_T$ where $\mathcal{T}_0(p)$ is defined, we have $\mathcal{T}_0(p) = (T_p^0, T_p^0)$ for some $T_p^0 \in [\alpha_p^s, \beta_p^s]$, and that all trains still have some remaining time before being blocked, i.e. $T_p^0 > 0$. We will denote by $Reach(C_0)$ the set of reachable configurations, i.e. $Reach(C_0) = \{(M, B, \mathcal{T}) \mid C_0 \xrightarrow{*} (M, B, \mathcal{T})\}$. We will say that a trajectory net is *bounded* iff, there exists an integer K such that for every configuration $(M, B, \mathcal{T})$ in $Reach(C_0)$, and for every place $p \in P_C$, $M(p) \leq K$.

We will address reachability problems, i.e. study whether a particular configuration $(M, B, \mathcal{T})$ is reachable. Now, asking whether a configuration $(M, B, \mathcal{T})$ is reachable refers to the exact position of objects, and is a too precise question. To cope with this problem, we transform contents of trajectory places into markings as follows: given a configuration $C = (M, B, \mathcal{T})$ we define a *complete marking* M_C that associates an integral number of tokens to each place in P, such that $M_C(p) = M(p)$ if $p \in P_C$, and $M_C(p) = 1$ if $p \in P_T$ and $\mathcal{T}(p)$ is defined or if $p \in B$. We then differentiate three decision problems:

- *exact reachability*: for a given configuration C, does $C \in Reach(C_0)$?
- *boolean reachability*: for a given configuration C, is there a configuration $C' \in Reach(C_0)$ such that $M_C = M_{C'}$?
- *coverability*: for a given configuration C, is there a configuration $C' \in Reach(C_0)$ such that $M_C \leq M_{C'}$?

Coverability can be used to check that at a given instant, two trains cannot occupy rail sections that require mutual exclusion. Note however that coverability is not fine enough to define safety properties involving distances. For instance, one cannot use coverability to check existence of configurations where two objects are simultaneously in a critical area (a red zone similar to those in the diagrams of Fig. 1). As explained at the beginning of this Section, this is a standard property to ensure in metro networks to guarantee passengers safety. In Sect. 6, we will show that we can address such properties involving distances.

Reachability is decidable for timed automata, using an abstraction of clock values and building a *region automaton* [3]. Reachability and coverability are undecidable in general for time Petri nets [16]. Coverability is decidable for timed-arc Petri nets [1,13], but reachability remains undecidable [23]. For a weak interpretation of TPNs semantics, [22] shows that the set of reachable markings of a TPN $\mathcal{N}$ coincides whith the set of reachable markings of an untimed version of $\mathcal{N}$. It is well known that coverability [21] and reachability [19] are decidable for Petri nets, and hence also for TPNs with a weak semantics.

Inspiring from [22], we can easily transform a trajectory net $\mathcal{N} = (P = P_C \cup P_T, T, F, I, H)$ and its initial configuration $C_0 = (M_0, B_0, \mathcal{T}_0)$ into a standard untimed Petri net $\mathcal{U}(\mathcal{N}) = (P', T, F)$ where $P' = P_C \cup \{p^U \mid p \in P_T\}$ replaces every trajectory place by a standard place, with initial marking $M_0^{\mathcal{U}} = M_{C_0}$. However, in general $\mathcal{U}(\mathcal{N})$ allows more markings and more runs than $\mathcal{N}$. Consider the example of Fig. 4, with configuration $C_0 = (M_0, B_0, \mathcal{T}_0)$ with $M_0(p_3) = 1, M_0(p_4) = M_0(p_5) = 0$, $\mathcal{T}_0(p_1) = (1,1)$, $\mathcal{T}_0(p_2) = (2,2)$, and its translation to a standard Petri net $\mathcal{U}(\mathcal{N})$ on the right of the Figure. One can see from this example that in the initial configuration C_0, transition σ_1 is firable, but transition σ_2 will never fire. On the other hand, in $\mathcal{U}(\mathcal{N})$, both transitions σ_1 and σ_2 can fire from M_{C_0}. This example shows that erasing time makes some new markings reachable. Hence we cannot rely on a simple untiming of a trajectory net to address reachability or coverability. We can further show that trajectory nets are powerful enough to model a two-counter machine, yielding undecidability of most problems.

Theorem 1. *Reachability, boolean reachability and coverability are undecidable for trajectory nets.*

Proof (Sketch). We can simulate the behavior of an unbounded two-counter machine with an unbounded trajectory net. Control places can represent counters. From a configuration, urgency can be used to distinguish two behaviors: fire immediately a particular transition if a counter place is filled, or wait and fire a different transition if this counter place is empty. This is sufficient to encode a zero test [22]. The complete encoding is provided in [15]. □

A standard way to recover decidability of reachability and coverability in timed extensions of Petri nets is to restrict to *bounded* nets, and define a symbolic abstraction of timing information allowing a finite partition of the space of configurations. For time Petri nets, where time is measured by clocks attached to enabled transitions, [5] defines an abstraction called state classes, that are equivalence classes for sets of configurations with identical markings and equivalent constraints on the values of clocks. These constraints are called *domains*, and can be compared to zones or regions of timed automata [3]. In the next section, we consider state classes and domains for trajectory nets, and give a sound abstraction of continuous values appearing in configurations.

4 Domains

In this section we define domains for trajectory nets. Domains are a way to define symbolically the value of initial and remaining trajectory durations with positive real valued variables. For a given configuration $C = (M, B, \mathcal{T})$, we have two types of trajectories: blocked trajectories, and progressing ones. For blocked trajectories, the remaining running time is known (it is 0), and the initial time is of no use to define a position of an object: it is at distance 0 w.r.t the end of the represented space. For progressing trajectories, i.e. trajectories in a place $p \in P_T$ for which $\mathcal{T}(p)$ is defined, we have $\mathcal{T}(p) = (T_p, t_p)$. As already mentioned in Sect. 3, we only need these two variables T_p and t_p to compute the speed or the current position of an object, which are important information when considering some safety properties.

Definition 3 (Domains). *Let $\mathcal{N}$ be a trajectory net, with set of trajectory places P_T. Let $P \subseteq P_T$ represent places with progressing trajectories. Then, a domain for $\mathcal{N}$ with progressing trajectories in P is a set of inequalities D over variables $V_D = \{T_i, t_i \mid i \in P\}$, of the form:*

$$\alpha_i^1 \leq T_i \leq \beta_i^1 \textit{ for all } i \in P \tag{1}$$

$$\alpha_i^2 \leq t_i \leq \beta_i^2 \textit{ for all } i \in P \tag{2}$$

$$t_i - t_j \leq \gamma_{ij}^3 \textit{for all } i \neq j \tag{3}$$

$$\alpha_i^4 \leq T_i - t_i \leq \beta_i^4 \textit{for all } i \in P \tag{4}$$

$$\alpha_{ij}^5 \leq T_i - t_i + t_j \leq \beta_{ij}^5 \textit{for all } i \neq j \tag{5}$$

$$-T_i + t_i + T_j - t_j \leq \gamma_{ij}^6 \textit{for all } i \neq j \tag{6}$$

where each type of inequality $(1-6)$ *appears* ***exactly*** *once in D for each $i \in P$ and for each pair of distinct places $i, j \in P$, and $\alpha_i^1, \alpha_i^2, \alpha_i^4, \alpha_i^5$. $\beta_i^1, \beta_i^2, \beta_i^4, \beta_i^5$ $\gamma_{ij}^3, \beta_{ij}^5, \gamma_{ij}^6$ are either constant values, $-\infty$, or $+\infty$.*

Domains in Definition 3 are systems of linear inequalities, but inequalities of type (5) involve expressions with three variables, and inequalities of type (6) use expressions with four variables. Consequently, our domains are not two-dimensional, and differ from the domains proposed by [5]. Further, they **cannot be encoded using DBMs**. We will however show in the rest of the paper that these domains can be efficiently manipulated in polynomial time. Following the definitions of Sect. 2, we will say that a valuation $\mu : V_D \to \mathbb{R}$ for V_D is a *solution* for D iff replacing variables T_i, t_i in V_D by their values $\mu(T_i), \mu(t_i)$ yields a tautology, and denote by $[\![D]\!]$ the set of all solutions for D. Slightly abusing our notation, we will write $\mathcal{T} \in [\![D]\!]$ when the valuation $\mu_{\mathcal{T}}$ that associates variables $\{T_i, t_i\}$ with their respective values in $\mathcal{T}$ is a solution of D. We will say that two domains D_1, D_2 are equivalent iff $[\![D_1]\!] = [\![D_2]\!]$. Even if two domains are equivalent, they may have different representations. Indeed, consider a single pair of variables $T_1 = t_1$ whose values lie in the interval $[3, 4]$. We can represent the constraint on T_1, t_1 as $D_1 = \{3 \leq T_1 \leq 12; 0 \leq T_1 - t_1 \leq 0; 0 \leq t_1 \leq 4\}$. However the domain $D_2 = \{0 \leq T_1 \leq 4; 0 \leq T_1 - t_1 \leq 0; 3 \leq t_1 \leq 20\}$ represents the same set of solutions. We can show that a *canonical form* for domains exists (Proposition 1), and can be computed in polynomial time (Proposition 2).

Definition 4. *Let $D = \{a_i \leq expr_i \leq b_i\}$ be a domain of the form given in Definition 3. Then, the* canonical form *for D is a domain $D^* = \{a_i^* \leq expr_i \leq b_i^*\}$, where a_i^* is the smallest value taken by $expr_i$ in $[\![D]\!]$, and b_i^* is the largest value taken by $expr_i$ in $[\![D]\!]$.*

Proposition 1. *The canonical form of a domain D is unique and preserves $[\![D]\!]$.*

Proof (Sketch). A domain D is a set of inequalities of the form $a_i \leq expr_i$, and $expr_j \leq b_j$. A solution $\mu_S \in [\![D]\!]$ is a map that associates a real value with every variable t_i (resp. T_i). Let $\mu_S(expr_i)$ be the value obtained by replacing every variable t_i by $\mu_S(t_i)$ and T_i by $\mu_S(T_i)$ in $expr_i$. The values $a_i^* = \min_{\mu_S \in [\![D]\!]} \mu_S(expr_i)$ and $b_j^* = \max_{\mu_S \in [\![D]\!]} \mu_S(expr_j)$ are unique, so D^* is uniquely defined. Every expression of the form $a_i \leq expr_i \leq b_i$ can be equivalently rewritten as $a_i^* \leq expr_i \leq b_i^*$ because $a_i^* \leq \mu(expr_i) \leq b_i^*$ for every $\mu \in [\![D]\!]$, and we have $a_i \leq a_i^* \leq b_i^* \leq b_i$. A complete proof is given in [15]. □

As all domains over a fixed set of progressing trajectories have the same types of inequalities, and differ only by the constants used, a direct consequence of Proposition 1 is that two domains D, D' are equivalent if and only if $D^* = D'^*$.

Proposition 2. *The canonical form for a domain D can be computed in PTIME.*

Proof. We perform the following linear transformation: $x_i = T_i - t_i$ and $y_i = -t_i$ to get a new set of inequalities:

$$\begin{array}{rclrcl} \alpha_i^1 \leq x_i - y_i \leq \beta_i^1 & & & \alpha_i^4 \leq & x_i & \leq \beta_i^4 \\ -\beta_i^2 \leq \quad y_i \quad \leq -\alpha_i^2 & & & \alpha_{ij}^5 \leq & x_i - y_j & \leq \beta_{ij}^5 \\ y_j - y_i \leq \gamma_{ij}^3 & & & & -x_i + x_j & \leq \gamma_{ij}^6 \end{array}$$

Notice that our linear transformation is bijective. Hence, for any solution $\mu : \{T_i, t_i\} \rightarrow \mathbb{R}$ in the original domain, there exists a **unique** solution μ' such that $\mu'(x_i) = \mu(T_i) - \mu(t_i), \mu'(y_i) = -\mu(t_i)$ and for any solution $\mu' : \{x_i, y_i\} \rightarrow \mathbb{R}$ in the new domain, there exists a **unique** solution μ such that $\mu(T_i) = \mu'(x_i) - \mu'(y_i)$, and $\mu(t_i) = -\mu'(y_i)$. Hence there is a bijection between the two domains.

Observe that this new domain is of dimension 2. It can hence be encoded as a DBM or a constraint graph, and finding a canonical form for this new domain can be done by computing the shortest paths in the constraint graph. The cost of this calculus is in $O(n^3)$ for n variables. The optimal bounds obtained for the domain over variables $\{x_i, y_i\}$ are bounds for expressions of the form $x_i - y_i$, $x_i - y_j$, etc. that directly encode expressions of the original domain D (for instance $x_i - y_i = T_i$, and $x_i - y_j = T_i - t_i + t_j$). The sharp bounds obtained for the new domain can hence be immediately used as optimal bounds for D, except for the expression of the form $-(\beta_i^2)^* \leq y_i \leq -(\alpha_i^2)^*$, where we need a sign inversion to obtain $(\alpha_i^2)^* \leq t_i \leq (\beta_i^2)^*$.

For a domain D addressing properties of k trajectories, the linear transformation of D is in $O(k^2)$, as we have $3 \cdot k + 3 \cdot k^2$ inequalities D, and performing the transformation for each inequality takes constant time. Computing the canonical form of the new domain can be done in $O(k^3)$ time using the Floyd-Warshall algorithm, as the new domain has $2 \cdot k$ variables. Hence, the canonical form of D can be computed in $O(k^3)$. Note that the constants obtained in the canonical form are linear combinations of α_i^s and β_i^s with integer coefficients. □

Let $\mathcal{N}$ be a trajectory net, and let $P \subseteq P_T$ be the set of places containing progressing trajectories in initial configuration C_0. The initial domain D_0 for $\mathcal{N}$ and P is the set:

$$D_0 = \begin{cases} T_i^0 \leq T_i \leq T_i^0 & \text{for all } i \in P \\ T_i^0 \leq t_i \leq T_i^0 & \text{for all } i \in P \\ t_i - t_j \leq \infty & \text{for all } i \neq j \\ 0 \leq T_i - t_i \leq 0 & \text{for all } i \in P \\ -\infty \leq T_i - t_i + t_j \leq \infty & \text{for all } i \neq j \\ -T_i + t_i + T_j - t_j \leq \infty & \text{for all } i \neq j \end{cases}$$

Let μ_0 be the valuation such that $\mu_0(T_p) = \mu_0(t_p) = T_p^0$ for every place where $\mathcal{T}_0$ is defined. Obviously, $[\![D_0]\!] = \{\mu_0\}$. The initial domain D_0 meets the requirements of Definition 3. To show that the form of domain of Definition 3 is sufficient to represent all domains of a net, it remains to show that the effect of a transition firing, or of a trajectory blocking after some delay δ as proposed in the semantics of Sect. 3 can be encoded through algebraic operations (variable changes, unions of inequalities and projections) that preserve the types of inequalities considered in Definition 3.

4.1 Successors After Firing a Transition

Let D be a domain with set of progressing trajectories P. We want to compute the set of constraints on variables attached to progressing trajectories of the net after firing a transition σ. Let $p = {}^{\bullet}(\sigma) \cap P_T$ and $p' = (\sigma)^{\bullet} \cap P_T$. First, σ can fire only if p is an empty place (a trajectory in p was formerly blocked), and p' is also empty. According to our semantics, adding a trajectory in p' means sampling a new trip duration $T_{p'} \in [\alpha_{p'}^s, \beta_{p'}^s]$ and adding in p' a new progressing trajectory $(T_{p'}, T_{p'})$. The sampled value is totally independent from the values of variables in D, so the new set of constraint on progressing trajectories after firing of σ is the set:

$$\begin{aligned} SuccF(D, \sigma) = D & \cup \{\alpha_{p'}^s \leq T_{p'} \leq \beta_{p'}^s\} \cup \{\alpha_{p'}^s \leq t_{p'} \leq \beta_{p'}^s\} \cup \{0 \leq T_{p'} - t_{p'} \leq 0\} \\ & \cup \{t_{p'} - t_i \leq \infty \mid i \in P\} \cup \{t_i - t_{p'} \leq \infty \mid i \in P\} \\ & \cup \{-\infty \leq T_{p'} - t_{p'} + t_i \leq \infty \mid i \in P\} \\ & \cup \{-\infty \leq T_i - t_i + t_{p'} \leq \infty \mid i \in P\} \\ & \cup \{-T_{p'} + t_{p'} + T_i + t_i \leq \infty \mid i \in P\} \\ & \cup \{-T_i + t_i + T_{p'} + t_{p'} \leq \infty \mid i \in P\} \end{aligned}$$

One can immediately notice that if D is a domain, then so is $SuccF(D, \sigma)$.

4.2 Successors After Blocking a Trajectory

Blocking a progressing trajectory $tr_p = (T_p, t_p)$ from a configuration occurs after elapsing $\delta = t_p$ time units, and is allowed if δ is the minimal duration among all progressing trajectories. We hence have to consider transformations on a domain D occurring after a sequence of timed and discrete moves of the form $C \xrightarrow{\delta} C' \xrightarrow{block\ p} C''$. Remark, from the semantics that $\delta = t_p$, so blocking of tr_p can occur only if $t_p = \min\{t_j \mid \exists p_j \in P_T, \mathcal{T}(p_j) = (T_j, t_j)\}$. This requirement can be easily translated into a new constraint: trajectory tr_p can be blocked from some configuration satisfying domain D iff $D^{p\leq *} ::= D \cup \{t_p \leq t_j \mid j \neq p \wedge (T_j, t_j) \text{ is a progressing trajectory}\}$ is satisfiable.

As a blocked trajectory does not constrain any more possible durations of other trajectories, the domain capturing the remaining constraints in configuration C'' is the projection on remaining variables once t_p time units have elapsed. To obtain this set of constraints, we proceed as follows:

- We make a variable change. Let t'_j denote a variable representing the new value of remaining travel time of trajectory j after elapsing t_p time units. Then we have $t'_j = t_j - t_p$. We hence replace every variable t_j by $t'_j + t_p$ in every inequality of $D^{p\leq *}$. Let us call D' this new domain.
- We eliminate variables T_p and t_p from domain D'. This elimination can be done in polynomial time using the well-known Fourier-Motzkin algorithm (see [15] and [9]).
- We replace every occurrence of a variable t'_j by an unprimed variable t_j to obtain a successor domain $SuccB(D, p)$, and we compute its canonical form.

Proposition 3. *Let D be a domain of a trajectory net $\mathcal{N}$, and let D' be a system of linear inequalities that is a successor of D via construction of $SuccB(D, p)$ or $SuccF(D, \sigma)$. Then D' is a domain of $\mathcal{N}$.*

Proof (Sketch). $SuccF(D, \sigma)$ trivially satisfies this property, as it only adds constraints of the form $\alpha_i^S \leq T_i \leq \beta_i^S$, $\alpha_i^S \leq t_i \leq \beta_i^S$, and $T_i = t_i$. The proof for $SuccB(D, p)$ is more involved, as it requires eliminating variables for the blocked trajectory. Yet, during elimination, some inequalities are unchanged. For other inequalities, for instance when eliminating t_p, combining expressions of the form $expr_j \leq t_p$ and $t_p \leq expr_k$ to obtain a new expression $expr_j \leq expr_k$ during the elimination process either produces tautologies, or new expressions that are of the form of inequalities in Definition 3. A complete proof is given in [15]. □

5 Soundness, Completeness, Finiteness

Now that we have defined domains for trajectory nets, and shown that we can effectively compute a canonical representation for $SuccF(D, \sigma)$ the set of constraints that hold after firing a transition σ and $SuccB(D, p)$ the constraints that hold after blocking a trajectory in place p when starting from a domain D, we can define state classes.

Definition 5. *A* state class *of a trajectory net $\mathcal{N}$ is a triple $SC = (M, B, D)$, where M is a marking, B is a subset of trajectory places with blocked trajectories, and D is a domain of $\mathcal{N}$ in canonical form.*

We can define a symbolic transition relation among state classes as follows:

- $(M, B, D) \xrightarrow{Block\ p}_S (M, B', D')$ if $B' = B \cup \{p\}$, and D' is the canonical representation of $SuccB(D, p)$, and
- $(M, B, D) \xrightarrow{\sigma}_S (M', B', D')$ if $M[\sigma\rangle M'$, $B' = B \setminus {}^\bullet(\sigma)$, and D' is the canonical representation of $SuccF(D, \sigma)$.

We will write $(M, B, D) \longrightarrow_S (M', B', D')$ if either $(M, B, D) \xrightarrow{Block\ p}_S (M', B', D')$ or $(M, B, D) \xrightarrow{\sigma}_S (M', B', D')$. We will denote by $Reach^S(M_0, B_0, D_0)$ the set of state classes that can be built inductively from the initial sate class D_0 by application of the symbolic transition relation $\longrightarrow_S$.

Definition 6. *The state class graph of a trajectory net $\mathcal{N}$ is the transition system $SC(\mathcal{N}) = \big(Reach^S(M_0, B_0, D_0), \longrightarrow_S, (M_0, B_0, D_0)\big)$.*

Notice that $SC(\mathcal{N})$ is defined even if $Reach^S(M_0, B_0, D_0)$ is not finite. We will say that a configuration $C = (M, B, \mathcal{T})$ *matches* with a state class $SC = (M', B', D)$ iff $M = M'$, $B = B'$ and $\mathcal{T} \in [\![D]\!]$.

Definition 7. *A* symbolic run *of $\mathcal{N}$ is a sequence of state classes of the form $\rho^S = (M_0, B_0, D_0) \xrightarrow{e_0} (M_1, B_1, D_1) \xrightarrow{e_1} \dots$ such that for every index $i \geq 0$, $e_i \in \{Block\ p_i, \sigma_i\}$ and $(M_i, B_i, D_i) \xrightarrow{e_i}_S (M_{i+1}, B_{i+1}, D_{i+1})$.*

Proposition 4 (Soundness). *Let $\rho^S = (M_0, B_0, D_0) \xrightarrow{e_0} (M_1, B_1, D_1) \xrightarrow{e_1} \dots$ be a symbolic run of a trajectory net $\mathcal{N}$. Then, there exists a run $\rho = (M_0, B_0, \mathcal{T}_0) \xrightarrow{\delta_0} (M_0, B_0, \mathcal{T}_0 + \delta_0) \xrightarrow{e_0} (M_1, B_1, \mathcal{T}_1) \dots$ of $\mathcal{N}$ such that for every $i \geq 0$, $(M_i, B_i, \mathcal{T}_i)$ matches with (M_i, B_i, D_i).*

Proposition 5 (Completeness). *Let $\rho = (M_0, B_0, \mathcal{T}_0) \xrightarrow{\delta_0} (M_0, B_0, \mathcal{T}_0 + \delta_0) \xrightarrow{e_0} (M_1, B_1, \mathcal{T}_1) \dots$ be a run of a trajectory net $\mathcal{N}$. Then, there exists a symbolic run $\rho^S = (M_0, B_0, D_0) \xrightarrow{e_0} (M_1, B_1, D_1) \xrightarrow{e_1} \dots$ of $\mathcal{N}$ such that for every $i \geq 0$, $(M_i, B_i, \mathcal{T}_i)$ matches with (M_i, B_i, D_i).*

The proofs for Propositions 4 and 5 are obtained by induction on the length of runs, and are detailed in [15].

Theorem 2. *Let $\mathcal{N}$ be a bounded trajectory net, with initial configuration C_0. Then the set of canonical domains that can be computed inductively from D_0^* is finite.*

Proof (sketch). A domain is defined by a set of inequalities. The number of inequalities depend only on the number of progressing trajectories, and these inequalities involve constants. We can prove that constants appearing in canonical domains of a trajectory net are linear combinations of constants appearing in

D_0 and the bounds of intervals $[\alpha_p^s, \beta_p^s]$. It was also proved (see [5]) that the number of linear combinations of a finite set of constants in a rational interval $[A, B]$ is finite. It remains to show that this type of bounding interval exists for the values of constants $\alpha_i^1, \beta_i^1, \alpha_i^2, \beta_i^2, \gamma_{ij}^3, \alpha_i^4, \beta_i^4, \alpha_{ij}^5, \beta_{ij}^5, \gamma_{ij}^6$ appearing in domains of trajectory nets. This interval is $[-2 \cdot C_{max}, 2 \cdot C_{max}]$, where C_{max} is the maximal value appearing in an interval $[\alpha_p^s, \beta_p^s]$. So, we get that constants in domains can take a finite number of values, and a finite number of inequalities can appear in domains. Let us denote by I_0 this set of possible inequalities. The set of possible domains is finite, since each domain is a subset of I_0 (all proof details are given in [15]). □

Following Theorem 2, we can give an upper bound on the number of state classes in $SC(\mathcal{N})$. Let us first compute the size of I_0. Assuming that we consider only domains in canonical form, every constraint in I_0 is of the form $a \leq expr \leq b$, with $-2 \cdot C_{max} \leq a$ and $b \leq 2 \cdot C_{max}$. Assuming that all α_i^s and β_i^s are rational numbers with a common denominator d, there exists at most $4 \cdot C_{max} . d$ possibilities for values of a and b in expressions. Similarly, expression $expr$ are of the form given in Definition 3, and there are hence $3 \cdot (|P_T| + |P_T|^2)$ expressions. The size of I_0 is hence $12 \cdot C_{max} \cdot d \cdot (|P_T| + |P_T|^2)$, and each domain is a subset of inequalities from I_0. This gives an upper bound on the number of domains, which is in $O(2^{|I_0|})$. In a state class, component B is a subset of trajectory places, and hence there are at most $2^{|P_T|}$ possible values for B. Last, for a K bounded trajectory net, the number of possible markings for control places is in $O(|P_C|^{K+1})$ [12]. The number of state classes is hence in $O(2^{|I_0|+|P_T|} \cdot |P_C|^{K+1})$.

6 Reachability, Coverability, Safety

An important property of the state class graph is that all solutions for domains that are reachable in $SC(\mathcal{N})$ are also reachable in $\mathcal{N}$. This immediately gives an algorithm to check coverability or reachability properties.

Theorem 3. *Given a state class (M_n, B_n, D_n) reachable from initial state class (M_0, B_0, D_0), and a solution $\mathcal{T}_n \in [\![D_n]\!]$, there exists a run in the original trajectory net that ends in configuration $(M_n, B_n, \mathcal{T}_n)$.*

Proof (Sketch). As for Proposition 4), we can use an induction on the length of paths in the state class graph. (See details in [15]). □

Theorem 4. *Reachability, boolean reachability, and boolean coverability are decidable in PSPACE for bounded nets*

Proof. These problems can be solved by a non-deterministic exploration of the state class graph. Let us first consider reachability of a given configuration $(M, B, \mathcal{T})$. Assume that a state class (M, B, D) such that $\mathcal{T} \in [\![D]\!]$ is reachable in $SC(\mathcal{N})$. Then, according to Theorem 3, there exists a run of $\mathcal{N}$ reaching $(M, B, \mathcal{T})$. Consider now the boolean reachability and coverability problems. Let

$sc=(M,B,D)$ be a reachable state class. One can notice that the boolean marking M_C is identical for every configuration C matching sc. We hence denote this marking by M_{sc}, and compute it by assigning a token to a place $p_i \in P_T$ iff T_i, t_i are variables used in D or if $p_i \in B$. Then, deciding reachability (resp. coverability) of a marking M consists in finding a state class sc such that $M_{sc} = M$ (resp. $M_{sc} \geq M$).

As shown in Sect. 5, the size of the state class graph is exponential w.r.t. the number of places and w.r.t. the value on constants appearing in intervals attached to trajectory places. Encoding a state class can be done in $\log(|SC(\mathcal{N})|)$ hence in polynomial space, and reachability questions can be addressed in nlogspace w.r.t. the size of the graph. At each step of an exploration reaching a particular state class $SC_i = (M_i, B_i, D_i)$, checking $M = M_i$, $B = B_i$, $M = M_{SC_i}$ or $M \leq M_{SC_i}$ can be done in linear time w.r.t the number of places, and checking $\mathcal{T} \in [\![D_i]\!]$ can be done in linear time w.r.t the number of inequalities in D by replacing every variable by its value in each inequality. As the number of inequalities in canonical domains is quadratic w.r.t. the number of trajectory places, checking $\mathcal{T} \in [\![D_i]\!]$ can be done in PTIME. Hence, reachability boolean reachability, and coverability can be checked in NPSPACE, which is equivalent to PSPACE by Savitch's theorem [24]. □

Remark 1. Notice that a trajectory net without trajectory places is a Petri net (transitions can fire as soon as their preset is filled, after any delay). Hence, reachability and coverability in trajectory nets are at least as hard as reachability and coverability in 1-safe Petri nets. These problems are known to be PSPACE-Complete [7,11].

Corollary 1. *Reachability, boolean reachability, and boolean coverability for bounded nets are PSPACE-Complete.*

Proof. PSPACE membership is proved by Theorem 4. As highlighted in Remark 1, a reachability or a coverability problem for 1-safe nets is also a reachability/coverability problem for trajectory nets (with empty set of trajectory places). As these two problems are PSPACE-Complete [7,11], we get the result. □

6.1 Extending Coverability to Safety Properties

Reachability is often a too precise question and one is usually interested in properties that address ranges of values for positions of objects. Let us get back to the case study introduced in Sect. 3, namely avoidance of simultaneous dangerous situations in a metro network. Metro networks can be easily represented by trajectory nets: trajectory places represent track portions between two stations, or a finer partition of a physical network into track portions called *blocks*, transitions symbolize departures, arrivals etc. Obviously, metro networks have very strict safety requirements that must be guaranteed by physical equipments such as signals and brakes. Safety issues also appear at the operational level. At any instant, evacuation of passengers should be feasible with the lowest risks, and for that reason, operators want to avoid situations where more than K trains are in tunnels or on bridges.

Addressing such safety properties is neither a reachability nor a coverability question: it requires that *at any instant*, all trajectories of trains avoid a *set of unsafe positions*. Let $p \in P_T$ be a trajectory place, of length $H(p)$, and assume that the track portion represented by p contains a tunnel. We can easily define two values d_p^s and d_p^e defining respectively the position of the entry and exit of the tunnel in that track (for instance, in Fig. 3, we have $d_p^s = 600$ and $d_p^e = 400$). Let $\mathcal{T}p = (T_p, t_p)$. The function gives us the initial duration T_p of a trip from a station to the next one, the remaining time t_p before the end of this trip, but we can also compute the position d_p of the considered train on the track. We have assumed that all objects moving in a trajectory net have a constant speed during the whole duration of a trajectory, sampled when the object enters a place. Hence, $\frac{H(p)}{T_p} = \frac{d_p}{t_p}$ at any instant, and a train in place p is in a tunnel iff the following property $Tunnel(p)$ is satisfied:

$$Tunnel(p) ::= d_p^e \leq \frac{H(p) \cdot t_p}{T_p} \leq d_p^s$$

Now assume that we want to avoid a situation where trains in a set of places $X = \{p_1, \dots, p_K\} \subseteq P_T$ are in tunnels at the same instant. It means that we have to avoid any configuration that satisfies the property $Unsafe(X)$, where $Unsafe(X) ::= \bigcup_{p_i \in X} Tunnel(p_i)$.

The domains computed in the state class graph $SC(\mathcal{N})$ are symbolic representations of configurations reached immediately after discrete moves. It can be the case that, after each discrete move, all trains are located before a tunnel in their respective track segment. Verifying safety of a train network does not amount to verifying that $[\![D \cup Unsafe(X)]\!] = \emptyset$ for sets of places X containing tunnels and for every reachable domain D. We need to consider how configurations depicted in D evolve when elapsing time, i.e., build symbolic representations of configurations reached an arbitrary duration after a discrete move. Hence, we introduce time closure $S_\downarrow = (M, B, D_\downarrow)$ of a state class $S = (M, B, D)$.

Definition 8 (Time Closure). *Let $S = (M, B, D)$ be any state class having a set of places with progressing trajectories $P_{pr} \subseteq P_T$. We introduce variables δ (to represent the timed move) and t_i' for all $i \in P_{pr}$ (to represent time remaining in trajectories after a timed move of duration δ). The* time closure *of S is a 3-tuple $S_\downarrow = (M, B, D_\downarrow)$ with $D_\downarrow$ defined as:*

Case I: No transition is firable from the given marking M and set of blocked trajectories B, and hence timed moves are allowed by the semantics of trajectory nets. We have $D_\downarrow = D \cup \{0 \leq \delta \leq t_i \mid i \in P_{pr}\} \cup \{t_i' = t_i - \delta \mid i \in P_{pr}\}$

Case II: There exists a firable transition for the given marking M and set of blocked trajectories B, and hence timed moves are not allowed. We hence have $D_\downarrow = D \cup \{\delta = 0\} \cup \{t_i' = t_i \mid i \in P_{pr}\}$.

The time closure of a state class $S = (M, B, D)$ is a symbolic representation of possible configurations reachable after timed moves of arbitrary duration

δ, including the configurations in domain D (i.e., when $\delta = 0$). As explained above, a property of interest for metro networks is that no more than K trains are in a tunnel at any given instant. We denote by $Unsafe'(X)$ the set of inequalities obtained by replacing, for every place $p \in X$ variable t_p by t'_p in $Unsafe(X)$. A state class $S = (M, B, D)$ is *safe* for places $p_1, \dots p_K$ iff $[\![D_\downarrow \cup Unsafe'(p_1, \dots p_K)]\!] = \emptyset$. Verifying that a state class is safe for every subset of places of size K amounts to asking that $[\![D_\downarrow \cup Unsafe'(X)]\!] = \emptyset$ for every subset $X \subseteq P_T$ of places of size K containing tunnels. Notice that the set of all X's can be enumerated in $\log(|P_T|)$ space.

Remark 2. Non-emptiness of $D_\downarrow \cap Unsafe'(X)$ implies existence of a configuration violating the safety property, because all configurations in a state class D are reachable, and hence all configurations in $D_\downarrow$ too. Hence, checking safety for all state classes of $SC(\mathcal{N})$ guarantees that the trajectory net does not violate the safety property. This gives us a method to check safety of a metro network modeled with a trajectory net using its state class graph.

Let $X = \{p_1, \dots, p_K\}$. The constraint $Unsafe'(X)$ can be rewritten as $\left\{d^e_{p_i} \leq \frac{H(p_i) \cdot t'_i}{T_i} \leq d^s_{p_i} \mid p_i \in X\right\}$ and as every T_i is a positive value, simplified to get $\left\{d^e_{p_i} \cdot T_i \leq H(p_i) \cdot t'_i \leq d^e_{p_i} \cdot T_i \mid p_i \in X\right\}$. One can immediately observe that this set contains only linear inequalities of dimension 2 involving T_i and t'_i. Let us now consider $D_\downarrow$, defined for a set of places P of progressing trajectories. It obtained by replacing t_i by $t'_i + \delta$. It is hence a set of constraints of the form:

$$\begin{aligned}
\alpha^1_i \leq T_i \leq \beta^1_i &\text{ for all } p_i \in P \\
\alpha^2_i \leq t'_i + \delta \leq \beta^2_i &\text{ for all } p_i \in P \\
t'_i - t'_j \leq \gamma^3_{ij} &\text{for all } p_i, p_j \in P \text{ with } i \neq j \\
\alpha^4_i \leq T_i - t'_i + \delta \leq \beta^4_i &\text{for all } p_i \in P \\
\alpha^5_{ij} \leq T_i - t'_i + t'_j \leq \beta^5_{ij} &\text{for all } p_i, p_j \in P \text{ with } i \neq j \\
-T_i + t'_i + T_j - t'_j \leq \gamma^6_{ij} &\text{for all } p_i, p_j \in P \text{ with } i \neq j
\end{aligned}$$

Remark 3. Checking emptiness of $D_\downarrow \uplus \left\{d^e_{p_i} \cdot T_i \leq H(p_i) \cdot t'_i \leq d^s_{p_i} \cdot T_i \mid p_i \in X\right\}$ can be done by elimination of variables one after another, and stopping as soon as an inequality is unsatisfiable, or when all variables are eliminated. We can use a variable change as in Proposition 2, and get an equivalent system of dimension 2. Hence, checking satisfiability of $D_\downarrow \uplus Unsafe'(p_1, \dots, p_K)$ can be done in PTIME.

Proposition 6. *Checking a safety property for a bounded trajectory net is PSPACE-complete.*

Proof. $\mathcal{N}$ violates a safety property of the form "no more than K trains in a tunnel" iff there exists a reachable configuration $C = (M, B, \mathcal{T})$ such that $\mathcal{T} \models Unsafe(X)$ for some subset of places $X = \{p_1, \dots p_K\}$ with tunnels. According to Remark 2, this holds only if there exists a reachable state class

$S = (M, B, D)$ such that $[\![D_\downarrow \cup Unsafe'(X)]\!] \neq \emptyset$. We know from the decision procedures for reachability and coverability that exploration of all state classes can be done in PSPACE. Then, for each state class $S = (M, B, D)$ reached, we need to compute $D_\downarrow$ enumerate all subsets X of K places containing tunnels and with a progressing trajectory, and check emptiness of $D_\downarrow \cup Unsafe'(p_1, \dots, p_K)$. Enumeration of subsets of places of size K can be done in $\log(P_T)$ space. Following Remark 3, (un)satisfiability of $D_\downarrow \cup Unsafe'(p_1, \dots p_K)$ can be verified in PSPACE. The hardness comes from a reduction of the coverability problems for 1-safe nets: if C is a set of places to cover in a 1-safe Petri net $\mathcal{N}$, one can design a trajectory net $\mathcal{N}_T$ with one additional trajectory places $p_{T,c}$ per place in C and such that ${}^\bullet(p_{T,c}) = {}^\bullet(c)$, $(p_{T,c})^\bullet = (c)^\bullet$ and set zones to avoid as the whole length of these places. Reaching a configuration with objects in $X = \{p_{T,c} \mid c \in C\}$ in $\mathcal{N}_T$ amounts to covering C in $\mathcal{N}$. □

7 Conclusion

We have considered an extension of Petri nets enhanced with time and linear functions depicting trajectories of moving objects. Most problems for this model are undecidable in general. However, as soon as the control part is bounded, the behaviour of the model can be abstracted to a finite state class graph, and coverability, reachability, and safety properties addressing distance issues can be decided in PSPACE. Finiteness of the state class graph of trajectory nets comes from bounds on the values of variables, and from the particular structure of domains, that are conjunctions of linear inequalities with at most 4 variables, and coefficients in $\{1, -1\}$. This structure is preserved by projection, and hence by the successor relation among state classes. Preliminary work shows that domains for trajectory nets are a form of regular polyhedra, that can be encoded by Totally Unimodular Matrices (TUM). This needs to be formally proved, but it would explain why our domains have interesting closure and algorithmic properties. An interesting research direction is to consider extensions of trajectory nets that preserve this nice and efficient domain structure.

As future work, several extensions of the model can be considered. We have defined a restricted model, mainly tailored to represent metro networks. The first restriction is that transitions have at most one trajectory place in their preset and in their postset. This last restriction was used to simplify notations and reading, and can be easily relaxed: one can imagine firing rules consuming several blocked trajectories, and similarly producing several new trajectories in the postset of a transition. The techniques shown in this paper easily adapt. It should also be possible to consider trajectories in 2D or 3D spaces.

The second restriction imposes that trajectories are simple linear functions, i.e. the speed of an object is sampled once, and remains constant until a trajectory gets blocked. One could easily improve trajectories using piecewise linear functions depicting behavior of objects moving at varying speeds (this is the model proposed in [17], Chapter 5). This would be useful to describe acceleration and braking phases. We conjecture that this extension still allows to work with TUMs, as long as trajectory places contain at most one trajectory.

More involved improvements would be to consider trajectories specified by polynomials, or to allow more than one trajectory per place. The single trajectory restriction makes sense when modeling metro networks for instance, as many cities implement a fixed block traffic management policy, where tracks are partitioned in exclusive blocks that must be occupied by a single train at any instant. However, this type of management is progressively replaced by moving block policies, that allow several trains in a track segment provided they maintain safety distances. Relaxing the single trajectory restriction is also needed to model other situations such as road traffic. First experiments seem to show that a definition of domains for these two extensions require polynomial inequalities, i.e. expressions of the form $P(X) \leq c$, where $P(X)$ is a multivariate polynomial. It is known that such domains are closed under projection [25]. However, variable elimination has a doubly exponential complexity [8]. Further, we conjecture that finiteness of domains does not hold any more in this new setting.

References

1. Abdulla, P.A., Nylén, A.: Timed petri nets and BQOs. In: Colom, J.-M., Koutny, M. (eds.) ICATPN 2001. LNCS, vol. 2075, pp. 53–70. Springer, Heidelberg (2001). https://doi.org/10.1007/3-540-45740-2_5
2. Alur, R., et al.: The algorithmic analysis of hybrid systems. TCS **138**(1), 3–34 (1995)
3. Alur, R., Dill, D.L.: A theory of timed automata. TCS **126**(2), 183–235 (1994)
4. Alur, R., et al.: The algorithmic analysis of hybrid systems. Theor. Comput. Sci. **138**(1), 3–34 (1995)
5. Berthomieu, B., Diaz, M.: Modeling and verification of time dependent systems using time Petri nets. IEEE Trans. Softw. Eng. **17**(3), 259–273 (1991)
6. Chandru, V.: Variable elimination in linear constraints. Comput. J. **36**(5), 463–472 (1993)
7. Cheng, A., Esparza, J., Palsberg, J.: Complexity results for 1-safe nets. TCS **147**(1&2), 117–136 (1995)
8. Collins, G.E.: Quantifier elimination for real closed fields by cylindrical algebraic decompostion. In: Automata Theory and Formal Languages, pp. 134–183 (1975)
9. Dantzig, G., Eaves, B.C.: Fourier-Motzkin elimination and its dual. J. Comb. Theory, Ser. A **14**(3), 288–297 (1973)
10. Dill, D.L.: Timing assumptions and verification of finite-state concurrent systems. In: Sifakis, J. (ed.) CAV 1989. LNCS, vol. 407, pp. 197–212. Springer, Heidelberg (1990). https://doi.org/10.1007/3-540-52148-8_17
11. Esparza, J.: Decidability and complexity of Petri net problems — an introduction. In: Reisig, W., Rozenberg, G. (eds.) ACPN 1996. LNCS, vol. 1491, pp. 374–428. Springer, Heidelberg (1998). https://doi.org/10.1007/3-540-65306-6_20
12. Esparza, J.: Petri nets lecture notes. Technical report, 2019
13. Escrig, D.F., Ruiz, V.V., Alonso, O.M.: Decidability of properties of timed-arc petri nets. In: Nielsen, M., Simpson, D. (eds.) ICATPN 2000. LNCS, vol. 1825, pp. 187–206. Springer, Heidelberg (2000). https://doi.org/10.1007/3-540-44988-4_12
14. Hélouët, L., Agrawal, P.: Waiting nets. In: Bernardinello, L., Petrucci, L. (eds.) Application and Theory of Petri Nets and Concurrency. PETRI NETS 2022. LNCS, vol. 13288, pp. 67–89. Springer, Cham (2022). https://doi.org/10.1007/978-3-031-06653-5_4

15. Hélouët, L., Contractor, P.: Symbolic domains and reachability for nets with trajectories (long version). Technical report, Inria, 2024. https://inria.hal.science/hal-04528235
16. Jones, N.D., Landweber, L.H., Lien, Y.E.: Complexity of some problems in Petri nets. Theor. Comput. Sci. **4**(3), 277–299 (1977)
17. Kecir, K.: Performance Evaluation of Urban Rail Traffic Management Techniques. (Évaluation de Performances pour les Techniques de Régulation du Trafic Ferroviaire Urbain). PhD thesis, University of Rennes 1, France, 2019
18. Lime, D., Roux, O.H.: Model checking of time Petri nets using the state class timed automaton. Discret. Event Dyn. Syst. **16**(2), 179–205 (2006)
19. Mayr, E.W.: An algorithm for the general Petri net reachability problem. In: Proceedings of the 13th Annual ACM Symposium on Theory of Computing, 11–13 May 1981, Milwaukee, Wisconsin, USA, pp. 238–246. ACM (1981)
20. Merlin, P.M.: A Study of the Recoverability of Computing Systems. PhD thesis, University of California, Irvine, CA, USA (1974)
21. Rackoff, C.: The covering and boundedness problems for vector addition systems. Theor. Comput. Sci. **6**, 223–231 (1978)
22. Reynier, P.-A., Sangnier, A.: Weak time Petri nets strike back! In: Bravetti, M., Zavattaro, G. (eds.) CONCUR 2009. LNCS, vol. 5710, pp. 557–571. Springer, Heidelberg (2009). https://doi.org/10.1007/978-3-642-04081-8_37
23. Ruiz, V.V., Gomez, F.C., de Frutos-Escrig, D.: On non-decidability of reachability for timed-arc Petri nets. In: PNPM, p. 188. IEEE Computer Society (1999)
24. Savitch, W.J.: Relationships between nondeterministic and deterministic tape complexities. J. Comput. Syst. Sci. **4**(2), 177–192 (1970)
25. Tarski, A.: A decision method for elementary algebra and geometry. Technical report, RAND Corporation, 1957

Verification and Model Checking

Symbolic Model Checking Using Intervals of Vectors

Damien Morard(✉), Lucas Donati, and Didier Buchs

Computer Science Department, Faculty of Science, University of Geneva, Geneva, Switzerland
{damien.morard,didier.buchs}@unige.ch, lucas.donati@etu.unige.ch

Abstract. Model checking is a powerful technique for software verification. However, the approach notably suffers from the infamous state space explosion problem. To tackle this, in this paper, we introduce a novel symbolic method for encoding Petri net markings. It is based on the use of generalised intervals on vectors, as opposed to existing methods based on vectors of intervals such as Interval Decision Diagrams. We develop a formalisation of these intervals, show that they possess homomorphic operations for model checking CTL on Petri nets, and define a canonical form that provides good performance characteristics. Our structure facilitates the symbolic evaluation of CTL formulas in the realm of global model checking, which aims to identify every state that satisfies a formula. Tests on examples of the model checking contest (MCC 2022) show that our approach yields promising results. To achieve this, we implement efficient computations based on saturation principles derived from other symbolic model checking techniques.

1 Introduction

In the ever-evolving landscape of computer science and software engineering, the correctness and reliability of software systems remain paramount concerns. In response to these challenges, the discipline of model checking has emerged as a powerful tool in the arsenal of software verification and validation.

Model checking [8] is a formal verification technique used to ensure that a system adheres to its specifications and requirements. However, although model checking is a trustworthy and robust method, it is confronted with a variety of challenges. One of the most significant ones is the state space explosion problem [11,18]. This problem arises when dealing with complex software or hardware systems, where the number of possible states and transitions within a model grows exponentially with its size. The task becomes more challenging when the objective is to identify all states that adhere to a given property, as opposed to merely determining the validity of the property within a specific configuration. Moreover, our objective is to address the issue of *global model checking*, namely, to identify all states that fulfill a CTL (Computation Tree Logic) [9] formula.

While various techniques have been developed to mitigate the state space explosion problem, such as Decision diagrams [1,13,14], abstractions [10],

L. M. Kristensen and J. M. van der Werf (Eds.): PETRI NETS 2024, LNCS 14628, pp. 269–290, 2024.
https://doi.org/10.1007/978-3-031-61433-0_13

partial order reduction [20], and many others, it has become evident that this challenge will persist as models become more sophisticated. Therefore, the quest for innovative methods to address this issue remains a compelling area of research. As always, each new approach inevitably brings its own set of advantages and disadvantages compared to existing techniques.

This paper focusses on a symbolic method inspired by [21] that proposes to encode sets of markings in Petri net models with vectors that act as boundaries on the markings, similar to intervals. As markings are essentially vectors of values, and fireability of transitions are lower constraints and their negation are upper constraints, intervals seem to be a suitable choice for representing the satisfiable states of a CTL formula. However, expressing intervals on vectors is complicated, since vectors are not always comparable. We introduce such notions and demonstrate that, for non-total orders, we require additional information such as a set of lower and upper bounds, rather than a single bound as in intervals of natural numbers. Merely efficiently representing sets of markings is insufficient; we therefore also define homomorphic operations (w.r.t. union) on the symbolic structure to directly perform model checking on it.

Thanks to this encoded structure and operations, we can efficiently maintain the same structure and remain in the symbolic domain, without the need to decode it to perform computation over the encoded values. Furthermore, we introduce a canonical form incorporating optimisations to make it efficient to use in the context of CTL model checking. We tested our approach on examples from the model checking contest [16], yielding very promising results.

The paper is organised as follows: firstly, in Sect. 2, we provide an overview of our project. Next, in Sect. 3, we present the complete formalism of our symbolic structure. In Sect. 4, we demonstrate the connection between this symbolic structure and Petri nets to apply CTL model checking. Finally, Sect. 5 presents our results using this technique.

2 Informal Presentation

In the context of Petri nets, as illustrated in Fig. 1, the states of a model are represented using *markings*. A marking can be viewed as a vector where each place is associated with a natural number that indicates the number of resources. In our example, $(3, 1)$ is the current marking of the Petri net.

Model checking Petri nets consists in being able to answer queries such as "How many resources are required at each place to make t_0 fireable ?". Two perspectives on the problem are possible from this question. The first seeks all states that satisfy the query, namely *global model checking* [22]. The second checks whether a specific state satisfies the query, corresponding to *local model checking*. Obviously, because it computes the entire state space for a query, global model checking also provides an answer for any given state, making it more general than the local approach. In particular, such a space is often infinite, meaning that a simple explicit enumeration of all states is impossible, as would be the case for the property of making t_0 fireable.

Our approach uses a new structure called *symbolic vector*, described by a couple (a, b), where a and b represent sets of markings. To belong to (a, b), a given marking must include (given a partial order relation on vectors) all the markings in a and none of the markings in b. This is a generalisation of the interval concept to vectors. Indeed, membership in a natural number interval $[a, b]$ can be expressed as $\forall x \in \mathbb{N}, a \leq x \leq b$, which can equivalently be formulated as $\forall x \in \mathbb{N}, a \leq x \wedge b \nleq x$. The latter definition aligns with the conceptual framework of symbolic vectors, albeit in this case, the focus is on vectors rather than natural numbers. The latter definition shares the same idea as symbolic vector, excepts that we work on vectors instead of naturals. In the example in Fig. 1, the symbolic vector encoding all solutions of t_0 being fireable is $(\{(2,0)\}, \varnothing)$. This couple includes an infinite number of markings, such as $(2, 0)$ or $(3, 4)$ and so on. Let us assume that we do not want to include markings greater than or equal to $(8, 9)$; we would get the symbolic vector $(\{(2,0)\}, \{(8,9)\})$. In this refined version, valid markings could be $(7, 7)$ or $(10, 8)$, whereas invalid markings could be $(8, 9)$ or $(10, 9)$. Therefore, based on an inclusion and exclusion set, we can encapsulate sets of markings. In addition to this structure, we leverage homomorphic operations to perform computations directly on the symbolic structure, rather than on each element of the set individually. For example, the action of adding 1 to p_0 for all markings in $(\{(2,0)\}, \varnothing)$ can be done in a single step, leading to $(\{(3,0)\}, \varnothing)$, rather than having to iterate over each element in the set to increment it.

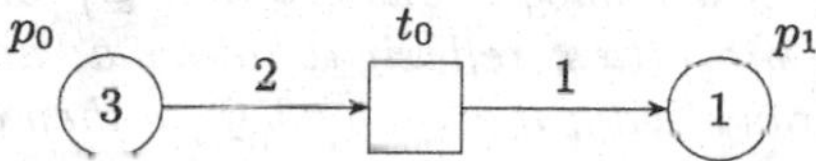

Fig. 1. Petri net with two places and one transition

Although symbolic vectors are a first step in encoding sets of markings, they are not sufficient to represent all possible sets, similar to intervals of natural numbers that cannot encode all sets of natural numbers. To address this limitation, a set of intervals is required. For instance, the set of naturals $\{1, 2, 3, 7, 8\}$ is encoded as $\{[1, 3], [7, 8]\}$, necessitating two intervals. However, constructing a symbolic version for sets poses a greater challenge. Indeed, a symbolic representation is often linked to the task of finding a canonical form, as failing to do so may result in redundancy and other issues that diminish the efficiency of the approach. For example, the set $\{[1, 4], [2, 5], [3, 6]\}$ could be reduced to $\{[1, 6]\}$ in natural numbers. Due to the non-explicit representation, the uniqueness set rule is not sufficient to preserve the uniqueness of the symbolic set. In fact, $[1, 4] \neq [2, 5]$ and $[1, 4] \cap [2, 5] \neq \varnothing$. Each interval in this example is distinct, but all share values with the others. The standard approach to constructing a canonical form in this context is to merge intervals whenever possible.

Unlike sets of intervals, creating a symbolic representation for sets of markings poses a more substantial challenge, primarily due to the non-strict partial order on its elements. Although all intervals can be compared to each other, the same is not true for markings. Besides, a symbolic vector itself is an intricate structure that also needs to be canonised. Consequently, additional constraints are necessary to manage canonicity of a *symbolic vector set.*

By combining a symbolic encoding for sets of markings and encoding the evaluation process to be applied to symbolic vector sets, we are able to compute CTL formulas for global model checking, i.e. to determine all the markings of a net that satisfy a formula. It should be noted that the necessary operations for doing CTL model checking are mainly set operations and a *pre* operation. All of them are defined as homomorphisms on symbolic vector sets.

3 Symbolic Structure for Sets of Vectors

This section formalises the structures known as *symbolic vector* and *symbolic vector set.* Symbolic vectors serve as the initial encoding layer for Petri net markings, whereas sets of symbolic vectors constitute the set that incorporates them. We present all of their respective definitions, properties, and canonical form.

3.1 Symbolic Vectors

Definition 1 (Non-strict partial order). *Let $a, b \in Q$ be two elements. A non-strict partial order is a homogeneous relation $\subseteq_f$ on a set Q, such that it verifies the following properties; reflexivity: $a \subseteq_f a$; antisymmetry: if $a \subseteq_f b \wedge b \subseteq_f a$ then $a = b$; transitivity: if $a \subseteq_f b \wedge b \subseteq_f c$ then $a \subseteq_f c$.*

Notice that two values can potentially not be compared, i.e. neither $a \subseteq_f b$ nor $b \subseteq_f a$ are in the relation.

Definition 2 (Inclusion relation on vectors). *Let $q_a, q_b \in Q_n$ be two elements such that Q_n is the set of vectors of size $n \in \mathbb{N}$. The non-strict partial order on two vectors of the same size, namely the inclusion of vectors, is a relation defined as $q_a \subseteq_f q_b \Leftrightarrow \forall p \in P, q_a(p) \leq q_b(p)$.*

We use the notation $\forall p \in P, q(p)$ where $q \in Q_n$ to iterate over the elements of our tuple. This parallels a function structure, where each tuple location serves as the domain and the associated value functions as its codomain. From the previous definition, the derivations of the definitions for $\not\subseteq_f, =_f, \neq_f$ can be inferred. In addition, for the remainder of the article, we assume the same definition of Q_n.

Definition 3 (Symbolic vector). *Let Q_n be a set of vectors such that there exists a non-strict partial order $\subseteq_f$ on its elements. A symbolic vector is defined as: $(a, b) \in \mathbb{SV}$, where $a, b \in \mathcal{P}(Q_n)$. A vector that belongs to it, noted $\in_{sv}$: $Q_n \times \mathbb{SV}$, is defined by the relation $(q \in_{sv} (a, b)) \Leftrightarrow \forall q_a \in a, q_a \subseteq_f q \wedge \forall q_b \in b, (q_b \not\subseteq_f q \wedge q_b \neq_f q)$. We assume that $\notin_{sv}$ is its negation.*

A symbolic vector relies on set theory and the definition of a non-strict partial order relation for elements inside it. For a vector q to be accepted by such a structure, all elements of a must be included in q and none of the elements of b must be included in it. Note that the whole relation may be seen as a sequence of conjunctions ($\wedge$) where each predicate must be true.

Remark 1. Note that the theory of symbolic vectors could be generalised to any structure with a non-strict partial order. However, for simplicity and a clear focus on Petri nets, we will describe it directly for vectors, the structure used to represent markings. Moreover, one advantage of vectors of totally ordered values is that there exists a lexicographic order on them that can be used as a total order. Nevertheless, it can help construct the canonical form, but this order alone is not useful in establishing the fireability condition.

Definition 4 (Underlying set of vectors). *Let $sv \in \mathbb{SV}$ be a symbolic vector. The set of underlying vectors of a symbolic vector is a function $uf : \mathbb{SV} \to \mathcal{P}(Q_n)$ such that $uf(sv) = \{q \in Q_n \mid q \in_{sv} sv\}$.*

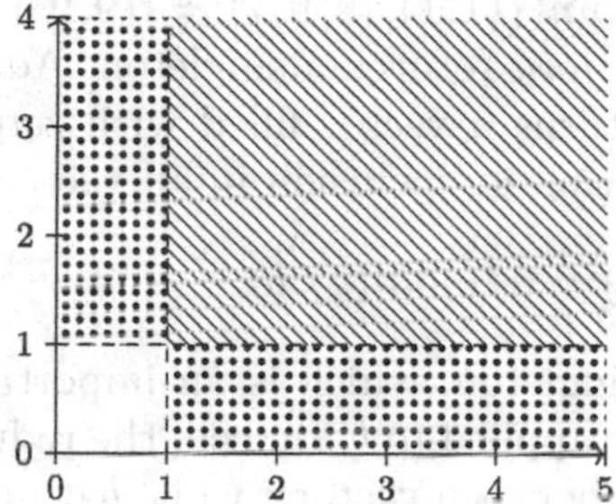

Fig. 2. Visualisation of the symbolic vector $(\{(1,0),(0,1)\},\varnothing)$.

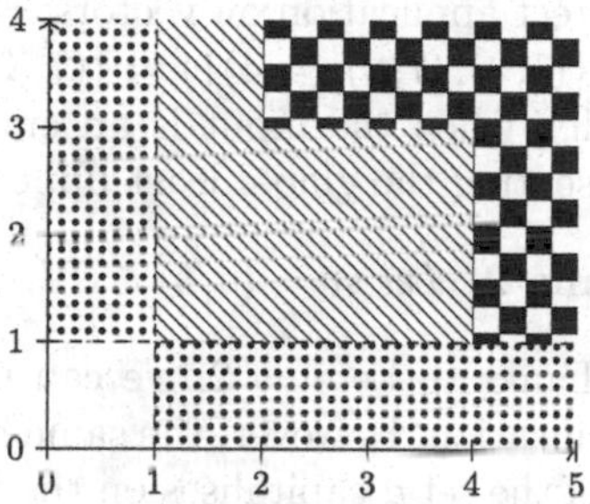

Fig. 3. Visualisation of the symbolic vector $(\{(1,0),(0,1)\},\{(2,3),(4,1)\})$.

In Fig. 2, we present an initial example visualising the underlying set of vectors from $(\{(1,0),(0,1)\},\varnothing)$. Vectors in the dot pattern satisfy only one of the conditions, while those in the hatch pattern satisfy both simultaneously. This illustrates that the left vectors must all be included, denoted by the "*and*" operator. In contrast, when adding multiple constraints, as shown in Fig. 3 with the chequerboard pattern, the "*or*" operator comes into play, making a solution invalid if it includes $(2,3)$, $(4,1)$, or both simultaneously. Therefore, the valid vectors are those belonging to the hatch pattern. Note that, while we bounded the domain size for the graphical representation, the hatch pattern symbolises an infinite number of solutions. Moreover, while this visualisation is effective for a vector of size 2, it needs adaptation for varying vector lengths. Nevertheless, the fundamental concept remains consistent across different vector sizes.

Definition 5 (Intersection of two symbolic vectors). *Let $sv = (a,b)$, $sv' = (c,d) \in \mathbb{SV}$ be two symbolic vectors. The intersection between two symbolic vectors, noted $\cap_{sv} : \mathbb{SV} \times \mathbb{SV} \to \mathbb{SV}$, is defined as: $sv \cap_{sv} sv' = (a \cup c, b \cup d)$.*

The intersection takes the conditions of each symbolic vector to combine them. Thus, to belong to their intersection, a new vector must belong to a and c, but not to b and d. Furthermore, the intersection is the only operation on symbolic vectors that gives a symbolic vector as an output by combining two of them. Similarly to intervals, the difference or union may return a set of intervals instead of a single interval.

Lemma 1. *Let $sv, sv' \in \mathbb{SV}$, then $uf(sv \cap_{sv} sv') = uf(sv) \cap uf(sv')$.*

In the following, we assume that $f_\varepsilon \in Q_n$ is the *zero vector* such that all its values are set to 0.

Definition 6 (Join). *Let us assume $p \in \mathcal{P}(Q)$ be a set of vectors. The join function, which finds the least upper bound of a set of vectors, written as $join : \mathcal{P}(Q_n) \to \mathcal{P}(Q_n)$, is defined as $join(\varnothing) = \{f_\varepsilon\}$ and $join(p) = \{q_{join}\}, p \neq \varnothing$, such that $\forall q \in p, (q \subseteq_f q_{join} \wedge \nexists q' \in \mathcal{P}(Q), q' \subseteq_f q_{join} \wedge q_{join} \neq_f q' \wedge q \subseteq q')$.*

The *join* operation has its origins in lattice theory and has been adapted for direct application on vectors. For example, $join(\{(1,1),(9,9)\}) = \{(9,9)\}$ or $join(\{(1,3),(7,2),(4,4)\}) = \{(7,4)\}$. Note that *join* returns a singleton. Additionally, when the input is an empty set, it returns a vector filled with zeros, representing the global least upper bound, ensuring its inclusion in all vectors.

Lemma 2. *Let $sv = (a,b) \in \mathbb{SV}$. Then, $uf(a,b) = uf(join(a),b)$.*

Thanks to Lemma 2, we can treat a as a singleton, which is an important simplification. However, the same idea cannot be applied to b. Indeed, the reduction of the set a capitalises on the '*and*' relation between vectors, while b cannot achieve the same simplification due to the '*or*' relation.

Definition 7 (Empty symbolic vector). *Let $sv = (a,b) \in \mathbb{SV}$ be a symbolic vector and $\{q_a\} = join(a)$[1]. An empty symbolic vector is defined as $sv \in \varnothing_{sv} \Leftrightarrow \exists q_b \in b, q_b \subseteq_f q_a$, where $\varnothing_{sv}$ is the set of all empty symbolic vectors. Furthermore, when expressing $sv \notin \varnothing_{sv}$, we assume its negation.*

In fact, addressing emptiness in symbolic vectors is a crucial challenge. Given their infinite number, there is considerable potential for redundancy, necessitating the ability to identify them for the development of a future canonical form. For example, $uf(\{(3,3)\},\{(1,1)\}) = uf(\{(0,0)\},\{(0,0)\}) = \varnothing$.

Definition 8 (Canonicity of symbolic vectors). *Let $sv = (a,b) \in \mathbb{SV}$ be a symbolic vector. sv is canonical if and only if:*

1. *$a = \{q_a\}$ is a singleton.*
2. *$sv \notin \varnothing_{sv} \vee sv = (\{f_\varepsilon\},\{f_\varepsilon\})$.*
3. *$\forall q_b, q'_b \in b, q_b \neq q'_b, q_b$ and q'_b are not comparable.*
4. *$\forall q_b \in b, \forall p \in P, q_a(p) \leq q_b(p)$.*

[1] If $a = \varnothing$, then $join(a) = \{f_\varepsilon\}$, where q_a is included in all markings of b.

To ensure the canonicity of a symbolic vector, several conditions must be satisfied, each contributing to its unique representation. Firstly, we rely on the join function described in Definition 6 as the initial condition, which ensures that a is always expressible as a singleton. Secondly, addressing the issue of multiple possible expressions for an empty symbolic vector, a deliberate decision has been made to maintain a single representation, namely $(\{f_\varepsilon\}, \{f_\varepsilon\})$. Thirdly, the vectors in the set b must not be encapsulated within a vector in the same set, since the most inclusive vector already encapsulates all constraints. For example, $uf(\{(1,1)\}, \{(4,4),(5,5)\}) = uf(\{(1,1)\}, \{(4,4)\})$, because $(4,4) \subseteq_f (5,5)$. Lastly, the fourth condition ensures that each value of q_a serves as a minimum bound for each value of each vector in b.

Example 1. $uf(\{(2,0,5)\}, \{(4,0,2)\}) = uf(\{(2,0,5)\}, \{(4,0,5)\})$. Both vectors have the same underlying representation. Although $(4,0,2) \neq_f (4,0,5)$, the minimum vector is $(2,0,5)$, which forces each component to be greater than or equal to it to be accepted. Therefore, as long as the value of the third component is less than or equal to 5 for the right part, the underlying set remains equivalent. Furthermore, by imposing the minimum value of q_a in b, all vectors in b are inherently included in a, i.e. $\forall q_b \in b, q_a \subseteq_f q_b$.

Theorem 1 (Unicity of the representation). *Let $sv, sv' \in \mathbb{SV}$ be two canonical symbolic vectors. Then, $uf(sv) = uf(sv') \Leftrightarrow sv = sv'$.*

3.2 Symbolic Vector Sets

As mentionned before when using intervals, union is not an internal operation of intervals. The same occurs for symbolic vectors. We then need the following definitions and properties that are based on the previous definition of symbolic vectors and can be seen as extensions that work on sets.

Definition 9 (Symbolic vector sets). *A symbolic vector set, belonging to $\mathcal{P}(\mathbb{SV})$, is a set containing exclusively symbolic vectors. A vector that belongs to it, noted $\in_{svs}$: $Q_n \times \mathcal{P}(\mathbb{SV})$, is defined by the relation $(q \in_{svs} svs) \Leftrightarrow \exists sv \in svs, q \in_{sv} sv$. We assume $\notin_{svs}$ as its negation. An empty symbolic vector set is defined as: $svs \in \varnothing_{svs} \Leftrightarrow \forall sv \in svs, sv \in \varnothing_{sv}$, where $\varnothing_{svs}$ is the set of all empty symbolic vector sets. Besides, $\varnothing \in \varnothing_{svs}$.*

Definition 10 (Underlying vectors of set). *The set of underlying vectors of a symbolic vector set is a function, noted $uf_{svs} : \mathcal{P}(\mathbb{SV}) \rightarrow \mathcal{P}(Q_n)$, such that:*

$$uf_{svs}(svs) = \bigcup_{sv \in svs} (uf(sv))$$

Similarly to uf, we unfold the result of each symbolic vector to obtain the final vector set.

Definition 11 (Union & intersection). *Let $svs, svs' \in \mathcal{P}(\mathbb{SV})$ be two symbolic vector sets. The union and intersection between two symbolic vector sets, noted $\cup_{svs}, \cap_{svs} : \mathcal{P}(\mathbb{SV}) \times \mathcal{P}(\mathbb{SV}) \rightarrow \mathcal{P}(\mathbb{SV})$, are defined as: $svs \cup_{svs} svs' = svs \cup svs'$ and $svs \cap_{svs} svs' = \bigcup_{sv \in svs} \bigcup_{sv' \in svs'} sv \cap_{sv} sv'$.*

Union for symbolic vector sets is simply set union. The intersection of two symbolic vector sets is the combination of the intersection between all the symbolic vectors of sv and sv'.

Lemma 3 (Commutativity). *Let $svs, svs' \in \mathcal{P}(\mathbb{SV})$ be two symbolic vector sets and $* \in \{\cup_{svs}, \cap_{svs}\}$ Then, $svs * svs' = svs' * svs$.*

Before introducing the difference between two symbolic vector sets, we formalise the difference of two symbolic vectors.

Definition 12 (Difference of symbolic vectors). *Let $sv = (a, b), sv' = (c, d) \in \mathbb{SV}$ be symbolic vectors. The difference between two symbolic vectors, noted $\setminus_{sv} : \mathbb{SV} \times \mathbb{SV} \to \mathcal{P}(\mathbb{SV})$, is defined as $sv \setminus_{sv} sv' = \{sv_1 \in \mathbb{SV} \mid q_c \in c,\ sv_1 = (a, b \cup \{q_c\})\} \cup_{svs} \{sv_2 \in \mathbb{SV} \mid q_d \in d,\ sv_2 = (a \cup c \cup \{q_d\}, b)\}$.*

We remove all the values that are common to both symbolic vectors. c and d are used as upper and lower bounds, respectively. Note that in the construction of the new set we have $b \cup \{q_c\}$ and $a \cup c \cup \{q_d\}$ instead of $\{q_c\}$ and $\{q_d\}$. The omission of c in the construction of the latter could cause a problem when $q_c \not\leq q_d$. This is crucial to preserve the initial condition in addition to the new bounds, preventing acceptance of values that were not part of the original conditions.

Example 2. Let us illustrate with an example:

$$\begin{aligned}
&(\{(1,1)\}, \{(9,9)\}) \setminus_{sv} (\{(4,4)\}, \{(7,2),(5,6)\}) \\
&= \{(\{(1,1)\}, \{(9,9)\} \cup \{(4,4)\})\} \cup_{svs} \{(\{(1,1)\} \cup \{(4,4)\} \cup \{(7,2)\}, \{(9,9)\})\} \\
&\quad \cup_{svs} \{(\{(1,1)\} \cup \{(4,4)\} \cup \{(5,6)\}, \{(9,9)\})\} \\
&= \{(\{(1,1)\}, \{(4,4)\}), (\{(7,4)\}, \{(9,9)\}), (\{(5,6)\}, \{(9,9)\})\}
\end{aligned}$$

The last step involves simplifications, ensuring that each symbolic vector adheres to the canonical form. This primarily involves using *join* for the left component of each symbolic vector. Then, we verify that there are no comparable vectors in b, as exemplified by $\{(9,9)\} \cup \{(4,4)\}$, reduced to $\{(4,4)\}$. Note that this operation may introduce redundancy in the resulting symbolic vector set. The vector $(7,6)$ belongs to $(\{(7,4)\}, \{(9,9)\})$ and $(\{(5,6)\}, \{(9,9)\})$.

Lemma 4. *Let $sv, sv' \in \mathbb{SV}$. Then, $uf_{svs}(sv \setminus_{sv} sv') = uf(sv) \setminus uf(sv')$.*

Definition 13 (Difference). *Let $svs = \{sv_1, \ldots, sv_n\}, svs' \in \mathcal{P}(\mathbb{SV}), n > 1$ and $sv, sv' \in \mathbb{SV}$. The difference between two symbolic vector sets, noted $\setminus_{svs} : \mathcal{P}(\mathbb{SV}) \times \mathcal{P}(\mathbb{SV}) \to \mathcal{P}(\mathbb{SV})$, is defined inductively as:*

$$\begin{aligned}
svs \setminus_{svs} \emptyset_{svs} = svs,\ \emptyset_{svs} \setminus_{svs} svs &= \emptyset_{svs} \\
\{sv\} \setminus_{svs} \{sv'\} &= sv \setminus_{sv} sv' \\
\{sv\} \setminus_{svs} \{sv_1, \ldots, sv_n\} &= (sv \setminus_{sv} sv_1) \setminus_{svs} \{sv_2, \ldots, sv_n\} \\
\{sv_1, \ldots, sv_n\} \setminus_{svs} svs' &= \{sv_1\} \setminus_{svs} svs' \cup_{svs} \{sv_2, \ldots, sv_n\} \setminus_{svs} svs'
\end{aligned}$$

The difference is carried out recursively by taking each left symbolic vector and applying the difference with each right symbolic vector. Moreover, since the result of $\setminus_{sv}$ is a new set, we need to invoke $\setminus_{svs}$.

Definition 14 (Negation). *Let* $svs \in \mathcal{P}(\mathbb{SV})$ *be a symbolic vector set and* $sv \in \mathbb{SV}$ *be a symbolic vector. The respective negation of both structures, noted* $\neg_{svs} : \mathcal{P}(\mathbb{SV}) \to \mathcal{P}(\mathbb{SV})$ *and* $\neg_{sv} : \mathbb{SV} \to \mathcal{P}(\mathbb{SV})$, *are defined as:* $\neg_{sv}(sv) = (\{f_\varepsilon\}, \varnothing) \setminus_{sv} sv$ *and* $\neg_{svs}(svs) = \{(\{f_\varepsilon\}, \varnothing)\} \setminus_{svs} svs$.

Lemma 5. *Let* $svs, svs' \in \mathcal{P}(\mathbb{SV})$ *and* $* \in \{\cup, \cap, \setminus\}$ *be operations. Then,* $uf_{svs}(svs) * uf_{svs}(svs') = uf_{svs}(svs *_{svs} svs')$ *and* $uf_{svs}(\neg_{svs} svs) = \neg uf_{svs}(svs)$.

Definition 15 (Symbolic vector set relations). *Let* $svs, svs' \in \mathcal{P}(\mathbb{SV})$ *be symbolic vector sets and* $q \in Q_n$ *be a vector. Each of the relations* $\subseteq_{svs}, \not\subseteq_{svs}, =_{svs}, \neq_{svs} \subseteq \mathcal{P}(\mathbb{SV}) \times \mathcal{P}(\mathbb{SV})$ *and* $\in_{svs} \subseteq \mathbb{SV} \times \mathcal{P}(\mathbb{SV})$ *are defined as follows:*

- $svs \subseteq_{svs} svs' \Leftrightarrow svs \setminus_{svs} svs' \in \varnothing_{svs}$
- $svs \not\subseteq_{svs} svs' \Leftrightarrow svs \setminus_{svs} svs' \notin \varnothing_{svs}$
- $svs =_{svs} svs' \Leftrightarrow (svs \subseteq_{svs} svs') \wedge (svs' \subseteq_{svs} svs)$
- $svs \neq_{svs} svs' \Leftrightarrow (svs \not\subseteq_{svs} svs') \vee (svs' \not\subseteq_{svs} svs)$
- $q \in_{svs} svs \Leftrightarrow \exists sv \in svs, q \in_{sv} sv$

The general set relations are expanded to directly work on sets of symbolic vectors. Their constructions are based on the operations described previously.

3.3 Canonicity of Symbolic Vector Sets

Establishing a canonical form for symbolic vector sets involves several conditions, some of which mirror those seen in interval sets. Among these conditions is the aim to prevent overlap between intervals, which corresponds to avoiding the inclusion of the same value in two distinct symbolic vector sets. Additionally, another scenario may arise when two intervals share no values, but can be merged into a single one. For instance, the two intervals [1,4] and [5,7] in natural numbers can be combined into [1,7]. Therefore, symbolic vector sets must take into account these conditions.

However, addressing the aforementioned cases is insufficient to guarantee future canonicity due to the non-strict partial order of its elements. Figure 4 depicts a scenario in which two symbolic vectors do not share values and cannot be merged. Despite this, two different sets exist, resulting in an equivalent final underlying set of vectors. The issue arises because the value within the symbolic vector $(\{(4,4)\}, \varnothing)$ may be included in both symbolic vectors. The visualisation in Fig. 4b offers an initial explanation of this phenomenon. The nodes are arranged according to their vectors, with the lowest at the bottom and the largest at the top. Vector (4,4) serves as a *join* point where the two symbolic vectors share the above values. Moreover, vector (4,4) can be moved over both excluding sets to transfer the constraint, leading to two different representations for the same underlying set. Thus, it is imperative to establish a deterministic rule for selecting the symbolic vector that should encompass the constraint. To address this, we introduce a total order on symbolic vectors.

$uf_{svs}(\{(\{(2,4)\},\varnothing),(\{(4,2)\},\{(4,4)\})\})$
$=$
$uf_{svs}(\{(\{(2,4)\},\{(4,4)\}),(\{(4,2)\},\varnothing)\})$

(a) Two symbolic vector sets with the same underlying set of vectors.

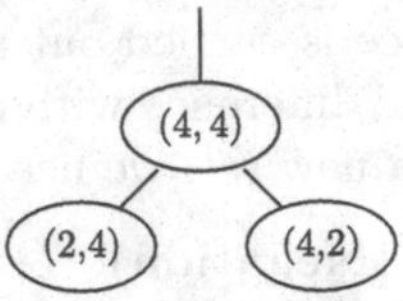

(4, 4)
(2,4) (4,2)

(b) Visualisation of the two symbolic vector sets.

Fig. 4. Illustration of the canonical issue encountered with the non-strict partial order on its vectors, where $(\{(4,4)\},\varnothing)$ is potentially contained in both symbolic vectors.

Definition 16 (Lexicographic ordering). *Let* $q = (x_1,\ldots,x_n), q' = (y_1,\ldots,y_n) \in Q_n$. *The lexicographic ordering, noted* $\leq_f$: $Q_n \times Q_n \to \mathbb{B}$, *is defined as:*

$$() \leq_f () = true$$

$$(x_1,\ldots,x_n) \leq_f (y_1,\ldots,y_n) = \begin{cases} true & x_1 < y_1 \\ false & y_1 < x_1 \\ (x_2,\ldots,x_n) \leq_f (y_2,\ldots,y_n) & x_1 = y_1 \end{cases}$$

Lemma 6. $\leq_f$ *defines a total order on* Q_n.

Using this new relation, we can determine the symbolic vector that should incorporate the constraint. Referring to the previous illustration in Fig. 4, we opted to retain the constraint in the larger symbolic vector. Consequently, the selected form among these two alternatives is $\{(\{(2,4)\},\varnothing),(\{(4,2)\},\{(4,4)\})\}$.

This relation enables us to establish the criteria that determine when two symbolic vectors are deemed *shareable*, implying the capability to transfer the constraint to the right symbolic vector if necessary.

Definition 17. *Let* $sv = (a,b), sv' = (c,d) \in \mathbb{SV}$ *be two symbolic vectors, and* $\{q_a\} = join(a), \{q_c\} = join(c)$ *be the simplifications of a and c. Additionally, let* $\{q_{max}\} = join(\{q_a, q_c\})$ *denote the join vector of* q_a *and* q_c. *The shareable function on two symbolic vectors, noted* $shareable_{sv} : \mathbb{SV} \times \mathbb{SV} \to \mathbb{B}$, *is defined as:*

$$\textbf{\textit{if}}\ sv \in \varnothing_{sv} \vee sv' \in \varnothing_{sv},\ shareable_{sv}(sv, sv') = true$$
$$\textbf{\textit{if}}\ sv \notin \varnothing_{sv} \wedge sv' \notin \varnothing_{sv}$$

$$shareable_{sv}(sv, sv') = \begin{cases} \textbf{\textit{if}} & q_a \leq_f q_c \\ & \begin{cases} \textbf{\textit{if}} & \nexists q_b \in b, (q_b \subseteq_f q_{max} \wedge q_b \neq_f q_{max}) \\ & \wedge\ \nexists q_d \in d, q_d \subseteq_f q_{max} \\ & true \\ \textbf{\textit{else}} & false \end{cases} \\ \textbf{\textit{else}} & Symmetrical\ case \end{cases}$$

The function $shareable_{sv}$ is designed to determine when two symbolic vectors have a shareable component. Checking shareability involves examining whether there exists a gap between two symbolic vectors. By building q_{max}, we seek the first value that is potentially shared between two symbolic vectors and will then be the minimum vector of the shared symbolic vector. Therefore, if a tuple of b or d is lower than q_{max}, this implies that sharing is not possible. This is similar to having a gap between two intervals such as $[1, 4]$ and $[7, 9]$.

As illustrated in Fig. 4a, a scenario arises in which two symbolic vectors are shareable. In this case, the shareability should be considered *true* only when the shared component is not on the correct symbolic vector. The condition $q_b \neq_f q_{max}$ guarantees this, ensuring that the vector with the lowest lexicographic order does not contain the constraint. In other words, only the largest symbolic vector will be constrained by q_{max}. This prevents overlap between the two structures.

The purpose of $shareable_{sv}$ extends beyond the mentioned scenario. These conditions also encompass situations of *overlapping* and *mergeability*. In essence, when the function evaluates to true, it signifies that the symbolic vector set is non-canonical.

Example 3. Let us look at the following examples:

$$shareable((\{(1,1)\},\{(5,5)\}),(\{(3,3)\},\{(7,7)\})) = true \quad (1)$$
$$shareable((\{(1,1)\},\{(3,3)\}),(\{(3,3)\},\{(7,7)\})) = true \quad (2)$$
$$shareable((\{(1,1)\},\{(3,3)\}),(\{(5,5)\},\{(7,7)\})) = false \quad (3)$$

In case (1), the overlap is evident, while in case (2) there is no overlap, but it could still be subject to merging. Both cases return *true* when invoking $shareable_{sv}$, whereas case (3) does not allow for any simplification.

We have all the necessary elements to formally define the canonical form of a symbolic vector set, as follows.

Definition 18 (Canonicity of symbolic vector sets). *Let $svs \in \mathcal{P}(\mathbb{SV})$ be a symbolic vector set. svs is canonical if and only if:*

1. $\forall sv \in svs, sv$ *is canonical.*
2. $svs \notin \varnothing_{svs} \vee svs = \varnothing$
3. $\forall sv, sv' \in svs, sv \neq sv', shareable_{sv}(sv, sv') = false$

Based on this definition, three conditions must be met to guarantee canonicity. The initial condition stipulates that all elements within a symbolic vector set must be canonical. The second condition requires a unique form for the empty symbolic vector set. Lastly, the condition we examined previously becomes significant. When $shareable_{sv}$ is evaluated to be true, a portion of the largest symbolic vector can be transferred to the smaller one between sv and sv'. However, it is important to note that the shared part might already partially or entirely exist within the smaller symbolic vector. This could occur in cases where one symbolic vector overlaps another. In such instances, the larger symbolic vector would be either entirely removed or adjusted to exclude the portion already present in the smaller symbolic vector.

Theorem 2 (Unicity of the representation). *Let* $svs, svs' \in \mathcal{P}(\mathbb{SV})$ *be two canonical symbolic vector sets. Then,* $uf_{svs}(svs) = uf_{svs}(svs') \Leftrightarrow svs = svs'$.

Remark 2. For the sake of brevity and simplicity, we omit the formal description of canonical homomorphisms for $\cup_{svs}$, $\cap_{svs}$, $\setminus_{svs}$, and $\neg_{svs}$. We assume their existence because, formally, they can be viewed as a canonisation of classical set operations and are intended to be used implicitly. Note that this work is a synthesis of the thesis in [19]. Furthermore, the complete theory is described there, including additional definitions and properties, as well as all the canonical operations and proofs related to them.

4 Petri Net Model Checking Using Symbolic Vector Sets

This section aims to describe the relation and application between CTL model checking and Petri nets using symbolic vector sets.

4.1 Encoding Markings Using Symbolic Vector Sets

Definition 19 (Capacity Petri net). *A Petri net with capacities is a tuple* $\Sigma = \langle P, T, in, out, w, k, m_0 \rangle$ *where* P *is the set of finite places,* T *is the set of finite transitions,* $in \subseteq (P \times T)$ *and* $out \subseteq (T \times P)$ *are the sets of input and output arcs linking places and transitions,* $w : (in \cup out) \to \mathbb{N}^+$ *is the function labeling each arc with a weight,* $k : P \to \mathbb{N}^+ \cup \{\infty\}$ *binds each place with a value that cannot be exceeded, and* $m_0 \in M$ *represents the initial marking, belonging to the set of all markings* M. *We assume that* w *is total, returning* 0 *when the relations in or out are not defined. Formally,* $w(p,t) = 0$ *if* $(p,t) \notin in$ *and* $w(t,p) = 0$ *if* $(t,p) \notin out$. *Furthermore,* $\mathbb{N}^+ = \mathbb{N} \setminus \{0\}$. *Operators and relations on* $\mathbb{N}$ *are extended to* $\mathbb{N}^+ \cup \{\infty\}$. *A marking is denoted by a vector* $(n_1, \ldots, n_m)$ *where* $n_1, \ldots, n_m \in \mathbb{N}, m = |P|$, *in which each tuple location is connected to a place that indicates the number of resources. Vectors can be seen as functions of finite domain* $m : P \to \mathbb{N}$. *A transition* $t \in T$ *is enabled for the marking* $m \in M$ *iff* $\forall p \in P, w(p,t) \leq m(p) \leq k(p) - w(t,p)$, *where* $\infty - n = \infty, n \in \mathbb{N}$ *and* $l \leq \infty$ *is true for* $l \in \mathbb{N} \cup \{\infty\}$. *When a transition is enabled, it may be fired, resulting in:* $fire(m,t) = m' \Leftrightarrow \forall p \in P, m'(p) = m(p) - w(p,t) + w(t,p)$.

We define a function, $\lambda_{in} : T \to M$, *called input marking, such that* $\forall p \in P, \lambda_{in}(t)(p) = w(p,t)$. λ_{in} *returns a marking that contains the required number of tokens to enable a transition. In the following, the definition* $\Sigma = \langle P, T, in, out, w, k, m_0 \rangle$ *is assumed for a Petri net.*

To apply CTL model checking on a model, a particular operation called *pre* is necessary. This operation is used to calculate the execution trace of a model backwards, step by step [25]. In the context of a Petri net, the *pre* operation serves as the inverse or opposite of the $fire$ operation. This operation computes the set of predecessor markings for a given transition, representing the state space before the firing of the transition. In our approach, encoding the operation *pre* is crucial to construct the state space symbolically. The choice of Petri nets with finite capacity is crucial to ensure the decidability of our system.

Definition 20 (Reversible). *Let $t \in T$ be a transition and $m \in M$ be a marking. The reversibility of a transition t for a marking m is defined as $rev(t, m) \Leftrightarrow \forall p \in P, m(p) \leq k(p) - w(p, t)$. In addition, the operation that computes its application, noted $pre_t : M \times T \to M$, is defined as:*

$$\forall p \in P, pre_t(m, t)(p) = \begin{cases} w(p, t) & \textbf{\textit{if}}\ m(p) \leq w(t, p) \\ m(p) + w(p, t) - w(t, p) & \textbf{\textit{else}} \end{cases}$$

pre_t may be seen as the *reverse fire* operation for a given transition. However, even if there are not enough tokens in the post-places, the operation is still applied, and the minimum amount of tokens is put in the *in*-places. Hence, the targeted transition is always fireable the next time allowing to explore all satisfiable states, related to global model checking. Our objective is not to compute the reachable configurations from a given marking, but to capture all the initial configurations satisfying a property. Thus, even if a transition does not influence a place, it should still be considered in the list of all valid initial configurations. To define the general *pre* operation that operates on a set of markings, it involves applying the pre_t operation to each marking with every transition and combining the results using the union operation.

Definition 21 (Operation pre_{sv}). *Let $sv = (a, b) \in \mathbb{SV}$ be a symbolic vector. The operation $pre_{sv} : \mathbb{SV} \to \mathcal{P}(\mathbb{SV})$ is defined as:*

$$pre_{sv}(a, b) = \begin{cases} \varnothing\ \textbf{\textit{if}}\ \exists t \in T, \exists q \in (a \cup b), \neg rev(t, m) \\ \bigcup_{t \in T}\ (pre_i(a, t), pre_e(b, t))\ \textbf{\textit{else}} \end{cases}$$

$$pre_i, pre_e : \mathcal{P}(M) \times T \to \mathcal{P}(M)$$

$$pre_i(\varnothing, t) = \{\lambda_{in}(t)\},\ pre_e(\varnothing, t) = \varnothing$$

$$pre_i(a, t) = pre_e(a, t) = \bigcup_{m\ \in\ M} \{pre_t(m, t)\},\ \textbf{\textit{if}}\ a \neq \varnothing$$

The operation pre_{sv} is designed to operate on each marking within the symbolic vector individually, computing its corresponding pre_t. Note that if one of the markings is not reversible ($\neg rev(t, m)$), the entire operation is not applied and returns $\varnothing$. Moreover, it preserves the capacity of Petri nets.

Definition 22 (Operation pre_{svs}). *Let $svs \in \mathcal{P}(\mathbb{SV})$. Operation pre_{svs} : $\mathcal{P}(\mathbb{SV}) \times T \to \mathcal{P}(\mathbb{SV})$ is defined as: $pre_{svs}(svs) = \bigcup_{sv\ \in\ svs} pre_{sv}(sv)$.*

Lemma 7. *Let $svs \in \mathcal{P}(\mathbb{SV})$ be a symbolic vector set and $sv \in \mathbb{SV}$ be a symbolic vector. Additionally, we assume the decoded version of the pre operation on markings, where pre_t is computed directly on each marking for each transition. Then, $pre(uf(sv)) = uf(pre_{sv}(sv))$ and $pre(uf_{svs}(svs)) = uf_{svs}(pre_{svs}(svs))$.*

Remark 3. Our approach is adaptable to models beyond Petri nets. For models other than Petri nets, encoding "*pre*" is necessary to apply model checking for CTL. It remains an open question whether this is always possible with symbolic vector sets. In some cases, it can be necessary to have an infinite number of symbolic vectors, such as a Petri net representing even or odd numbers of tokens.

4.2 Global Model Checking of Petri Nets with Symbolic Vector Sets

In tackling the global model checking problem, our focus has been on addressing CTL formulas. We briefly introduce the corresponding formalism and the satisfaction relation.

Definition 23 (CTL definition and satisfaction). *Let AP be a set of atomic propositions and $ap \in AP$. The set of CTL formulas over AP, noted $\phi \in CTL$, is defined as: $\phi ::= ap \mid \neg\phi \mid \phi \vee \phi \mid \mathbf{EX}\ \phi \mid \mathbf{EG}\ \phi \mid \mathbf{E}[\phi\ \mathbf{U}\ \phi]$. We assume the usual CTL extended syntax: $\mathbf{AX}\ \phi$, $\mathbf{EF}\ \phi$, $\mathbf{AF}\ \phi$, $\mathbf{AG}\ \phi$, and $\mathbf{A}[\phi\ \mathbf{U}\ \phi]$.*

Let $\Sigma = \langle P, T, in, out, w, k, m_0\rangle$ be a Petri net, $m \in M$ be a marking and $t \in T$ be a transition. The satisfaction relation is denoted as $\models \subseteq \Sigma \times M \times CTL$ and its definition based on the standard literature [9] (Σ is omitted in the final notation). Furthermore, the set of atomic propositions for our Petri nets is $AP = \{isFireable(t) \mid t \in T\}$ such that: $m \models isFireable(t) \Leftrightarrow t$ is enabled for $m \Leftrightarrow \lambda_{in}(t) \subseteq_f m$.

$isFireable(t)$ is intended to observe the firing of transitions in Petri nets. We have assembled all the necessary components to provide the evaluation semantics of CTL formulas as the computation of a symbolic vector set. This semantics has been revisited on the basis of [21].

Definition 24 (Evaluation of a CTL formula). *Let $\phi, \psi \in CTL$, $ap \in \{isFireable(t) \mid t \in T\}$, and $\Sigma = \langle P, T, pre, post, w, k, m_0\rangle$ be a Petri net. The evaluation of a CTL formula as a symbolic vector set is noted $[\![\ ]\!] : CTL \to \mathcal{P}(\mathbb{SV})$. The function is defined as follows:*

$$\widetilde{pre}_{svs}([\![\phi]\!]) = \neg_{svs} pre_{svs}(\neg_{svs}[\![\phi]\!]) \qquad [\![\neg\phi]\!] = \neg_{svs}[\![\phi]\!]$$
$$[\![true]\!] = \{(\{f_\varepsilon\}, \varnothing)\} \qquad [\![\phi \vee \psi]\!] = [\![\phi]\!] \cup_{svs} [\![\psi]\!]$$
$$[\![isFireable(t)]\!] = \{(\{pre_t(t)\}, \varnothing)\} \qquad [\![\mathbf{EX}\ \phi]\!] = pre_{svs}([\![\phi]\!])$$
$$[\![\mathbf{EG}\ \phi]\!] = \nu Y.[\![\phi]\!] \cap_{svs} (pre_{svs}(Y) \cup_{svs} \widetilde{pre}_{svs}(Y))$$
$$[\![\mathbf{E}[\phi\ \mathbf{U}\ \psi]]\!] = \mu Y.[\![\psi]\!] \cup_{svs} ([\![\phi]\!] \cap_{svs} pre_{svs}(Y))$$

Remark 4. The semantics of evaluation closely resemble those of a Kripke structure. However, due to the left-total relation on Kripke structure arcs (i.e., each state has at least one arc starting from it), sink states are nonexistent. The semantics need to be adapted to account for this, expanding its scope. For instance, the evaluation of **EG** in a Kripke structure could be simplified as follows: $\nu Y.[\![\phi]\!] \cap pre_{svs}(Y)$, which does not require the computation of $\widetilde{pre}_{svs}(Y)$ and the union application. Furthermore, a Petri net without sink states could also benefit from this advantage.

Theorem 3. *Let $\phi \in CTL$ be a CTL formula and $m \in M$ a marking. Then, $m \models \phi \Leftrightarrow m \in_{svs} [\![\phi]\!]$.*

The theorem asserts that the satisfaction of a CTL formula for a given marking can be determined by evaluating the formula and verifying whether the marking belongs to it. This feature enables us to perform exhaustive computations to identify all states that satisfy a given formula. The perspective of computing $[\![\phi]\!]$ aligns with the challenge of *global model checking*, where the focus extends beyond a single configuration. This approach guides the remainder of the article.

4.3 Use Case: Mutual Exclusion

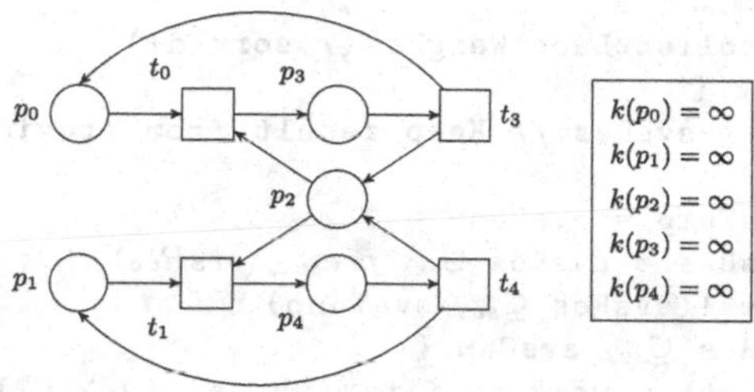

Fig. 5. Petri net modelling the mutual exclusion problem.

Figure 5 illustrates a Petri net example containing the well-known *mutual exclusion problem*. Markings are displayed using the order: $(p_0, p_1, p_2, p_3, p_4)$. Avoiding mutual exclusion in this Petri net can be expressed by the property: $\forall m \in M, m \models \neg(\mathbf{EF}\ isFireable(t_3) \wedge isFireable(t_4))$, computed as follows:

$$\begin{aligned}
[\![\mathbf{EF}\ isFireable(t_3) \wedge isFireable(t_4)]\!] &= \mu Y.\{(\{(0,0,0,1,1)\}, \varnothing)\} \cup_{svs} pre_{svs}(Y) \\
(1) &= \{(\{(0,0,0,1,1)\}, \varnothing)\} \\
(2) &= \{(\{(0,0,0,1,1)\}, \varnothing), (\{(0,1,1,1,0)\}, \varnothing), (\{(1,0,1,0,1)\}, \varnothing)\} \\
(3) &= \{(\{(0,0,0,1,1)\}, \varnothing), (\{(0,1,1,1,0)\}, \varnothing), (\{(1,0,1,0,1)\}, \varnothing), \\
&\quad (\{(1,1,2,0,0)\}, \varnothing), (\{(0,1,0,2,0)\}, \varnothing), (\{(1,0,0,0,2)\}, \varnothing)\}
\end{aligned}$$

In the fixpoint computation of **EF**, we have omitted the detailed steps and provided the final result of each step. The result does not change after step (3). Hence, a marking that is part of $[\![\mathbf{EF}\ isFireable(t_3) \wedge isFireable(t_4)]\!]$ is considered not safe from the mutual exclusion problem. For example, $(1,1,1,0,0)$ is a valid marking, whereas $(1,1,2,0,0)$ is not. This result also yields an infinite number of valid markings. Note that when the capacity of places is unbounded, the computation may not terminate if the solutions diverge. Conversely, if all places are bounded, the computation always concludes.

4.4 Optimisation Through Saturation

Similarly to methods such as decision diagrams that encounter a *peak* [7,13] effect, the construction of symbolic vector sets is subject to the same challenge.

While computing the solution iteratively through the fixed-point, intermediate solutions are generated. However, among these steps, some of the intermediate solutions will converge to the same solution. Thus, we end up working with more objects than necessary to obtain the final solution.

To address this issue, we propose a solution called *saturation*, which effectively mitigates the *peak* effect during computation.

```
1  func evalEF(ϕ: CTL) -> SVS {
2      n = 1 // The current capacity
3      svsRes = ⟦ϕ⟧svs
4      weights = collectLabelWeights().sorted()
5      while n ≤ k {
6          svsCap = svsRes // Keep result from previous capacity
7          do {
8              svsTemp = svsRes
9              svsRes = svsRes ∪svs pre_svs_n(svsRes)
10         } while !(svsRes ⊆svs svsTemp)
11         if svsRes ⊆svs svsCap {
12             weight = weights.first(where: {w > n})
13             if weight.isNil() { break }
14             n = weight
15         } else { n = n + 1 } // Increment current capacity
16     }
17     return svsRes
18 }
```

Fig. 6. Saturated algorithm of the evaluation of **EF** ϕ

Figure 6 is the pseudo-code summarising this notion for the CTL temporal operator **EF**. *svsRes* is the symbolic vector set storing the solution that will evolve, initialised with the previous evaluation of ϕ (line 3). *weights* is an array that compiles all weight labels from a Petri net, sorted from the smallest to the largest (line 4). The functions *collectLabelWeights* and *sorted* are given.

To streamline our approach, we consider the capacity k as a uniform bound for all places. The idea is to iterate on k (using n) starting from 1 (line 2) and compute the intermediate results of the CTL formula for a lower k. Subsequently, the result of the previous step is leveraged to compute the next one, creating an iterative process (from line 5 to 16).

From line 7 to 10, we have the computation of fixed-point for **EF**. The main difference from the common evaluation is the use of pre_{svs_n}. This function defined earlier in Definition 22 has been complemented by the index n. This index specifies the current capacity used when computing pre_{svs}. It ensures that symbolic vectors cannot contain markings with a value greater than n.

Upon completing the computation of the fixed-point for a given capacity, we compare the new result *svsRes* with the previous one stored in *svsCap* (line 6). Furthermore, if the result of two consecutive iterations for two different capacities remains unchanged (line 11), we inspect the weights collected previously (line 4) to determine whether any of them is capable of generating a marking greater than the current capacity n but less than k (line 12). The function $first$ serves as a filter that attempts to extract a weight from the array $weights$ such that its value is greater than n. Here, w is the variable that iterates through each value of $weights$. If the function fails to find such a value, it returns nil. This means that no transition in the net is capable of altering the final result. Then, the $break$ statement (line 13) is executed to exit the loop and return $svsRes$ (line 17). If such a weight is found, we update the corresponding capacity (line 14), allowing the potential discovery of new symbolic vectors in the next iteration. On the other hand, if condition line 11 does not hold, n is incremented by one (line 15). Note that if the value of $weight$ is greater than k, the condition of the loop $while$ will also terminate. For example, consider an arc of a transition that requires 6 tokens with $k = 9$. If $svsRes$ remains unchanged for $n = 2$ and $n = 3$, and the only other transition available also demands 6 tokens for an arc, we can update n to 6. In fact, if no new transition is available under such conditions, the creation of new markings is not possible.

To make this pseudo-code work for other CTL temporal operators, it is enough to update line 9 for the corresponding CTL operation. In addition, line 10 and 11 should be modified for the operations **EG** and **AG** due to the greatest fixed-point, requiring the elements of both relations to be swapped.

5 Benchmarks and Results

This section is dedicated to examining the results obtained with our tool, namely *SVSKit*, which is a Swift library available on GitHub. Our library implements the theory developed in this article. All the showcased examples are sourced directly from the model checking contest for the year 2022. However, our tool is presented from the perspective of global model checking. All tests were conducted on a MacBook Pro computer equipped with a 3.2 GHz CPU and 32 GB of RAM.

5.1 Circadian Clock: A General Model

The Petri net of the circadian clock model, originally defined in [23], consists of 14 places, 16 transitions, and 58 arcs. Several places in this net are associated with a variable value N, which can be adjusted according to specific objectives. In our system, N can be considered as the number that determines the capacity of each place. In addition, increasing this value raises the system's complexity.

In Table 1, we compare the calculation of a reachability property both with and without saturation, for various capacities. Saturation optimisation is instrumental in making symbolic vector sets practical for Petri nets of reasonable size. It is worth noting that without saturation, the time complexity grows exponentially, reaching over 5 h for a capacity of 6. In contrast, with saturation enabled, the time remains constant starting from capacity 2 and above. The use of saturation allows us to build symbolic vector sets step by step with the minimum

Table 1. Comparison of a reachability property computation (**EF** ϕ) for the circadian clock model, with and without the use of saturation.

Capacity	Saturated	Time (s)	Peak SV number	Final SV number
1	*true*	1.9	29	26
	false	1.9	29	
2	*true*	3.2	29	
	false	45	86	
6	*true*	3.2	29	
	false	19000	954	

of information required. This can be observed in the *Peak SV number* column, where the saturated solution remains consistently below 29 and does not exhibit significant growth. Moreover, the final solution consistently comprises 26 symbolic vectors for all capacities, indicating that increasing it should not significantly alter the computational complexity.

We have implemented the *query reduction* optimisation discussed in [5]. In fact, because the CTL formulas for the competition are generated randomly, we are interested in simplifying them before evaluating them. Furthermore, only the reduction rules concerning global model checking have been reused, which does not require any initial marking.

Table 2. Progression of the prior **EF** reachable property from Table 1 across varying capacities, assuming the saturation by default.

Capacity	Marking number (~)	Final SV number	Time (s)
100	1.1×10^{28}	26	3.2
10000	1.0×10^{56}		
1000000	1.0×10^{84}		
∞	∞		

In Table 2, we observe the behaviour when we assume saturation by default and only vary the capacity. This result aligns with the previous finding, indicating that changing the capacity no longer significantly impacts the efficiency of the computation. The primary result affected by this change is the number of markings encoded by the symbolic vector set. Although the count of symbolic vectors remains consistent across all rows, the decoded version can still change. This is because capacity is not taken into account in the creation of symbolic vectors, and will only appear when decoding the set of symbolic vectors.

In addition, the table offers an approximate count of markings, calculated without the need to decode the structure. Note that this number is related to the number of initial markings satisfying the CTL formula and not to the number of reachable markings from a given configuration. Thus, we can efficiently handle markings until, potentially, infinity if the result converges. The earlier convergence occurs, the more efficient the computation becomes, enabling the capture of the Petri net's evolution at the earliest possible stage thanks to saturation.

Table 3. Computation of CTL fireability formulas for the circadian clock model with $N = 100\,000$. Each CTL formula is numbered in # from 00 to 15, following the order provided in competition resources. $*$ column contains the local answer of each formula of the competition, where F and T stand for *false* and *true*, respectively. "nb" is the abbreviation for "number".

#	Time(s)	SV nb	Marking nb	*	#	Time(s)	SV nb	Marking nb	*
00	11	1	2.0×10^{60}	F	08	3.8	3	1.0×10^{60}	F
01	1.7	0	0	F	09	0.15	17	1.0×10^{70}	T
02	1.7	1	3.0×10^{60}	F	10	4	20	1.0×10^{60}	F
03	0.06	0	0	F	11	1350	1	1.0×10^{84}	F
04	20.5	17	1.0×10^{70}	T	12	0.28	6	1.1×10^{70}	T
05	2.4	1	1.0×10^{70}	T	13	1.4	9	2.0×10^{65}	T
06	418	686	1.0×10^{84}	T	14	11.4	29	1.0×10^{70}	T
07	6.6	4	1.0×10^{70}	F	15	23.8	40	1.0×10^{60}	F

In the context of the model checking contest, the same model is provided with varying initial markings, each time increasing the number of tokens. The most challenging scenario for the circadian clock model is when N is set to 100 000, which means that several places in the Petri net contain such a number of tokens.

In the *CTL fireability* category, three tools participated in 2022: *GreatSPN* [3], *ITS-Tools* [26], and *Tapaal* [15]. None of these tools managed to produce responses to all queries within the 60-min time limit, underscoring the challenging nature of the task. The best tool for this execution was Tapaal, which was able to answer 9 out of 16 queries[2]. In contrast, our results in Table 3 show that our symbolic technique successfully addressed all queries in approximately 30 min. Furthermore, we computed the complete set of valid solutions and checked whether the initial marking of the model belongs to the resulting symbolic vector set. By focussing on a specific configuration, new optimisations become available, such as the *on-the-fly* technique [4], *structural reduction* [6], or *stubborn reduction* [28]. It should be noted that while our tool was not tested under identical conditions, the setup was not significantly different[3].

Nevertheless, our present implementation of canonical symbolic vector sets faces scalability challenges. Efficiency hinges on the implementation of canonical operations, exhibiting a worst-case complexity of $\mathcal{O}(n!)$. Our implementation strategy involves moving each element from one set to the other. Throughout

[2] The results can be found here, by looking for the *CircadianClock* table.

[3] Each execution on the virtual machine for the MCC was limited to a maximum of 16 GB, 2.4 GHz, and 4 cores.

this process, we must ensure that each moved element does not conflict with elements in the other set. Additionally, when conflicts arise, the lexicographical order determines the symbolic vector that will contain the shared portion. Obtaining this shared part entails merging it with the lowest symbolic vector and subtracting it from the greatest symbolic vector. This subtraction may result in the creation of new symbolic vectors, requiring the canonical union to be reapplied to each of them, causing a deeper level of recursion.

In addition, most operations rely on canonical union. Hence, we restricted our testing to a relatively small-sized Petri net due to these limitations. Despite the inherent complexity, our method allows us to compute formulas that even the most proficient tools find challenging. This approach already demonstrates its viability in specific scenarios, and we are confident that further refinement can address the mentioned issue and enhance its capabilities.

6 Background and Related Works

The original framework, initially introduced as *Predicate structures* in [21], remained stagnant for almost 30 years without further development. In its initial version, performance limitations rendered it nearly unusable or, at the very least, non-competitive. Building on this foundation, we have expanded and enhanced it by introducing new operations, a novel canonical form, and optimisations.

In the realm of techniques addressing the state-space explosion problem in model checking, our method resembles symbolic model checking approaches. Decision diagrams [1,12–14,27] are the prevailing methods in symbolic model checking, employing various types of data encoding and optimisations [2,7,17, 24]. These representations enhance the efficiency of data representation and leverage homomorphisms for symbolic processing. In contrast to our approach, based on encapsulation akin to intervals, decision diagrams exploit shared representations, allowing common components to be shared among different entities when encoding a set. Moreover, our representation can handle certain infinite representations, a capability not universally shared by decision diagrams. Among the big family of decision diagrams, Interval decision diagrams (IDDs) [24] appear to be the most closely related to our work. However, symbolic vector sets generalise intervals for vectors, while IDDs consist of vectors of intervals.

7 Conclusion

In this paper, we introduce a novel structure termed *symbolic vector set*. Alongside this structure, we present homomorphic operations, canonical form, and saturation optimisation to enhance computational efficiency. Using this approach, we encode Petri net markings and employ the symbolic representation for CTL formula evaluation. Our initial results are promising, indicating potential avenues for further explorations and analysis. We believe that our current implementation can be improved, particularly by enhancing the canonical operations.

Our work opens up several research directions. Exploring our approach within the realm of local model checking holds promise. Additionally, exploring models beyond Petri nets is another line of research, contingent upon our ability to construct the *pre* operation symbolically for the model. One of the most promising directions for our research is to integrate our method with other symbolic model checking methods, such as decision diagrams. We believe that both approaches are orthogonal and could mutually benefit from each other.

References

1. Akers, S.B.: Binary decision diagrams. IEEE Trans. Comput. **27**(06), 509–516 (1978)
2. Amparore, E.G., Donatelli, S., Beccuti, M., Garbi, G., Miner, A.: Decision diagrams for petri nets: a comparison of variable ordering algorithms. In: Koutny, M., Kristensen, L.M., Penczek, W. (eds.) Transactions on Petri Nets and Other Models of Concurrency XIII. LNCS, vol. 11090, pp. 73–92. Springer, Heidelberg (2018). https://doi.org/10.1007/978-3-662-58381-4_4
3. Babar, J., Beccuti, M., Donatelli, S., Miner, A.: GreatSPN enhanced with decision diagram data structures. In: Lilius, J., Penczek, W. (eds.) PETRI NETS 2010. LNCS, vol. 6128, pp. 308–317. Springer, Heidelberg (2010). https://doi.org/10.1007/978-3-642-13675-7_19
4. Bhat, G., Cleaveland, R., Grumberg, O.: Efficient on-the-fly model checking for CTL. In: Proceedings of Tenth Annual IEEE Symposium on Logic in Computer Science, pp. 388–397. IEEE (1995)
5. Bønneland, F., Dyhr, J., Jensen, P.G., Johannsen, M., Srba, J.: Simplification of CTL formulae for efficient model checking of petri nets. In: Khomenko, V., Roux, O.H. (eds.) PETRI NETS 2018. LNCS, vol. 10877, pp. 143–163. Springer, Cham (2018). https://doi.org/10.1007/978-3-319-91268-4_8
6. Bønneland, F.M., Dyhr, J., Jensen, P.G., Johannsen, M., Srba, J.: Stubborn versus structural reductions for petri nets. J. Log. Algebr. Methods Program. **102**, 46–63 (2019)
7. Ciardo, G., Lüttgen, G., Siminiceanu, R.: Saturation: an efficient iteration strategy for symbolic state—space generation. In: Margaria, T., Yi, W. (eds.) TACAS 2001. LNCS, vol. 2031, pp. 328–342. Springer, Heidelberg (2001). https://doi.org/10.1007/3-540-45319-9_23
8. Clarke, E.M.: Model checking. In: Ramesh, S., Sivakumar, G. (eds.) FSTTCS 1997. LNCS, vol. 1346, pp. 54–56. Springer, Heidelberg (1997). https://doi.org/10.1007/BFb0058022
9. Clarke, E.M., Emerson, E.A.: Design and synthesis of synchronization skeletons using branching time temporal logic. In: Workshop on Logic of Programs. Carnegie Mellon University (1981)
10. Clarke, E.M., Grumberg, O., Long, D.E.: Model checking and abstraction. ACM Trans. Program. Lang. Syst. (TOPLAS) **16**(5), 1512–1542 (1994)
11. Clarke, E.M., Klieber, W., Nováček, M., Zuliani, P.: Model checking and the state explosion problem. In: Meyer, B., Nordio, M. (eds.) LASER 2011. LNCS, vol. 7682, pp. 1–30. Springer, Heidelberg (2012). https://doi.org/10.1007/978-3-642-35746-6_1
12. Couvreur, J.-M., Encrenaz, E., Paviot-Adet, E., Poitrenaud, D., Wacrenier, P.-A.: Data decision diagrams for petri net analysis. In: Esparza, J., Lakos, C. (eds.) ICATPN 2002. LNCS, vol. 2360, pp. 101–120. Springer, Heidelberg (2002). https://doi.org/10.1007/3-540-48068-4_8

13. Hamez, A., Thierry-Mieg, Y., Kordon, F.: Hierarchical set decision diagrams and automatic saturation. In: van Hee, K.M., Valk, R. (eds.) PETRI NETS 2008. LNCS, vol. 5062, pp. 211–230. Springer, Heidelberg (2008). https://doi.org/10.1007/978-3-540-68746-7_16
14. Hostettler, S., Marechal, A., Linard, A., Risoldi, M., Buchs, D.: High-level petri net model checking with alpina. Fund. Inform. **113**(3–4), 229–264 (2011)
15. Jensen, J.F., Nielsen, T., Oestergaard, L.K., Srba, J.: TAPAAL and reachability analysis of P/T nets. In: Koutny, M., Desel, J., Kleijn, J. (eds.) Transactions on Petri Nets and Other Models of Concurrency XI. LNCS, vol. 9930, pp. 307–318. Springer, Heidelberg (2016). https://doi.org/10.1007/978-3-662-53401-4_16
16. Kordon, F., et al.: Complete Results for the 2022 Edition of the Model Checking Contest, June 2022. http://mcc.lip6.fr/2022/results.php
17. López Bóbeda, E., Colange, M., Buchs, D.: StrataGEM: a generic petri net verification framework. In: Ciardo, G., Kindler, E. (eds.) PETRI NETS 2014. LNCS, vol. 8489, pp. 364–373. Springer, Cham (2014). https://doi.org/10.1007/978-3-319-07734-5_20
18. McMillan, K.L.: Symbolic Model Checking. In: Symbolic Model Checking, pp. 25–60. Springer, Berlin, Heidelberg (1993). https://doi.org/10.1007/978-1-4615-3190-6
19. Morard, D.: Global Symbolic Model Checking based on Generalised Intervals. Ph.D. thesis, University of Geneva (2024)
20. Peled, D.: Combining partial order reductions with on-the-fly model-checking. In: Dill, D.L. (ed.) CAV 1994. LNCS, vol. 818, pp. 377–390. Springer, Heidelberg (1994). https://doi.org/10.1007/3-540-58179-0_69
21. Racloz, P., Buchs, D.: Properties of petri nets modellings: the temporal way. In: 7th International Conference on Formal Description Techniques for Distributed Systems Communications Protocols. Services, Technologies (1994)
22. Schuele, T., Schneider, K.: Global vs. local model checking: a comparison of verification techniques for infinite state systems. In: Proceedings of the Second International Conference on Software Engineering and Formal Methods, 2004. SEFM 2004, pp. 67–76. IEEE (2004)
23. Schwarick, M., Heiner, M.: CSL model checking of biochemical networks with interval decision diagrams. In: Degano, P., Gorrieri, R. (eds.) CMSB 2009. LNCS, vol. 5688, pp. 296–312. Springer, Heidelberg (2009). https://doi.org/10.1007/978-3-642-03845-7_20
24. Schwarick, M., Rohr, C., Liu, F., Assaf, G., Chodak, J., Heiner, M.: Efficient unfolding of coloured petri nets using interval decision diagrams. In: Janicki, R., Sidorova, N., Chatain, T. (eds.) PETRI NETS 2020. LNCS, vol. 12152, pp. 324–344. Springer, Cham (2020). https://doi.org/10.1007/978-3-030-51831-8_16
25. Sifakis, J.: A unified approach for studying the properties of transition systems. Theor. Comput. Sci. **18**(3), 227–258 (1982)
26. Thierry-Mieg, Y.: Symbolic model-checking using ITS-tools. In: Baier, C., Tinelli, C. (eds.) TACAS 2015. LNCS, vol. 9035, pp. 231–237. Springer, Heidelberg (2015). https://doi.org/10.1007/978-3-662-46681-0_20
27. Tovchigrechko, A.: Efficient symbolic analysis of bounded Petri nets using Interval Decision Diagrams. Ph.D. thesis, BTU Cottbus-Senftenberg (2009)
28. Valmari, A.: Stubborn sets for reduced state space generation. In: Rozenberg, G. (ed.) ICATPN 1989. LNCS, vol. 483, pp. 491–515. Springer, Heidelberg (1991). https://doi.org/10.1007/3-540-53863-1_36

Safety Verification of Wait-Only Non-Blocking Broadcast Protocols

Lucie Guillou[1(✉)], Arnaud Sangnier[2(✉)], and Nathalie Sznajder[3(✉)]

[1] IRIF, CNRS, Université Paris Cité, Paris, France
guillou@irif.fr
[2] DIBRIS, Università di Genova, Genoa, Italy
arnaud.sangnier@unige.it
[3] Sorbonne Université, CNRS, LIP6, 75005 Paris, France
nathalie.sznajder@lip6.fr

Abstract. We study networks of processes that all execute the same finite protocol and communicate synchronously in two different ways: a process can broadcast one message to all other processes or send it to at most one other process. In both cases, if no process can receive the message, it will still be sent. We establish a precise complexity class for two coverability problems with a parameterised number of processes: the state coverability problem and the configuration coverability problem. It is already known that these problems are Ackermann-hard (but decidable) in the general case. We show that when the protocol is *Wait-Only*, i.e., it has no state from which a process can send and receive messages, the complexity drops to P and PSPACE, respectively.

Keywords: Parameterised Networks · Broadcast · Verification

1 Introduction

Verification of Distributed Systems. The ubiquity of distributed and concurrent systems in nowadays applications leads to an increasing need to ensure their correct behaviour. Over the last two decades, the verification of such systems has become a crucial research direction in the field of computer science. Indeed, analysing distributed systems has proven to be challenging. One difficulty is due to the numerous interleavings caused by the concurrent behaviour of the system entities, that make the design and modelling of these systems very complex. Moreover, the number of agents is often not known a priori; in that case, verifying all possible behaviours of such a system amounts to analyse it for any number of agents, i.e. an infinite number of times. The unpredictability of the number of participants in a system makes classical techniques such as model-checking impractical and requires some new techniques.

Parameterised Verification. Addressing the challenge of unbounded entities involves designing schematic programs or protocols intended for implementation

L. M. Kristensen and J. M. van der Werf (Eds.): PETRI NETS 2024, LNCS 14628, pp. 291–311, 2024.
https://doi.org/10.1007/978-3-031-61433-0_14

by multiple identical processes and parameterised by the number of entities. While in general parameterised verification is undecidable [2], several realistic restrictions enable automatic verification. Among them, one can highlight systems where entities have no identity, and systems with simple communication mechanism. Several papers have considered synchronous communication means, as rendez-vous [3,4,9,10,12] and broadcast [5,8]. Note that, surprisingly, parameterised verification, when decidable, is sometimes significantly easier than the same problem with a fixed number of entities [6]. In all those models, all the entities execute the same program which is modelled as a finite-state automaton.

Wait-Only Non-Blocking Broadcast Protocols. In [10], the authors have studied the complexity of several parameterised verification problems in the context of non-blocking rendez-vous. This communication mechanism, motivated by Java Threads programming, involves at most two processes: when a process sends a message, it is received by at most one process ready to receive the message, and both processes jointly change their local state. However, when no process is ready to receive the message, the message is sent anyway and lost, and only the sender changes its local state. This is in contrast with classical rendez-vous as studied for instance in [9], where a sender is prevented to send a message if no process can receive it. The model proposed in [10] allows to capture some behaviour of the Threads: when a Thread is suspended in a waiting state, it can be woken up upon the reception of a `notify` message sent by another Thread, but the sender is not blocked if no Thread is suspended; it simply continues its execution and the `notify` message is lost. However, this fails to capture the behaviour of what occurs when a Thread sends a `notifyAll` message that will be received by all the suspended Threads waiting for that message. This, as already highlighted in [4], is modelled by the broadcast mechanism, in which a message sent by a process will be received by all the processes ready to receive it. Observe that broadcast is also a non-blocking means of communication. In this work we consider Non-Blocking Broadcast protocols, that allow for both broadcast and non-blocking rendez-vous. One important problem in parameterised verification is the coverability problem: is it possible that, starting from an initial configuration, (at least) one process reaches a bad state? With classical rendez-vous mechanism, this problem is in P [9], while with non-blocking rendez-vous, it is EXPSPACE-complete [10]. For protocols enabling both broadcast and non-blocking rendez-vous, the problem is decidable [7] and Ackermann-hard [1,8,15]. In this work we study the coverability problem for a syntactic restriction of the protocols, introduced in [10], namely Wait-Only protocols, in which there is no state from which a process can both send and receive a message. In this context, when processes communicate with non-blocking rendez-vous only, the coverability problem is in P [10].

Our Contributions. We show that the coverability problem for Wait-Only Non-Blocking Broadcast protocols is P-complete, and that the *configuration coverability problem* is PSPACE-complete. This last problem asks whether it is possible to cover a given configuration (and not simply a bad state) from an initial state. Note that both problems are in P when forbidding broadcasts [10], however, in this work, the P-membership proof is less involved and the PSPACE-membership proof uses a different technique. Due to lack of space, some proofs can be found on the extended version available in [11].

2 Model and Verification Problems

We denote by $\mathbb{N}$ the set of natural numbers. For a finite set E, the set $\mathbb{N}^E$ represents the multisets over E. For two elements $s, s' \in \mathbb{N}^E$, we denote by $s + s'$ the multiset such that $(s+s')(e) = s(e) + s'(e)$ for all $e \in E$. We say that s' is bigger than s, denoted $s \leq s'$ if and only if $s(e) \leq s'(e)$ for all $e \in E$. If $s \leq s'$, then $s' - s$ is the multiset such that $(s' - s)(e) = s'(e) - s(e)$ for all $e \in E$. Given a subset $E' \subseteq E$ and $s \in \mathbb{N}^E$, we denote by $||s||_{E'}$ the sum $\Sigma_{e \in E'} s(e)$ of elements of E' present in s. The size of a multiset s is given by $||s|| = ||s||_E$. For $e \in E$, we use sometimes the notation e for the multiset s verifying $s(e) = 1$ and $s(e') = 0$ for all $e' \in E \setminus \{e\}$ and, to represent for instance the multiset with four elements a, b, b and c, we will also use the notations $\wr a, b, b, c \int$ or $\wr a, 2 \cdot b, c \int$.

2.1 Networks of Processes Using Rendez-Vous and Broadcast

We now present the model under study in this work. We consider networks of processes where each entity executes the same protocol given by a finite state automaton. Given a finite alphabet Σ of messages, the transitions of a protocol are labelled with four types of actions that can be executed by the processes of the network. For $m \in \Sigma$ a process can (1) send a (non-blocking) rendez-vous over the message m with $!m$, (2) send a broadcast over m with $!!m$, (3) receive a rendez-vous or a broadcast over m with $?m$ and (4) perform an internal action with τ (assuming $\tau \notin \Sigma$). In order to refer to these different actions, we denote by $!\Sigma$ the set $\{!m \mid m \in \Sigma\}$, by $!!\Sigma$ the set $\{!!m \mid m \in \Sigma\}$ and by $?\Sigma$ the set $\{?m \mid m \in \Sigma\}$. Finally, we use the notation Op_Σ to represent the set of labels $!!\Sigma \cup !\Sigma \cup ?\Sigma \cup \{\tau\}$ and Act_Σ to represent the set of actions $!!\Sigma \cup !\Sigma \cup \{\tau\}$.

Definition 1. *A* Non-Blocking Broadcast protocol P *(NB-Broadcast protocol) is a tuple* (Q, Σ, q_{in}, T) *such that* Q *is a finite set of states,* Σ *is a finite alphabet,* q_{in} *is an initial state, and* $T \subseteq Q \times \mathsf{Op}_\Sigma \times Q$ *is the transition relation.*

In this work, we are in particular interested in studying some syntactical restrictions on such protocols. We say that a protocol is *Wait-Only* when for all $q \in Q$, either $\{q' \mid (q, \alpha, q') \in T \text{ with } \alpha \in ?\Sigma\} = \emptyset$, or $\{q' \mid (q, \alpha, q') \in T \text{ with } \alpha \in !!\Sigma \cup !\Sigma \cup \{\tau\}\} = \emptyset$. We call a state respecting the first or both conditions an *active* state and a state respecting the second condition a *waiting* state. In the following, we denote by Q_A the set of active states of P and Q_W its set of waiting states.

If the protocol does not contain any broadcast transition of the form $(q, !!m, q')$, we call it a *Non-Blocking Rendez-vous protocol* (NB-Rendez-vous protocol).

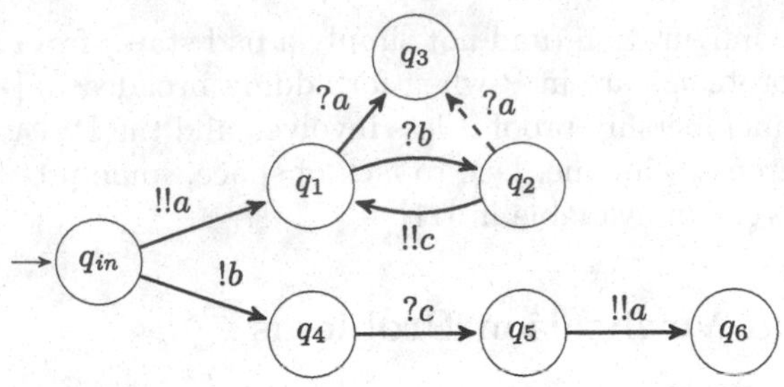

Fig. 1. Example of a protocol denoted P_{dashed} (we note P the protocol P_{dashed} without the dashed arrow between q_2 and q_3)

Example 1. An example of protocol is depicted on Fig. 1. We name P the protocol drawn without the dashed arrow between q_2 and q_3, and P_{dashed} the complete protocol. Note that P is a *Wait-Only* protocol, indeed each state is either an active state (q_{in}, q_2, q_3, q_5 and q_6), or a waiting state, (q_1 and q_4). However, P_{dashed} is not a Wait-Only protocol, since q_2 is neither an active state nor a waiting state as it has an outgoing transition labelled with an action $!!c$, and an outgoing transition labelled with an action $?a$.

We shall now present the semantics associated to protocols. Intuitively, we consider networks of processes, each process being in a state of the protocol and changing its state according to the transitions of the protocol with the following assumptions. A process can perform on its own an internal action τ and this does not change the state of the other processes. When a process sends a broadcast with the action $!!m$, then all the processes in the network which are in a state from which the message m can be received (i.e. with an outgoing transition labelled by $?m$) have to take such a transition. And when a process sends a rendez-vous with the action $!m$, then *at most* one process receives it: in fact, if there is at least one process in a state from which the message m can be received, then exactly one of these processes has to change its state, along with the receiver (while the other processes do not move), but if no process can receive the message m, only the sender performs the action $!m$. This is why we call this communication mechanism a *non-blocking* rendez-vous.

We move now to the formal definition of the semantics. Let $P = (Q, \Sigma, q_{in}, T)$ be a protocol.

A *configuration* C over P is a non-empty multiset over Q, it is *initial* whenever $C(q) = 0$ for all $q \in Q \setminus \{q_{in}\}$. We note $\mathcal{C}$ the set of all configurations over P, and $\mathcal{I}$ the set of all initial configurations over P.

For $q \in Q$, we denote by $R(q)$ the set $\{m \in \Sigma \mid \text{there exists } q' \in Q, (q, ?m, q') \in T\}$ of messages that can be received when in the state q. Given a transition $t = (q, \alpha, q') \in T$, we define the relation $\xrightarrow{t} \subseteq \mathcal{C} \times \mathcal{C}$ as follows: for two configurations C, C' we have $C \xrightarrow{t} C'$ iff one of the following conditions holds:

(a) $\alpha = \tau$, and $C(q) > 0$ and $C' = C - \{\!\{q\}\!\} + \{\!\{q'\}\!\}$;
(b) $\alpha = !!m$, and $C = \{\!\{q_1, q_2, \ldots, q_n, q\}\!\}$ for some $n \in \mathbb{N}$, and $C' = \{\!\{q'_1, q'_2, \ldots, q'_n, q'\}\!\}$ where for all $1 \leq i \leq n$, either $m \notin R(q_i)$ and $q'_i = q_i$, or $(q_i, ?m, q'_i) \in T$;

(c) α=!m, and $C(q) > 0$, and $(C - \llbracket q \rrbracket)(p) = 0$ for all $p \in Q$ such that $m \in R(p)$, and $C' = C - \llbracket q \rrbracket + \llbracket q' \rrbracket$;

(d) α=!m and $C(q) > 0$ and there exists $p \in Q$ such that $(C - \llbracket q \rrbracket)(p) > 0$ and $(p, ?m, p') \in T$ for some $p' \in Q$, and $C' = C - \llbracket p, q \rrbracket + \llbracket q', p' \rrbracket$.

Observe that when $C \xrightarrow{t} C'$, we necessarily have $||C|| = ||C'||$.

The case (a) corresponds to the internal action of a single process, the case (b) to the emission of a broadcast hence all the processes that can receive the message have to receive it. The case (c) corresponds to the case where a process sends a rendez-vous and there is no process to answer to it, hence only the sender changes its state. The case (d) corresponds to a classical rendez-vous where a process sends a rendez-vous and another process receives it. Note that for both the broadcast and the rendez-vous, the absence of a receiver does not prevent a sender from its action. We call non-blocking our semantics because of the case (c), which contrasts with the broadcast model of [8] for instance, where this case is not possible.

We write $C \to C'$ whenever there exists $t \in T$ such that $C \xrightarrow{t} C'$, and denote by $\to^*$ [resp. $\to^+$] the reflexive and transitive [resp. transitive] closure of $\to$. An execution ρ is then a finite sequence of the form $C_0 \xrightarrow{t_1} C_1 \xrightarrow{t_2} \dots \xrightarrow{t_n} C_n$, it is said to be initialized when C_0 is an initial configuration in $\mathcal{I}$.

Example 2. We consider the protocol P of Fig. 1. We have then the following execution starting at the initial configuration $\llbracket q_{in}, q_{in}, q_{in} \rrbracket$ with three processes:

$$\llbracket q_{in}, q_{in}, q_{in} \rrbracket \xrightarrow{(q_{in},!!a,q_1)} \llbracket q_1, q_{in}, q_{in} \rrbracket \xrightarrow{(q_{in},!b,q_4)} \llbracket q_2, q_4, q_{in} \rrbracket \xrightarrow{(q_{in},!b,q_4)} \llbracket q_2, q_4, q_4 \rrbracket$$
$$\xrightarrow{(q_2,!!c,q_1)} \llbracket q_1, q_5, q_5 \rrbracket \xrightarrow{(q_5,!!a,q_6)} \llbracket q_3, q_6, q_5 \rrbracket.$$

It corresponds to the following sequence of events: one of the agents broadcasts message a (not received by anyone), then another agent sends message b which leads to a rendez-vous with the first agent on q_1, the last agent sends message b which is not received by anyone (the sending is possible thanks to the non-blocking semantics), the agent in state q_2 broadcasts message c which is received by the two other agents, and finally, one of the agents in q_5 broadcasts letter a which is received by the process on q_1.

Remark 1. Observe that internal transitions labelled by τ can be replaced by broadcast transitions of the form $!!\tau$. Since no transition is labelled by $?\tau$, when τ is broadcasted, no process is ready to receive it and the semantics is equivalent to the one of an internal transition. Observe also that since $\tau \in \mathsf{Act}_\Sigma$, transforming internal transitions into broadcasts keeps a protocol Wait-Only.

Following this remark, we will omit internal transitions in the rest of this work.

2.2 Verification Problems

We present now the verification problems we are interested in. Both these problems consist in ensuring a safety property: we want to check that, no matter the number of processes in the network, a configuration exhibiting a specific pattern can never be reached. If the answer to the problem is positive, it means in our context that the protocol is not safe.

The state coverability problem StateCover is stated as follows:

StateCover
Input: An NB-Broadcast Protocol P and a state $q_f \in Q$;
Question: Do there exist $C \in \mathcal{I}$ and $C' \in \mathcal{C}$ such that $C \rightarrow^* C'$, and $C'(q_f) > 0$?

When the answer is positive, we say that q_f is *coverable* by P. The second problem, called the configuration coverability problem ConfCover, is a generalisation of the first one where we look for a multi-set to be covered.

ConfCover
Input: An NB-Broadcast Protocol P and a configuration $C_f \in \mathcal{C}$;
Question: Do there exist $C \in \mathcal{I}$ and $C' \in \mathcal{C}$ such that $C \rightarrow^* C'$, and $C_f \leq C'$?

Remark 2. Note that if P is a Wait-Only protocol and its initial state q_{in} is a waiting state, then no state besides q_{in} is coverable and the only coverable configurations are the initial ones. Hence, when talking about Wait-Only protocols, we assume in the rest of this work that the initial state q_{in} is always an *active state*.

Example 3. In the protocol P of Fig. 1, configuration $\wr q_3, q_6 \int$ is coverable as $\wr q_{in}, q_{in}, q_{in} \int \rightarrow^* \wr q_3, q_6, q_5 \int$ (see Example 2) and $\wr q_3, q_6 \int \leq \wr q_3, q_6, q_5 \int$.

Results. We summarize in Table 1 the results on NB-Broadcast protocols, where our results appear in red. Note that for what concerns the lower bounds for NB-Broadcast protocols, they have been proved in [15] [Fact16, Remark 17] for Broadcast protocols with a classical "blocking" rendez-vous semantics, i.e. where a process requesting a rendez-vous cannot take the transition if no process answers the rendez-vous. However, it is possible to retrieve the lower bound for NB-Broadcast protocols without rendez-vous by using the fact that "blocking" rendez-vous can be simulated by broadcast as shown in [1,8].

Table 1. Coverability in NB-Broadcast protocols

Type of protocols	StateCover	ConfCover
NB-Broadcast	Decidable [7] and Ackermann-hard [1,8,15]	
NB-Rendez-vous	ExpSpace-complete [10]	
Wait-Only NB-Rendez-vous	in P [10], P-hard	
Wait-Only NB-Broadcast	P-complete	PSpace-complete

In the rest of this work, we will focus on Wait-Only NB-Broadcast protocols which we name Wait-Only protocols for ease of notation.

3 Preliminary Properties

Wait-Only protocols enjoy a nice property on coverable states. The property makes a distinction between active states and waiting states. First, we show that when an active state is coverable, then it is coverable by a number of processes as big as one wants, whereas this is not true for a waiting state. Indeed, it is possible that a waiting state can be covered by exactly one process at a time, and no more. However, we show that if two active states, or if an active state and a waiting state are coverable, then there is an execution that reaches a configuration where they are both covered.

This property relies on the fact that once the active state has been covered in an execution, it will not be emptied while performing the sequence of actions allowing to cover the second (waiting state), since no reception of message can happen in such a state. As we will see, this phenomenon can be generalised to a subset of active states.

Example 4. Going back to the protocol P of Fig. 1, consider the active state q_2. It is coverable as shown by the execution $\wr q_{in}, q_{in} \int \xrightarrow{(q_{in},!!a,q_1)} \wr q_1, q_{in} \int \xrightarrow{(q_{in},!b,q_4)} \wr q_2, q_4 \int$. From this execution, for any integer $n \in \mathbb{N}$, one can build an execution leading to a configuration covering $\wr n \cdot q_2 \int$. For instance, for $n = 2$, we build the following execution:

$$\wr q_{in}, q_{in}, q_{in}, q_{in} \int \xrightarrow{(q_{in},!!a,q_1)} \wr q_1, q_{in}, q_{in}, q_{in} \int \xrightarrow{(q_{in},!b,q_4)} \wr q_2, q_4, q_{in}, q_{in} \int$$
$$\xrightarrow{(q_{in},!!a,q_1)} \wr q_2, q_4, q_1, q_{in} \int \xrightarrow{(q_{in},!b,q_4)} \wr q_2, q_4, q_2, q_4 \int.$$

Furthermore each coverable waiting state is coverable by a configuration that also contains q_2. For instance, $\wr q_2, q_4 \int$ is coverable as shown by the above execution.

Note that when considering P_{dashed}, which is not Wait-Only, such an execution is not possible as the second broadcast of a should be received by the process on q_2. In fact, q_2 is coverable by only one process and no more. This is because q_1 is coverable by at most one process at a time; every new process arriving in state q_1 will do so by broadcasting a, message that will be received by the process already in q_1. Then any attempt to send two processes in q_2 requires a broadcast of a, hence the reception of a by the process already in q_2. For the same reason, in P_{dashed}, $\wr q_1, q_2 \int$ is not coverable whereas q_1 is a coverable waiting state and q_2 a coverable active state.

Before stating the main lemma of this section (Lemma 1), we need an additional definition. For each *coverable* state $q \in Q$, let min_q be the minimal number of processes needed to cover q. More formally, $\mathsf{min}_q = \min\{n \mid n \in \mathbb{N}$, there exists $C \in \mathcal{C}$ s. t. $\wr n.q_{in} \int \rightarrow^* C$ and $C(q) > 0\}$. Note that min_q is defined only when q is coverable.

Lemma 1. *Let $P = (Q, \Sigma, q_{in}, T)$ be a Wait-Only protocol, $A = \{q_1, \ldots, q_n\} \subseteq Q_A$ a subset of* coverable *active states and $p \in Q_W$ a* coverable *waiting state. Then, for all $N \in \mathbb{N}$, there exists an execution $C_0 \rightarrow^* C_m$ such that $C_0 \in \mathcal{I}$, and $\{N \cdot q_1, \ldots, N \cdot q_n, p\} \leq C_m$. Moreover, $||C_0|| = N.\sum_{i=1}^{n} \mathsf{min}_{q_i} + \mathsf{min}_p$.*

The proof of this lemma relies on the two following properties on executions of Wait-Only protocols:

Lemma 2. *Given an initialized execution* $C_0 \xrightarrow{t_1} C_1 \xrightarrow{t_2} \dots\dots \xrightarrow{t_k} C_k$ *and another initial configuration* C'_0, *we can build an execution* $\widehat{C_0} \xrightarrow{t_1} \widehat{C_1} \xrightarrow{t_2} \dots\dots \xrightarrow{t_k} \widehat{C_k}$ *with* $\widehat{C_0} = C_0 + C'_0$. *For all* $0 \leq i \leq k$, $\widehat{C_i}(q) = C_i(q)$ *for all* $q \in Q \setminus \{q_{in}\}$, *and* $\widehat{C_i}(q_{in}) = C_i(q_{in}) + C'_0(q_{in})$.

This property comes from the fact that q_{in} is an active state. Hence, if we start from a bigger configuration, we can take exactly the same transitions as in the initial execution, the additional processes will stay in the initial state.

Lemma 3. *Given an initialized execution* $C_0 \xrightarrow{t_1} C_1 \xrightarrow{t_2} \dots\dots \xrightarrow{t_k} C_k$, *given some* $M \geq 1$, *for all configurations (not necessarily initial)* $\widetilde{C}_0$ *such that* $\widetilde{C}_0(q_{in}) \geq M.C_0(q_{in})$, *we have the execution* $\widetilde{C}_0 \xrightarrow{t_1} \dots \xrightarrow{t_k} \widetilde{C}_k$ *in which, for all* $0 \leq i \leq k$: $\widetilde{C}_i(q) \geq \widetilde{C}_0(q) + C_i(q)$ *for all* $q \in Q_A \setminus \{q_{in}\}$, $\widetilde{C}_i(q) \geq C_i(q)$ *for all* $q \in Q_W$ *and* $\widetilde{C}_i(q_{in}) \geq (M-1).C_0(q_{in}) + C_i(q_{in})$.

This last property states that, if one mimicks an initialized execution from another (non initial) configuration, the processes already present in *active* states (different from the initial state) will not move during the execution.

Proof of Lemma 1. Let $N \in \mathbb{N}$. Using these two properties, we can now prove the lemma. We start by proving that there exists an execution $C_0 \to^* C_m$ such that for all $q \in A$, $C_m(q) \geq N$ and $||C_0|| = N.\sum_{i=1}^{n} \mathsf{min}_{q_i}$. We prove it by induction on the size of A. If $A = \emptyset$, the property is trivially true. Let $n \in \mathbb{N}$, and assume the property to hold for all subsets $A \subseteq Q_A$ of size n. Take $A = \{q_1, q_2, \dots, q_{n+1}\} \subseteq Q_A$ of size $n+1$ such that all states $q \in A$ are coverable and let $A' = A \setminus \{q_1\}$. Let $C_0 \xrightarrow{t_1} C_1 \xrightarrow{t_2} \dots\dots \xrightarrow{t_k} C_k$ be an execution covering q_1 with $||C_0|| = \mathsf{min}_{q_1}$. Let $C'_0 \xrightarrow{t'_1} C'_1 \xrightarrow{t'_2} \dots \xrightarrow{t'_m} C'_m$ be an execution such that for all $q' \in A'$, $C'_m(q') \geq N$ and $||C'_0|| = N.\sum_{i=2}^{n+1} \mathsf{min}_{q_i}$ (it exists by induction hypothesis). We let $C_0^N = \wr(N.\mathsf{min}_{q_1}) \cdot q_{in} \int$ and $C''_0 = C'_0 + C_0^N$. Thanks to Lemma 2, we can build an execution $C''_0 \xrightarrow{t'_1} C''_1 \xrightarrow{t'_2} \dots \xrightarrow{t'_m} C''_m$, with $C''_m(q) = C'_m(q)$ for all $q \in Q \setminus \{q_{in}\}$ and $C''_m(q_{in}) = C'_m(q_{in}) + C_0^N(q_{in}) = C'_m(q_{in}) + N.\mathsf{min}_{q_1}$. So, for all $q' \in A'$, $C''_m(q') = C'_m(q') \geq N$ and $||C''_0|| = ||C'_0|| + ||C_0^N|| = N.\sum_{i=2}^{n+1} \mathsf{min}_{q_i} + N.\mathsf{min}_{q_1}$.

Now that we have shown how to build an execution that leads to a configuration with more than N processes on all states in A' and enough processes in the initial state, we show that mimicking N times the execution allowing to cover q_1 allows to obtain the desired result. Let $C_{0,1} = C''_m$. We know that for all $q' \in A'$, $C_{0,1}(q') \geq N$, and $C_{0,1}(q_{in}) \geq N.\mathsf{min}_{q_1}$. Since $||C_0|| = \mathsf{min}_{q_1}$, using Lemma 3, we can build the execution $C_{0,1} \xrightarrow{t_1} \dots \xrightarrow{t_k} C_{k,1}$ with $C_{k,1}(q_{in}) \geq (N-1).\mathsf{min}_{q_1}$, $C_{k,1}(q') \geq C_{0,k}(q') + C_k(q') \geq N$ for all $q' \in A'$ and $C_{k,1}(q_1) \geq C_{0,k}(q_1) + C_k(q_1) \geq 1$. Iterating this construction and applying each time Lemma 3, we obtain that there is an execution $C_{0,1} \xrightarrow{t_1} \dots \xrightarrow{t_k} C_{k,1} \xrightarrow{t_1} \dots \xrightarrow{t_k} C_{k,2} \dots \xrightarrow{t_1} \dots \xrightarrow{t_k}$

$C_{k,N-1} \xrightarrow{t_1} \ldots \xrightarrow{t_k} C_{k,N}$ with $C_{k,i}(q_{in}) \geq (N-i).\mathsf{min}_{q_1}$, $C_{k,i}(q') \geq N$ for all $q' \in A'$ and $C_{k,i}(q_1) \geq C_{k,i-1}(q_1) + 1 \geq i$. Observe that to obtain that $C_{k,i}(q_1) \geq i$ from Lemma 3, we use the fact that $q_1 \in Q_A$. Hence, $C_{k,N}(q_1) \geq N$ and $C_{k,N}(q') \geq N$ for all $q' \in A'$ and we have build an execution where $C_{k,N}(q) \geq N$ for all $q \in A$ and $||C_{k,N}|| = ||C''_0|| = N.\sum_{i=1}^{|A|} \mathsf{min}_{q_i}$, as expected.

At last, take a subset $A=\{q_1,\ldots,q_n\} \subseteq Q_A$ of coverable active states. Let $C_0 \to^* C_m$ be an execution such that $C_m(q) \geq N$ for all $q \in A$ and $||C_0|| = N.\sum_{i=1}^{n} \mathsf{min}_{q_i}$. Let $p \in Q_W$ a coverable state and $C'_0 \to^* C'_k$ such that $C'_k(p) \geq 1$ and $||C'_0|| = \mathsf{min}_p$. By Lemma 2, we let $\widehat{C}_0 = C_0 + C'_0$ and we have an execution $\widehat{C}_0 \to^* \widehat{C}_m$ with $\widehat{C}_m(q) = C_m(q)$ for all $q \in Q \setminus \{q_{in}\}$, and $\widehat{C}_m(q_{in}) = C_m(q_{in}) + C'_0(q_{in})$. Hence, $\widehat{C}_m(q) \geq N$ for all $q \in A$ and $\widehat{C}_m(q_{in}) \geq C'_0(q_{in})$, and note that $||\widehat{C}_m|| = ||\widehat{C}_0|| = ||C_0|| + ||C'_0|| = N.\sum_{i=1}^{n} \mathsf{min}_{q_i} + \mathsf{min}_p$. Then, with $\widetilde{C}_0 = \widehat{C}_m$, by Lemma 3, we have an execution $\widetilde{C}_0 \to^* \widetilde{C}_k$ with $\widetilde{C}_k(q) \geq \widetilde{C}_0(q) + C_k(q) \geq \widetilde{C}_0(q) \geq N$ for all $q \in A$, and $\widetilde{C}_k(p) \geq C_k(p) \geq 1$, and $||\widetilde{C}_0|| = ||\widehat{C}_0|| = N.\sum_{i=1}^{n} \mathsf{min}_{q_i} + \mathsf{min}_p$. □

4 StateCover for Wait-Only Protocols is P-Complete

4.1 Upper Bound

We present here a polynomial time algorithm to solve the state coverability problem when the considered protocol is Wait-Only. Our algorithm computes in a greedy manner the set of coverable states using Lemma 1.

Given a Wait-Only protocol $P = (Q, \Sigma, q_{in}, T)$, we compute iteratively a set of states $S \subseteq Q$ containing all the states that are coverable by P, by relying on a family $(S_i)_{i \in \mathbb{N}}$ of subsets of Q formally defined as follows (we recall that $\mathsf{Act}_\Sigma = !!\Sigma \cup !\Sigma$):

$$
\begin{aligned}
S_0 &= \{q_{in}\} \\
S_{i+1} &= S_i \cup \{q \mid \text{there exists } q' \in S_i, (q', \alpha, q) \in T, \alpha \in \mathsf{Act}_\Sigma\} \\
&\cup \{q'_2 \mid \text{there exist } q_1, q_2 \in S_i, q'_1 \in Q, a \in \Sigma \text{ s. t. } (q_1, !a, q'_1) \in T \text{ and } (q_2, ?a, q'_2) \in T\} \\
&\cup \{q'_2 \mid \text{there exist } q_1, q_2 \in S_i, q'_1 \in Q, a \in \Sigma \text{ s. t. } (q_1, !!a, q'_1) \in T \text{ and } (q_2, ?a, q'_2) \in T\}
\end{aligned}
$$

Intuitively at each iteration, we add some control states to S_{i+1} either if they can be reached from a transition labelled with an action (in Act_Σ) starting at a state in S_i or if they can be reached by two transitions corresponding to a communication by broadcast or by rendez-vous starting from states in S_i. We then define $S = \bigcup_{n \in \mathbb{N}} S_n$. Observe that $(S_i)_{i \in \mathbb{N}}$ is an increasing sequence such that $|S_i| \leq |Q|$ for all $i \in \mathbb{N}$. Then we reach a fixpoint $M \leq |Q|$ such that $S_M = S_{M+1} = S$. Hence S can be computed in polynomial time.

The two following lemmas show correctness of this algorithm. We first prove that any state $q \in S$ is indeed coverable by P. Moreover, we show that min_q the minimal number of processes necessary to cover $q \in Q$ is smaller than $2^{|Q|}$.

Lemma 4. *If $q \in S$, then there exists $C \in \mathcal{I}$ and $C' \in \mathcal{C}$ such that $C \to^* C'$, $C'(q) > 0$ and $||C|| \leq 2^{|Q|}$.*

Proof. Let $M \in \mathbb{N}$ be the first natural such that $S_M = S_{M+1}$. We have then $S_M = S$ and $M \leq |Q|$. We prove by induction that for all $0 \leq i \leq M$, for all $q \in S_i$, there exists $C \in \mathcal{I}$ and $C' \in \mathcal{C}$ such that $C \rightarrow^* C'$, $C'(q) > 0$ and $||C|| \leq 2^i$.

As $S_0 = \{q_{in}\}$, the property trivially holds for $i=0$, since $\wr q_{in} \wr \in \mathcal{I}$ and $\wr q_{in} \wr (q_{in}) > 0$.

Assume now the property to be true for $i < M$ and let $q \in S_{i+1}$. If $q \in S_i$, then by induction hypothesis, we have that there exists $C \in \mathcal{I}$ and $C' \in \mathcal{C}$ such that $C \rightarrow^* C'$, $C'(q) > 0$ and $||C|| \leq 2^i < 2^{i+1}$. We suppose that $q \notin S_i$ and proceed by a case analysis on the way q has been added to S_{i+1}.

1. there exists $q' \in S_i$ and $t = (q', \alpha, q) \in T$ with $\alpha \in \mathsf{Act}_\Sigma$. By induction hypothesis, there exists an execution $C \rightarrow^* C'$ such that $C'(q') > 0$ and $||C|| \leq 2^i$. But we have then $C' \xrightarrow{t} C''$ with $C''(q) > 0$, and consequently as well $C \rightarrow^* C''$. This is true because of the "non-blocking" nature of both broadcast and rendez-vous message in this model. Hence there is no need to check for a process to receive the message to ensure the execution $C \rightarrow^* C''$.
2. there exist $q_1, q_2 \in S_i$ and $q_1' \in Q$ and there exists $a \in \Sigma$ such that $(q_1, !a, q_1'), (q_2, ?a, q) \in T$. By induction hypothesis, we have that there exists $C_1, C_2 \in \mathcal{I}$ and $C_1', C_2' \in \mathcal{C}$ such that $C_1 \rightarrow^* C_1'$ and $C_2 \rightarrow^* C_2'$ and $C_1'(q_1) > 0$ and $C_2'(q_2) > 0$ and $||C_1|| \leq 2^i$ and $||C_2|| \leq 2^i$. Note furthermore that by definition q_1 is in Q_A and as $(q_2, ?a, q) \in T$, q_2 do not belong to Q_A. Hence $q_1 \neq q_2$. By Lemma 1, we know that there exist $C \in \mathcal{I}$ and $C' \in \mathcal{C}$ such that $C \rightarrow^* C'$ and $C'(q_1) > 0$ and $C'(q_2) > 0$. Furthermore, recall that min_{q_i} for $i \in \{1, 2\}$ is the minimal number of processes needed to cover q_i, by Lemma 1, $||C|| \leq \mathsf{min}_{q_1} + \mathsf{min}_{q_2}$. By induction hypothesis, $\mathsf{min}_{q_1} + \mathsf{min}_{q_2} \leq 2^i + 2^i$, hence $||C|| \leq 2^{i+1}$.

   We then have $C' \xrightarrow{(q_1, !a, q_1')} C''$ with $C'' = C' - \wr q_1, q_2 \wr + \wr q_1', q \wr$. Hence $C \rightarrow^* C''$ with $C''(q) > 0$.
3. there exist $q_1, q_2 \in S_i$ and $q_1' \in Q$ and there is some $a \in \Sigma$ such that $(q_1, !!a, q_1'), (q_2, ?a, q) \in T$. As above, we obtain the existence of an execution $C \rightarrow^* C'$ with $C'(q_1) > 0$ and $C'(q_2) > 0$, and $||C|| \leq 2^{i+1}$. Then $C' = \wr q_1, q_2, \ldots, q_k \wr$ and $C' \xrightarrow{(q_1, !!a, q_1')} C''$ with $C'' = \wr q_1', q, \ldots, q_k' \wr$ with, for all $3 \leq j \leq k$, either $a \notin R(q_j)$ and $q_j = q_j'$ or $(q_j, ?a, q_j') \in T$. In any case, we have $C \rightarrow^* C''$ with $C''(q) > 0$.

So, for any $q \in S$, $q \in S_M$ and we have an execution $C \rightarrow^* C'$ with $C \in \mathcal{I}$ such that $C'(q) > 0$ and $||C|| \leq 2^M \leq 2^{|Q|}$. □

We now prove the completeness of our algorithm by showing that every state coverable by P belongs to S.

Lemma 5. *If there exists $C \in \mathcal{I}$ and $C' \in \mathcal{C}$ such that $C \rightarrow^* C'$ and $C'(q) > 0$, then $q \in S$.*

Proof. We consider the initialized execution $C_0 \xrightarrow{t_1} C_1 \xrightarrow{t_2} \ldots \xrightarrow{t_n} C_n$ with $C = C_0$ and $C_n = C'$. We will prove by induction on $0 \leq i \leq n$ that for all q such that $C_i(q) > 0$, we have $q \in S_i$.

For $i = 0$, we have $C_0 = \wr ||C|| \cdot q_{in} \wr$, and $S_0 = \{q_0\}$. Hence the property holds.

Assume the property to be true for $i < n$, and let $q \in Q$ such that $C_{i+1}(q) > 0$. If $C_i(q) > 0$, then by induction hypothesis we have $q \in S_i$ and since $S_i \subseteq S_{i+1}$, we deduce that $q \in S_{i+1}$. Assume now that $C_i(q) = 0$. We proceed by a case analysis.

1. $t_{i+1}=(q', !a, q)$ or $t_{i+1}=(q', !!a, q)$ for some $a \in \Sigma$ and $q' \in Q$. Since $C_i \xrightarrow{t_{i+1}} C_{i+1}$, we have necessarily $C_i(q')>0$. By induction hypothesis, $q' \in S_i$, and by construction of S_{i+1}, we deduce that $q' \in S_{i+1}$.
2. $t_{i+1} = (q_1, !a, q_1')$ or $t_{i+1} = (q_1, !!a, q_1')$ with $q_1' \neq q$. Since $C_i(q) = 0$ and $C_{i+1}(q) > 0$, there exists a transition of the form $(q_2, ?a, q)$ with $q_1 \neq q_2$ (because $q_1 \in Q_A$ and $(q_2, ?a, q) \in T$ hence $q_2 \notin Q_W$). Consequently, we know that we have $C_i(q_1) > 0$ and $C_i(q_2)>0$. By induction hypothesis q_1, q_2 belong to S_i and by construction of S_{i+1} we deduce that $q \in S_{i+1}$. □

The two previous lemmas show the soundness and completeness of our algorithm to solve StateCover based on the computation of the set S. Since this set of states can be computed in polynomial time, we obtain the following result.

Theorem 1. StateCover *is in* P *for Wait-Only protocols.*

Furthermore, completeness of the algorithm along with the bound on the number of processes established in Lemma 4 gives the following result.

Corollary 1. *Given a Wait-Only protocol* $P=(Q, \Sigma, q_{in}, T)$, *for all* $q \in Q$ *coverable by* P, *then* min_q *the minimal number of processes necessary to cover* q *is at most* $2^{|Q|}$.

4.2 Lower Bound

We show that StateCover for Wait-Only protocols is P-hard. For this, we provide a reduction from the Circuit Value Problem (CVP) which is known to be P-complete [14]. CVP is defined as follows: given an acyclic Boolean circuit with n input variables, one output variable, m boolean gates of type *and*, *or*, *not*, and a truth assignment for the input variables, is the value of the output equal to a given boolean value? Given an instance of the CVP, we build a protocol in which the processes broadcast variables (input ones or associated with gates) along with their boolean values. These broadcasts will be received by other processes that will use them to compute boolean value of their corresponding gate, and broadcast the obtained value. Hence, different values are propagated through the protocol representing the circuit, until the state representing the output variable value we look for is covered.

Take for example a CVP instance C with two variables v_1, v_2, and two gates: one *not* gate on variable v_1 denoted $g_1(v_1, \neg, o_1)$ (where o_1 stands for the output variable of g_1), and one *or* gate on variable v_2 and o_1 denoted $g_2(o_1, v_2, \vee, o_2)$ (where o_2 stands for the output variable of gate g_2). Assume the input boolean value for v_1 [resp. v_2] is $\top$ [resp. $\perp$]. The protocol associated to C is displayed on Fig. 2. Assume the output value of C is o_2, we will show that $q^2_\top$ [resp. $q^2_\perp$] is coverable if and only if o_2 evaluates to $\top$ [resp. $\perp$]. Note that with the truth assignment depicted earlier, o_2 evaluates to $\bot$, and indeed one can build an execution covering $q^2_\bot$ with three processes:

$$\langle 3.q_{in} \rangle \rightarrow \langle 2.q_{in}, q^2_0 \rangle \rightarrow \langle q_{in}, q^1_0, q^2_0 \rangle \xrightarrow{(q_{in}, !!(v_1,\top), q_{in})} \langle q_{in}, q^1_\bot, q^2_0 \rangle$$
$$\xrightarrow{(q^1_\bot, !!(o_1,\bot), q^1_\bot)} \langle q_{in}, q^1_\bot, q^2_1 \rangle \xrightarrow{(q_{in}, !!(v_2,\bot), q_{in})} \langle q_{in}, q^1_\bot, q^2_\bot \rangle$$

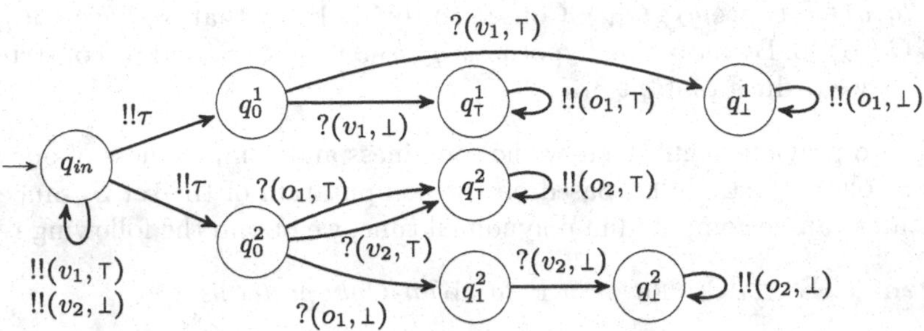

Fig. 2. Protocol for a CVP instance with two variables v_1, v_2, two gates $g_1(\neg, v_1, o_1)$ and $g_2(\vee, o_1, v_2, o_2)$, and input $\top$ for v_1 and $\bot$ for v_2 and output variable o_2. Depending on the truth value of o_2 to test, the state we ask to cover can be $q^2_\bot$ or $q^2_\top$.

Together with Theorem 1, we get the following theorem.

Theorem 2. STATECOVER *for Wait-Only protocols is P-complete.*

This reduction can be adapted to Wait-Only NB-Rendez-vous protocols, which leads to the following theorem, proving that the upper bound presented in [10] is tight.

Theorem 3. STATECOVER *for Wait-Only NB-Rendez-vous protocols is P-hard.*

5 ConfCover for Wait-Only Protocols is PSpace-Complete

We present here an algorithm to solve the configuration coverability problem for Wait-Only protocols in polynomial space.

5.1 Main Ideas

For the remaining of the section, we fix a Wait-Only protocol $P = (Q, \Sigma, q_{in}, T)$ and a configuration $C_f \in \mathcal{C}$ to cover, and we let $K = ||C_f||$. The intuition is the following: we (only) keep track of the K processes that will cover C_f. Of course, they might need other processes to reach the desired configuration, if they need to receive messages. That is why we also maintain the set of reachable states along the execution. An abstract configuration will then be a multiset of K

states (concrete part of the configuration) and a set of all the reachable states (abstract part of the configuration). Lemma 1 ensures that it is enough to know which active states are reachable to ensure that both the concrete part and the active states of the abstract part are coverable at the same time. However, there is a case where this abstraction would not be enough: assume that one of the K processes has to send a message, and this message *should not* be received by the other $K - 1$ processes. This can happen when the message is received by a process in the part of the configuration that we have abstracted away. In that case, even if the (waiting) state is present in the set of reachable states, Lemma 1 does not guarantee that the entire configuration is reachable, so the transition to an abstract configuration where none of the $K - 1$ processes has received the message might be erroneous. This is why in that case we need to precisely keep track of the process that will receive the message, even if in the end it will not participate in the covering of C_f. This leads to the definition of the $\Rightarrow_{\mathsf{switch}}$ transition below.

This proof is structured as follows: we present the formal definitions of the abstract configurations and semantics in Sect. 5.2. In Sect. 5.3, we present the completeness proof, Sect. 5.4 is devoted to prove the soundness of the construction. In the latter, we also give some ingredients to prove an upper bound on the number of processes needed to cover the configuration. In Sect. 5.5 one can find the main theorem of this section: it states that the ConfCover problem is in PSpace and if the configuration is indeed coverable, it presents an upper bound on the number of processes needed to cover it. In Sect. 5.6, we prove that this lower bound is tight as the problem is PSpace-hard.

5.2 Reasoning with Abstract Configurations

We present the abstract configurations we rely on. Let us fix $K = ||C_f||$. An *abstract configuration* γ is a pair (M, S) where M is a configuration in $\mathcal{C}$ such that $||M||=K$ and $S \subseteq Q$ is a subset of control states such that $\{q \in Q \mid M(q)>0\} \subseteq S$. We call M the M-part of γ and S its S-part. We denote by Γ the set of abstract configurations and by γ_{in} the initial abstract configuration $\gamma_{in} = (\wr K \cdot q_{in} \int, \{q_{in}\})$. An abstract configuration $\gamma=(M, S)$ represents a set of configurations $[\![\gamma]\!] = \{C \in \mathcal{C} \mid M \leq C$ and $C(q) > 0$ implies $q \in S\}$. Hence in $[\![\gamma]\!]$, we have all the configurations C that are bigger than M as long as the states holding processes in C are stored in S (observe that this implies that all the states in M appear in S).

We now define an abstract transition relation for abstract configurations. For this matter, we define three transition relations $\Rightarrow_{\mathsf{step}}$,$\Rightarrow_{\mathsf{ext}}$ and $\Rightarrow_{\mathsf{switch}}$ and let $\Rightarrow$ be defined by $\Rightarrow_{\mathsf{step}} \cup \Rightarrow_{\mathsf{ext}} \cup \Rightarrow_{\mathsf{switch}}$. Let $\gamma = (M, S)$ and $\gamma' = (M', S')$ be two abstract configurations and $t = (q, \alpha, q')$ be a transition in T with $\alpha = !a$ or $\alpha = !!a$. For $\kappa \in \{\mathsf{step}, \mathsf{ext}, \mathsf{switch}\}$, we have $\gamma \overset{t}{\Rightarrow}_\kappa \gamma'$ iff all the following conditions hold:

- $S \subseteq S'$, and,
- for all $p \in S' \setminus S$, either $p = q'$ or there exist $p' \in S$ and $(p', ?a, p)$ in T, and,
- one of the following cases is true:

- κ = step and $M \xrightarrow{t} M'$. This relation describes a message emitted from the M-part of the configuration;
- κ = ext and $q \in S$ and $M + \wr q \wr \xrightarrow{t} M' + \wr q' \wr$;
- κ = ext and there exists $(p, ?a, p')$ in T such that $q, p \in S$ and $M + \wr q, p \wr \xrightarrow{t} M + \wr q', p' \wr$ (note that in that case $M = M'$). The relation ext hence describes a message emitted from the S-part of the configuration;
- κ = switch and $q \in S$ and α=!a and there exists $t' = (p, ?a, p') \in T$ such that $\wr p \wr \leq M$ and $\wr q' \wr \leq M'$ and $M - \wr p \wr = M' - \wr q' \wr$ and $M + \wr q \wr \xrightarrow{t} M' + \wr p' \wr$. This relation describes a sending from a state in the S-part of the abstract configuration leading to a rendez-vous with one process in the M-part, and a"switch" of processes: we remove the receiver process of the M-part and replace it by the sender.

Note that in any case, q, the state from which the message is sent, belongs to S. We then write $\gamma \overset{t}{\Rightarrow} \gamma'$ whenever $\gamma \overset{t}{\Rightarrow}_{\kappa} \gamma'$ for $\kappa \in \{\text{step}, \text{ext}, \text{switch}\}$ and we do not always specify the used transition t (when omitted, it means that there exists a transitions allowing the transition). We denote by $\Rightarrow^*$ the reflexive and transitive closure of $\Rightarrow$.

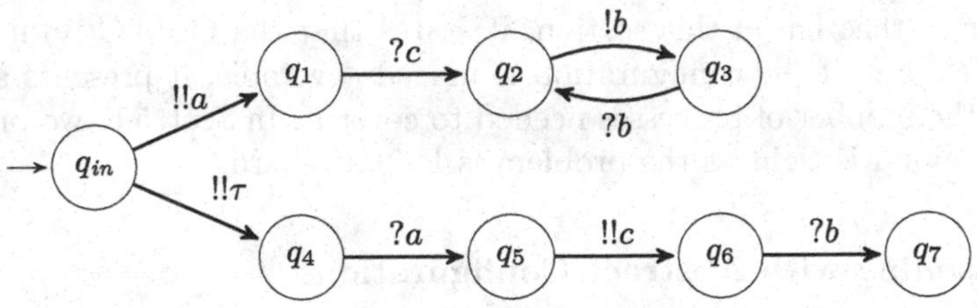

Fig. 3. A Wait-Only protocol P'.

Example 5. We consider the Wait-Only protocol P' depicted on Fig. 3 with set of states Q'. We want to cover $C_f = \wr q_3, q_3, q_6 \wr$ (hence $K = 3$). In this example, the abstract configuration $\gamma = (\wr q_2, q_2, q_4 \wr, \{q_{in}, q_1, q_2, q_4, q_5, q_6\})$ represents all the configurations of P' with at least two processes on q_2 and one on q_4, and no process on q_3 nor q_7.

Considering the following abstract execution, we can cover C_f:

$$\gamma_{in} \xRightarrow{(q_{in},!!\tau,q_4)}_{\text{step}} (\wr q_4, q_{in}, q_{in} \wr, \{q_{in}, q_4\}) \xRightarrow{(q_{in},!!\tau,q_4)}_{\text{step}} (\wr q_4, q_4, q_{in} \wr, \{q_{in}, q_4\})$$
$$\xRightarrow{(q_{in},!!a,q_1)}_{\text{step}} (\wr q_5, q_5, q_1 \wr, \{q_{in}, q_1, q_4, q_5\}) \xRightarrow{(q_5,!!c,q_6)}_{\text{ext}} (\wr q_5, q_5, q_2 \wr, Q' \setminus \{q_3, q_7\})$$
$$\xRightarrow{(q_2,!b,q_3)}_{\text{step}} (\wr q_5, q_5, q_3 \wr, Q') \xRightarrow{(q_5,!!c,q_6)}_{\text{step}} (\wr q_6, q_5, q_3 \wr, Q')$$
$$\xRightarrow{(q_2,!b,q_3)}_{\text{switch}} (\wr q_3, q_5, q_3 \wr, Q') \xRightarrow{(q_5,!!c,q_6)}_{\text{step}} (\wr q_3, q_6, q_3 \wr, Q')$$

It corresponds for instance to the following concrete execution:

$$\wr q_{in}, q_{in}, q_{in} \wr + \wr q_{in}, q_{in} \wr \rightarrow^{+} \wr q_4, q_4, q_{in} \wr + \wr q_4, q_{in} \wr \xrightarrow{(q_{in},!!a,q_1)} \wr q_5, q_5, q_1 \wr + \wr q_5, q_{in} \wr$$
$$\xrightarrow{(q_{in},!!a,q_1)} \wr q_5, q_5, q_1 \wr + \wr q_5, q_1 \wr \xrightarrow{(q_5,!!c,q_6)} \wr q_5, q_5, q_2 \wr + \wr q_6, q_1 \wr$$
$$\xrightarrow{(q_2,!b,q_3)} \wr q_5, q_5, q_3 \wr + \wr q_7, q_1 \wr \xrightarrow{(q_5,!!c,q_6)} \wr q_6, q_5, q_3 \wr + \wr q_7, q_2 \wr$$
$$\xrightarrow{(q_2,!b,q_3)} \wr q_3, q_5, q_3 \wr + \wr q_7, q_7 \wr \xrightarrow{(q_5,!!c,q_6)} \wr q_3, q_6, q_3 \wr + \wr q_7, q_7 \wr$$

The M-part of our abstract configuration $\wr q_6, q_5, q_3 \wr$ reached just before the $\Rightarrow_{\mathsf{switch}}$ transition does not correspond to the set of processes that finally cover C_f, at this point of time, the processes that will finally cover C_f are in states q_2, q_5, and q_3. But here we ensure that the process on q_6 will actually receive the b sent by the process on q_2, leaving the process on q_3 in its state. Once this has been ensured, process on q_6 is not useful anymore, and instead we follow the process that was on q_2 before the sending, hence the $\Rightarrow_{\mathsf{switch}}$ transition.

The algorithm used to solve ConfCover, is then to seek in the directed graph $(\Gamma, \Rightarrow)$ if a vertex of the form (C_f, S) is reachable from γ_{in}.

Before proving that this algorithm is correct, we establish the following property.

Lemma 6. *Let* (M, S) *and* (M', S') *be two abstract configurations and* $\widetilde{S} \subseteq Q$ *such that* $S \subseteq \widetilde{S}$. *We have:*

1. $[\![(M, S)]\!] \subseteq [\![(M, \widetilde{S})]\!]$.
2. *If* $(M, S) \Rightarrow (M', S')$ *then there exists* $S'' \subseteq Q$ *such that* $(M, \widetilde{S}) \Rightarrow (M', S'')$ *and* $S' \subseteq S''$.

Proof. The first point is a direct consequence of the definition of $[\![]\!]$. For the second point, it is enough to take $S'' = S' \cup \widetilde{S}$ and apply the definition of $\Rightarrow$. □

5.3 Completeness of the Algorithm

In this subsection we show that if C_f can be covered then there exists an abstract configuration $\gamma = (C_f, S)$ such that $\gamma_{in} \Rightarrow^* \gamma$. We use $\mathcal{C}_{\geq K}$ to represent the set $\{C \in \mathcal{C} \mid ||C|| \geq K\}$ of configurations with at least K processes and $\mathcal{C}_{=K}$ the set $\{C \in \mathcal{C} \mid ||C|| = K\}$ of configurations with exactly K processes. This first lemma shows the completeness for a single step of our abstract transition relation (note that we focus on the M-part, as it is the one witnessing C_f in the end).

Lemma 7. *Let* $C, C' \in \mathcal{C}_{\geq K}$ *and* $t \in T$ *such that* $C \xrightarrow{t} C'$. *Then for all* $M' \in \mathcal{C}_{=K}$ *such that* $M' \leq C'$, *there exists* $M \in \mathcal{C}_{=K}$ *and* $S' \subseteq Q$ *such that* $(M, S) \Rightarrow (M', S')$ *with* $S = \{q \in Q \mid C(q) > 0\}$, $C \in [\![(M, S)]\!]$ *and* $C' \in [\![(M', S')]\!]$.

Proof. Let $M' \in \mathcal{C}_{=K}$ such that $M' \le C'$. We assume that $t = (q, \alpha, q')$ with $\alpha \in \{!a, !!a\}$. We let $S = \{p \in Q \mid C(p) > 0\}$ and $S' = S \cup \{p \in Q \mid C'(p) > 0\}$. By definition of S and S', for all $p \in S' \setminus S$, either $p = q'$ or there exists $p' \in S$ and $(p', ?a, p)$ in T. In fact, let $p \in S' \setminus S$ such that $p \not= q'$. Since $C \xrightarrow{t} C'$, we have necessarily that there exist $p' \in Q$ such that $C(p') > 0$ (hence $p' \in S$), and $(p', ?a, p)$ in T. We now reason by a case analysis to determine $M \in \mathcal{C}_{=K}$ such that $(M, S) \Rightarrow (M', S')$ and $C \in [\![(M, S)]\!]$. The different cases are: (i) α=!!a, (ii) α=!a and the message is *not received*, (iii) α=!a and the message is received by a process. Because of space constraints, we present here only case (iii) as it exhibits the most different abstract behaviours. The two other cases can be found on [11].

(iii) if α=!a and the message is received by a process (i.e. it is a rendez-vous), denote by $(p, ?a, p')$ the reception transition issued between C and C'. Using the definition of $\rightarrow$, we get $C' = C - \{\!\{q, p\}\!\} + \{\!\{q', p'\}\!\}$. We consider the four following disjoint cases:
- $M'(q')$=0 and $M'(p')$=0. Since $\{\!\{q', p'\}\!\} \le C'$ and $M' \le C'$, we get that $C' = M' + \{\!\{q', p'\}\!\} + M_2$ for some multiset M_2. We deduce that $C = M' + \{\!\{q, p\}\!\} + M_2$. This allows us to deduce that $M' \le C$ and consequently $C \in [\![(M', S)]\!]$. Moreover, $M' + \{\!\{p, q\}\!\} \xrightarrow{t} M' + \{\!\{q', p'\}\!\}$ and $q, p \in S$. Hence $(M', S) \overset{t}{\Rightarrow}_{\mathsf{ext}} (M', S')$.
- $M'(q') = 0$ and $M'(p') > 0$. In that case $C' = M' + \{\!\{q'\}\!\} + M_2$ for some multiset M_2. Let $M = M' - \{\!\{p'\}\!\} + \{\!\{p\}\!\}$. We have then $C = C' + \{\!\{q, p\}\!\} - \{\!\{q', p'\}\!\} = M' + \{\!\{q'\}\!\} + M_2 + \{\!\{q, p\}\!\} - \{\!\{q', p'\}\!\} = M' + M_2 - \{\!\{p'\}\!\} + \{\!\{p\}\!\} + \{\!\{q\}\!\} = M + \{\!\{q\}\!\} + M_2$. This allows us to deduce that $M \le C$ and consequently $C \in [\![(M, S)]\!]$. Furthermore $q \in S$ and $M + \{\!\{q\}\!\} \xrightarrow{t} M' + \{\!\{q'\}\!\}$. Hence $(M, S) \overset{t}{\Rightarrow}_{\mathsf{ext}} (M, S')$.
- $M'(q') > 0$ and $M'(p') = 0$. In that case $C' = M' + \{\!\{p'\}\!\} + M_2$ for some multiset M_2. Let $M = M' - \{\!\{q'\}\!\} + \{\!\{p\}\!\}$. We have then $C = C' + \{\!\{q, p\}\!\} - \{\!\{q', p'\}\!\} = M' + \{\!\{p'\}\!\} + M_2 + \{\!\{q, p\}\!\} - \{\!\{q', p'\}\!\} = M' + M_2 - \{\!\{q'\}\!\} + \{\!\{p\}\!\} + \{\!\{q\}\!\} = M + \{\!\{q\}\!\} + M_2$. This allows us to deduce that $q \in S$ and $M \le C$ and consequently $C \in [\![(M, S)]\!]$. We also have $\{\!\{p\}\!\} \le M$ and $\{\!\{q'\}\!\} \le M'$ and $M - \{\!\{p\}\!\} = M' - \{\!\{q'\}\!\}$ and $M + \{\!\{q\}\!\} \xrightarrow{t} M' + \{\!\{p'\}\!\}$. Hence $(M, S) \overset{t}{\Rightarrow}_{\mathsf{switch}} (M', S')$. Observe that we need to use the $\Rightarrow_{\mathsf{switch}}$ transition relation in this case. Assume that $C(s) > 0$ for some state $s \in S$ such that $(s, ?a, s') \in T$, and that any configuration in $\mathcal{C}_{=K}$ such that $M \le C$ contains such state s. Then, applying $\Rightarrow_{\mathsf{step}}$ to such a multiset M will take away the process on state s and will lead to an abstract configuration with $M' \not\le C'$.
- $M'(q') > 0$ and $M'(p') > 0$. In that case $C' = M' + M_2$ for some multiset M_2. Let $M = M' - \{\!\{p', q'\}\!\} + \{\!\{p, q\}\!\}$. We have then $C = C' + \{\!\{q, p\}\!\} - \{\!\{q', p'\}\!\} = M + M_2$. This allows us to deduce that $M \le C$ and consequently $C \in [\![(M, S)]\!]$ and that $M \xrightarrow{t} M'$. Hence $(M, S) \overset{t}{\Rightarrow}_{\mathsf{step}} (M', S')$. □

The two previous lemmas allow us to establish completeness of the construction, by a simple induction on the length of the considered execution.

Lemma 8. *Let $C_{in} \in \mathcal{I}$ and $C \in \mathcal{C}_{\ge K}$ such that $C_{in} \rightarrow^* C$. For all $M \in \mathcal{C}_{=K}$ such that $M \le C$ there exists $S \subseteq Q$ such that $C \in [\![(M, S)]\!]$ and $\gamma_{in} \Rightarrow^* (M, S)$.*

5.4 Soundness of the Algorithm

We now prove that if we have $\gamma_{in} \Rightarrow^* (M, S)$ then the configuration M can be covered. We first establish that the S-part of a reachable abstract configuration stores only states that are reachable in a concrete execution.

Lemma 9. *If $\gamma = (M, S)$ is an abstract configuration such that $\gamma_{in} \Rightarrow^* \gamma$, then all states $q \in S$ are coverable.*

Proof. We suppose that we have $\gamma_{in} = \gamma_0 \Rightarrow \gamma_1 \Rightarrow \ldots \Rightarrow \gamma_n = (M, S)$ and we prove this lemma by induction on n, the length of the abstract execution.

Case $n = 0$: In that case $(M, S) = \gamma_0 = (K \cdot \wr q_{in} \int, \{q_{in}\})$, as q_{in} is trivially coverable, the property holds.

Case $n > 0$: We assume that the property holds for all $0 \leq m < n$ and consider the abstract execution $\gamma_0 \overset{t_1}{\Rightarrow} \gamma_1 \overset{t_2}{\Rightarrow} \ldots \overset{t_n}{\Rightarrow} \gamma_n$ where $\gamma_0 = \gamma_{in}$ and $\gamma_i = (M_i, S_i)$ for all $0 \leq i \leq n$. Let $p \in S_n$. If $p \in S_{n-1}$, then by induction hypothesis, p is coverable. Otherwise, $p \in S_n \setminus S_{n-1}$, and let $t_n = (q, \alpha, q')$ with $\alpha \in \{!a, !!a \mid a \in \Sigma\}$. By definition of $\Rightarrow$, $q \in S_{n-1}$ and

- either $q' = p$, and by induction hypothesis, there exists an initialized execution $C_0 \to^* C$ with $C(q) > 0$ and in that case, $C_0 \to^* C \xrightarrow{t} C'$ for a configuration C' such that $C'(p) > 0$ and p is coverable.
- or $\alpha \in \{!a, !!a\}$ for some $a \in \Sigma$ and $(p', ?a, p) \in T$, with $p' \in S_{n-1}$. By induction hypothesis, both q and p' are coverable, with $q \in Q_A$ and $p \in Q_W$. By Lemma 1, there exists an execution $C_0 \to^* C$ such that $C(q) \geq 1$ and $C(p') \geq 1$. We then have $C \xrightarrow{t_n} C'$ with $C'(p) > 0$ (if α=$!a$ then the process on p' can receive the message a and move to p, and if α=$!!a$ then the process on p' will necessary receive the broadcast and move to p), and p is coverable. □

The next lemma establishes soundness of the algorithm. Moreover, it gives an upper bound on the minimal number of processes needed to cover a configuration.

Lemma 10. *Let (M, S) be an abstract configuration such that $\gamma_{in} \Rightarrow \gamma_1 \Rightarrow \ldots \Rightarrow \gamma_n = (M, S)$. Then, there exist $C_{in} \in \mathcal{I}$, $C \in \mathcal{C}$ such that $M \leq C$ and $C_{in} \to^* C$. Moreover, $||C_{in}|| = ||C|| \leq K + 2^{|Q|} \times n$.*

Proof. We reason by induction on n, the length of the abstract execution.

Case $n = 0$: The property trivially holds for $C = \wr K \cdot q_{in} \int$.

Case $n > 0$: We assume that the property holds for all $0 \leq m < n$ and consider the abstract execution $\gamma_0 \overset{t_1}{\Rightarrow} \gamma_1 \overset{t_2}{\Rightarrow} \ldots \overset{t_n}{\Rightarrow} \gamma_n$ where $\gamma_0 = \gamma_{in}$ and $\gamma_i = (M_i, S_i)$ for all $0 \leq i \leq n$. By induction hypothesis, we know that there exist $C_{in} \in \mathcal{I}$, $C_{n-1} \in \mathcal{C}$ such that $M_{n-1} \leq C_{n-1}$ and $C_{in} \to^* C_{n-1}$ and $||C_{in}|| = ||C_{n-1}|| \leq K + 2^{|Q|} \times (n-1)$. If $M_n = M_{n-1}$ then the property holds. Assume now that $M_n \not\models M_{n-1}$. We let $t_n = (q, \alpha, q')$ with α=$!a$ or α=$!!a$. By definition of $\Rightarrow$, we know that $S_{n-1} \subseteq S_n$ and that $q \in S_{n-1}$. Thanks to Lemma 9, q is coverable. We now perform a case analysis:

– Assume $\gamma_{n-1} \overset{t_n}{\Rightarrow}_{\mathsf{step}} \gamma_n$. Then $M_{n-1} \xrightarrow{t_n} M_n$, and since $M_{n-1} \le C_{n-1}$, we have $C_{n-1} = M_{n-1} + M$ for some multiset M.
  - If α=!!a, then $M_{n-1} + M = \wr q_1, \dots q_{K-1}, q \wr + \wr p_1, \dots, p_L \wr$ and $M_n = \wr q'_1, \dots q'_{K-1}, q' \wr$ where for all $1 \le i \le K-1$, either $(q_i, ?a, q'_i) \in T$ or $a \notin R(q_i)$ and $q_i = q'_i$. For each $1 \le i \le L$, define p'_i as $p'_i = p_i$ if $a \notin R(p_i)$ or p'_i is such that $(p_i, ?a, p'_i) \in T$. If we let $M' = \wr p'_1, \dots, p'_L \wr$ and $C_n = M_n + M'$, we have by definition that $C_{n-1} \xrightarrow{t_n} C_n$ with $M_n \le C_n$.
  - If α=!a and $(M_{n-1} - \wr q \wr)(p) > 0$ for some $p \in Q$ such that $(p, ?a, p') \in T$ (i.e., a rendez-vous occurred), it holds that $M_{n-1} + M \xrightarrow{t} M_n + M$ and we choose $C_n = M_n + M$.
  - If α=!a and $(M_{n-1} - \wr q \wr)(p) = 0$ for all $p \in Q$ such that $a \in R(p)$ (i.e. it was a non-blocking sending of a message), then either there exists $(p, ?a, p') \in T$ such that $M(p) > 0$, either for all $p \in Q$ such that $M(p) > 0$, $a \notin R(p)$. In the first case, a rendez-vous will occur in the execution of t_n over C_{n-1}, and we have $M_{n-1} + M \xrightarrow{t_n} M_n + M - \wr p \wr + \wr p' \wr$. We then let $C_n = M_n + M - \wr p \wr + \wr p' \wr$. In the latter case, $M_{n-1} + M \xrightarrow{t_n} M_n + M$ and with $C_n = M_n + M$. In both cases, we have , $C_{n-1} \xrightarrow{t} C_n$ and $M_n \le C_n$.

  In all cases, we have $C_{in} \to^* C_{n-1} \to C_n$ and $||C_{in}|| = ||C_n|| = ||C_{n-1}|| \le K + 2^{|Q|} \times (n-1) \le K + 2^{|Q|} \times n$.

– Assume $\gamma_{n-1} \overset{t_n}{\Rightarrow}_{\mathsf{ext}} \gamma_n$ or $\gamma_{n-1} \overset{t_n}{\Rightarrow}_{\mathsf{switch}} \gamma_n$. As q is coverable, from Lemma 1, there exists an execution $C^q_{in} \to^* C^q$ such that $C^q_{in} \in \mathcal{I}$ and $C^q(q) > 0$ and $||C^q_{in}|| \le 2^{|Q|}$. From Lemma 2, we have the following execution: $C_{in} + C^q_{in} \to^* C_{in} + C^q$. Next, from Lemma 3, by taking $M = 1$ and $\tilde{C}_0 = C_{in} + C^q$, we have an execution $C_{in} + C^q \to^* C_{n-1} + C'^q$ where $C'^q(q) > 0$. We deduce that $C_{n-1} + C'^q = M_{n-1} + \wr q \wr + M$ for some multiset M. We now proceed with a case analysis:
  - Case $\gamma_{n-1} \overset{t_n}{\Rightarrow}_{\mathsf{ext}} \gamma_n$ where $M_{n-1} + \wr q \wr \xrightarrow{t_n} M_n + \wr q' \wr$. By definition of $\to$, we also have $M_{n-1} + \wr q \wr + M \to M_n + \wr q' \wr + M'$ for some multiset M'. Letting $C_n = M_n + \wr q' \wr + M'$ gives us that that $C_{in} + C^q_{in} \to^* M_{n-1} + \wr q \wr + M \to C_n$.
  - Case $\gamma_{n-1} \overset{t_n}{\Rightarrow}_{\mathsf{switch}} \gamma_n$: by definition of $\Rightarrow_{\mathsf{switch}}$, we know that there exists $(p, ?a, p') \in T$ with $p \in S_{n-1}$ such that $M_{n-1} = M' + \wr p \wr$ and that $M' + \wr p \wr + \wr q \wr \xrightarrow{t_n} M' + \wr q' \wr + \wr p' \wr$ and $M_n = M' + \wr q' \wr$. Furthermore, by definition of $\to$, we have $M' + \wr p, q \wr + M \to M' + \wr p', q' \wr + M$. Hence setting $C_n = M' + \wr p', q' \wr + M = M_n + \wr p' \wr + M$ gives us that that $C_{in} + C^q_{in} \to^* M_{n-1} + \wr q \wr + M \to C_n$, with $M_n \le C_n$.

  In both cases, we have shown that there exists C_n such that $M_n \le C_n$ and $C_{in} + C^q_{in} \to^* C_n$. Furthermore we have that $||C_n|| = ||C_{in} + C^q_{in}|| \le K + 2^{|Q|} \times (n-1) + 2^{|Q|} \le K + 2^{|Q|} \times n$. □

5.5 Upper Bound

Using Lemmas 8 and 10, we know that there exists $C \in \mathcal{I}$ and $C' \in \mathcal{C}$ such that $C \to^* C'$ and $C_f \le C'$ iff there exists an abstract execution $\gamma_{in} \Rightarrow \gamma_1 \Rightarrow \cdots \Rightarrow \gamma_n$ with $\gamma_n = (C_f, S)$ for some $S \subseteq Q$, hence the algorithm consisting in deciding reachability of a vertex of the form (C_f, S) from γ_{in} in the finite graph $(\Gamma, \Rightarrow)$ is correct. Note furthermore that the number of abstract configurations $|\Gamma|$ is bounded by $|Q|^{||C_f||} \times 2^{|Q|}$. As the reachability of a vertex in a graph is NL-complete, this gives us a NPSPACE procedure, which leads to a PSPACE procedure thanks to Savitch's theorem.

Theorem 4. CONFCOVER *is in* PSPACE.

Remark 3. Thanks to Lemma 10, we know that $\gamma_{in} \Rightarrow \gamma_1 \Rightarrow \cdots \Rightarrow \gamma_n$ with $\gamma_n = (C_f, S)$ iff there exists $C_{in} \in \mathcal{I}$, $C \in \mathcal{C}$ such that $C_f \leq C$ and $C_{in} \rightarrow^* C$ and $||C_{in}|| = ||C|| \leq K + 2^{|Q|} \times n$. But due to the number of abstract configurations, we can assume that $n \leq 2^{|Q|} \times |Q|^{||C_f||}$ as it is unnecessary in the abstract execution $\gamma_{in} \Rightarrow \gamma_1 \Rightarrow \cdots \Rightarrow \gamma_n$ to visit twice the same abstract configuration. Hence the configuration C_f is coverable iff there is $C \in \mathcal{I}$ and $C' \in \mathcal{C}$ such $C \rightarrow^* C'$ and $C_f \leq C'$ and $||C|| = ||C'|| \leq K + 2^{|Q|} \times 2^{|Q|} \times |Q|^{||C_f||}$.

5.6 Lower Bound

To prove PSPACE-hardness of the CONFCOVER problem for Wait-Only protocols, we reduce the intersection non-emptiness problem for deterministic finite automata, which is known to be PSPACE-complete [13]. The PSPACE-hardness in fact holds when considering Wait-Only protocols without any (non-blocking) rendez-vous transitions, i.e. transitions of the form $(q, !a, q')$.

Let $\mathcal{A}_1, \ldots, \mathcal{A}_n$ be a list of deterministic finite and *complete* automata with $\mathcal{A}_i = (\Sigma, Q_i, q_i^0, \{q_i^f\}, \Delta_i)$ for all $1 < i \leq n$. Observe that we restrict our reduction to automata with a unique accepting state, which does not change the complexity of the problem. We note Σ^* the set of words over the finite alphabet Σ and Δ_i^* the function extending Δ_i to Σ^*, i.e., for all $q \in Q_i$, $\Delta_i^*(q, \varepsilon) = q$, and for all $w \in \Sigma^*$ and $a \in \Sigma$, $\Delta_i^*(q, wa) = \Delta_i(\Delta_i^*(q, w), a)$.

We build the protocol P with set of states Q, displayed in Fig. 4 where P_i for $1 \leq i \leq n$ is a protocol mimicking the behaviour of the automaton $\mathcal{A}_i$: $P_i = (Q_i, \Sigma, q_i^0, T_i)$, with $T_i = \{(q, ?a, q') \mid (q, a, q') \in \Delta_i\}$. Moreover, from any state $q \in \bigcup_{1 \leq i \leq n} Q_i$, there is an outgoing transition $(q, ?go, q_{fail})$. These transitions are depicted by the outgoing transitions labelled by $?go$ from the orange rectangles.

Note that P is Wait-Only as all states in P_i for all $1 \leq i \leq n$ are waiting states and the only active states are q_{in} and q_s. We show that $\bigcap_{1 \leq i \leq n} L(\mathcal{A}_i) \neq \emptyset$ if and only if there is an initial configuration $C \in \mathcal{I}_P$ and a configuration $C' \in \mathcal{C}_P$ such that $C \rightarrow^* C'$ and $C_f \leq C'$ with $C_f = \wr q_1^f, \ldots, q_n^f \int$.

The idea is to synchronize (at least) n processes into simulating the n automata. To this end, we need an additional (leader) process that will broadcast a message go, which will be received by the n processes, leading each of them to reach a different automaton initial state. Then, the leader process will broadcast a word letter by letter. Since the automata are all complete, these broadcast will be received by all the processes that simulate the automata, mimicking an execution. If the word belongs to all the automata languages, then each process simulating the automata ends the simulation on the unique final state of the

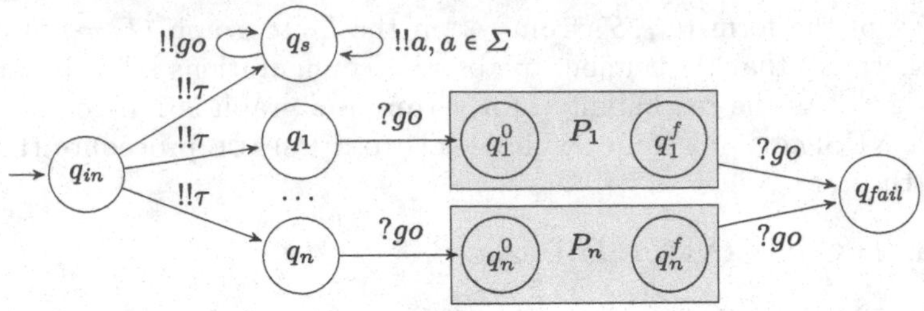

Fig. 4. Protocol P for PSPACE-hardness of CONFCOVER.

automaton. Note that if the leader process broadcasts the message *go* a second time, then all the processes simulating the automata stop their simulation and reach the state q_{fail}.

6 Conclusion

We have proved that when extending the model presented in [10] with broadcasts, STATECOVER for Wait-Only protocols remains in P, and is even P-complete. We also explained how to retrieve a P lower bound for the model of [10] for both problems restricted to Wait-Only protocols. However CONFCOVER for Wait-Only protocols is now PSPACE-complete. In the future, we wish to study not only coverability problems but extend the analysis of this model to liveness properties. We also wish to expand this model with dynamic creations of messages and processes in order to take a step closer to the modelling of Java Threads programming, where Threads can dynamically create new objects in which they can synchronize with notify and notifyAll messages.

References

1. Aminof, B., Rubin, S., Zuleger, F.: On the expressive power of communication primitives in parameterised systems. In: Davis, M., Fehnker, A., McIver, A., Voronkov, A. (eds.) LPAR 2015. LNCS, vol. 9450, pp. 313–328. Springer, Heidelberg (2015). https://doi.org/10.1007/978-3-662-48899-7_22
2. Apt, K.R., Kozen, D.C.: Limits for automatic verification of finite-state concurrent systems. Inf. Process. Lett. **22**(6), 307–309 (1986)
3. Balasubramanian, A.R., Esparza, J., Raskin, M.A.: Finding cut-offs in leaderless rendez-vous protocols is easy. In: Kiefer, S., Tasson, C. (eds.) FOSSACS 2021. LNCS, vol. 12650, pp. 42–61. Springer, Cham (2021). https://doi.org/10.1007/978-3-030-71995-1_3
4. Delzanno, G., Raskin, J.F., Begin, L.V.: Towards the automated verification of multithreaded java programs. In: Katoen, J.P., Stevens, P. (eds.) TACAS'02. LNCS, vol. 2280, pp. 173–187. Springer, Heidelberg (2002). https://doi.org/10.1007/3-540-46002-0_13

5. Delzanno, G., Sangnier, A., Traverso, R., Zavattaro, G.: On the complexity of parameterized reachability in reconfigurable broadcast networks. In: FSTTCS 2012. LIPIcs, vol. 18, pp. 289–300. Schloss Dagstuhl - Leibniz-Zentrum für Informatik (2012)
6. Durand-Gasselin, A., Esparza, J., Ganty, P., Majumdar, R.: Model checking parameterized asynchronous shared-memory systems. Formal Methods Syst. Des. **50**(2–3), 140–167 (2017)
7. Emerson, E.A., Kahlon, V.: Model checking guarded protocols. In: (LICS 2003, pp. 361–370. IEEE (2003)
8. Esparza, J., Finkel, A., Mayr, R.: On the verification of broadcast protocols. In: LICS 1999, pp. 352–359. IEEE Computer Soc. Press (1999)
9. German, S.M., Sistla, A.P.: Reasoning about systems with many processes. J. ACM **39**(3), 675–735 (1992)
10. Guillou, L., Sangnier, A., Sznajder, N.: Safety analysis of parameterised networks with non-blocking rendez-vous. In: CONCUR'23. LIPIcs, vol. 279, pp. 7:1–7:17. loss Dagstuhl - Leibniz-Zentrum für Informatik (2023)
11. Guillou, L., Sangnier, A., Sznajder, N.: Safety verification of wait-only non-blocking broadcast protocols (2024). https://doi.org/10.48550/arXiv.2403.18591
12. Horn, F., Sangnier, A.: Deciding the existence of cut-off in parameterized rendez-vous networks. In: CONCUR'20. LIPIcs, vol. 171, pp. 46:1–46:16. Schloss Dagstuhl - Leibniz-Zentrum für Informatik (2020)
13. Kozen, D.: Lower bounds for natural proof systems. In: FOCS 1977, pp. 254–266. IEEE Computer Society (1977)
14. Ladner, R.E.: The circuit value problem is log space complete for P. SIGACT News **7**(1), 18–20 (1975)
15. Schmitz, S., Schnoebelen, P.: The power of well-structured systems. In: D'Argenio, P.R., Melgratti, H. (eds.) CONCUR 2013. LNCS, vol. 8052, pp. 5–24. Springer, Heidelberg (2013). https://doi.org/10.1007/978-3-642-40184-8_2

Modular State Spaces - A New Perspective

Julian Gaede, Sophie Wallner, and Karsten Wolf(✉)

University of Rostock, Schwaansche Str. 2, 18055 Rostock, Germany
{julian.gaede,sophie.wallner,karsten.wolf}@uni-rostock.de

Abstract. A *modular Petri net* is built from individual Petri nets, the instances, which have disjoint sets of internal transitions and interface transitions. Whereas internal transitions represent the internal behavior of an instance, interface transitions are used to synchronize behavior between instances. For a modular Petri net, we can use the *modular state space* as an implicit representation of its reachability graph. This concept has also been examined in [3] and [6]. The modular state space is the entirety of *local reachability graphs* that present the internal behavior of the instances and a *synchronization graph* that keeps record of the synchronized behavior. In this paper we present a data structure and a construction algorithm for the modular state space. Our conceptualization is a generalization of the work of [3], and we have emphasized the differences in our approach. Furthermore, we describe how reachability-set-based properties can be verified in the modular state space. For the evaluation of properties, we aggregate information from the local reachability graphs and then combine the aggregated information when inspecting the synchronization graph. This way, we can evaluate deadlock freedom more efficiently and can for the first time propose a method to evaluate the reachability of state predicates. In addition, we emphasize the capability of modular state spaces to cope with modular Petri nets where some modules are structurally equal. In fact, when a module is used in more than one instance, its local reachability graph needs to be represented only once. This leads to a state space reduction that is not fully covered by the exploitation of symmetries.

Keywords: Petri nets · Modular · Verification and Model Checking using Nets · Regular Paper

1 Introduction

Large Petri net models are typically constructed as a composition of *modules*, place-disjoint subnets composed by transition fusion. For the composed system, we can build the *modular state space* [3,6], an implicit representation of the reachability graph. The exploration of the local state spaces of the modules (*local reachability graphs*) is coordinated by a global synchronization structure (*synchronization graph*) such that exactly the reachable behavior of the Petri

L. M. Kristensen and J. M. van der Werf (Eds.): PETRI NETS 2024, LNCS 14628, pp. 312–332, 2024.
https://doi.org/10.1007/978-3-031-61433-0_15

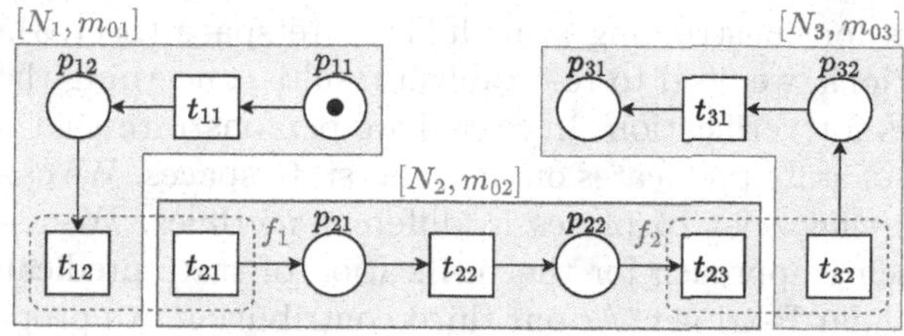

Fig. 1. Modular Version of a Channel Model.

net is depicted. The modular state space is always finite if the composed Petri net is bounded. For a first impression, consider the following example for which the notation is presented hereafter.

Example 1. Figure 1 describes a modular version of a channel model. On the left, $[N_1, m_{01}]$ describes a sender that executes some local activity t_{11} before passing the token into the channel, $[N_2, m_{02}]$. As transition t_{12} and t_{21} are fused, their firing is synchronized and the token is only produced on p_{21}, if it was present on p_{12} before. On the right, the receiver $[N_3, m_{03}]$ can execute an internal activity t_{31} after the token arrived by firing fused transitions t_{23} and t_{32}.

In verification, we can benefit from the modular structure of a Petri net - an idea that has been intensely studied as *compositional verification* [1,5,7,12]. Here, the local state spaces of the modules have to be given and finite. Then, compositional minimization tries to condense the local state spaces individually before aggregating them into some explicit or implicit representation of the global state space of the composed system. It has been observed [4] that the local state spaces may suffer from their own state explosion whereas sometimes only a tiny fraction is realizable in the composed system. However, for Petri nets, we may observe an extreme case of this phenomenon: Modules with an infinite local state space can be composed into a Petri net with a finite number of states. Consider $[N_2, m_{02}]$ of the just introduced Example 1. As transition t_{21} has no preplace in $[N_2, m_{02}]$, the module by oneself has an infinitely many states. The limitation to finite state modules is a severe burden if we try to automatically separate a Petri net into modules to make compositional verification applicable.

In this paper, we reconsider modular state spaces and make the following contributions. First, we introduce a fresh conceptualization of the modular state space in differentiation to existing work (Sect. 3). The synchronization graph refers to *segments* of the local reachability graphs that abstract local markings that are only connected with internal transitions. In [3], a segment is constructed as the forward closure only regarding internal transitions from a *generator*; a set of local markings. They claim the generator to be strongly connected via internal transitions. We drop the requirement that a generator needs to be strongly connected. This way, segments become larger, and we obtain a smaller synchronization graph whereas our local reachability graphs are equally large as in [3]. As verification algorithms mainly refer to the synchronization graph, a smaller synchronization graph increases efficiency. Additionally, we propose an interleaved algorithm for constructing a modular state space (Sect. 3). Second, due to our modifications, we need to re-establish results concerning the use of modular state spaces for verification. In Sect. 4 we demonstrate that we can verify the reachability of state predicates on modular state spaces. We can handle even state predicates that refer to places in different modules. We also present an efficiency-enhancing approach for the verification of state predicates and deadlocks in the modular Petri net. As our third contribution, we propose to exploit the fact that a composition may involve several instances of structurally identical modules. We propose to compute only one local state space in this case, yielding additional reduction. We outline our approach to replicated modules in Sect. 5.

2 Preliminaries

We build the modular state space for a Petri net system that contains multiple Petri net systems as components. We define a *module* as a template for those components.

Definition 1 (Module). *A* module *is a place/transition Petri net* $N_i = [P_i, T_i, F_i, W_i]$ *for some* $i \in \mathbb{N}$ *where the set of transitions* T_i *is partitioned into the subsets* $T_{i|internal}$ *of internal transitions and* $T_{i|interface}$ *of interface transitions.*

As all Petri nets, modules are only structures without behavior. Based on this we introduce *instances* as Petri net systems with behavior.

Definition 2 (Instance). *An* instance *is a Petri net system* $[N_j, m_{0j}]$ *where* N_j *is a module and* m_{0j} *is the initial marking of* N_j *for some* $j \in \mathbb{N}$. *A marking in general is a mapping* $m : P_j \to \mathbb{N}$.

Instances form the components of our composition. Whenever we refer to constituents of an instance we refer to the constituents of the included module. A marking is denoted as a multiset of marked places in this work. Interface transitions of the instances describe actions that may be synchronized with those of other instances. We define *fusion vectors* that formalize those synchronizations.

Definition 3 (Fusion Vector, Support). *Let* $\{[N_1, m_{01}], \ldots, [N_\ell, m_{0\ell}]\}$ *be a set of instances. A* fusion vector $f \in (T_{1|interface} \cup \{\bot\}) \times \ldots \times (T_{\ell|interface} \cup \{\bot\})$ *is a vector of interface transitions of the instances or* $\bot$. *For* $j \in \{1, \ldots, \ell\}$, *instance* $[N_j, m_{0j}]$ *participates in fusion vector* f *with interface transition* t, *if* $f[j] = t$ *for* $t \in T_{j|interface}$. *If* $f[j] = \bot$, $[N_j, m_{0j}]$ *does not participate in the fusion. For fusion vector* f, *the* support $supp(f) = \{t \mid f[j] = t\}$ *of* f *is the nonempty set of contained interface transitions.*

This definition ensures that a fusion vector contains at most one transition per instance. Next, we introduce the *modular structure* as a blueprint for our composition.

Definition 4 (Modular Structure). *A modular structure is a tuple* $\mathcal{M} = [\mathcal{I}, \mathcal{F}]$, *where* $\mathcal{I} = \{[N_1, m_{01}], \ldots, [N_\ell, m_{0\ell}]\}$ *is a set of instances with pairwise disjoint modules and* $\mathcal{F} \subseteq (T_{1|interface} \cup \{\perp\}) \times \ldots \times (T_{\ell|interface} \cup \{\perp\})$ *is a set of fusion vectors.*

With reference to Example 1, the modular structure $\mathcal{M} = [\mathcal{I}, \mathcal{F}]$ of the depicted Petri net has the set of instances $\mathcal{I} = \{[N_1, m_{01}], [N_2, m_{02}], [N_3, m_{03}]\}$ and the set of fusion vectors $\mathcal{F} = \{f_1 = [t_{12}, t_{21}], f_2 = [t_{23}, t_{32}]\}$. A modular structure provides all necessary information we need to build a *modular Petri net* that is basically a Petri net system composed of instances guided through fusion vectors. To implement the synchronization by fusion vectors properly, we introduce a fresh *fusion transition* for each fusion vector that inherits the environments of its support transitions.

Definition 5 (Modular Petri Net, Fusion Transition). *Let* $\mathcal{M} = [\mathcal{I}, \mathcal{F}]$ *be a modular structure.*

From $\mathcal{M}$*, we can derive a Petri net system* $N = [P, T, F, W, m_0]$*, where*

- $P = \bigcup_{j \in \{1,\ldots,\ell\}} P_j$,
- $T = \bigcup_{j \in \{1,\ldots,\ell\}} T_{j|internal} \cup \{t_f \mid f \in \mathcal{F}\}$, *where* t_f *is the* fusion transition *for* $f \in \mathcal{F}$,
- $F = \bigcup_{j \in \{1,\ldots,\ell\}} (F_j \cap (P_j \times T_{j|internal} \cup T_{j|internal} \times P_j)) \cup \{(p, t_f) \mid f \in \mathcal{F}, (p, f[j]) \in F_j\} \cup \{(t_f, p) \mid f \in \mathcal{F}, (f[j], p) \in F_j\}$
- $W(t,p) = \begin{cases} W_j(t,p) \text{ for } (t,p) \in F_j \text{ and } t \in T_{j|internal} \\ W_j(t^*,p) \text{ for } (t^*,p) \in F_j \text{ and } t^* \in supp(f), \\ \qquad f \in \mathcal{F}, t = t_f, j \in \{1,\ldots,\ell\} \end{cases}$
- $W(p,t) = \begin{cases} W_j(p,t) \text{ for } (p,t) \in F_j \text{ and } t \in T_{j|internal} \\ W_j(p,t^*) \text{ for } (p,t^*) \in F_j \text{ and } t^* \in supp(f), \\ \qquad f \in \mathcal{F}, t = t_f, j \in \{1,\ldots,\ell\} \end{cases}$
- $m_0 = \bigcup_{j \in \{1,\ldots,\ell\}} m_{0j}$

We call N *the* modular Petri net *for* $\mathcal{M}$.

Note that interface transitions that do not occur in the support of any fusion vector are no constituent of N. However, if we insist an interface transition $t \in T_{j|interface}$ to be a constituent of N anyway, we add fusion vector f with $supp(f) = \{t\}$ to $\mathcal{F}$. An interface transition can be part of more than one fusion set. This results in multiple fusion transitions that have the same influence locally, but may cause different effects in other instances. The initial marking of N is well-defined since the domains of all initial markings m_{0j} are pairwise disjoint. The weight function is also well-defined since a fusion vector permits at most one transition t^* per instance. Consider again Example 1. The interface transitions t_{12} and t_{21} are fused to fusion transition t_{f1} that inherits pre- and postplaces of t_{21} and t_{21}. The procedure is similar for t_{23} and t_{32}. The set of places and the internal transitions of an instance stay unaffected. The behavior of a Petri net system $N = [P, T, F, W, m_0]$, whether instance or modular Petri net, can be described as follows.

Definition 6 (Activation of a Transition). *Let m be a marking and $t \in T$ be a transition of N. Marking m* activates *transition t (denoted as $m \xrightarrow{t}$) iff, $\forall p \in P : W(p,t) \leq m(p)$. If m does not activate t we denote this with $m \not\xrightarrow{t}$.*

Note that a fusion transition is activated if and only if all support transitions are activated themselves.

Definition 7 (Transition Rule). *For $t \in T$ of N let Δt as $\Delta t(p) = W(t,p) - W(p,t)$, for all $p \in P$ be a $|P|$-indexed vector. Transition t* fires *in marking m and leads to marking m' ($m \xrightarrow{t} m'$) if $m \xrightarrow{t}$ and $m' = m + \Delta t$.*

We can extend the transition rule to a sequence of transitions $\omega \in T^*$, i.e. $m \xrightarrow{\epsilon} m$ and $m \xrightarrow{\omega t} m''$ if $m \xrightarrow{\omega} m' \xrightarrow{t} m''$. A marking m' is *reachable* from marking m, if we can find a sequence $\omega \in T^*$ such that $m \xrightarrow{\omega} m'$, denoted as $m \xrightarrow{*} m'$. Two markings m, m' are *strongly connected*, if and only if $m \xrightarrow{*} m'$ and $m' \xrightarrow{*} m$. In short, we write $m \leftrightarrow m'$. For Petri net system N and a set M of markings, we can calculate the *reachability set* $RS(M)$ of markings that are reachable from any marking in M. We define $RS(M) = \{m' \mid m \xrightarrow{*} m', m \in M\}$.

Definition 8 (Reachability Graph). *The* reachability graph *of N is the directed, labeled graph $R = [V^R, E^R]$, where $V = RS(\{m_0\})$ and $(m,t,m') \in E^R$ iff $m \xrightarrow{t} m'$ for $t \in T$.*

Based on the strong connectedness of markings, the reachability graph decays into *strongly connected components (SCCs)*. We call an SCC *terminal*, if it has no outgoing edges.

3 Modular State Space

During model checking on modular Petri net N the well-known state explosion problem also may occur. Our goal is to introduce the *modular state space* as an implicit representation of the state space of N that is based on the component structure and permits verification algorithms to take advantage of that. An intuitive way of building the modular state space is to generate the state spaces of the instances independently and then remove the actually non-reachable behavior due to composition. As instances become a component of a modular Petri net, their interface transition behavior can be narrowed down by fusion with other interface transitions. This is basically the proceeding compositional verification [1,5,7,12]. However, potentially unnecessary calculation are at the expense of runtime and even worse: Components may have an infinite number of states. With compositional model checking methods, we chance to get lost in unbounded behavior that cannot occur in the context of the composition. We present an approach for generating a modular state space avoiding this.

3.1 Conceptualization of the Modular State Space

In the following, let $R = [V^R, E^R]$ be the reachability graph of the modular Petri net $N = [P, T, F, W, m_0]$ based on $\mathcal{M} = [\mathcal{I}, \mathcal{F}]$ and $R_j = [V_j^R, E_j^R]$ be the reachability graph of instance $[N_j, m_{0j}]$ for $j \in \{1, \ldots, \ell\}$ according to Definition 8, where $E_j^R = E_{j|internal}^R \cup E_{j|interface}^R$ such that an edge is in $E_{j|internal}^R$ resp. in $E_{j|interface}^R$ if the labelling transition is internal resp. interface.

A modular state space contains two main concepts: the *local reachability graph* and the *synchronization graph.* A local reachability graph represents the behavior of an instance in the context of a modular structure. It contains projections of the reachable markings of N to the places of the according instance.

Definition 9 (Projection). *Let $m \in RS(\{m_0\})$. For instance $[N_j, m_{0j}]$ for $j \in \{1, \ldots, \ell\}$ of N, let $\pi_j(m) = m \cap (P_j \times \mathbb{N})$ be the* projection *of marking m to instance $[N_j, m_{0j}]$.*

We call those projections *local markings* of an instance. We build one local reachability graph per instance.

Definition 10 (Local Reachability Graph). *Let $[N_j, m_{0j}]$ be an instance for $j \in \{1, \ldots, \ell\}$. The* local reachability graph *is defined as $L_j = [V_j^L, E_j^L]$ where*

- $V_j^L = \{\pi_j(m) \mid m \in V\}$ *and*
- $E_j^L = E_{j|internal}^L \cup E_{j|interface}^L$ *with*
  - $(\pi_j(m), t, \pi_j(m')) \in E_{j|internal}^L$ *iff* $(m, t, m') \in E^R, t \in T_{j|internal}$
  - $(\pi_j(m), t, \pi_j(m')) \in E_{j|interface}^L$ *iff* $(m, t_f, m') \in E^R, f[j] = t, f \in \mathcal{F}$

The local reachability graph decays into *locally strongly connected components* (LSCCs), equivalence classes of local markings that are reachable from each other via internal transitions. The reachability graph of an instance is an overapproximation of its local reachability graph.

Lemma 1 (Relation between R_j and L_j). *Let $R_j = [V_j^R, E_j^R]$ be the reachability graph and $L_j = [V_j^L, E_j^L]$ be the local reachability graph of instance $[N_j, m_{0j}]$ for $j \in \{1, \ldots, \ell\}$. The local reachability graph is a subgraph of the reachability graph, i.e. $V_j^L \subseteq V_j^R$, $E_{j|internal}^L = E_{j|internal}^R$ and $E_{j|interface}^L \subseteq E_{j|interface}^R$.*

The proof follows from the construction. For firing a fusion transition the information about the activation of its support transitions must be shared between the instances. As local markings that are connected with internal transitions only have the same set of activatable interface transitions, we abstract them into *segments*.

Definition 11 (Segment). *In $L_j = [V_j^L, E_j^L]$ of $[N_j, m_{0j}]$ for $j \in \{1, \ldots, \ell\}$, a* segment *is a set of local markings $O \subseteq V_j^L$, which is forwardly closed in the following way: If $m \in O$ and $(m, t, m') \in E_{j|intern}^L$, then $m' \in O$.*

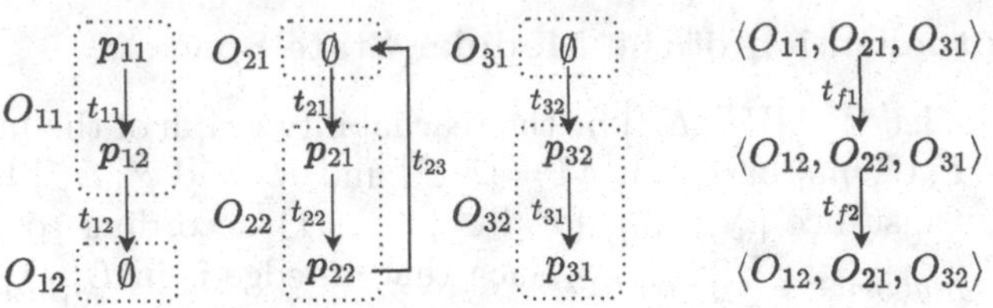

Fig. 2. State Space L_1, L_2, L_3 and S for a Channel Model (cf. Fig. 1).

For a set of local markings $M \subseteq V_j^L$, let $\overset{\bullet}{M}$ be the smallest segment that contains M, i.e. the *closure* of M regarding internal transitions. We call set M a *generator* of $\overset{\bullet}{M}$. Markings in a segment do not necessarily need to be connected or strongly connected, as the elements of the generator are not required to be connected. We define the successor segment for a fixed segment and a fixed interface transition.

Definition 12 (Successor Segment). *Let O be a segment of L_j of $[N_j, m_{0j}]$ for $j \in \{1, \ldots, \ell\}$ and $t \in T_{j|interface}$ be an interface transition. The* successor segment *is defined as $O^{+t} = \overset{\bullet}{M}$, $M = \{m' \mid m \in O, (m, t, m') \in E^L_{j|interface}\}$.*

As we only permit one interface transition per instance in a fusion vector, the successor segment is unambiguous. We may abuse this notation for successor segments of fusion transitions, i.e. $O_j^{+t_f} = O^{+t}$ for $t_f \in T$ where $f \in \mathcal{F}$ and $f[j] = t$. For $f[j] = \perp$, $O^{+t_f} = O$. After abstracting local markings to segments, we lift the activation of interface transitions from a local marking to a tuple of segments.

Definition 13 (Global Activation). *Let O be a segment of L_j of $[N_j, m_{0j}]$ with $j \in \{1, \ldots, \ell\}$. An interface transition $t \in T_{j|interface}$ is activated in O (denoted by $O \xrightarrow{t}$), if there is $m \in O$ such that $m \xrightarrow{t}$. Let t_f the corresponding fusion transition for $f \in \mathcal{F}$. Fusion transition t_f is* globally activated *in $<O_1, \ldots, O_\ell>$, if for all instances $[N_j, m_{0j}]$ with $f[j] = t$, it holds that $O_j \xrightarrow{t}$ for $j \in \{1, \ldots, \ell\}$.*

A vertex in the synchronization graph is an ℓ-tuple of segments from the local reachability graphs $L_1, \ldots, L_\ell$. Edges are drawn in case of globally activated fusion transitions.

Definition 14 (Synchronization Graph). *The* synchronization graph *$S = [V^S, E^S]$ is inductively defined as follows:*
Base: $<\{\overset{\bullet}{m}_{01}\}, \ldots, \{\overset{\bullet}{m}_{0\ell}\}> \in V^S$
Step: if $v = <O_1, \ldots, O_\ell> \in V^S$ and t_f is globally activated in v, then $v' = <O_1^{+t_f}, \ldots, O_\ell^{+t_f}> \in V^S$ and $(v, t_f, v') \in E^S$.

Algorithm 1. Generating the Segment for a Given Set of Markings

procedure BUILDCLOSURE(j: instance index, M: set of markings)
  Unprocessed $\leftarrow M, O = \emptyset$
  while Unprocessed $\neq \emptyset$ **do**
    $m \leftarrow$ choose from Unprocessed
    Unprocessed $\leftarrow$ Unprocessed $\setminus \{m\}$
    $O \leftarrow O \cup \{m\}$
    $V_j^L \leftarrow V_j^L \cup \{m\}$ ▷ all explored markings are added to L_j
    for all $t \in T_{j|internal} : m \xrightarrow{t}$ **do** ▷ explore internal transitions
      $m' \leftarrow m + \Delta t$
      $E_j^L \leftarrow E_j^L \cup \{(m, t, m')\}$
      if $m' \notin O$ **then**
        Unprocessed $\leftarrow$ Unprocessed $\cup \{m'\}$
      end if
    end for
  end while
  return O
end procedure

As an example, consider the modular state space in Fig. 2 for the modular Petri net depicted in Example 1. Note, that if we would calculate the local reachability graphs separately, L_2 is infinite due to transition t_{21} without preplace in $[N_2, m_{02}]$. To avoid this, we now provide a procedure to calculate modular state space interleaved such that the above-mentioned definitions hold. The main procedure *computeModularStateSpace* (Algorithm 3) calls two subprocedures, *buildClosure* (Algorithm 1) and *processInterfaceTransition* (Algorithm 2) that are used to build the local reachability graphs for all instances.

Algorithm 2. Firing an Interface Transition

procedure PROCESSINTERFACETRANSITION(j: instance index, O: segment, t: interface transition)
  $M \leftarrow \emptyset$ ▷ M is the generator of O^{+t}
  for all $m \in O$ **do**
    if $m \xrightarrow{t}$ **then**
      $m' \leftarrow m + \Delta t$
      $M \leftarrow M \cup \{m'\}$ ▷ adding m' to M
      $V_j^L \leftarrow V_j^L \cup \{m'\}, E_j^L \leftarrow E_j^L \cup \{(m, t, m')\}$
    end if
  end for
  return M
end procedure

Algorithm 3. Computing the Modular State Space

```
procedure COMPUTEMODULARSTATESPACE(M = [I, F]: modular structure)
    V^S ← ∅, E^S ← ∅                                   ▷ initialize synchronization graph
    for all j ∈ {1, ..., ℓ} do
        V_j^L ← ∅, E_j^L ← ∅                          ▷ initialize local reachability graphs
    end for
    for all j ∈ {1, ..., ℓ} do
        O_j ← buildClosure(j, {m_{0j}})                ▷ compute initial segments
    end for
    Unprocessed ← {<O_1, ..., O_ℓ>}
    V^S ← {<O_1, ..., O_ℓ>}                            ▷ initial vertex of synchronization graph
    while Unprocessed ≠ ∅ do
        <O_1, ..., O_ℓ> ← choose from Unprocessed
        Unprocessed ← Unprocessed \ <O_1, ..., O_ℓ>
        for all f ∈ F : t_f is globally activated in <O_1, ..., O_ℓ> do
            for all j ∈ {1, ..., ℓ} do                 ▷ compute successor segments
                if f[j] = t then                       ▷ if instance participates in f
                    newGenerator ← processInterfaceTransition(j, O_j, t)
                    O_j^{+t_f} ← buildClosure(j, newGenerator)
                else                                   ▷ instance does not participate in f
                    O_j^{+t_f} ← O_j
                end if
            end for
            E^S ← E^S ∪ {(<O_1, ..., O_ℓ>, t_f, <O_1^{+t_f}, ..., O_ℓ^{+t_f}>)}
            if < O_1^{+t_f}, ..., O_ℓ^{+t_f} > ∉ V^S then
                V^S ← V^S ∪ {< O_1^{+t_f}, ..., O_ℓ^{+t_f} >}
                Unprocessed ← Unprocessed ∪{< O_1^{+t_f}, ..., O_ℓ^{+t_f} >}
            end if
        end for
    end while
end procedure
```

3.2 Differentiation from Existing Approaches

The concept of modular state spaces is not new. In [3], the definition of the modular state space seems similar to ours, however there are dissimilarities that we want to state in this subsection. In [3] the set of local markings that generates a segment is locally strongly connected. This property leads to the fact that less local markings can be used as one generator. As a result, [3] might get more and smaller generators and consequently more and smaller segments as we might generate with our more liberal approach. The number of segments has a direct influence on the size of the synchronization. By dropping the requirement of locally strongly connectedness to our generators we usually generate fewer, but never more segments. This results in a smaller synchronization graph than [3] while keeping our local reachability graphs equally large. The following example demonstrates how non-strongly connected generators affect size and structure of the synchronization graph. For the comparison, we have adapted the notation

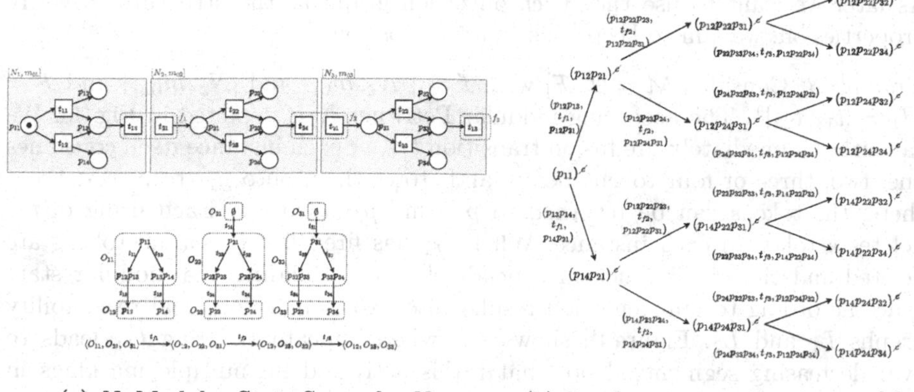

(a) N, Modular State Space for N. (b) Synchronization Graph as in [3].

Fig. 3. Influence of Strongly Connectedness on Synchronization Graph Size.

from [3] for their solution: Given a marking $m = m_1 \dots m_\ell$, $m^{\not{c}}$ denotes the cross product of LSCCs for every m_j with $j \in \{1, \dots, \ell\}$.

Example 2. Consider $\mathcal{M} = [\mathcal{I}, \mathcal{F}]$ with $\mathcal{I} = \{[N_1, m_{01}], [N_2, m_{02}]$ and $[N_3, m_{03}]\}$ and $\mathcal{F}$ for that the modular Petri net N is depicted in Fig. 3a. This token from p_{11} is passed from instance to instance, but each instance has two competing transitions for the incoming token. In both cases the token is passed further to the next instance, but the decision is recorded by producing an additional token. Those tokens have no effect on the further behavior of N; unless they ensure, that the terminal segments of each local reachability graph have non-locally-strongly connected generator elements. According to our definition, those generators form can one segment together. The modular state space of N is shown in Fig. 3a. If we build the modular state space according to [3], non-locally-strongly connected generator markings lead to different segments. In the worst case, the strongly-connectedness property leads to an exponential number of states in the synchronization graph. As you can see in Fig. 3b, firing three fusion transitions results into 13 states in the synchronization graph, whereas in our notation the synchronization graph only has 4 vertices.

According to our definition of the synchronization graph $S = [V^S, E^S]$, a synchronization graph edge $(v, t_f, v') \in E^S$ is labeled with the name of the according fusion transition t_f. This edge occurs once; regardless of which and how many local markings of the segments activate t_f. The definition of the synchronization graph according to [3] is in contrast to this. The edges in the synchronization graph are labeled with a triple (m, t_f, m'), where m and m' are source and target marking of firing t_f. If there would be another marking in the segment or successor segment for an instance that activated t_f, this would result in an additional edge. The following example shows how the synchronization graph edge notation of [3] can lead to a swiftly increasing number of edges.

As later we want to use the synchronization graph as the structure to verify properties on, we aim to keep it as small as possible.

Example 3. Consider $\mathcal{M} = [\mathcal{I}, \mathcal{F}]$ with $\mathcal{I} = \{[N_1, m_{01}]$ and $[N_2, m_{02}]\}$ and $\mathcal{F} = \{f_1 = [t_{13}, t_{23}]\}$. The according modular Petri net N is depicted in Fig. 4a. We can either immediately fire fusion transition t_{f1}, or each instance itself consumes one, two, three or four tokens before and stores them onto p_{11} resp. p_{21}. From there, the tokens can be returned to p_{11} and p_{21} anytime. Each firing of t_{f1} deletes a token in each instance. When t_{f1} has fired five times, all tokens are deleted and the net is stuck in a deadlock state. Building the modular state space according to our definition results into two isomorphic local reachability graphs L_1 and L_2. Figure 4b shows L_1, where repeatedly firing t_{f1} leads to ever decreasing segments. Note that t_{f1} is activated in multiple markings in a segment. Our synchronization graph has a linear structure, as you can see in Fig. 4c. The synchronization graph in style of [3] that is depicted in Fig. 4d. Their local reachability graphs will look the same as ours, whereas no interface edges are recorded. In initial synchronization graph vertex $(5p_{12}5p_{22})^{\not c}$ there exist 25 markings that activate t_{f1}, i.e. all combinations of local markings in the segments O_{11} and O_{21}. So, from $(5p_{12}5p_{22})^{\not c}$ we initiate 25 edges. This pattern continues through the whole synchronization graph. So, in total this synchronization graph contains of 55 edges compared to our synchronization graph with 5 edges.

3.3 Modular State Space vs. Actual State Space

Now we demonstrate that the reachability graph of a modular Petri net can be fully reconstructed from the modular state space. Before the main theorem for completeness we present two lemmata that help us to formulate the proof. The first one states that we can deduce the reachability of segment markings from the reachability of generators given.

Lemma 2 (Generator-Induced Reachability). *Let $L_j = [V_j^L, E_j^L]$ be the local reachability graph of instance $[N_j, m_{0j}]$ and $M_j \subseteq V_j^L$ be given for $j \in \{1, \ldots, \ell\}$. If all markings in $M_1 \times \ldots \times M_\ell$ are reachable, so are all markings in $\overset{\bullet}{M}_1 \times \ldots \times \overset{\bullet}{M}_\ell$.*

The proof of this lemma follows directly from the fact that building the closure $\overset{\bullet}{M}_j$ only uses internal transitions of $[N_j, m_{0j}]$ that are independent of each other, as their pre- and postplaces are disjoint. Sequences of internal transitions can arbitrarily be interleaved without affecting the resulting marking.

The second lemma explains how conclusions on the reachability of markings can be drawn in the synchronization graph. It benefits from the fact, that although the interface transitions of a fusion vectors are fused, their individual influence is restricted to their respective instances.

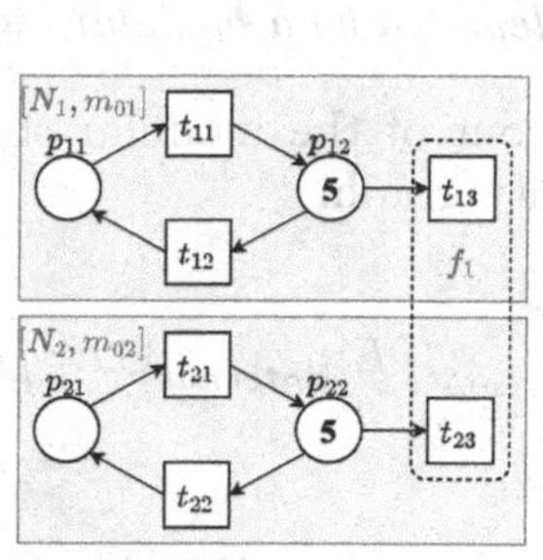

(a) Modular Petri Net N.

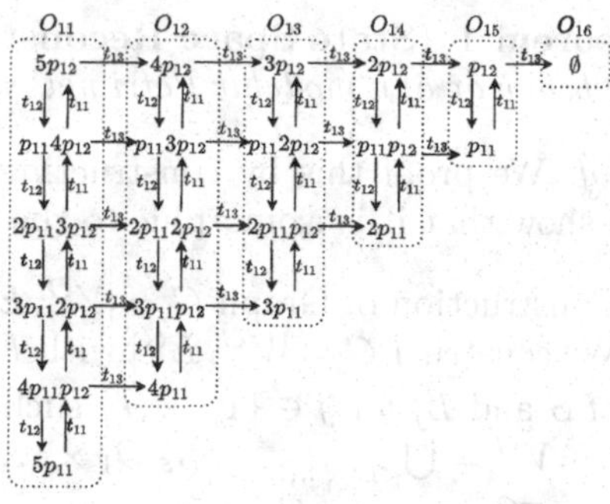

(b) Local Reachability Graph L_1.

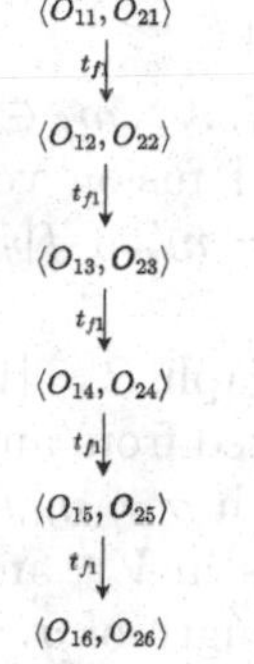

(c) Synchronization Graph S.

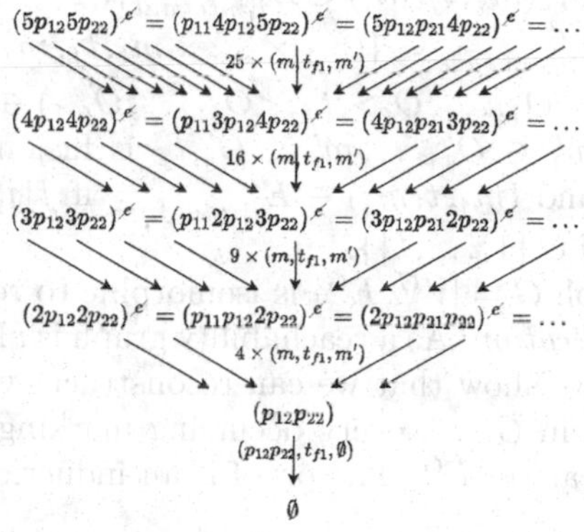

(d) Synchronization Graph S as in [3].

Fig. 4. Modular Petri Net N and Modular State Space.

Lemma 3 (Successor Segment Reachability). *Let $S = [V^S, E^S]$ be a synchronization graph. Let $v = <O_1, \ldots, O_\ell> \in V^S$ be a vertex in S. If all markings in $O_1 \times \ldots \times O_\ell$ are reachable and $(v, t_f, v') \in E^S$, then all markings in $v' = O_1' \times \ldots \times O_\ell'$ are reachable.*

Proof. Let $f \in \mathcal{F}$ be the fusion vector to fusion transition t_f. Firing t_f in $<O_1, \ldots, O_\ell> \in V^S$ requires global activation of t_f. This implies that for every instance $[N_j, m_{0j}]$ for all $j \in \{1, \ldots, \ell\}$ with $f[j] = t$, $m \xrightarrow{t} m'$ with $m \in O_j$ and $m' \in O_j'$ is possible (cf. Algorithm 2). For instance $[N_j, m_{0j}]$ we define M_j as a set of markings generated in this manner - the generator of O_j'. As the effect of processing t is restricted to $[N_j, m_{0j}]$, all markings in $M_1 \times \ldots \times M_\ell$ are reachable. The just introduced Lemma 2 on generator-induced reachability justifies that all markings in $O_1' \times \ldots \times O_\ell'$ are reachable as $O_j' = \overset{\bullet}{M_j}$.

With those two lemmata we now proceed with the following theorem:

Theorem 1 (State Space Reconstruction from Modular State Space). *The behavior of a modular Petri net N is fully depicted by a modular state space.*

Proof. We proof this by constructing a graph out of the modular state space and show that it is isomorphic to the reachability graph.

1. Construction of Graph $G = [V^G, E^G]$:
   We construct $G = [V^G, E^G]$ with $E^G = \bigcup_{j\in\{1,\ldots,\ell\}} E^G_{j|internal} \cup E^G_{interface}$ out of S and L_j for $j \in \{1,\ldots,\ell\}$ such that
   - $V^G = \bigcup_{<O_1,\ldots,O_\ell>\in V^S} O_1 \times \ldots \times O_\ell$;
   - $E^G_{j|internal} = \{(m_1 \times \ldots \times m_{j-1} \times m_j \times m_{j+1} \times \ldots \times m_\ell, t_j, m_1 \times \ldots \times m_{j-1} \times m'_j \times m_{j+1} \times \ldots \times m_\ell) \mid <O_1,\ldots,O_\ell> \in V^S, m_1 \in O_1,\ldots,m_\ell \in O_\ell, (m_j, t_j, m'_j) \in E^L_{j|internal}\}$;
   - $E^G_{interface} = \{(m_1 \times \ldots \times m_\ell, t_f, m'_1 \times \ldots \times m'_\ell) \mid (<O_1,\ldots,O_\ell>, t_f, <O'_1,\ldots,O'_\ell>) \in E^S, m_1 \in O_1,\ldots,m_\ell \in O_\ell, m'_1 \in O'_1,\ldots,m'_\ell \in O'_\ell, t_f$ is fusion transition of fusion vector $f \in \mathcal{F}$, and $(m_j, t, m'_j) \in E^L_{j|interface}$ if $f[j] = t$ or $m_j = m'_j$ if $f[j] = \perp$ for all $j \in \{1,\ldots,\ell\}\}$.
2. Graph $G = [V^G, E^G]$ is isomorphic to reachability graph $R = [V^R, E^R]$:
   Implication: As a reachability graph is always connected from an initial marking we show that we can reconstruct every finite path $\sigma = m_0t_1m_1\ldots t_nm_n$ of R in G, i.e. every occurring marking also appears in V^G and every edge appears in E^G. The proof is an induction over the length of σ.

   Base: m_0.

By construction every projection of the initial marking to an instance $\pi_j(m_0)$ appears in the initial segment of L_j for all $j \in \{1,\ldots,\ell\}$. These initial segments form the initial vertex the synchronization graph V^S and thus, m_0 appears in V^G.

   Step: For $0 \leq i \leq n$, assume m_{i-1} appears in V^G.

If $m_{i-1} \in V^G$, then there is a vertex $<O_1,\ldots,O_\ell> \in V^S$, such that $m_{i-1} \in O_1 \times \ldots \times O_\ell$. Now, let $m_{i-1}t_im_i$ be part of σ. Transition t_i might be

a) an internal transition to instance $[N_j, m_{0j}]$:
   As all of its preplaces are located in P_j, the influence of t_i is restricted to N_j. For $j' \neq j \in \{1,\ldots,\ell\}$ it holds that $\pi_{j'}(m_i) = \pi_{j'}(m_{i-1})$. Firing of t_i in m_{i-1} implies that $(\pi_j(m_{i-1}), t_i, \pi_j(m_i)) \in E^L_{j|internal}$ and $\pi_j(m_i) \in V^L_j$ (cf. Algorithm 1). By construction of G, $(m_{i-1}, t_i, m_i) \in E^G$ and $m_i \in V^G$.
b) a fusion transition for fusion vector $f \in \mathcal{F}$:
   Firing of t_i in m_{i-1} implies that all support interface transitions of f are activated in the projections $\pi_j(m_{i-1})$ for instance N_j. Therefore, we can execute Algorithm 2, the successor markings $\pi_j(m_i)$ form the new generator of the successor segment and are added to V^L_j. The edges $(\pi_j(m_{i-1}), t_i, \pi_j(m_i))$ are added to E^L_j as well. Thus, marking m_i appears in V^G and the edge (m_{i-1}, t_i, m_i) in E^G.

Replication: We need to show that the graph G does not contain unreachable behavior, i.e. every behavior in G can somehow be found in R. Starting with the initial vertex $<\{\overset{\bullet}{m}_{01}\},\ldots,\{\overset{\bullet}{m}_{0\ell}\}>$ of S, $\{m_{0j}\}$ with $j \in \{1,\ldots,\ell\}$ is the generator of the initial segment $\overset{\bullet}{m}_{0j}$ of every local reachability graph L_j. The only marking in $\{m_{01}\} \times \ldots \times \{m_{0\ell}\}$ is m_0 per definition. With Lemma 2 we can deduce that all markings in $\{\overset{\bullet}{m}_{01}\} \times \ldots \times \{\overset{\bullet}{m}_{0\ell}\}$ are reachable markings of the modular Petri net, thus occur in R. Algorithm 1 ensures that only the according internal edges are recorded in E_j^L for instance $[N_j, m_{0j}]$ and nothing else. The other vertices of S emerge from building the closure of generators produced by Algorithm 2 for each instance. In this way, only realizable interface edges are recorded in E_j^L for each instance $[N_j, m_{0j}]$. Additionally, fusion transition edges are recorded in E_S. Assume now to have a vertex $v \in V^S$ that only covers reachable markings and an edge $(v, t_f, v') \in E^S$. Then, Lemma 3 describes that only reachable markings evolve that all occur in R.

Concluding, the modular state is a complete representation of the state space of N.

4 Verification in the Modular State Space

In the following, we want to verify standard properties in the modular state space of modular Petri net $N = [P, T, F, W, m_0]$ based on a modular structure $\mathcal{M} = [\mathcal{I}, \mathcal{F}]$. The modular state space with S and L_j for $j \in \{1,\ldots,\ell\}$ is built as just described. We consider properties that can be verified regarding the set of reachable markings only. First, we check the existence of a reachable marking that fulfills some *state predicate*, a comparison of a weighted marking sum and some integer. The reachability of a specific marking can easily be expressed as the reachability of a state predicate that describes this marking.

Definition 15 (State Predicate, Support). *Let N be a Petri net system with the place set $P = \{p_1,\ldots,p_n\}$. We define $\varphi : a_{p1}m(p_1)+\ldots+a_{p_k}m(p_k) \leq a$ where $a_{p_1},\ldots,a_{p_k}, a \in \mathbb{Z}$ as an* atomic state predicate. *An atomic state predicate is true for a marking m, if the inequality is holds for m. For atomic state predicate φ the* support *supp(φ) includes all occurring places in φ. A* state predicate *is a conjunction of atomic state predicates $\varphi : \bigwedge_{i=1}^{n} \varphi_i$ and is true, if every atomic state predicate φ_i is true. For short, within a state predicate φ the atomic state predicate φ_i can be written as $s_i(m) \leq a_i$.*

This definition gives a canonical form for state predicates in which every kind of state predicate can be transformed. Verifying a state predicate φ in the modular state space depends on the instance affiliation of the occurring places in formal sums. Let $\varphi_i : s_i \leq a_i$ be an atomic state predicate with $i \in \{1,\ldots,n\}$.

I. In φ_i, all occurring places are from the same instance: Assume *supp*(φ_i) $\subseteq P_j$ for instance $[N_j, m_{0j}]$, $j \in \{1,\ldots,\ell\}$. The verification of φ_i can be accomplished on the local reachability graph L_j as it contains all relevant

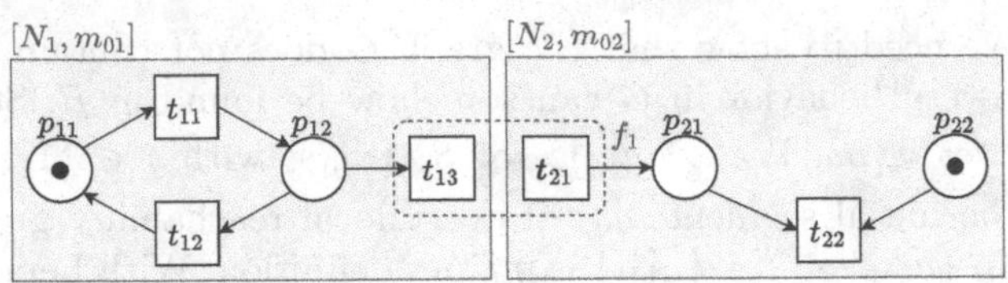

Fig. 5. Modular Petri net N.

places for φ_i. In state predicate $\varphi : \bigwedge_{i=1}^{n} \varphi_i$, multiple atomic state predicates may refer to the same instance. Therefore, we split φ such that $\varphi : \bigwedge_{j=1}^{\ell} \varphi^j$, where $\varphi^j : \bigwedge_{i=1, vis(\varphi_i) \subseteq P_j}^{n} \varphi_i$. While computing the modular state space, we check whether we can find a vertex of the synchronization graph that covers a marking that satisfies all atomic propositions. Consider vertex $v = <O_1, \ldots, O_\ell>$. For $j \in \{1, \ldots, \ell\}$, we mark segment O_j if it contains a marking m_j that satisfies φ^j. We also mark segment O_j if it is not relevant for any atomic state predicate. The state predicate is true, if and only if there exists a vertex $<O_1, \ldots, O_\ell> \in V^S$ where every segment O_j is marked.

II. In φ_i, all occurring places are not from the same instance: If φ_i is not restricted to one instance, we split the formal sum $s_i(m)$ according to the instance affiliation of the places: $s_i(m) = \sum_{j \in \{1,\ldots,\ell\}} s_{ij}(m_j)$ where $s_{ij}(m_j) = \sum_{p \in P_j} a_p m_j(p)$ is the subsum that regards only places in instance $[N_j, m_{0j}]$ and $m_j = \pi_j(m)$ for $j \in \{1, \ldots, \ell\}$. Then φ_i holds if and only if there is a marking m such that $\sum_{j \in \{1,\ldots,\ell\}} s_{ij}(m_j) \leq a_i$. While computing the modular state space, we calculate $s_{ij}(m_j)$ for every marking $m_j \in L_j$. Over all atomic state predicates $\varphi_1, \ldots, \varphi_n$ of φ, every marking $m_j \in V_j^L$ gives a vector $\vec{c}_{m_j} = (s_{1j}(m_j), \ldots, s_{nj}(m_j))$.

For the verification of the φ, we proceed as follows when traversing the synchronization graph: Consider vertex $<O_1, \ldots, O_\ell>$. For every segment $O_j \subseteq V_j^L$, for every marking $m_j \in O_j$, we collect the vectors $\vec{c}_{m_j}$ in $C_j = \{\vec{c}_{m_j} \mid m_j \in O_j\}$.

Subsequently, the set C_j can then be reduced by vectors that are larger than some other vector contained - if a marking of the modular Petri net containing $m_j \in V_j^L$ satisfies φ, this marking containing $m'_j \in V_j^L$ with $\vec{c}_{m'_j} \leq \vec{c}_{m_j}$ satisfies φ a fortiori. Vector $\vec{c}_{m'_j}$ contributes smaller (at least not bigger) values to every formal sum. Let $C_j^{min} = \{\vec{c}_{m_j} \in C_j \mid \forall \vec{c}_{m'_j} \in C_j : \vec{c}_{m'_j} \not< \vec{c}_{m_j}\}$ for segment O_j be the set of minimal vectors of sum values for every φ_i. Note that multiple vectors might produce the same value, and we cannot find the one smallest vector, as they are partially incomparable. If there exists some selection $\vec{c}_1 \in C_1^{min}, \ldots, \vec{c}_\ell \in C_\ell^{min}$ such that for all $i \in \{1, \ldots, n\}$ it holds that $\sum_{j \in \{1,\ldots,\ell\}} s_{ij} \leq a_i$, then the state predicate $\varphi : \bigwedge_{i=1}^{n} \varphi_i$ is true in this vertex $<O_1, \ldots, O_\ell>$. If we cannot find such a selection for any vertex of the synchronization graph, the state predicate is false. We will underline this with an example.

Example 4. Consider $\mathcal{M} = [\mathcal{I}, \mathcal{F}]$ with $\mathcal{I} = \{[N_1, m_{01}], [N_2, m_{02}]\}$ and $\mathcal{F} = \{f_1 = [t_{13}, t_{21}]\}$, where N is depicted in Fig. 5. We want to verify state predicate

$\varphi : p_{11} + p_{12} \leq 1 \wedge p_{12} + p_{21} \leq 0 \wedge p_{11} + p_{21} + p_{22} \leq 1$, containing three atomic state predicates. At first, we consider the initial vertex of the synchronization graph $<O_{11}, O_{21}>$ where $O_{11} = \overset{\bullet}{p_{11}} = \{p_{11}, p_{p12}\}$ and $O_{21} = \overset{\bullet}{p_{22}} = \{p_{22}\}$. For segment O_{11}, the set of minimal vectors is $C_1^{min} = \{(1,0,1),(1,1,0)\}$, as $(1,0,1)$ and $(1,1,0)$ are incomparable. Analogously, the $C_2^{min} = \{(0,0,1)\}$ for the only marking p_{22}. In $C_1^{min} \times C_2^{min}$, we cannot find a combination, i.e. marking of the modular Petri net, that fulfills all three atomic state predicates. So we proceed to $<O_{12}, O_{22}>$ and calculate $C_1^{min} = \{(0,0,0)\}$ and $C_2^{min} = \{(0,0,0)\}$ - the lowest vector of values the instance $[N_2, m_{02}]$ adds to every atomic state predicate. When we now sum the sums of the instances with each other, we get $(0,0,0)$ which fulfills all the state predicate.

A *deadlock* is a marking where no transition is activated. In a modular Petri net, we need to consider both, the local behavior of the instances and the synchronized behavior, to detect a deadlock. First, we define sets of activated transitions for a given local marking.

Definition 16 (Activated Transitions). *Let m_j be a local marking of instance $[N_j, m_{0j}]$ for $j \in \{1, \ldots, \ell\}$. The set $act_{j|internal}(m_j) \subseteq T_{j|internal}$ describes the internal transitions that are activated in m. Analogously, the set $act_{j|interface}(m_j) \subseteq T_{j|interface}$ describes the interface transitions that are activated in m_j.*

Theorem 2 (Deadlock). *In N, a deadlock is reachable if and only if the synchronization graph contains a vertex $<O_1, \ldots, O_\ell> \in V^S$, such that*

- *Segment O_j contains marking m_j with $act_{j|internal}(m_j) = \emptyset$ for all $j \in \{1, \ldots, \ell\}$ and*
- *$\bigcup_{j \in \{1,\ldots,\ell\}} act_{j|interface}(m_j)$ does not include supp(f) of any fusion vector $f \in \mathcal{F}$.*

Proof. Let m_d be a deadlock marking of N. As m_d is reachable, there exists $<O_1, \ldots, O_\ell> \in V^S$, such that $m_d \in O_1 \times \ldots \times O_\ell$. For every $j \in \{1, \ldots, \ell\}$ let $m_j = \pi_j(m_d) \in O_j$ in L_j. For all instances $[N_j, m_{0j}]$, no internal transition can be activated in m_j. Furthermore, $\bigcup_{j \in \{1,\ldots,\ell\}} act_{j|interface}(m_j)$ cannot include *supp*(*f*) of any fusion vector f, since otherwise t_f would be activated in m_d. Now assume that the two conditions are satisfied for a local marking $m_j \in O_j$ for all $j \in \{1, \ldots, \ell\}$ from vertex $<O_1, \ldots, O_\ell> \in V^S$. We can construct a deadlock marking such that $m_d = \cup_{j \in \{1,\ldots,\ell\}} m_j$ of N, where no internal transitions is activated, and every fusion vector contains a disabled interface transition in m_d.

Detecting deadlocks can easily be implemented in our concept of the modular state space. By exploring L_j, we can identify markings that do not activate internal transitions (*candidate markings*). Then, traversing the vertices of the synchronization graph, for a vertex $<O_1, \ldots, O_\ell>$, we check, whether each segment O_j contains at least one candidate. If we find such a vertex where each segment contains at least one candidate, we check all combinations of candidates, whether their activated interface transitions are the support of a fusion vector; firing a fusion transition would be the only way to escape the potential deadlock. For the latter check, we can again reduce the number of combinations that need to be checked. First, it is sufficient to traverse the set of activated interface transitions $act_{j|interface}(m_j)$ of any candidate $m_j \in O_j$. Second, if for two candidates of a segment $m_j, m'_j \in O_j$ for $j \in \{1, \ldots, \ell\}$ it holds that $act_{j|interface}(m_j) \subset act_{j|interface}(m'_j)$, we do not need to consider $act_{j|interface}(m'_j)$. The reason is that if any marking $m_1 \ldots m'_j \ldots m_\ell$ is a deadlock, so is $m_1 \ldots m_j \ldots m_\ell$, because it activates even less transitions. So, for every segment, we collect the smallest set of activated interface transitions $act_{j|interface}(m_j)$ of all candidates. Then, we join those sets and compare it with the given support of the fusion vectors of our modular Petri net. If the union does not cover the support of any fusion vector, the considered vertex is a deadlock vertex. We consider a transition as *dead*, if it is not activated in any reachable marking. For verification in the modular Petri net, we distinguish internal and fusion transitions. The deadness of a fusion transition can be analyzed based on the synchronization graph.

Theorem 3 (Dead Transitions). *A* fusion transition *is* dead *if and only if it does not appear in any edge of the synchronization graph. Let $t \in T_{j|internal}$ be an internal transition of $[N_j, m_{0j}]$ with $j \in \{1, \ldots, \ell\}$. Transition t is* dead, *if and only if for all $m \in V_j^L : m \overset{t}{\not\to}$.*

The proof follows from the construction. Knowing the maximum number of tokens that a place can hold, may minimize the storage space to store markings. For modular Petri nets, the main advantage for this property arises from the disjointness of the place sets of the instances, so the *bounds* of a place can be calculated based on the local reachability graph.

Theorem 4 (Boundedness). *Let $p \in P_j$ be a place of $[N_j, m_{0j}]$ for $j \in \{1, \ldots, \ell\}$. For p, the* upper bound *of tokens is $ub(p) = max\{m(p) \mid m \in V_j^L\}$ and the* lower bound *of tokens is $lb(p) = min\{m(p) \mid m \in V_j^L\}$.*

5 Handling Replications

Similar structures in a Petri net system cause similar behavior that does not need to be calculated multiple times. Exploiting this, the symmetry method [9] aims to find isomorphisms of the Petri net system and calculates their behavior only once. Therefore, net structures need to be identical. If structures are largely similar but differ in some nodes we cannot apply the symmetry method. Although the structures behave the same mostly, we need to calculate their state spaces individually - at the expense of efficiency. We want to introduce a more robust method to handle those similar structures with small deviations that avoids multi-calculations of the same states. Therefore, we consider similar structures as different instances of the same module. When building the modular state space, the local state space is then only explored once per module and shared between its instances. First, we need to adapt the Definition 2 of instances and assign an ID to an instance to distinguish between instances of the same module. Those IDs do not need to be consecutive for instances of the same module.

Definition 17 (R-Instance). *An* r-instance *is a Petri net system* $[N_i, j, m_{0j}]$ *of module* N_i*, where* N_i *is a module for* $i \in \mathbb{N}$*,* $j \in \mathbb{N}$ *is an instance identifier and* m_{0j} *is an initial marking of* N_i*.*

Two r-instances are *twins* if they uprise from the same module - they have the same module structure but may have different initial markings. Defining an r-modular structure and r-modular Petri net with twin r-instances is straightforward. For an r-modular structure, we do not require the modules to be disjoint as the unambiguous assignment of places and transitions to r-instances is realized through the instance identifiers. As announced above, we want to generate one local reachability graph for twin r-instances - The *local reachability graph of the module.* Before defining that, we need to make a small adjustment: Markings of twin r-instances are not defined over the same set of places, technically. To allow r-instances to still share markings we define a to-module-mapping that maps the marking of an r-instance place to a module place marking.

Definition 18 (To-Module-Mapping for R-Instances). *For r-instances* $\mathcal{I} = \{[N_{i_1}, 1, m_{01}], \ldots, [N_{i_\ell}, \ell, m_{0\ell}]\}$ *of an r-modular Petri net, we define a* to-module-mapping $\mu : ((P_{i_j} \times \{1, \ldots, \ell\}) \to \mathbb{N}) \to (P_{i_j} \to \mathbb{N})$ *for every* $j \in \{1, \ldots, \ell\}$*, such that* $\mu(m((p, j)) = m(p)$*.*

This to-module-mapping notation for markings is sometimes abused for the components of the r-instances following the same principle, i.e. $\mu((x, j)) = x$ for $x \in (T_{i_j} \cup P_{i_j})$. The local reachability graph of a module contains those μ-markings.

Definition 19 (Local Reachability Graph of a Module). *Let* N_i *be a module for* $i \in \mathbb{N}$ *that The* local reachability graph *of* N_i *is defined as* $L_i = [V_i^L, E_i^L]$*, where*

- $V_i^L = \{\mu(\pi_j(m)) \mid m \in V, [N_i, j, m_{0j}] \in \mathcal{I}\}$
- $E_i^L = E_{i|internal}^L \cup E_{i|interface}^L$ *with*
  - $(\mu(\pi_j(m)), \mu(t), \mu(\pi_j(m'))) \in E_{i|internal}^L$ *iff* $(m, t, m') \in E$ *for* $t \in \bigcup_{[N_i,j,m_{0j}] \in \mathcal{I}} T_{j|internal}$
  - $(\mu(\pi_j(m)), \mu(t), \mu(\pi_j(m'))) \in E_{i|interface}^L$ *iff* $(m, t_f, m') \in E$ *for* $f[j] = t, f \in \mathcal{F}, [N_i, j, m_{0j}] \in \mathcal{I}$*.*

By accumulating the local reachability graphs of twin r-instances, none of their behavior gets lost.

Corollary 1 (Relation between L_i and L_j). *Let* $L_i = [V_i^L, E_i^L]$ *be the local reachability graph of module* N_i *and* $L_j = [V_j^L, E_j^L]$ *be the local reachability graph of r-instance* $[N_i, j, m_{0j}]$ *for* $i \in \mathbb{N}$ *and* $j \in \{1, \ldots, \ell\}$*. The local reachability graph* L_j *is isomorphic to the induced subgraph of* L_i *by the set of nodes* $\{\mu(m_j) \mid m_j \in V_j^L\}$*. This subgraph is forwardly closed regarding internal transitions, i.e. for all* $t \in T_{i|internal}$*,* $(m_j, t, m_j') \in E_i^L$ *implies that* $(m_j, t, m_j') \in E_j^L$*.*

The proof for this lemma follows from the definitions and from the fact that internal transitions of distinct r-instances cannot prohibit each other from firing. Note, that the subgraph is only forwardly closed regarding internal transitions. Different r-instances of one module can take part in different fusion sets. So their interface behavior may differ, while when twin instances reach the same state, the upcoming internal behavior is equivalent. Building segments and successor segments for r-instances follows the same criteria as building them for instances described in Definition 11 and Definition 12. We also use the notation global activation Definition 13) here. A vertex in the synchronization graph remains an ℓ-tuple of segments, where ℓ is the number of r-instances. Finally, the synchronization graph $S = [V^S, E^S]$ is defined as in Definition 14. However, for r-instance $[N_i, j, m_{0j}]$ with $i \in \mathbb{N}$ and $j \in \{1, \ldots, \ell\}$ we can extract its segment structure from the local reachability graph L_i of its underlying module - Every segment of the local reachability graph L_j of $[N_i, j, m_{0j}]$ is as well a segment of L_i. Mind that not every segment of L_i necessarily is a segment of L_j as it contains all segments of all twin instances. The replicated modular state can be constructed as described in Sect. 3. Regarding the verification in Subsect. 4, we can transfer the results on reachability, state predicates, deadlock and dead fusion transitions directly into the replicated modular state space. The theorems of dead internal transitions and bounds need to be refined.

Theorem 5 (Dead Internal Transition in the Replicated Modular State Space). *Let* $[N_i, j, m_{0j}]$ *with* $i \in \mathbb{N}, j \in \{1, \ldots, \ell\}$ *be an r-instance of an r-modular structure. An internal transition* $t \in T_i \times \{j\}$ *is* dead to the r-instance *if for all* $m \in V_i^L : m \overset{t}{\not\to}$, *or if for all vertices* $<O_1, \ldots, O_\ell>$ *in the synchronization graph, it holds for all* $m \in O_j : m \overset{t}{\not\to}$.

The same argumentation and example fit for the upper and lower bounds of tokens. Bounds observed just in the local reachability graph of the module can be used as a first approach. For precise bounds, we need to consider the synchronization graph as well.

Definition 20 (Boundedness in the Replicated Modular State Space). *Let* $[N_i, j, m_{0j}]$ *with* $i \in \mathbb{N}, j \in \{1, \ldots, \ell\}$ *be an r-instance of a modular structure. Let* $p \in P_i \times \{j\}$ *be a place of* $[N_i, j, m_{0j}]$. *For p, the* upper bound *of tokens is* $ub(p) = max\{m(p) \mid m \in O_j, \forall <O_1, \ldots, O_\ell> \in V^S\}$ *and the* lower bound *of tokens is* $lb(p) = min\{m(p) \mid m \in O_j, \forall <O_1, \ldots, O_\ell> \in V^S\}$.

To demonstrate the advantages of replicated modularization consider the well known 10 dining philosophers. In our version, a philosopher takes the forks one by one but releases them synchronously. Nine of the philosophers start picking up their left fork first and their right fork afterwards. The tenth philosopher breaks the symmetry by taking the right fork before the left fork. For 10 dining philosophers, we build an r-modular structure $\mathcal{M} = [\mathcal{I}, \mathcal{F}]$ with $\mathcal{I} = \{[N_1, 1, m_{01}], [N_1, 2, m_{02}], [N_1, 3, m_{03}], [N_2, 1, m_{01}]\}$ of the modules depicted in Fig. 6 and $\mathcal{F} = \{f_1 = \{(tr.c, 1), (nt.t, 2)\}, f_2 = \{(r.c, 1), (nr.t, 2)\},$ $f_3 = \{(tr.c, 2), (nt.t, 3)\}, f_4 = \{(r.c, 2), (nr.t, 3)\}, f_5 = \{(tr.c, 3), (nt.s, 4)\}, f_6 =$

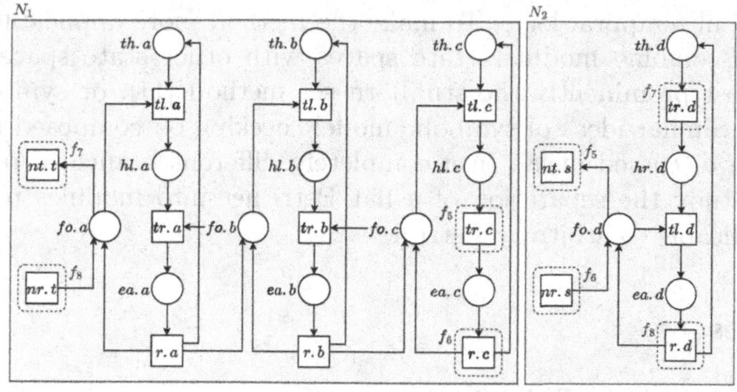

Fig. 6. R-Modular Petri Net Model for Dining Philosophers.

$\{(r.c, 3), (nr.s, 4)\}$, $f_7 = \{(tr.d, 4), (nt.t, 1)\}$, $f_8 = \{(r.d, 4), (nr.t, 1)\}\}$. The local reachability graph L_1 for N_1 consists of 24 markings, the local reachability graph L_2 for N_2 consists of 5 markings. The synchronization graph consists of 16 states. In total the modular state space has 45 vertices. In comparison, the full state space contains 5741 markings and can be reduced to 165 states if deadlock preserving stubborn sets [11] are used. The numbers correspond to the LoLA tool [10]. Symmetry reduction is not available since the single philosopher (modeled in N_2) breaks the symmetry.

6 Conclusion and Future Work

We proposed a new perspective and construction algorithm for the modular state space of a modular Petri net. The main difference to [3] is the nature of segments. Our segments are generated from a not necessarily locally strongly connected set of markings. This way, segments may become larger and the synchronization graph may get smaller. Nevertheless, the coarser structure still permits the verification of interesting properties.

We proposed a method for verifying the reachability of state predicates on a modular state space. For every segment of every local reachability graph, information is collected and then combined when the synchronization graph is inspected. This way, the verification approach corresponds to the structure of the modular state space. The local information can be computed independently (and even concurrently) for each module.

We showed that several instances of one and the same module may share a single local reachability graph. This yields a reduction that, in non-modular verification, would be addressed by the symmetry method. However, our approach to replication does not require the whole Petri net to be symmetric. Consequently, our method may very well be more robust than the symmetry method concerning non-symmetrical arrangement of identical instances to an overall system.

In future work, we need to re-establish results on the verification of more complex properties such as liveness, reversibility, or the verification of properties specified in temporal logic. To make the method more applicable, it is also necessary to combine modular state spaces with other state space reduction methods, most prominently the stubborn set method [11], or symbolic model checking [2]. Similar ideas of symbolic model checking on composed models has already been discussed in [8]. In a completely different branch of research, we propose to study the separation of a flat Petri net into modules to make the method applicable to arbitrary Petri nets.

References

1. Buchholz, P., Kemper, P.: Efficient computation and representation of large reachability sets for composed automata. Discret. Event Dyn. Syst. **12**(3), 265–286 (2002). https://doi.org/10.1023/A:1015669415634
2. Burch, J.R., Clarke, E.M., McMillan, K.L., Dill, D.L., Hwang, L.J.: Symbolic model checking: 1020 states and beyond. Inf. Comput. **98**(2), 142–170 (1992)
3. Christensen, S., Petrucci, L.: Modular analysis of Petri nets. Comput. J. **43**(3), 224–242 (2000). https://doi.org/10.1093/comjnl/43.3.224
4. Garavel, H., Lang, F., Mounier, L.: Compositional verification in action. In: Howar, F., Barnat, J. (eds.) FMICS 2018. LNCS, vol. 11119, pp. 189–210. Springer, Cham (2018). https://doi.org/10.1007/978-3-030-00244-2_13
5. Graf, S., Steffen, B.: Compositional minimization of finite state systems. In: Clarke, E.M., Kurshan, R.P. (eds.) CAV 1990. LNCS, vol. 531, pp. 186–196. Springer, Heidelberg (1991). https://doi.org/10.1007/BFb0023732
6. Latvala, T., Mäkelä, M.: LTL model checking for modular Petri nets. In: Cortadella, J., Reisig, W. (eds.) ICATPN 2004. LNCS, vol. 3099, pp. 298–311. Springer, Heidelberg (2004). https://doi.org/10.1007/978-3-540-27793-4_17
7. Le Cornec, Y.S.: Compositional analysis of modular Petri nets using hierarchical state space abstraction. In: Joint 5th International Workshop on Logics, Agents, and Mobility, LAM 2012, The 1st International Workshop on Petri Net-Based Security, WooPS 2012 and the 2nd International Workshop on Petri Nets Compositions, CompoNet 2012. CEUR Workshop Proceedings, Hamburg, Germany, vol. 853, pp. 119–133, June 2012
8. Miner, A.S., Ciardo, G.: Efficient reachability set generation and storage using decision diagrams. In: Donatelli, S., Kleijn, J. (eds.) ICATPN 1999. LNCS, vol. 1639, pp. 6–25. Springer, Heidelberg (1999). https://doi.org/10.1007/3-540-48745-X_2
9. Schmidt, K.: How to calculate symmetries of petri nets. Acta Inform. **36**(7), 545–590 (2000). https://doi.org/10.1007/S002360050002
10. Schmidt, K.: LoLA a low level analyser. In: Nielsen, M., Simpson, D. (eds.) ICATPN 2000. LNCS, vol. 1825, pp. 465–474. Springer, Heidelberg (2000). https://doi.org/10.1007/3-540-44988-4_27
11. Valmari, A.: Stubborn sets for reduced state space generation. In: Rozenberg, G. (ed.) ICATPN 1989. LNCS, vol. 483, pp. 491–515. Springer, Heidelberg (1991). https://doi.org/10.1007/3-540-53863-1_36
12. Valmari, A.: Compositional analysis with place-bordered subnets. In: Valette, R. (ed.) ICATPN 1994. LNCS, vol. 815, pp. 531–547. Springer, Heidelberg (1994). https://doi.org/10.1007/3-540-58152-9_29

Verifying Temporal Logic Properties in the Modular State Space

Lukas Zech and Karsten Wolf[(✉)]

University of Rostock, Schwaansche Str. 2, 18055 Rostock, Germany
{lukas.zech,karsten.wolf}@uni-rostock.de

Abstract. A *modular Petri net* is composed of multiple individual Petri nets, the modules, by fusing their *interface* transitions. *Internal* transitions are not related to other modules. Their behavior is recorded in *local reachability graphs* for each module. The behavior of interface transitions is recorded in a single *synchronization graph*, linking the local reachability graphs together to form the *modular state space*. Our notion of modular state spaces is similar to previous proposals [5,6,17] but drops a few assumptions for the sake of additional compression.

In this paper, we study the verification of temporal logic properties using the modular state space. For linear time properties, we re-establish a result from [10] for our revised concepts. In our proof, we use completely different arguments and generalize the result to non-regular linear time properties. Modular state spaces do not easily permit the verification of branching time properties and there is no existing result on this matter. We demonstrate, however, that CTL properties can very well be verified after applying a certain refinement to the modular state space.

Keywords: Petri nets · Modular · Model Checking · CTL · Computation Tree Logic · LTL · Linear Time · Verification and model checking using nets · Regular Paper

1 Introduction

A Modular Petri net is composed of multiple components, the modules. Modeling tools like CPN Tools [20] already support modular or hierarchical models. Translations from other models to Petri nets, e.g. [1,12] often compose small building blocks to larger units. For nets that do not have a user-defined modular structure, proposals exist to automatically separate them into modules [4]. In this paper, we assume transition fusion as the composition operation, similarly to [6]. We expect that instances of a module behave similarly, so we can have just one copy of the data structures related to a module, saving computation time. The modular structure of a Petri net introduces the modular state space [5,6,17]. The modular state space consists of local state spaces, one for each module, and a single global synchronization graph. By computing multiple smaller graphs instead of the potentially large reachability graph, the impact of the state explosion problem can be reduced. The local state spaces are basically the projection

L. M. Kristensen and J. M. van der Werf (Eds.): PETRI NETS 2024, LNCS 14628, pp. 333–354, 2024.
https://doi.org/10.1007/978-3-031-61433-0_16

of the reachability graph of the Petri net to a module. It differs from the reachability graph of the module in general. There exist bounded modular Petri nets that are composed of unbounded modules. By exploring the local state spaces and the synchronization graph together, the local state space is always finite as long as the actual reachability graph of the modular Petri net is also finite. This way of exploration is the main difference to compositional verification techniques [3,8,11,19], which explore local state spaces independently, assuming them to be finite, before composing them into a global structure.

In the local state space, *segments* can be identified as the internal behavior of a module between two consecutive firings of interface transitions. We generalize assumptions from [6], allowing for wider applicability of the method.

A vertex in the synchronization graph is a tuple of segments, one for each module. Unlike [6,10], segments in our approach are not necessarily connected from a single marking. An edge in the synchronization graph describes an interface transition. Thus, the synchronization graph describes how segments of modules synchronously move to another.

In the sequel, we analyze how temporal logic properties like *Linear Temporal Logic* (LTL) [14] or *Computation Tree Logic* (CTL) [7] can be verified in the modular state space. For this, we introduce updated definitions in Sect. 2. In Sect. 3, we re-establish the result of [10] concerning the verification of linear time properties. This new consideration is necessary due to changes in the basic data structure. Our result, however, is more general than [10]. While their proof relied on Büchi automata, restricting their result to regular linear time properties, our proof carries over to non-linear time properties. In Sect. 4, we discuss the verification of CTL properties. The authors of [13] also discussed this, but did not provide proofs for their results and do not go into detail. Our result concerns the preservation of CTL properties between the actual reachability graph of the modular Petri net and the synchronization graph.

2 Preliminaries

2.1 Net Components

Definition 1 (Module). *A place/transition Petri net $N = [P, T, F, W]$, consisting of places P, transitions T, arcs F and weight function W, where T is partitioned into subsets $T_{internal}$ of internal transitions and $T_{interface}$ of interface transitions is called a* module.

Contrary to a full Petri net definition, modules are only structures without behavior. For this, we introduce instances as Petri net systems with behavior.

Definition 2 (Instance). *An* instance *is a Petri net system $[N, m_0]$ where N is a module and m_0 is the initial marking of N. In general, a marking is a mapping $m: P \to \mathbb{N}$.*

Now, instances can be composed along their interface transitions, which is described through fusion vectors.

Definition 3 (Fusion Vector). *Let* $\{[N_1, m_{01}], \ldots, [N_\ell, m_{0\ell}]\}$ *be a set of instances. A* fusion vector $f \in (T_{1|interface} \cup \{\bot\}) \times \ldots \times (T_{\ell|interface} \cup \{\bot\})$ *is a vector of interface transitions of the instances or the* $\bot$*-symbol. If* $f[j] = t$ *for* $t \in T_{j|interface}$*, then instance* $[N_j, m_{0j}]$ *participates in this fusion with* t*. If* $f[j] = \bot$*, then instance* $[N_j, m_{0j}]$ *does not participate in this fusion.*

The above definition is similar to *fusion sets* defined in [6,10]. The definitions for modules and instances allow for multiple instances to be of the same underlying module. By using vectors instead of sets and explicitly marking non-participating instances with $\bot$, we remove ambiguities. We further rule out that multiple interface transitions of the same instance participate in the same fusion. With this, we can define our "blueprint" for a modular Petri net, a system composed of instances through fusion vectors.

Definition 4 (Modular Structure). *A* modular structure *is a tuple* $\mathcal{M} = [\mathcal{I}, \mathcal{F}]$*, where*

- $\mathcal{I} = \{[N_1, m_{01}], \ldots, [N_\ell, m_{0\ell}]\}$ *is a set of instances with disjoint modules and*
- $\mathcal{F} \subseteq (T_{1|interface} \cup \{\bot\}) \times \ldots \times (T_{\ell|interface} \cup \{\bot\})$ *is a set of fusion vectors.*

A modular Petri net is obtained by gluing together the ingredients of a modular structure to a flat Petri net.

Definition 5 (Modular Petri Net). *From a modular structure* $\mathcal{M} = [\mathcal{I}, \mathcal{F}]$*, we can derive a Petri net system* $N = [P, T, F, W, m_0]$*, where*

- $P = \bigcup_{j \in \{1,\ldots,\ell\}} P_j$,
- $T = \bigcup_{j \in \{1,\ldots,\ell\}} T_{j|internal} \cup \{t_f \mid f \in \mathcal{F}\}$*, where* t_f *is a freshly introduced* fusion transition *for fusion vector* f,
- $F = \bigcup_{j \in \{1,\ldots,\ell\}} (F_j \cap (P_j \times T_{j|internal} \cup T_{j|internal} \times P_j))$
  $\cup \{(p, t_f) \mid f \in \mathcal{F}, \exists j \in \{1, \ldots, \ell\} \colon f[j] = t, (p, t) \in F_j\}$
  $\cup \{(t_f, p) \mid f \in \mathcal{F}, \exists j \in \{1, \ldots, \ell\} \colon f[j] = t, (t, p) \in F_j\}$,
- $W(t,p) = \begin{cases} W_j(t,p) \text{ for } (t,p) \in F_j \text{ and } t \in T_{j|internal} \\ W_j(t^*,p) \text{ for } (t^*,p) \in F_j \text{ and } t^* = f[j], f \in \mathcal{F}, j \in \{1,\ldots,\ell\} \end{cases}$,
- $W(p,t) = \begin{cases} W_j(p,t) \text{ for } (p,t) \in F_j \text{ and } t \in T_{j|internal} \\ W_j(p,t^*) \text{ for } (p,t^*) \in F_j \text{ and } t^* = f[j], f \in \mathcal{F}, j \in \{1,\ldots,\ell\} \end{cases}$,
- $m_0 = \bigcup_{j \in \{1,\ldots,\ell\}} m_{0j}$.

We call N *the* modular Petri net *for* $\mathcal{M}$.

Interface transitions of the individual instances are combined into fusion transitions of the modular net. Interface transitions, which are not present in a fusion vector, are not represented in the modular Petri net. Their behavior can be neglected, since they do not participate in the composition. In theory, they can still be added by introducing a fusion vector in which they participate, but all other instances do not with any transition. Since only one transition participates in a fusion vector per instance, the weight function is well-defined. Further, the domains of all initial markings m_{0j} for $j \in \{1, \ldots, \ell\}$ are pairwise disjoint, making the initial marking also well-defined.

2.2 Modular Behavior

Naturally, we are interested in the semantics of a modular structure and how this relates to the behavior of the corresponding modular Petri net. In particular, we want to be able to describe the modular state space without constructing the composed state space in the background. For this, we introduce the notions of local reachability and synchronization.

Local Reachability. Since the modular structure is defined through a set of instances, we describe their behavior as *local reachability*. Since instances are Petri net systems at their core, we can use existing notions for Petri nets to describe their behavior. This typically includes the activation and firing of transitions.

Definition 6 (Transition Activation). *Let m be a marking and $t \in T$ be a transition of Petri net N. Transition t is* activated *or* fireable *in marking m (denoted by $m \xrightarrow{t}$) iff, $\forall p \in P$: $W(p,t) \leq m(p)$. If m does not activate t we denote this by $m \not\xrightarrow{t}$.*

Definition 7 (Firing Rule). *Let $t \in T$ be a transition of Petri net N. First, we define $\Delta_t \colon P \to \mathbb{N} \cup \{0\}$ as $\Delta_t(p) = W(t,p) - W(p,t)$ for all $p \in P$, where we assume $W(x,y) = 0$ for $(x,y) \notin F$ for convenience. t* fires *in m and leads to marking m' (denoted $m \xrightarrow{t} m'$) if $m \xrightarrow{t}$ and $m' = m + \Delta_t$.*

One can see that this firing rule can be extended to transition sequences $\omega \in T^*$, saying $m \xrightarrow{\epsilon} m$ for the empty sequence and $m \xrightarrow{\omega t} m''$, if $m \xrightarrow{\omega} m' \xrightarrow{t} m''$. Similarly, since we use $\perp$ in fusion vectors like transitions, we say $m \xrightarrow{\perp} m$. By extension, we define that marking m' is *reachable* from marking m (denoted $m \to m'$), if a sequence $\omega \in T^*$ exists such that $m \xrightarrow{\omega} m'$.

With this, we can define the state space of an instance. This follows similar notions regarding reachability graphs, but we make one restriction to produce the *local* reachability graph. First, we can calculate the *reachability set* $RS_N(M)$ for a given Petri net system N and a set of markings M. We define $RS_N(M) = \{m' \mid m \to m', m \in M\}$.

Definition 8 (Projection). *Let $m \in RS(\{m_0\})$ be a reachable marking of a modular Petri net N. For a given instance $[N_j, m_{0j}]$ for $j \in \{1, \ldots, \ell\}$ of N, let $\pi_j(m) = m \cap (P_j \times \mathbb{N})$ be the* projection *of m to $[N_j, m_{0j}]$.*

Definition 9 ((Local) Reachability Graph). *The* reachability graph *of a Petri net system $N = [P, T, F, W, m_0]$ is the directed labeled graph $R = [V^R, E^R]$, where $V^R = RS_N(\{m_0\})$ and $(m, t, m') \in E^R$ iff $m \xrightarrow{t} m'$ for some $t \in T$.*

Given a modular structure $\mathcal{M} = [\mathcal{I}, \mathcal{F}]$ and its corresponding modular Petri net N, we can define the local reachability graph *$L_j = [V_j^L, E_j^L]$ for an instance $[N_j, m_{0j}]$ with $j \in \{1, \ldots, \ell\}$ based on R:*

- $V_j^L = \{\pi_j(m) \mid m \in V^R\}$

- $E_j^L = E_{j|internal}^L \cup E_{j|interface}^L$ *with*
  - $(\pi_j(m), t, \pi_j(m')) \in E_{j|internal}^L$ *iff* $(m, t, m') \in E^R, t \in T_{j|internal}$
  - $(\pi_j(m), t, \pi_j(m')) \in E_{j|interface}^L$ *iff* $(m, t_f, m') \in E^R, f[j] = t, f \in \mathcal{F}$

Projecting the behavior of the modular Petri net to the instance restricts its behavior to just the one relevant to the composed system. Since we disallow arbitrary firing of interface transitions through the projection, we can also handle instances whose state space would be infinite in the standalone case, but finite in the composed case. Despite that, handling local reachability graphs that remain infinite after this restriction are still infeasible to handle, which is why we restrict ourselves to finite local reachability graphs.

Synchronization Graph. In addition to describing the behavior of singular instances we want to reason about the composed system without generating its full state space. As seen in Definition 5, the global behavior of the modular Petri net (the behavior not fully described through a single instance) is generated through fusion transitions. Per this same definition, a fusion transition can only fire if every interface transition in the corresponding fusion vector can fire. In a similar vein, an interface transition can therefore only appear in a local reachability graph if the corresponding fusion transition can fire. This means, at some point, every interface transition of a fusion vector needs to be activated for this to happen. These points of *synchronization* of the instances are captured in the *synchronization graph*.

At this point we can introduce a simple abstraction. For a single instance, it does not matter which marking from other instances activates their relevant interface transition, only that it can be independently activated. Independently here simply means without the help of other instances (through interface transitions). Therefore, we can abstract from single markings to sets of markings reachable through just internal transitions. We call those sets *segments*.

Definition 10 (Segment). *Let $L_j = [V_j^L, V_j^L]$ be the local reachability graph of instance $[N_j, m_{0j}]$ for $j \in \{1, \ldots, \ell\}$. A* segment *is a set of markings $O \subseteq V_j^L$, which is forwardly closed in the following way:*
If $m \in O$ and $(m, t, m') \in E_{j|internal}^L$, then $m' \in O$.

For a set of markings $M \subseteq V_j^L$, let $\mathring{M}$ be the smallest segment that contains M, i.e. the *closure* of M regarding internal transitions. We call set M a *generator* of $\mathring{M}$.

Markings in a segment are not necessarily connected or strongly connected, as generators of a segment are not required to be connected. This is the main difference from previous approaches [5,6,17], which require generators to be strongly connected. We drop this requirement, allowing for coarser segments. Segments can lie arbitrarily to each other, i.e. they can intersect, be disjoint or contain each other.

Segments describe internal behavior of an instance between occurrences of interface transitions. For a fixed interface transition and a fixed segment, we define the successor segment.

Definition 11 (Successor Segment). *Let O be a segment of the local reachability graph L_j of instance $[N_j, m_{0j}]$ and $t \in T_{j|interface}$ be an interface transition. The* successor segment *is defined as*

$$O^{+t} = \mathring{M}, \text{ where } M = \{m' \mid m \in O, (m, t, m') \in E^L_{j|interface}\}$$

As we permit only one interface transition per instance in a fusion vector, the successor segment is unambiguous. From time to time, we will abuse the notation for a fusion transition $t_f \in T$ where $f \in \mathcal{F}$ is a fusion vector as $O^{+t_f} = O^{+f[j]}$ for $f[j] \neq \bot$ and $O^{+t_f} = O$ for $f[j] = \bot$ for $j \in \{1, \ldots, \ell\}$.

After abstracting the markings of an instance to segments, the next step is to abstract the activation of interface transitions, from a single marking in the reachability graph of the modular Petri net to a tuple of segments for every instance.

Definition 12 (Global Activation). *A fusion transition t_f for fusion vector $f \in \mathcal{F}$ is* globally activated *in a tuple of segments $<O_1, \ldots, O_\ell>$ if for all instances $[N_j, m_{0j}]$ with $j \in \{1, \ldots, \ell\}$ and $f[j] \neq \bot$, there exists a $m_j \in O_j$ with $m_j \xrightarrow{f[j]}$.*

With this abstraction, we are ready to define the synchronization graph. A vertex in the synchronization graph is an ℓ-tuple of segments of the local reachability graphs, one per instance.

Definition 13 (Synchronization Graph). *The* synchronization graph *$S = [V^S, E^S]$ of modular structure $\mathcal{M} = [\mathcal{I}, \mathcal{F}]$ is inductively defined as follows:*
Base: $<\{\mathring{m}_{01}\}, \ldots, \{\mathring{m}_{0\ell}\}> \in V^S$
Step: if $v = <O_1, \ldots, O_\ell> \in V^S$ and fusion transition t_f for fusion vector $f \in \mathcal{F}$ is globally activated in v, then $v' = <O_1^{+t_f}, \ldots, O_\ell^{+t_f}> \in V^S$ and $(v, t_f, v') \in E^S$.

The graphs in Fig. 1b depict the above definitions applied to the modular structure in Fig. 1a. Markings are written in multiset notation, while segments are depicted through dotted borders.

Since we assumed local reachability graphs of the modules to be finite, the synchronization graph will in turn also be finite. Based on these definitions, the modular state space is well-defined for a given modular structure. Since the modular Petri net resulting from that modular structure is also well-defined (per Definition 5), a modular structure producing finite local reachability and synchronization graphs would also produce a finite reachability graph of the modular Petri net.

In the following we shall write $N = [P, T, F, W, m_0]$ as a modular Petri net, S for the synchronization graph of the underlying modular structure and R for the corresponding reachability graph. When relating markings m of R to vertices $v = <O_1, \ldots, O_\ell>$ of S in the following sections, we shall write $m \in v$ to mean $m \in O_1 \times \ldots \times O_\ell$.

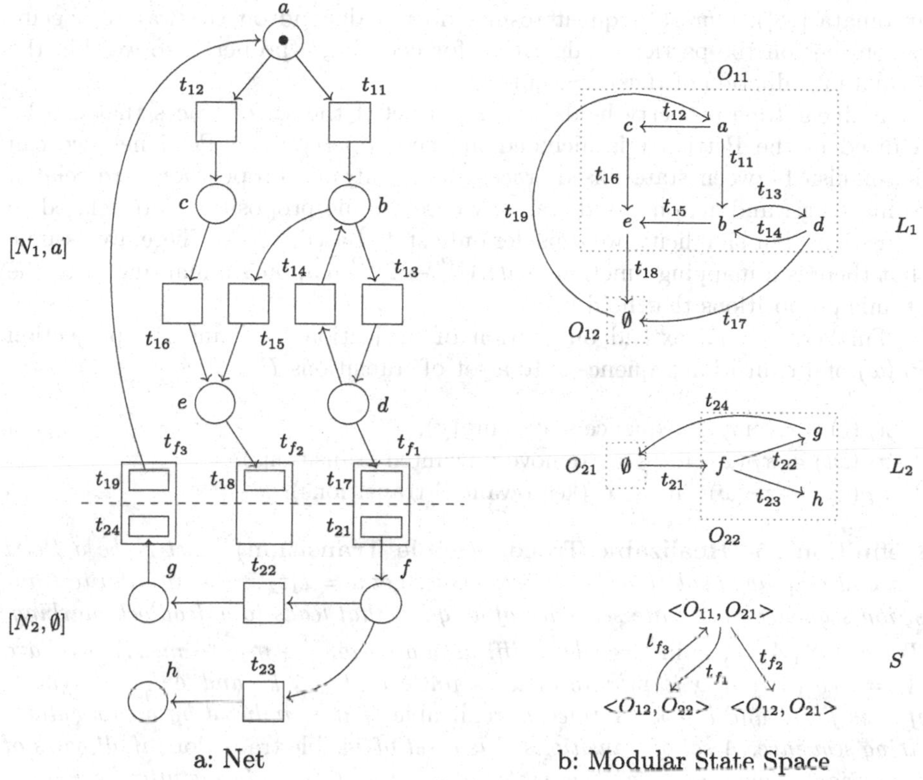

Fig. 1. Example net and modular state space

3 Verifying Linear Time Properties

A linear time property is defined as a set of traces. Intuitively, a trace may be seen as a run of the system, abstracted to values of a given set of atomic propositions.

Definition 14 (Traces, Reduced Form, Stuttering). *Let AP be a finite set of atomic propositions. A* trace *is an infinite sequence over the alphabet 2^{AP}. For a trace $A_0A_1A_2\ldots$, its* reduced form *is the finite or infinite sequence $A_0A_{i_1}A_{i_2}\ldots$ such that the values i_j are precisely the indices where $A_{i_j-1} \neq A_{i_j}$, in ascending order. A trace with a finite reduced form is called* diverging. *Two traces with the same reduced form are called* stutter equivalent. *A property where, for each contained trace, all stuttering equivalent traces are contained as well, is called* stutter invariant.

Linear time properties can be specified using linear time temporal logic (e.g. LTL [14]), or using one of various classes of automata that accept infinite sequences (e.g. Büchi automata, Streett automata, Rabin automata or parity automata [18]). Our subsequent results do not depend on the way of specification, nor on the particular algorithm for checking a property, so we skip the formal introduction of these concepts.

A linear time property holds in a Petri net if the set of traces that can be realized in the Petri net is included in the property. In a Petri net, we can distinguish between state based traces, where atomic propositions are related to markings, and action based traces, where atomic propositions are related to transitions. For simplicity, we consider only state based traces. There, we assume that there is a mapping function $eval\colon V^R \to 2^{AP}$ mapping a marking m to the atomic propositions that hold in m.

Further, we can extend our notion of projection to define the projection $\pi_\Gamma(\omega)$ of a transition sequence ω to a set of transitions $\Gamma \subseteq T$:

1. $\pi_\Gamma(\epsilon) = \epsilon$ (empty sequence stays empty),
2. $\pi_\Gamma(\omega t) = \pi_\Gamma(\omega)$, if $t \notin \Gamma$ (remove unwanted transitions),
3. $\pi_\Gamma(\omega t) = \pi_\Gamma(\omega)t$, if $t \in \Gamma$ (keep wanted transitions).

Definition 15 (Realizable Trace, Visible Transition). *Let N be a Petri net and mapping eval be as described above. Let $\omega = t_1 t_2 \dots$ be an infinite transition sequence or a finite sequence of length k that leads to a deadlock marking. Trace $A_0 A_1 A_2 \dots$ is* induced *by ω iff, assuming $m_0 \xrightarrow{t_0} m_1 \xrightarrow{t_1} m_2 \dots$, we have $A_i = eval(m_i)$ if ω is infinite or ω is finite and $i \leq k$, and $A_i = eval(m_k)$, if ω is finite and $i > k$. A trace is* realizable *if it is induced by an executable firing sequence. A set of transitions Γ is a set of* visible transitions *if all pairs of transitions sequences ω_1 and ω_2 with $\pi_\Gamma(\omega_1) = \pi_\Gamma(\omega_2)$ induce stutter equivalent traces.*

The intuition behind visible transitions is that their firing can directly change the value of atomic propositions. They therefore describe the "relevant" behavior of the system w.r.t. a property. Stutter equivalence describes this for traces, and the definition above translates this to transition sequences.

In the remainder of this section, we present results that relate realizable traces to stutter equivalent traces in the modular state space. This way, the modular state space can be used for verifying linear time properties. We avoid the term "trace equivalence" since the modular state space as a whole is not a transition system.

We develop our results bottom-up and start with a lemma concerning transitions in the synchronization graph.

Lemma 1 (Transition Lemma). *Let $v \xrightarrow{t_f} v'$ be a transition in S with $v = <O_1, \dots, O_\ell>$ and $v' = <O'_1, \dots, O'_\ell>$. Let $m'_j \in O'_j$ for $j \in \{1, \dots, \ell\}$. Then there exist markings $m_j \in O_j$ and a sequence $\omega \in T^*_{internal}$ such that $\bigcup_j m_j \xrightarrow{t_f \omega} \bigcup_j m'_j$.*

Proof. Consider instance $[N_j, m_{0j}]$ and some marking m'_j with $j \in \{1, \dots, \ell\}$. For $f[j] = \bot$, we can simply choose $m_j = m'_j$. Otherwise, O'_j is induced by firing $t = f[j]$ from O_j. By construction, we then have nonempty sets $Pre = \{m_{pre} \mid m_{pre} \in O_j, m_{pre} \xrightarrow{t}\}$ and $Post = \{m_{post} \mid \exists m_{pre} \in Pre, m_{pre} \xrightarrow{t} m_{post}\}$. Furthermore, O'_j is the closure of $Post$, so there exists a marking $m^*_j \in Post$ and a sequence $\omega_j \in T^*_{j|internal}$ of internal transitions with $m^*_j \xrightarrow{\omega_j} m'_j$. Let m_j be the marking in Pre holding $m_j \xrightarrow{t} m^*_j$. Since internal transitions of different instances are independent and t_f is enabled in $\bigcup_j m_j$, we have $\bigcup_j m_j \xrightarrow{t_f} \bigcup_j m^*_j \xrightarrow{\omega_1\omega_2\ldots\omega_\ell} \bigcup_j m'_j$. □

By induction, we can lift this lemma to finite transition sequences.

Corollary 1 (Finite Trace Lemma). *Let $t_1t_2\ldots t_n$ be a path in S starting in $<\{\mathring{m}_{01}\},\ldots,\{\mathring{m}_{0\ell}\}>$ leading to node v and let $m \in v$. Then there exist sequences $\omega_0,\ldots,\omega_n \in T^*_{internal}$ such that $m_0 \xrightarrow{\omega_0 t_1 \omega_1 t_2 \omega_2 \ldots t_n \omega_n} m$.*

Proof. (Sketch) We apply the transition lemma repeatedly. We start with vertex v and its predecessor, obtaining ω_n for t_n first. We proceed backwards until we finally derive ω_1 for t_1. Finally, we observe that every marking in the initial segment is reachable from the initial marking by internal transitions. This way, we obtain ω_0. □

Since the proof of the finite trace lemma proceeds backwards (starting from the final vertex), it cannot be trivially extended to infinite traces. For proving a result for infinite traces, we need to argue with the finiteness of a modular state space and apply an argument that resembles the proof of König's Graph Lemma [9].

Lemma 2 (Infinite Trace Lemma). *Let $t_1t_2\ldots$ be an infinite path in S starting in $<\{\mathring{m}_{01}\},\ldots,\{\mathring{m}_{0\ell}\}>$. Then there exist sequences $\omega_0,\omega_1,\ldots \in T^*_{internal}$ such that $m_0 \xrightarrow{\omega_0 t_1 \omega_1 t_2 \omega_2 \ldots}$.*

Proof. For every $i \in \mathbb{N}$, applying of the finite trace lemma to the prefix $t_1t_2\ldots t_i$ of the given infinite transition sequence asserts the existence of $\omega_{i0},\ldots,\omega_{ii}$ such that $m_0 \xrightarrow{\omega_{i0} t_1 \omega_{i1} \ldots t_i \omega_{ii}}$. Without loss of generality we may assume that every subsequence ω_{xy} is cycle-free, i.e. does not visit any marking twice. We now define, for every $k \in \mathbb{N} \cup \{-1\}$, an infinite set Ψ_k and a transition sequence ω_k.
Base: Let $\omega_{-1} = \epsilon$ (the empty sequence) and $\Psi_{-1} = \{(\omega_{i0}, \omega_{i1}\ldots\omega_{ii}) \mid i \in \mathbb{N}\}$. Ψ_{-1} is obviously infinite.
Step: Let $k \in \mathbb{N}$. We define ω_k and Ψ_k assuming that ω_j and Ψ_j are already defined for all $j < k$. Since we assume N to be bounded, there are only finitely many reachable markings and thus only finitely many cycle-free paths. Since Ψ_{k-1} is by assumption infinite, there exists a transition sequence ω_k such that $\omega_k = \omega_{ik}$ for infinitely many elements $(\omega_{i0}, \omega_{i1}, \ldots, \omega_{ik}, \ldots, \omega_{ii}) \in \Psi_{k-1}$. Fix such ω_k and let $\Psi_k = \{(\omega_{i0}, \omega_{i1}, \ldots, \omega_{ik}, \ldots, \omega_{ii}) \mid (\omega_{i0}, \omega_{i1}, \ldots, \omega_{ik}, \ldots, \omega_{ii}) \in \Psi_{k-1}, \omega_{ik} = \omega_k\}$. By choice of ω_k, Ψ_k is still infinite. Furthermore, our construction asserts that, for all $j < k$ and all $(\omega_{i0}, \omega_{i1}, \ldots, \omega_{ii}) \in \Psi_k$, $\omega_{ij} = \omega_j$. Since

all the ω_{xy} have been chosen according to the finite trace lemma, every element remaining in Ψ_k proves that $m_0 \xrightarrow{\omega_0 t_1 \omega_1 \ldots t_k \omega_k}$. For $k_1 < k_2$, trace $\omega_0 t_1 \omega_1 \ldots t_{k_1} \omega_{k_1}$ is a prefix of trace $\omega_0 t_1 \omega_1 \ldots t_{k_2} \omega_{k_2}$. An infinite sequence of ascending (w.r.t. prefix) finite transition sequences has a limit which is an infinite transition sequence. Consequently, the infinite sequence $\omega_0 t_1 \omega_1 t_2 \omega_2 \ldots$ is executable from m_0 in N as well. □

With the finite and infinite trace lemmas, we can map paths in the synchronization graph to paths in the reachability graph of a modular Petri net. The reverse mapping follows from the construction of a modular state space.

Lemma 3 (Projection Lemma). *Let ω be a finite or infinite sequence of transitions in N. If $m_0 \xrightarrow{\omega}$ then $\pi_{T_{fusion}}(\omega)$ is a path in S.*

We will not discuss the construction in this paper, but since Definitions 9 and 13 for the local reachability and synchronization graphs are similar to definitions for usual reachability graphs, the construction and mapping can easily be derived from there.

Next, we proceed to reestablish and generalize the main result of [10]. That result states that properties definable in the temporal logic LTL, where fusion transitions form a set of visible transitions, can be verified using the modular state space. We extend this result to arbitrary linear time properties, including properties that are not ω-regular (i.e. not recognizable by a finite Büchi automaton). LTL can only specify ω-regular properties. The next results use the assumption that fusion transitions form a set of visible transitions. As briefly described in the previous section, this assumption can be enforced by inserting new fusion vectors containing just a visible, internal transition.

Every executable infinite sequence ω in N with only finitely many occurrences of fusion transitions generates a diverging trace. Using the projection lemma, the fusion transitions form a finite path to a vertex $<O_1, \ldots, O_\ell>$ in the synchronization graph. The suffix of ω beyond the last occurrence of a fusion transition consists of internal transitions where those internal transitions that belong to instance $[N_j, m_{0j}]$ do not leave the corresponding segment O_j for $j \in \{1, \ldots, \ell\}$. Assuming boundedness of N, this can only happen if there is a cycle of internal transitions in at least one of the O_j, or there exist markings m_j in O_j such that $\bigcup_j m_j$ is a deadlock marking in N. This observation is formalized in the next definition.

Definition 16 (Diverging State). *A vertex $<O_1, \ldots, O_\ell>$ in S is called a* diverging state *if one of the following conditions holds:*

(1) There exist markings $m_j \in O_j$ and nonempty sequences $\omega_j \in T^+_{j|internal}$ such that $m_j \xrightarrow{\omega_j} m_j$ for $j \in \{1, \ldots, \ell\}$, or
(2) There exist markings $m_j \in O_j$ for $j \in \{1, \ldots, \ell\}$ such that $\bigcup_j m_j$ is a deadlock marking in N.

If only fusion transitions are visible, then all markings that correspond to the same vertex of the synchronization graph satisfy the same atomic propositions.

Lemma 4 (Uniformity Lemma). *Let $m, m' \in v$ for some vertex $v \in V^S$. Assuming visible transitions to only be fusion transitions, then m and m' satisfy the same atomic propositions, such is $eval_R(m) = eval_R(m')$.*

Proof. Since m and m' lie in the same vertex of the synchronization graph, applying the finite trace lemma to m and m' can obtain sequences ω and ω' such that $m_0 \xrightarrow{\omega} m$, $m_0 \xrightarrow{\omega'} m'$ and $\pi_{T_{fusion}}(\omega) = \pi_{T_{fusion}}(\omega')$. Since the fusion transitions are the only visible ones, $\pi_{T_{visible}}(\omega) = \pi_{T_{visible}}(\omega')$ holds. Per Definition 15, ω and ω' induce stutter equivalent traces. This results in $eval_R(m) = eval_R(m')$. □

The uniformity lemma justifies the next definition.

Definition 17 (Evaluation of Vertices in the Synchronization Graph). *The function $eval_S \colon V^S \rightarrow 2^{AP}$ maps vertices v to satisfied atomic propositions and is defined as $eval_R(m)$ for any marking $m \in v$.*

Using the extended evaluation function, we can define the set of traces that are realized in the modular state space.

Definition 18 (Traces Realized in the Modular State Space). *Let $N = [P, T, F, W, m_0]$ be a modular Petri net where the fusion transitions form a set of visible transitions with respect to a linear time property Θ. Let $\theta = A_0A_1A_2\ldots$ be a trace using the atomic propositions of Θ. θ is realizable in the modular state space of N if one of the following conditions holds:*

(1) There exists an infinite path $v_0v_1v_2\ldots$ in S such that $eval_S(v_i) = A_i$ for all i, or
(2) There exists a finite path $v_0v_1\ldots v_n$ in S such that v_n is a diverging state and $eval_S(v_i) = A_i$ for all i with $0 \leq i \leq n$ and $A_i = A_n$ for all i with $i > n$.

Now we are ready to prove that linear time properties are preserved in the modular state space.

Theorem 1 (Verification of Linear Time Properties). *Let N be a modular Petri net and Θ a linear time property where fusion transitions of N form a set of visible transitions. Then, any trace θ of Θ is realized in N iff a stutter equivalent trace θ' is realized in the modular state space of N.*

Proof. Implication: Let θ be realized in N, say, using an infinite or deadlocking sequence ω and let $\omega_{\mathcal{F}} = \pi_{T_{fusion}}(\omega)$. Using the projection lemma, $\omega_{\mathcal{F}}$ defines a path in the synchronization graph as well. By Definition 17, this sequence realizes a trace that is stutter equivalent to θ. If $\omega_{\mathcal{F}}$ is finite, it leads to a diverging state of the synchronization graph since it ends in a deadlock or an infinite sequence with only internal transitions. Again, $\omega_{\mathcal{F}}$ realizes a stutter equivalent trace.
Replication: Let, in the first case, ω be an infinite path in the synchronization graph. The infinite trace lemma yields a corresponding stutter equivalent transition sequence in N. In the second case, let ω be a finite sequence in the

synchronization graph that ends in a diverging state. Assume, in the first subcase, that divergence is caused by a deadlock state. Then the finite trace lemma yields a corresponding path in N that ends in that deadlock state too and thus realizes a stutter equivalent trace. If, in the second subcase, divergence is caused by an internal cycle in any instance, the finite sequence lemma yields a sequence to a marking that enters this cycle (and enters arbitrary markings in the other instances). The assumed cycling sequence can be executed indefinitely in N since internal transitions do not depend on other instances. Since internal transitions are all invisible by assumption, that infinite sequence stutters in the same way as in Definition 18. □

With Theorem 1, we can verify many linear time properties using the modular state space. The result does not depend on the particular verification algorithm to be used. In fact, we may turn the synchronization graph into a labeled transition system using the extended *eval* function. The conditions regarding diverging states can be integrated by adding a self-loop to every such vertex. The result in [10] relies on Büchi automata as representation of properties, so it does not extend to non-regular properties.

4 Verifying CTL Properties

We aim to find results about CTL [7] verification in the modular state space similar to the results obtained through Theorem 1 concerning linear time properties. To verify properties in the modular state space, we have to check how obtained results translate to the regular state space of the modular Petri net. Such preservation of CTL properties between systems is provided through *bisimulation* [2,16] relations. In the following, we take the liberty to define such relations on reachability graphs of Petri nets instead of general transition systems.

4.1 Relation Between Modular Petri Nets and Modular Structures for CTL

Normally, bisimulation relations are defined between two transition systems, our modular structure however is not such. Since we are interested in model checking the complete modular Petri net, we analyze CTL preservation between the reachability graph of the modular Petri net and its synchronization graph, since the latter represents the *global* behavior of the system.

As we will see in Fig. 2 below, bisimulation relations can not be found for our use case with sensible results. Therefore, we will instead focus on *weak bisimulations* [2,15,16], that preserve CTL_X, the subclass of CTL without the X operator. Since not every possible action in a transition system influences atomic propositions of a formula, this is not very restrictive in practical scenarios. At the basis of weak bisimulation again lie *visible* transitions. While properties in CTL are not directly defined over traces like for linear time properties, we can still apply that notion. When talking about "visible" and "invisible" transitions in the general sense in the following, we mean (in)visible in relation to a property Θ without explicitly naming Θ. Further, since the synchronization graph is akin to a reachability graph, we can also substitute it for the following definition.

Definition 19 (Weak Bisimulation). *Let R, R' be two reachability graphs for two Petri nets N, N' with corresponding functions eval and eval', where the nets share transition labels, i.e. $T_N = T_{N'}$. Let $T = T_N$ for simplicity. A relation $\sigma \subseteq V^R \times V^{R'}$ is a* weak bisimulation relation *iff*

i) $\sigma(m, m')$ *implies* $eval(m) = eval'(m')$,

ii) $\sigma(m, m')$ *and* $m \xrightarrow{t} m_1$ *for some* $t \in T$ *implies a* m'_1 *with* $m' \overset{t}{\Rightarrow} m'_1$ *and* $\sigma(m_1, m'_1)$,

iii) $\sigma(m, m')$ *and* $m' \xrightarrow{t} m'_1$ *for some* $t \in T$ *implies a* m_1 *with* $m \overset{t}{\Rightarrow} m_1$ *and* $\sigma(m_1, m'_1)$,

where $\overset{t}{\Rightarrow}$ *denotes* $\xrightarrow{\omega_1} \ldots \xrightarrow{t} \ldots \xrightarrow{\omega_2}$ *when t is visible and* $\xrightarrow{\omega}$ *when t is invisible for* $\omega, \omega_1, \omega_2 \in T^*_{invisible}$.

Two reachability graphs R, R' are further *weakly bisimular* iff a weak bisimulation relation σ exists, such that $\sigma(m_0, m'_0)$ holds.

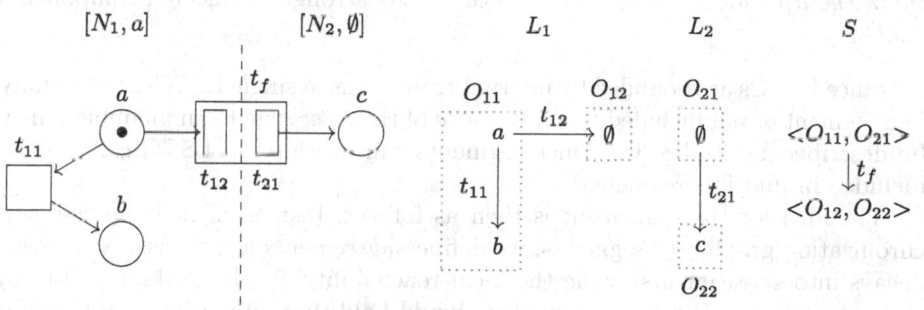

Fig. 2. Counterexample for (weak) bisimulation

Like in the previous section, we restrict visible transitions to simply be fusion transitions of the modular Petri net. Unfortunately, this restriction alone is not enough to satisfy Definition 19. This can be seen through Fig. 2. There, the modular Petri net is made up of two instances $[N_1, a]$ and $[N_2, \emptyset]$ with one fusion vector $f = (t_{12}, t_{22})$. Figure 2 also shows the local reachability graphs and the synchronization graph. We choose t_f to be a visible transition to have some relevant behavior in the net. Let $\sigma \subseteq V^R \times V^S$ now describe a relation where $\sigma(a, <O_{11}, O_{21}>)$ holds. Since t_{11} is activated in a, $<O_{11}, O_{21}>$ also needs a valid transition sequence in the synchronization graph in order to satisfy Definition 19 for σ. Since t_{11} is invisible (and t_f is not), the only valid transition sequence is the empty sequence, implying $\sigma(b, <O_{11}, O_{21}>)$. But t_f is activated in $<O_{11}, O_{21}>$, while b is a deadlock marking. Therefore, Definition 19 cannot hold for σ. Ignoring our restriction from before, when every transition is visible, weak bisimulation is equivalent to normal bisimulation. Since weak bisimulation does not hold in the general case, bisimulation also does not hold, supporting our earlier argument. To deal with weak bisimulation in the general case, we will discuss a possible refinement of the synchronization graph in the following section with the aim to satisfy Definition 19.

4.2 Refinement in the Synchronization Graph

As implied by Fig. 2, weak bisimulation requires $\sigma(m, v)$ for any marking m and vertex v with $m \in v$. Since every $m \in v$ therefore needs to be able to activate visible transitions activated in v, weak bisimularity generally does not hold, since our definition of segments does not require that. We aim to refine the synchronization graph to allow this based on local strong connectivity.

Definition 20 (Local Strong Connectivity). *Let $L_j = [V_j^L, E_j^L]$ be the local reachability graph of an instance $[N_j, m_{0j}]$ for $j \in \{1, \ldots, \ell\}$. Two markings $m, m' \in V_j^L$ are* locally strongly connected *(we write $m \sim_{L_j} m'$), iff $m \xrightarrow{\omega} m'$ and $m' \xrightarrow{\omega'} m$, where $\omega, \omega' \in T^*_{j|internal}$. Relation $\sim_{L_j}$ is an equivalence relation. The equivalence classes are named* locally strongly connected components (LSCCs).

Since LSCCs are bounded by internal transitions, a single LSCC is either fully in a segment or not included at all. Because of this, a segment can unambiguously be described by its LSCCs. Since segments can overlap, an LSCC can also be included in multiple segments.

The idea for the refinement is then as follows: Instead of defining the synchronization graph over segments, we define *subsegments* as its base. A segment decays into subsegments, while the local reachability graph can be partitioned into subsegments. For a subsegment it should hold that all markings in the subsegment can activate the same interface transitions, perhaps with some internal transitions firing before. Additionally, firing an interface transition from a subsegment should always lead to the same successor subsegment. This is needed, since generators of segments are not required to be strongly connected, therefore not every marking of a generator has to be in the same subsegment. This restriction should similarly hold for internal transitions that leave a subsegment. Since internal transitions are assumed to be invisible anyway, we can however weaken the restriction here: When leaving a subsegment through firing an internal transition, the successor subsegment should be reachable from all markings in the subsegment through an arbitrary sequence of internal transitions.

Definition 21 (Subsegment). *Let $L = L_j$ for $j \in \{1, \ldots, \ell\}$ be the local reachability graph of some instance. Let $\{O_1, \ldots, O_r\}$ be the set of segments of L. Every Segment O_x for $x \in \{1, \ldots, r\}$ decays into a set of q_x* subsegments *$\{U_{x1}, \ldots, U_{xq_x}\}$, such that the following holds:*

i) $\{U_{x1}, \ldots, U_{xq_x}\}$ *is a partition of* O_x

ii) Let m *be a marking of subsegment* $U_{xa} \subseteq O_x$ *for* $a \in \{1, \ldots, q_x\}$. *If there is a fusion vector* $f \in \mathcal{F}$ *such that* $m \xrightarrow{\omega_1 f[j] \omega_2} m_1$ *where* $m_1 \in U_{x'b} \subseteq O_{x'}$ *with* $x' \in \{1, \ldots, r\}$, $b \in \{1, \ldots, q_{x'}\}$ *and* $\omega_1, \omega_2 \in T^*_{j|internal}$, *then for every marking* $m' \in U_{xa}$, *there is a* $m'_1 \in U_{x'b}$ *and* $\omega'_1, \omega'_2 \in T^*_{j|internal}$ *such that* $m' \xrightarrow{\omega'_1 f[j] \omega'_2} m'_1$.

iii) Let m *be a marking of subsegment* $U_{xa} \subseteq O_x$ *for* $a \in \{1, \ldots, q_x\}$. *If* $m \xrightarrow{\omega} m_1$ *where* $m_1 \in U_{xb} \subseteq O_x$ *with* $b \in \{1, \ldots, q_x\}$ *and* $\omega \in T^*_{j|internal}$, *then for every marking* $m' \in U_{xa}$, *there is* $m'_1 \in U_{xb}$ *and* $\omega' \in T^*_{j|internal}$ *such that* $m' \xrightarrow{\omega'} m'_1$.

In the above, Definition 21ii deals with the first restriction on subsegments, that every fireable interface transition should be fireable from every marking in the subsegment. Definition 21iii deals with the second restriction, that successor subsegments reached through internal transitions should also be reachable from every marking of the subsegment, albeit not necessarily through the same sequence.

Even with this restriction, subsegments are not equal to segments in [5,6,17]. We form different requirements for subsegments. Therefore, subsegments may not generally be coarser than segments defined in [5,6,17] and vice versa.

Lemma 5 (LSCCs are Subsegments). *Let* L_j *for* $j \in \{1, \ldots, \ell\}$ *be the local reachability graph of an instance* $[N_j, m_{0j}]$. *Any LSCC of* L_j *is a valid subsegment.*

Proof. The proof for this lemma is trivial. Local strong connectivity partitions the local reachability graph. If an interface transition is activated in one marking m, it is fireable in every marking strongly connected to m, so we can perform the transition to the successor subsegment. The same holds for internal transitions. □

We can advance this fact to the following.

Lemma 6 (Closure of Subsegments). *Let* L_j *for* $j \in \{1, \ldots, \ell\}$ *be the local reachability graph of an instance* $[N_j, m_{0j}]$. *Subsegments of* L_j *are closed with respect to local strong connectivity.*

Proof. Let t be an interface transition or an internal transition that occurs between two subsegments. Assume marking $m \xrightarrow{\omega t}$ with $\omega \in T^*_{j|internal}$ and another marking m' that is locally strong connected to m. We can extend ω to $\omega'' = \omega'\omega$ for $\omega' \in T^*_{j|internal}$ with $m' \xrightarrow{\omega'} m$ and thus, m' belongs to the same subsegment as m. □

This means that an LSCC either lies completely within a subsegment or not at all. Based on this, we can partition every local reachability graph of the instances into subsegments. As allured to before, the set of LSCCs is a valid partition as well. In the worst case, the refined synchronization graph is the LSCC graph of the modular Petri net, where a vertex constitutes an LSCC. Even though this is intuitive, this might cause too many and too small subsegments. More partitions lead to a more complex synchronization structure that is at the expense of the efficiency of the verification. Therefore, the subsegment partitions should be as coarse as possible.

We reuse the wording of global activation (cf. Definition 12) for subsegments and eliminate misunderstandings by adding the context of a global activation in a tuple of segments for the synchronization graph and a tuple of subsegments for the refined synchronization graph respectively. A fusion transition t_f is globally activated in a tuple of subsegments $<U_1, \ldots, U_\ell>$, if for all instances $[N_j, m_{0j}]$ with $f[j] \neq \perp$ and $j \in \{1, \ldots, \ell\}$, there exists a $m_j \in U_j$ such that $m_j \xrightarrow{f[j]}$.

With the concept of subsegments we can now introduce the refined synchronization graph. A vertex in the refined synchronization graph is an ℓ-tuple of subsegments of the local reachability graphs, one per instance.

Definition 22 (Refined Synchronization Graph). *Based on a subsegmental partitioning, the* refined synchronization graph $S^* = [V^{S^*}, E^{S^*}]$ *is inductively defined as follows:*
Base: $<U(\{m_{01}\}), \ldots, U(\{m_{0\ell}\})> \in V^{S^*}$, *where* $U(\{m_{0j}\}) \subseteq \{\overset{\circ}{m_{0j}}\}$ *is the subsegment containing* m_{0j} *for* $j \in \{1, \ldots, \ell\}$.
Step: Let $v = <O_1, \ldots, O_\ell>, v' = <O_1^{+t_f}, \ldots, O_\ell^{+t_f}> \in V^S$ *be vertices and* $(v, t_f, v') \in E^S$ *an edge of* S *(Definition 13). Further, let* $u = <U_1, \ldots, U_\ell> \in V^{S^*}$ *be a vertex of* S^* *such that* $U_j \subseteq O_j$ *for* $j \in \{1, \ldots, \ell\}$. *Assuming that*

i) Fusion transition t_f *for fusion vector* $f \in \mathcal{F}$ *is globally activated in* u *and for all* $j \in \{1, \ldots, \ell\}$, $U'_j \subseteq O_j^{+t_f}$ *is a subsegment with* $m \in U_j$ *and* $m' \in U'_j$ *such that* $m \xrightarrow{f[j]} m'$. *Then,* $u' = <U'_1, \ldots, U'_\ell> \in V^{S^*}$ *and* $(u, t_f, u') \in E^{S^*}$.
ii) For instance $[N_j, m_{0j}]$ *with* $j \in \{1, \ldots, \ell\}$ *there is a transition* $t \in T_{j|internal}$, *subsegment* $U'_j \subseteq O_j$ *and markings* $m \in U_j$ *and* $m' \in U'_j$ *such that* $m \xrightarrow{t} m'$. *Then* $u' = <U_1, \ldots, U_{j-1}, U'_j, U_{j+1}, \ldots, U_\ell> \in V^{S^*}$ *and* $(u, \tau, u') \in E^{S^*}$.

In the definition above, subpart 22i conserves all the edges of the synchronization graph in the refined version. The new vertices contain the subsegments reached by the interface transitions of every instance. Subpart 22ii adds the edges between subsegments that result from splitting segments to the refined synchronization graph. Here, only one subsegment changes, namely the one where the corresponding segment was partitioned. The edge is then labeled with τ, since subpart 21iii does not define *one* single internal transition.

Figure 3 shows the refinement applied to the simple modular structure depicted in Fig. 1a. Subsegments are drawn using dashed lines. While many subsegments are comprised of single LSCCs, U_{11} and U_{12} show that dropping the restriction for strongly connected segments can allow for coarser refinement.

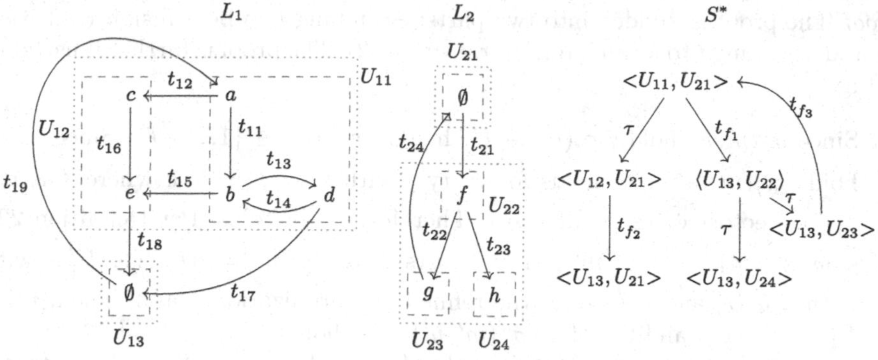

Fig. 3. Refinement applied to example from Fig. 1

4.3 Relation Between Modular Petri Nets and the Refined Modular Structure for $\mathrm{CTL}_{(X)}$

With the definitions introduced in the previous section, we can discuss if and how they allow for CTL preservation between a modular Petri net and its modular structure. As in Sect. 4.1, we already find that CTL cannot be preserved, so we now discuss the subclass CTL_X. The goal then becomes finding a relation between the reachability graph of a modular Petri net and the refined synchronization graph of the underlying modular structure that proves weak bisimularity.

Definition 23 (Relation Between R **and** S^***).** *Let* $u \in V^{S^*}$ *and* $m \in V^R$. *Then* $\sigma^* \subseteq V^R \times V^{S^*}$ *is a relation between* S^* *and* R *defined as* $\sigma^*(m, u) \Leftrightarrow m \in u$.

In the following, we will show how this relation proves weak bisimularity between the two graphs, assuming visible transitions of formulae to be fusion transitions only. We will discuss if the refined synchronization graph can imitate transition firings of the reachability graph and vice versa in terms of weak bisimularity for Definition 19.

Lemma 7 (Reachability Imitation Lemma). *Let* $u = <U_1, \ldots, U_\ell> \in V^{S^*}$ *and* $m \in V^R$ *with* $\sigma^*(m, u)$ *according to Definition 23. Let* $m \xrightarrow{t} m'$ *be a transition in* R, *then there exists a* $u' = <U'_1, \ldots, U'_\ell> \in V^{S^*}$ *where the following holds:*

- $u \xrightarrow{t} u'$ *if* t *is a fusion transition*
- $u \xrightarrow{\tau} u'$ *or* $u' = u$ *if* t *is an internal transition*
- $\sigma^*(m', u')$

Proof The proof is divided into two parts: Assuming t to be a fusion transition (1) and assuming t to be an internal transition (2). The proof is further illustrated in Fig. 4.

(1) Since $\sigma^*(m,u)$ holds, $\pi_j(m) \in U_j$ holds for all $j \in \{1,\ldots,\ell\}$. Since $m \xrightarrow{t}$ holds, $\pi_j(m) \xrightarrow{t_j}$ also holds for every j with $t_j = f[j] \neq \bot$ where f is the fusion vector corresponding to t. Therefore, $u \xrightarrow{t}$ holds. Per Definition 22i, since $\pi_j(m) \xrightarrow{f[j]} \pi_j(m')$, there exists a vertex $u' = <U'_1,\ldots,U'_\ell>$ with $\pi_j(m') \in U'_j$ and $u \xrightarrow{t} u'$ in the refined synchronization graph. Then, $m' \in U'_1 \times \ldots \times U'_\ell$, and therefore $\sigma^*(m',u')$, also holds.

(2) Similarly to above, $\pi_j(m) \in U_j$ holds for $j \in \{1,\ldots,\ell\}$. Since t is internal to just one instance, we can fix j to where $t \in T_{j|internal}$ holds. Assuming $\pi_j(m') \in U_j$, we can simply set $u' = u$ and $\sigma^*(m',u')$ holds because of the definition of σ^* (cf. Definition 23). Assuming $\pi_j(m') \in U'_j$ where $U_j \neq U'_j$, a vertex $u' = <U_1,\ldots,U_{j-1},U'_j,U_{j+1},\ldots,U_\ell>$ exists in the refined synchronization graph with $u \xrightarrow{\tau} u'$ per Definition 22ii. Further, $\sigma^*(m',u')$ holds since $m' \in u'$. □

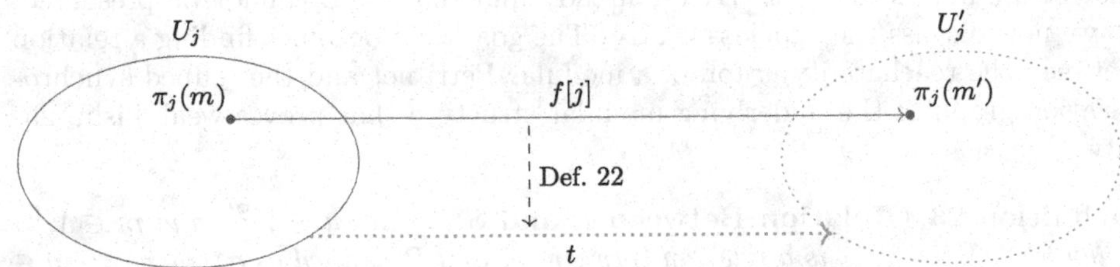

Fig. 4. Illustration for Lemma 7

Figure 4 illustrates the above proof. The imitated transition in the reachability graph and other assumptions from the lemma are seen through solid lines. The proof matter is noted through dotted lines. Dashed lines illustrate how we can derive the latter from the former.

Similarly to the above, we can find a lemma regarding the imitation of actions of the refined synchronization graph through the reachability graph.

Lemma 8 (Refined Imitation Lemma). *Let $u = <U_1,\ldots,U_\ell> \in V^{S^*}$ and $m \in V^R$ with $\sigma^*(m,u)$ according to Definition 23. Let $u \xrightarrow{t} u'$ be a transition in S^* with $u' = <U'_1,\ldots,U'_\ell>$, then there exists a marking $m' \in V^R$ where the following holds:*

- $m \xrightarrow{\omega_1 t \omega_2} m'$ *if t is a fusion transition with $\omega_1,\omega_2 \in T^*_{internal}$*
- $m \xrightarrow{\omega} m'$ *if t is an internal transition with $\omega \in T^*_{internal}$*
- $\sigma^*(m',u')$

Proof. This proof is also divided into two parts: Assuming t to be a fusion transition (1) and assuming t to be an internal transition (2). The proof is further illustrated in Fig. 5.

(1) First, $\pi_j(m) \in U_j$ holds for all $j \in \{1, \ldots, \ell\}$. Since t is globally enabled in u, there exist markings $m_{1j} \in U_j$ and $m'_{1j} \in U'_j$ with $m_{1j} \xrightarrow{f[j]} m'_{1j}$ for fusion vector f corresponding to t per Definition 22i. According to Definition 21ii, there exists a marking $m'_j \in U'_j$ with $\pi_j(m) \xrightarrow{\omega_{1j} f[j] \omega_{2j}} m'_j$ and $\omega_{1j}, \omega_{2j} \in T^*_{j|internal}$. Since internal transitions of different instances do not interact with each other, the transition sequence $\omega_{11} \ldots \omega_{1\ell} t \omega_{21} \ldots \omega_{2\ell}$ leads to $m' = \bigcup_j m'_j$ starting in m. Since $m' \in u'$, $\sigma^*(m', u')$ holds.

(2) For an internal transition t, $u \xrightarrow{T} u'$ holds in the refined synchronization graph and $u' = <U_1, \ldots, U_{j-1}, U'_j, U_{j+1}, \ldots, U_\ell>$ for some $j \in \{1, \ldots, \ell\}$. Per Definition 22ii, U_j and U'_j lie in the same segment and there exist markings $m_{1j} \in U_j$ and $m'_{1j} \in U'_j$ such that $m_{1j} \xrightarrow{t} m'_{1j}$. Per Definition 21iii, there exists a marking $m'_j \in U'_j$ such that $\pi_j(m) \xrightarrow{\omega} m'_j$ with $\omega \in T^*_{j|internal}$. Since ω does not interact with other instances, ω leads to $m' = m'_j \cup \bigcup_{a \neq j} \pi_a(m)$ for $a \in \{1, \ldots, \ell\}$. Therefore, $m' \in u'$ and $\sigma^*(m', u')$ holds. □

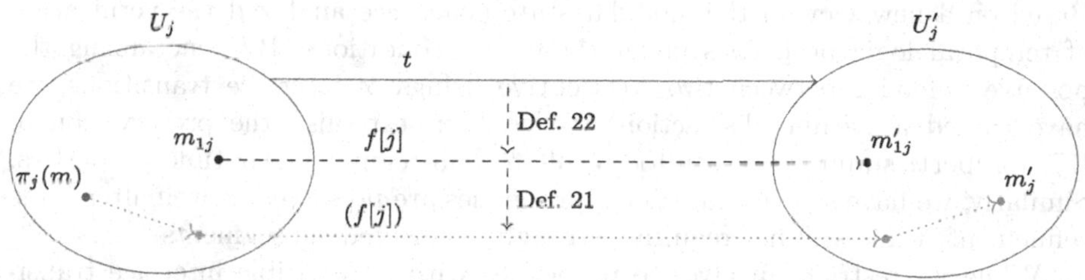

Fig. 5. Illustration for Lemma 8

Similarly to before, in Fig. 5 solid lines indicate assumptions from the lemma, while dotted lines indicate proof matter and dashed lines derivatives from the assumptions through definitions.

The previous lemmas deal with transition imitation of the reachability graph through the refined synchronization graph and vice versa, which align with the requirements for weak bisimulation for Definition 19. In addition to that, $eval_{S^*}$ needs to be defined. This can be done akin to $eval_S$ in Definition 17. While that definition only applies to the unrefined synchronization graph, it can easily be extended to the refined synchronization graph. Since $U_j \subseteq O_j$ for a given vertex in the refined synchronization graph $<U_1, \ldots, U_\ell>$ and its corresponding vertex in the synchronization graph $<O_1, \ldots, O_\ell>$ for $j \in \{1, \ldots, \ell\}$, one can also define $eval_{S^*}$ like $eval_S$. Finally, weak bisimularity is then provided through Definitions 22 and 23 for S^* and σ^* respectively. This leads to the following theorem.

Theorem 2 (Weak bisimularity through σ^*). *Let $R = [V^R, E^R]$ be the reachability graph of a modular Petri net and let $S^* = [V^{S^*}, E^{S^*}]$ be the refined synchronization graph of the underlying modular structure. Weak bisimularity holds through σ^* for a formula $\Theta \in CTL_X$, whose visible transitions are only fusion transitions.*

Corollary 2 (CTL_X preservation through the refined synchronization graph). *Let $N = [P, T, F, W, m_0]$ be a modular Petri net with its reachability graph R based on modular structure $\mathcal{M} = [\mathcal{I}, \mathcal{F}]$ and let S^* be the refined synchronization graph of $\mathcal{M}$. Then R and S^* satisfy the same formulae from CTL_X, i.e. for some formula $\Theta \in CTL_X$*

$$R \models \Theta \Leftrightarrow S^* \models \Theta$$

holds.

Proof. The proof follows from Theorem 2 and the proof for CTL_X preservation from [2]. □

5 Conclusion

Based on a new view of the modular state space, we analyzed the verification of temporal logic properties under these new conditions. By generalizing the possible behavior between two consecutive firings of interface transitions, we have gained a greater abstraction. We could re-establish the preservation of LTL properties and generalize the result to non-regular linear time properties. Similarly, we have shown that CTL_X properties are preserved as well after some refinement, while still not requiring strongly connected subsegments.

While we restrict ourselves to properties with only visible interface transitions, properties where internal transitions are visible can be verified as well. We simply need to declare these transitions as interface transitions and introduce singleton fusion vectors. Of course, the additional interface transitions may cause a combinatorial explosion in the synchronization graph. It is therefore desirable to have results that permit the verification of such properties without changing the set of interface/fusion transitions. Such results are a natural field for future research.

The obvious next step would be to implement the methods. While the preservation allows standard model checking algorithms to be used, the construction of the modular state space is not implemented yet. Further, a method to split a plain Petri net into modules would allow for broader applicability.

References

1. Best, E., Devillers, R., Koutny, M.: The box algebra – a model of nets and process expressions. In: Donatelli, S., Kleijn, J. (eds.) ICATPN 1999. LNCS, vol. 1639, pp. 344–363. Springer, Heidelberg (1999). https://doi.org/10.1007/3-540-48745-X_21
2. Browne, M.C., Clarke, E.M., Grümberg, O.: Characterizing finite Kripke structures in propositional temporal logic. Theoret. Comput. Sci. **59**(1), 115–131 (1988). https://doi.org/10.1016/0304-3975(88)90098-9
3. Buchholz, P., Kemper, P.: Efficient computation and representation of large reachability sets for composed automata. Discret. Event Dyn. Syst. **12**(3), 265–286 (2002). https://doi.org/10.1023/A:1015669415634
4. Buchholz, P., Kemper, P.: Hierarchical reachability graph generation for Petri nets. Formal Methods Syst. Des. **21**(3), 281–315 (2002). https://doi.org/10.1023/A:1020321222420
5. Christensen, S., Petrucci, L.: Modular state space analysis of coloured Petri Nets. In: De Michelis, G., Diaz, M. (eds.) ICATPN 1995. LNCS, vol. 935, pp. 201–217. Springer, Heidelberg (1995). https://doi.org/10.1007/3-540-60029-9_41
6. Christensen, S., Petrucci, L.: Modular analysis of Petri nets. Comput. J. **43**(3), 224–242 (2000). https://doi.org/10.1093/comjnl/43.3.224
7. Clarke, E.M., Emerson, E.A., Sistla, A.P.: Automatic verification of finite-state concurrent systems using temporal logic specifications. ACM Trans. Program. Lang. Syst. **8**(2), 244–263 (1986). https://doi.org/10.1145/5397.5399
8. Graf, S., Steffen, B.: Compositional minimization of finite state systems. In: Clarke, E.M., Kurshan, R.P. (eds.) CAV 1990. LNCS, vol. 531, pp. 186–196. Springer, Heidelberg (1991). https://doi.org/10.1007/BFb0023732
9. König, D.: Theorie der endlichen und unendlichen Graphen: Kombinatorische Topologie der Streckenkomplexe, vol. 16. Akademische Verlagsgesellschaft, Leipzig (1936)
10. Latvala, T., Mäkelä, M.: LTL model checking for modular Petri nets. In: Cortadella, J., Reisig, W. (eds.) ICATPN 2004. LNCS, vol. 3099, pp. 298–311. Springer, Heidelberg (2004). https://doi.org/10.1007/978-3-540-27793-4_17
11. Le Cornec, Y.S.: Compositional analysis of modular Petri nets using hierarchical state space abstraction. In: Joint 5th International Workshop on Logics, Agents, and Mobility, LAM 2012, The 1st International Workshop on Petri Net-Based Security, WooPS 2012 and the 2nd International Workshop on Petri Nets Compositions, CompoNet 2012. CEUR Workshop Proceedings, Hamburg, Germany, vol. 853, pp. 119–133, June 2012. https://hal.archives-ouvertes.fr/hal-00785569
12. Lohmann, N.: A feature-complete Petri net semantics for WS-BPEL 2.0. In: Dumas, M., Heckel, R. (eds.) WS-FM 2007. LNCS, vol. 4937, pp. 77–91. Springer, Heidelberg (2008). https://doi.org/10.1007/978-3-540-79230-7_6
13. Lounnaci, F.O., Ioualalen, M., Boukala, M.C.: Verification of modular systems. In: IT4OD, pp. 240–245 (2014). https://www.researchgate.net/publication/269094681_IT4OD_2014_Proceedings_Tebessa
14. Manna, Z., Pnueli, A.: Temporal Verification of Reactive Systems. Springer, New York (1995). https://doi.org/10.1007/978-1-4612-4222-2
15. Milner, R. (ed.): A Calculus of Communicating Systems. LNCS, vol. 92, p. 20. Springer, Heidelberg (1980). https://doi.org/10.1007/3-540-10235-3
16. Milner, R.: Communication and Concurrency, vol. 84. Prentice Hall, Englewood Cliffs (1989)

17. Petrucci, L.: Modularity and Petri nets. In: 7th International Symposium on Programming and Systems (ISPS 2005), Alger, Algeria, pp. 7–8 (2005). https://hal.archives-ouvertes.fr/hal-00012214
18. Thomas, W.: Automata on infinite objects. In: Van leeuwen, J. (ed.) Formal Models and Semantics. Handbook of Theoretical Computer Science, pp. 133–191. Elsevier, Amsterdam, January 1990. https://doi.org/10.1016/B978-0-444-88074-1.50009-3
19. Valmari, A.: Compositional analysis with place-bordered subnets. In: Valette, R. (ed.) ICATPN 1994. LNCS, vol. 815, pp. 531–547. Springer, Heidelberg (1994). https://doi.org/10.1007/3-540-58152-9_29
20. Westergaard, M.: CPN tools 4: multi-formalism and extensibility. In: Colom, J.-M., Desel, J. (eds.) PETRI NETS 2013. LNCS, vol. 7927, pp. 400–409. Springer, Heidelberg (2013). https://doi.org/10.1007/978-3-642-38697-8_22

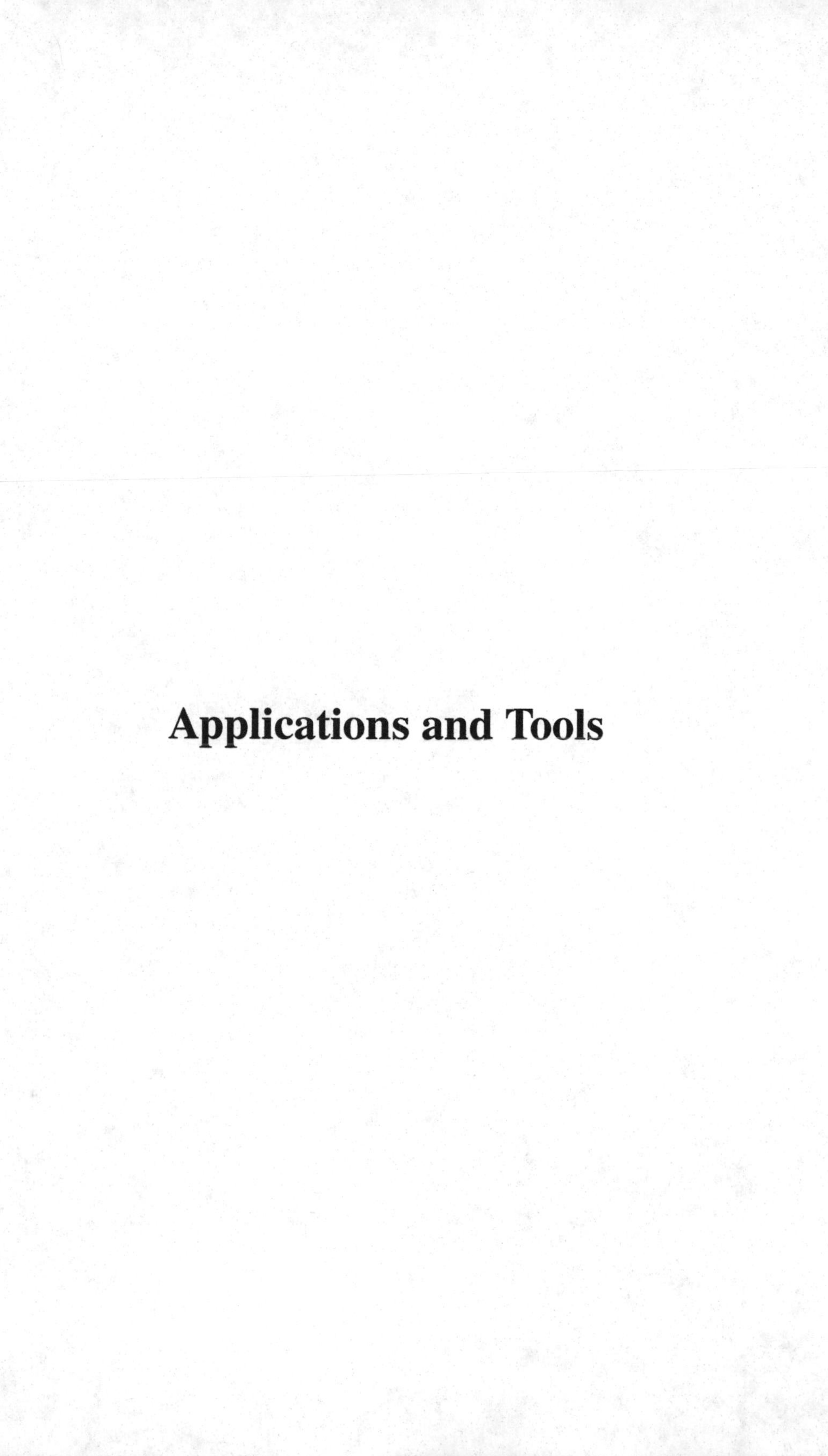

Applications and Tools

Design of Event-Driven Tsetlin Machines Using Safe Petri Nets

Alex Chan[1(✉)], Adrian Wheeldon[2], Rishad Shafik[1], and Alex Yakovlev[1]

[1] Newcastle University, Newcastle upon Tyne NE1 7RU, UK
{alex.chan,rishad.shafik,alex.yakovlev}@newcastle.ac.uk
[2] Literal Labs, Newcastle upon Tyne NE1 5JE, UK
adrian.wheeldon@newcastle.ac.uk

Abstract. In the last decade, there has been a significant shift towards the use of machine learning (ML) within the technology industry. One prominent ML algorithm is the Tsetlin Machine (TM), where it uses a collection of learning automata to learn new patterns through propositional logic. While TMs are considered computationally simpler and more efficient than neural networks (NNs), there is difficulty in how TMs can be better understood by industrial practitioners. Although many approaches help demonstrate the benefits of TMs, there is however no approach that helps better explain the behaviour of TMs, e.g. how the TM's decision is influenced by the initial states of their learning automata and how the TM's learning is determined by the calculations made from its inference and feedback components. In this paper, we present the concept of event-driven TMs, where we model the complete behaviour of TMs using 1-safe Petri nets. The key aspects of Petri nets are their flexibility to model many types of specifications including distributed systems and concurrent systems, and their rich support from many well-established tools including Petrify, MPSat, and Workcraft. To highlight the benefits of our approach, we conduct a simple experiment where we showcase our Petri net specifying the complete behaviour of a TM, analyse its behaviour through a set number of epochs, and most importantly evaluate its accuracy.

Keywords: Petri nets · Tsetlin Machine · Learning Automata · Workcraft · Machine Learning

1 Introduction

The use of Artificial Intelligence (AI) has recently seen a significant increase within many areas of application ranging from large classification problems, such as management of high-dimensional data, to automated control processes and synthetic mediums, such as self-driving cars and audio synthesis respectively. Although AI is shown to be beneficial, the energy cost required to use this technology is extremely high and requires a large amount of resources. As a

L. M. Kristensen and J. M. van der Werf (Eds.): PETRI NETS 2024, LNCS 14628, pp. 357–378, 2024.
https://doi.org/10.1007/978-3-031-61433-0_17

result, it is paramount that the designer not only improves the accuracy of their AI solutions like machine learning (ML) algorithms, but also optimises its performance and reduces its energy costs as a whole.

One prominent ML algorithm is the Tsetlin Machine (TM) [9], which is a type of learning automaton collective that uses a team of Tsetlin Automata (TAs) [19] to learn new patterns through propositional logic. With recent developments in TMs, they show great promise as they have high accuracy with relatively good performance and low energy costs, while solving several large classification problems [2,3,5] and seeing use in areas like circuit design [12,14] and financing [4].

However, despite the TM's growing potential, there is still difficulty in how they can be understood by practitioners from industry, when compared to more traditional ML algorithms like neural networks (NNs). Although the aforementioned works help demonstrate the benefits of TMs, there is currently no approach that helps explain the behaviour of TMs, e.g. how the initial states of TAs affect the calculations of the TM's clauses and how the TM's guess impacts the feedback process that influences the training process of its TAs.

In this paper, we introduce the concept of event-driven TMs by proposing the use of Petri nets [16] to model the complete behaviour of TMs. Here, we surmise why Petri nets are a good tool for modelling TMs and why TMs are a good application for Petri nets to enter the world of ML.

Firstly, let us consider why Petri nets for modelling TMs. Petri nets are a very flexible model that can express many types of specifications including distributed systems, concurrent systems, asynchronous circuits, and even potentially ML algorithms. One key aspect as to why Petri nets should be used is due to the TM's training process that is based on Finite State Machines (FSMs) [8], which are naturally discrete-event based. In particular, the events in FSMs are based on rules that derive from Boolean logic, where they are similar to the ML algorithm's inference procedure that is also based on Boolean logic. Additionally, some operations like the accumulation of votes in class sums can be easily represented using operation semantics, which are also natural for Petri nets. In fact, the whole process of computing the classification and reinforcing the TA states is essentially discrete-event driven, as the idea of multiple features, multiple clauses being calculated independently, and the ensemble of TAs that are evolving in parallel, all naturally extends to the concepts of concurrency and causality that can be easily captured by Petri nets.

Now, let us consider why TMs for enabling ML-based Petri nets. What makes TMs a particularly interesting model for Petri nets is how they establish the necessary bridge for Petri nets to enter the world of ML, due to the above points. Additionally, the relatively transparent behavioural discrete-event semantics of TMs makes them a convenient application use case for Petri nets, and the level of detail in modelling TMs using Petri nets is significantly higher than what can be achieved with modelling deep NNs, due to the latter essentially operating more like a "black" box with some potentially obscure internal behaviour that cannot be easily captured with the standard Petri net representation.

By considering the above points and their impact on modelling ML processes, one can see that the challenge of an explainable ML algorithm or AI can be potentially resolved using the various automation and visualisation tools that are available in the arsenal of Petri net analysis.

Thus, the main contribution of this work is as follows:

- A step-by-step approach on how event-driven TMs can be modelled using Petri nets (Sect. 3) where we provide the blueprint Petri nets of the TM's inference component (Sect. 3.2) and feedback component (Sect. 3.3).
- An analysis of the TM Petri net's size and behaviour including a visual methodology of using another Petri net specifying a measurement recorder to collect the analytical data during simulation in WORKCRAFT (Sect. 3.4).
- A simple experiment that demonstrates how the TM Petri net learns the pattern of a given dataset before we analyse its behaviour and evaluate its average running accuracy against a reference TM between each epoch, with discussion on some potential benefits of our approach (Sect. 4).

2 Preliminaries

Before we show how event-driven TMs can be modelled using Petri nets, it is important that we first cover the necessary background in TMs and some of the available features in the WORKCRAFT toolset that are used in this work.

2.1 Tsetlin Machines

TM [9] is a ML algorithm that uses a collection of learning automata to learn new patterns through propositional logic. Like many other ML algorithms, there are two major stages that contribute to the TM's learning: inference and feedback. The former is the process of running data points into the ML algorithm to calculate some output, e.g. a single numerical score, while the latter is the process of how the ML algorithm learns from this output based on the user's desired outcome before adjusting its parameters for the next calculation.

To help understand the procedures involved in the TM's inference and feedback components, Fig. 1 illustrates a Petri net specifying a simplified overview of the multi-class TM architecture. Note that a more detailed overview of the multi-class TM architecture can be found in Sect. 3.

Firstly, before inferencing, a set of datapoints is fed into the TM as shown with the transition labelled "Load data". Note that these datapoints may be in raw data format (i.e. raw features) meaning they must be converted into Boolean literals, which are required by TMs as their input. To convert these raw features (either integer or floating point values), they are passed through a Booleaniser component that compares it with a user specified threshold to generate a Boolean feature (i.e. a single bit Boolean value of 0 or 1). These Boolean features are then converted into Boolean literals to also contain their complement, as using both the feature and complement allows the Boolean literal space to represent every possible value that each Boolean feature can acquire [13].

Once the TM begins inferencing as shown with the transition labelled "Begin Inference", the Boolean literals are fed into every class, such that each clause receives this literal and calculates some output until all clauses have calculated

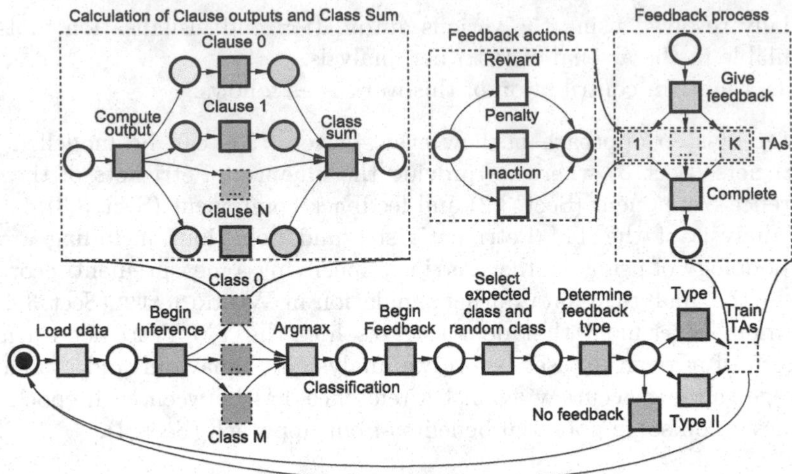

Fig. 1. Simple Petri net view of the Multiclass TM's procedure.

an output before the class sum is calculated. These are shown in the dashed area labelled "Calculation of Clause outputs and Class sum" that comprises the class' operation, where the transition labelled "Compute output" produces a token to each transition labelled "Clause 1..N" before merging at the transition labelled "Class sum". When the class sums of every class is calculated, they are fed into an argmax component as shown with the transition labelled "Argmax" to calculate the final classification that is used to determine whether the class(es) will receive feedback (i.e. a penalty or reward action) or not (i.e. an inaction).

After the final classification has been calculated, the TM can begin feedback. Here, the feedback process begins by selecting the expected class and a random class to receive feedback, as shown with the transition labelled "Select expected class and random class". Note that the chance of these classes receiving feedback depends on a probability check based on their class sum, where, if feedback is given, we iterate through each clause of the class and determine the type of feedback to be given depending on whether the class is the expected one, the polarity of the clause, and the current state of the corresponding TA.

2.2 Tsetlin Automata

TAs are a class of the finite reinforcement automaton [19], where it produces an exclude output (i.e. 0) for states below the midpoint and an include output (i.e. 1) for states above the midpoint as shown in Fig. 2. For TAs to transition between states, it must receive either a penalty action or a reward action from the TM's feedback component. Continued reward actions in the end states (i.e. state 1 and state $2n$) causes the TA to saturate, while penalty actions in one of the midstates (i.e. state n or state $n+1$) causes the TA to transition across the decision boundary and invert its output from exclude to include or vice versa [20]. Note that the TM's team of TAs comprises the TAs of each Boolean literal and are randomly initialised to one of the midstates.

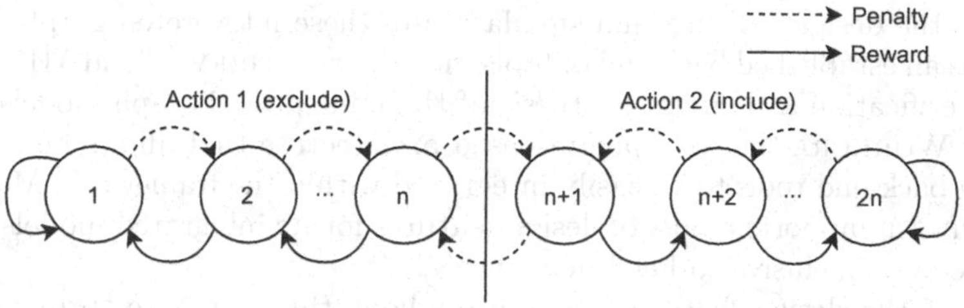

Fig. 2. Overview of a TA.

The logic of the TM's feedback can be decomposed into three stages: FB1, FB2 and FB3. FB1 focuses on feedback at the TM level, while FB2 focuses on feedback at the clause level and FB3 focuses on feedback at the TA level [20]. FB1 and FB2 are assigned either types none, T1 and T2, where the latter two are based on the Type I and Type II feedbacks that are shown as the transitions labelled "Type I" and "Type II" respectively in Fig. 1, while FB3 determines whether a TA receives a penalty, reward, or inaction as shown in the dashed area labelled "Feedback actions" comprising the corresponding transitions. Additionally, each feedback stage's output is the next stage's input (e.g. FB1's output is FB2's input) until the FB3's output is the input of the corresponding TA.

By referring to the above paragraph and the TM's payoff matrix [9], we can create the following truth table [20] shown in Table 1 where it includes the feedback type (i.e. the output of FB2), the TA's state, the clause's output, the literal's value and the TA's next action (i.e. the output of FB3). The values in this truth table can be defined as T1 being Type I feedback, T2 being Type II feedback, inc(0) and inc(1) being exclude and include respectively, c(0) and c(1) being clause output 0 and clause output 1 respectively, x(0) and x(1) being literal value 0 and literal value 1 respectively, $p_s = 0 \equiv \frac{1}{s}$ and $p_s = 1 \equiv \frac{s-1}{s}$, I being inaction, R being reward, P being penalty, and × being a don't care.

Table 1. TM's Payoff matrix

FB2	none	T1	T1	T1	T1	T1	T1	T1	T1	T1	T1	T2	T2	T2
inc	×	1	1	1	1	0	0	0	0	0	0	1	0	0
c	×	0	0	1	1	0	0	1	1	1	1	×	1	0
x	×	×	×	×	×	×	×	0	0	1	1	×	0	×
p_s	×	0	1	0	1	0	1	1	0	0	1	×	×	×
FB3	I	P	I	I	R	R	I	I	R	I	P	I	P	I

2.3 WORKCRAFT Toolset

WORKCRAFT [1] is a visual framework that provides rich support for the design automation of many interpreted graph models including FSMs, Petri nets, and Signal Transition Graphs [6,18]. WORKCRAFT has a graphical front-end that supports the design, editing, and simulation of these interpreted graph models, as well as an established back-end of tools including PETRIFY [7] and MPSAT [11] for the verification and circuit synthesis of these integrated graph models. Additionally, WORKCRAFT uses a plugin-based architecture that allows new plugins and new back-end tools to be easily integrated within the framework. Moreover, WORKCRAFT supports a host of design features for its integrated models, where some are even exclusive to Petri nets and STGs.

One of the design features used throughout this work are "proxy places" or proxies for short. Proxies are places that have been "graphically collapsed", where their original arc connections that connect to and from transitions are condensed and are replaced with a simple text block of the place's name as shown in Fig. 3. Note that proxies are purely visual and do not change the behaviour of the Petri net, as the original place still exists semantically and is only visually disconnected from the rest of the Petri net. This allows the designer to easily model their Petri net without creating too many overlapping arc connections, and to easily keep track of specific places without traversing the whole Petri net, e.g. a proxy corresponding to when a system enters its critical section. Moreover, proxies can even simplify the design of a Petri net to ensure that they remain comprehensible, and enables the ability to analyse several metrics of the Petri net, e.g. the total number of tokens gained by firing a repeated action.

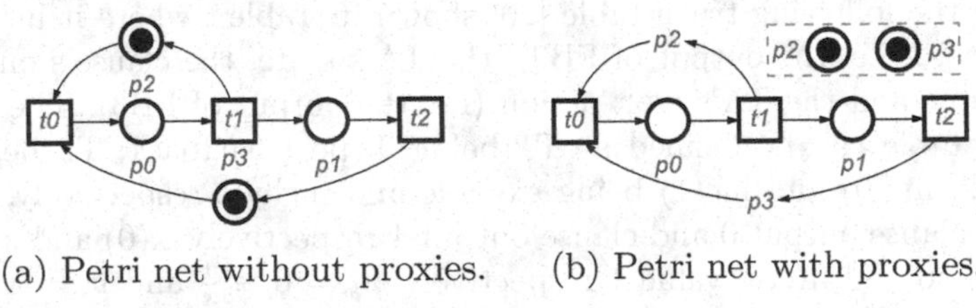

(a) Petri net without proxies. (b) Petri net with proxies.

Fig. 3. Visual comparison of non-proxies with proxies.

3 Design of Event-Driven Tsetlin Machines

In this section, we present how event-driven TMs can be modelled using Petri nets. Some initial work on representing asynchronous behaviour of TMs using Petri nets has been carried out in [20], where these Petri nets superficially modelled partial tiles that comprise the TM's clause calculation and the TM's class sum, before their results are merged to generate the final classification output.

For this work, our Petri nets are modelled to the level of multi-class TMs [9, 13], where we focus on how the TM's inference and feedback components can be designed hierarchically by modelling each operation as smaller segments before composing them [15] to the component level and subsequently to the TM level.

We then analyse the size and behaviour of our TM Petri net, where for the former we calculate the sizes of each Petri net segment and compare them to used segments of our TM Petri net, and for the latter we create another Petri net that specifies a measurement recorder to capture analytical behaviour of the TM Petri net during simulation in WORKCRAFT.

For simplicity purposes, our example will be based on a two-class TM learning the pattern of a two-input XOR gate using the dataset {00, 01, 10, 11} where each TM contains two clauses, each clause contains four TAs, and each TA contains six states (i.e. three exclude states and three include states).

3.1 Architecture of Multi-class Tsetlin Machine

To help understand how each TM component can be modelled as a Petri net, Fig. 4 provides a more detailed overview of the multi-class TM interface.

Figure 4(a) shows the loading of datapoints into the TM, where these datapoints may be raw features that must be converted into Boolean features through a Booleaniser component and then subsequently into Boolean literals to produce the feature's binary and complement values. Note that Boolean literals are required as inputs for the TM as explained in Sect. 2.1.

Figure 4(b) shows how the Booleanised data and the state values of the team of TAs are used to calculate the outputs of each TM's clauses, where an OR expression of each Boolean literal and the value of the literal's corresponding TA is first composed before an AND expression of every OR expression's value is composed to produce the clause output.

Figure 4(c) shows the team of TAs, where these TAs produce an exclude output for states below the midpoint and an include output for states above the midpoint. These TAs may transition between states by receiving either a penalty action or a reward action from the feedback component, where continued reward actions in the end states cause the TAs to saturate and penalty actions in one of the midstates cause the TAs to transition across the decision boundary and invert its output from exclude to include or vice versa [20].

Figure 4(d) shows how a TM class comprises many clauses that produce votes, which are split into a group of positive clauses and a group of negative clauses, such that the former votes in favour of the class and the latter votes against the class. Note that a majority vote gives an indication of class confidence, which is used to classify the input data and influence future decisions of the TAs through the feedback component [9], and that the composition of each clause is controlled by a vector of exclude bits, where these bits are parameters that are learned by the TAs. Additionally, the inclusion of inhibition in the voting system enables non-linear decision boundaries in the inference process.

Figure 4(e) shows how the class sums are fed into the argmax mechanism to determine the class with the most votes to be subsequently used in the classification process, where the winning class is used as the TM's guess to be compared with the expected value by the feedback component.

Figure 4(f) shows the whole feedback procedure, where Fig. 4(f)(i) shows the selection process of the TM classes to receive feedback, Fig. 4(f)(ii) shows the determination process of the feedback type, Fig. 4(f)(iii) shows the decision tree of Type I feedback to combat false negatives, and Fig. 4(f)(iv) shows the decision tree of Type II feedback to combat false positives.

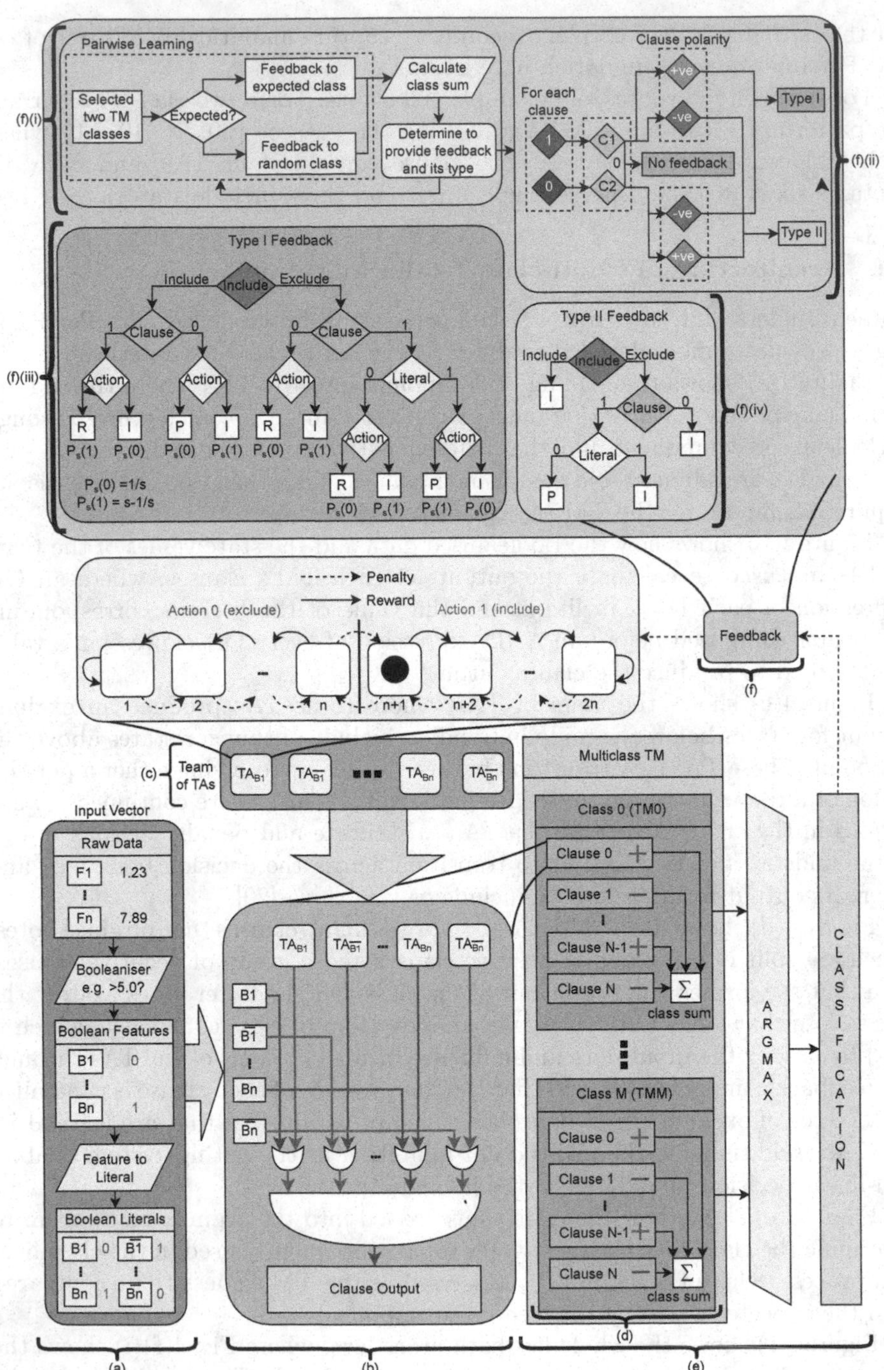

Fig. 4. Detailed overview of the Multi-class TM interface. (a) Data Vector. (b) Clause computation. (c) Team of TAs. (d) Single class TM. (e) Multiclass TM. (f). Feedback component where (i) Select classes for learning. (ii) Determine feedback type. (iii) Type I decision tree. (iv) Type II decision tree.

3.2 Inference Component

By referring to Fig. 4, let us first model the TM's inference component starting from the loading of data to the argmax of the class sums. For simplicity, we will assume that the data fed into the TM has already been converted into Boolean features through the Booleaniser component [13].

Firstly, by following Fig. 4(a), we can model the loading of datapoints as individual transitions labelled 00, 01, 10 and 11, where we then connect them to and from the places that determine the order of how datapoints are loaded (i.e. from 00 to 11), when the next datapoint gets loaded (i.e. the place named "loadVector") and the expected value (e.g. firing transition 01 marks a token at the place corresponding to the expected value 1). We then connect each "loading" transition to the two places that represent the Boolean features $bf1$ and $bf2$, which are connected to the transitions that subsequently convert them into the Boolean literals $b1$ and $b2$ (i.e. $litX1$ and $litX2$ respectively) and the complements $\bar{b1}$ and $\bar{b2}$ (i.e. $litCX1$ and $litCX2$ respectively). This results in the Petri net segment shown in Fig. 5, where we can fire the transitions labelled 00, 01, 10 and 11 to mark a token at the places that correspond to each feature's value, and convert them into literals by firing the subsequent transitions that mark a token at the places corresponding to the feature's value and complement.

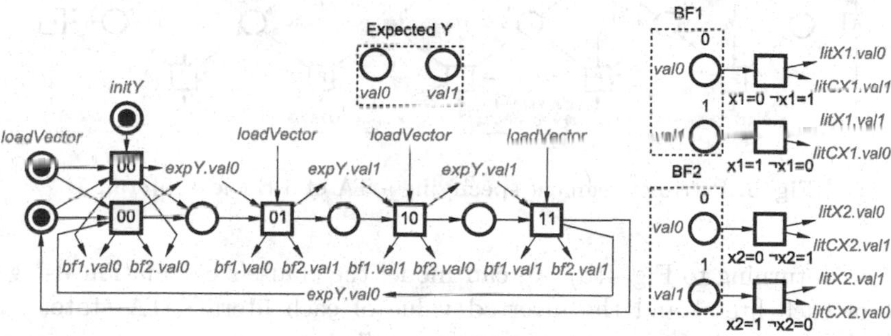

Fig. 5. Petri net segment specifying the loading and conversion of datapoints.

Next, by following Fig. 4(b), we can model the clauses of a TM. For simplicity, we will only show the Petri net segment for one clause, as this segment can be repeated for every clause with the exception that each clause's polarity alternates from their neighbour, e.g. if the polarity of clause i is positive (negative) then the polarity of clause j is negative (positive).

As part of the clause design, we must first follow Fig. 4(c) and create the TAs for every Boolean literal, where each TA has exactly three exclude states and three include states. This can be modelled by creating the same number of places to represent the exclude and include states, and marking a token to either places $exc3$ and $inc4$ to determine whether the TA is in state n or state $n+1$. Note that this initialisation is completed by another set of transitions before the datapoints are loaded, such that the set of transitions contains all possible combinations of every TAs' initial state. Additionally, the penalty and reward actions can be modelled by creating several transitions labelled penalty and reward, and connecting them to the places that lead closer to a decision boundary change (i.e. from $exc3$ to $inc4$ and vice versa) and saturation of rewards (i.e. from $exc3$ to $exc1$ for excludes and from $inc4$ to $inc6$ for includes) respectively. Moreover, the TA's inaction can be explicitated by creating a transition labelled inaction.

Note that places $fb3\_in$ and $fb3\_out$ are used to determine whether the TA receives a reward, penalty or inaction based on the provided feedback and are connected to and from their respective transitions, while places $inExc$ and $inInc$ are connected to and from the mid-state places $exc3$ and $inc4$ and are used to conveniently determine the TA's current state for clause calculation.

This results in the Petri net segment shown in Fig. 6, where we model the TA for the Boolean literal $x1$, such that it is initialised to be in the include state as indicated by the token marked at place $inc4$.

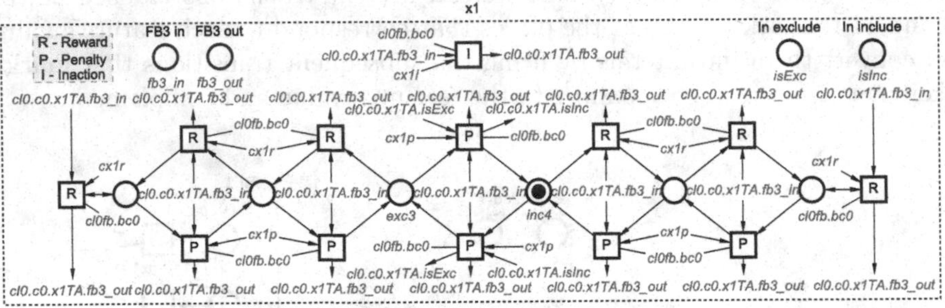

Fig. 6. Petri net segment specifying a TA of a Boolean literal.

Now, returning to Fig. 4(b), we can model the clause's calculation using the value of each literal and the inverted value of each literal's TA state, which results in the Petri net segment shown in Fig. 7.

Figure 7(a) shows the OR computation block for every Boolean literal comprising the literal's value and the literal's TA value, which has been inverted through the transition labelled INV. This OR computation block contains the total number of possible combinations (i.e. four transitions) where each transition is connected to and from the respective places that produce the values of an OR's truth table, e.g. an OR-transition connected from places $x1\_i0$ and $NOT(x1.TA) = 0$ leads to a token being marked at place $x1\_o0$.

Figure 7(b) shows the AND computation block involving the four OR computation blocks that comprise the truth table of all possible combinations leading to 0 (i.e. fifteen transitions), while Fig. 7(c) shows another AND computation block involving the four OR computation blocks that comprise the truth table of all possible combinations leading to 1 (i.e. one transition).

After repeating the above steps to create n clauses, we follow Fig. 4(d) and create the transitions that correspond to the class sum by totalling the clause

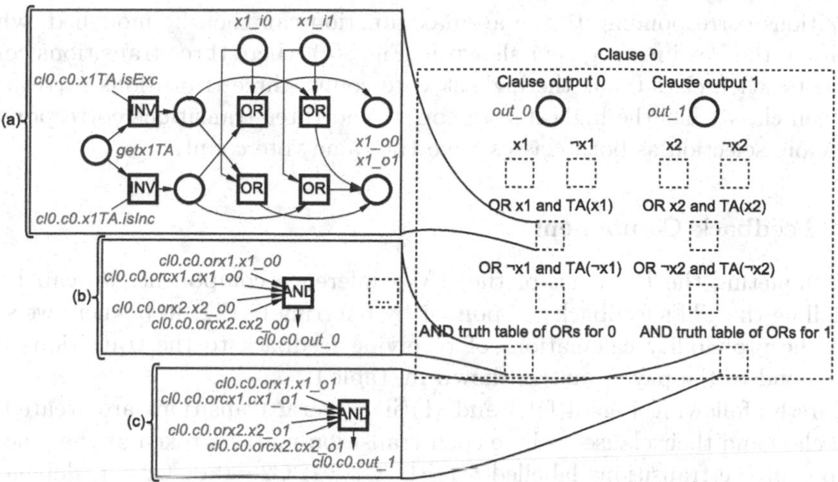

Fig. 7. Simplified view of a Petri net segment specifying a clause. (a) OR computation block for literals and their respective TA's inverted value. (b) AND computation block for OR calculations resulting 0. (c) AND computation block for OR calculations resulting 1.

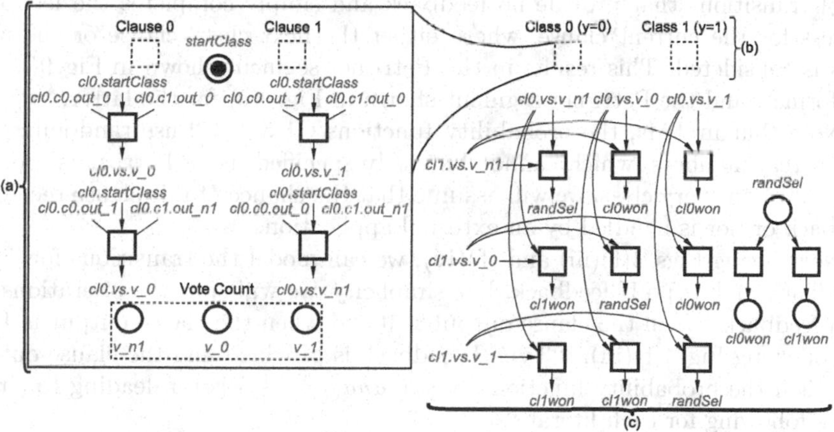

Fig. 8. Simplified view of Petri net segment specifying a multiclass TM. (a) Structure of a singular TM class. (b) Structure of a two-class TM with a collapsed view of classes. (c) Argmax transitions comprising the truth table of all class vote possibilities.

calculations, which leads to a vote count between $-n/2$ to $n/2$. This results in the Petri net segment shown in Fig. 8(a) where the outputs of our two clauses are connected to the four class sum transitions that add a token to either places v_n1 (for -1), v_0 (for 0) and v_1 (for 1).

Finally, by following Fig. 4(e), we can repeat the above step to create m TM classes, which results in the Petri net segment shown in Fig. 8(b). The set of transitions corresponding to the argmax function can then be modelled, which results in the Petri net segment shown in Fig. 8(c) where three transitions correspond to when class 0 has the highest vote count, three transitions correspond to when class 1 has the highest vote count, and three transitions correspond to a random selection as both classes have the same vote count.

3.3 Feedback Component

By completing the Petri net of the TM's inference component, we can begin modelling the TM's feedback component by referring to Fig. 4(f), where we start from the probability calculations of receiving feedback to the transitions that correspond to the payoff matrix shown in Table 1.

Firstly, following Figs. 4(f)(i) and (f)(ii), several transitions are created for each class and their clauses, where each transition marks a token at the place to fire one of the transitions labelled C1 = 0, C1 = 1, C2 = 0 or C2 = 1, depending on if the class is expected (i.e. C1) or random (i.e. C2) and if it receives feedback (i.e. 1) or not (i.e. 0). This results in the Petri net segment shown in Fig. 9(a).

If feedback is provided then this leads to the set of transitions, which comprise all the possibilities of whether Type I feedback or Type II feedback is provided to the clause, depending on its polarity. Otherwise, this leads to the set of transitions that provide no feedback and simply completes the feedback process for the current clause, where either the next class' clause or the next class is considered. This results in the Petri net segment shown in Fig. 9(b) for the former, and the Petri net segment shown in Fig. 9(c) for the latter.

Note that in TMs, the probability functions C1 and C2 use randomly generated real numbers, which cannot be easily specified using Petri nets. So, for each clause in every class, we will assume that the chance of this clause receiving feedback or not is handled by an external application.

Next, using Figs. 4(f)(iii) and (f)(iv), we can model the transitions for Type I feedback and Type II feedback. For simplicity, we will split the operations for both feedbacks when the clause output is 0 and when the clause output is 1.

Following Fig. 4(f)(iii), if Type I feedback is provided and the clause output is 0, then the probability function $S1 = (rand() \leqslant (\frac{1}{s}))$ is ran leading to either of the following for each literal:

- If the literal's TA is in the include state then the probability for the TA to receive a penalty is $p_s = 0$ and the probability to receive an inaction is $p_s = 1$, as shown in Fig. 10(a).
- If the literal's TA is in the exclude state then the probability for the TA to receive a reward is $p_s = 0$ and the probability to receive an inaction is $p_s = 1$, as shown in Fig. 10(b).

Note that places $cx1r$, $cx2r$, $cnx1r$, $cnx2r$, $cx1p$, $cx2p$, $cnx1p$, $cnx2p$, $cx1i$, $cx2i$, $cxn1i$ and $cxn2i$ are all common places used by the transitions corresponding to Type I and Type II feedbacks, where a token is marked to determine a reward, penalty or inaction for the TAs in the current clause.

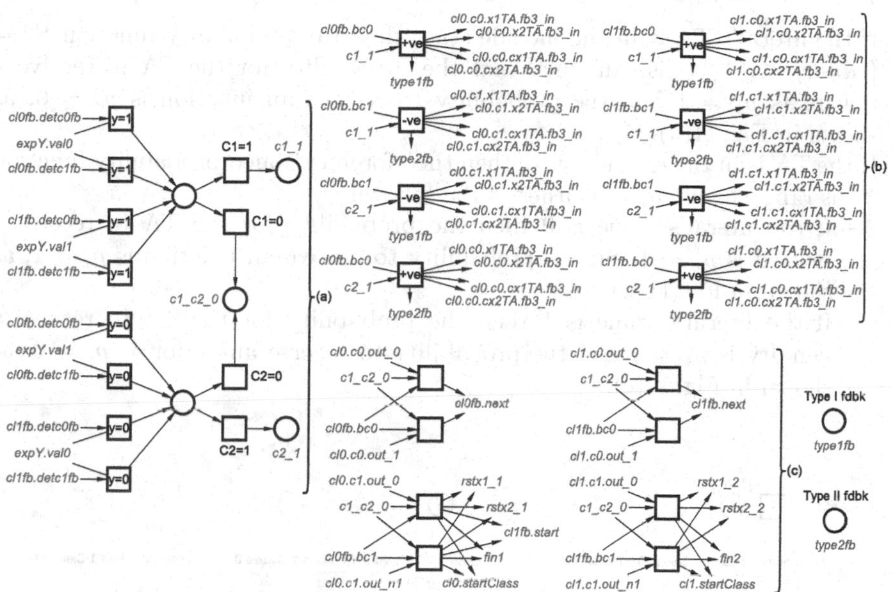

Fig. 9. Petri net segment specifying the probability calculations of the TM's feedback component. (a) Transitions corresponding to the probability functions C1 and C2. (b) Transitions corresponding to the operations of C1 = 1 and C2 = 1. (c) Transitions corresponding to the operations of C1 = 0 and C2 = 0.

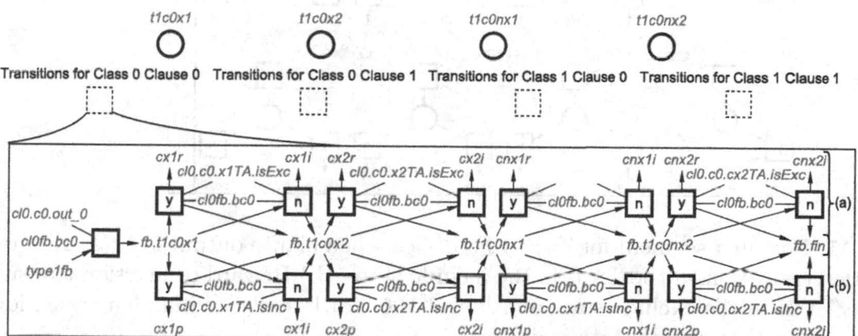

Fig. 10. Petri net segment for Type I Feedback when clause output is 0. (a) Reward or inaction transitions if the TA is in the exclude state. (b) Penalty or inaction transitions if the TA is in the include state.

Again, following Fig. 4(f)(iii), if Type I feedback is provided and the clause output is 1, then any of the following happens for each literal:

1. If the literal's TA is in the include state then the probability function $S2 = (rand() \leqslant (\frac{s-1}{s}))$ is ran, such that the probability for the TA to receive a reward is $p_s = 1$ and the probability to receive an inaction is $p_s = 0$, as shown in Fig. 11(a).
2. If the TA is in the exclude state then the aforementioned probability function $S1$ is ran, which leads to either of the following:
   - If the literal's value is 0 then the probability for the TA to receive a reward is $p_s = 0$ and the probability to receive an inaction is $p_s = 1$, as shown in Fig. 11(b).
   - If the literal's value is 1 then the probability for the TA to receive a penalty is $p_s = 1$ and the probability to receive an action is $p_s = 0$, as shown in Fig. 11(c).

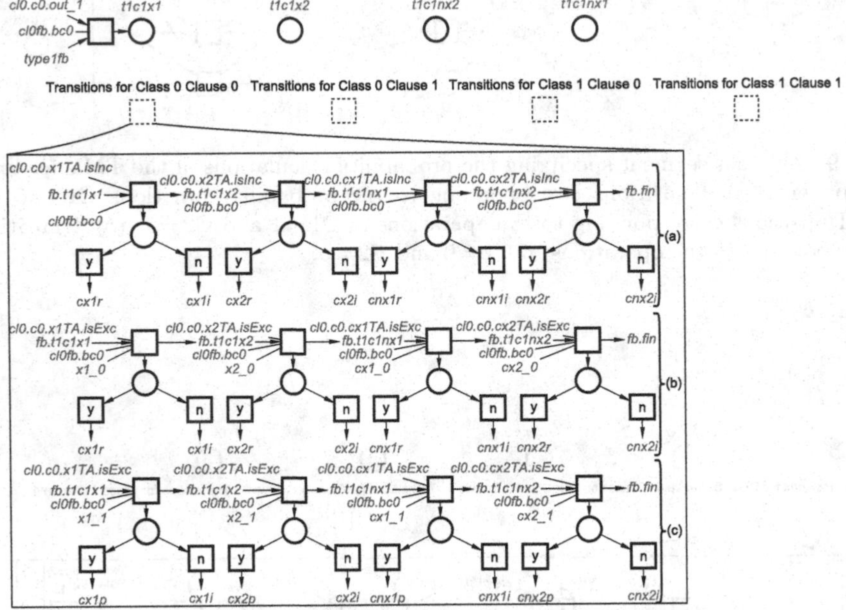

Fig. 11. Petri net segment for Type I Feedback when clause output is 1. (a) Reward or inaction transitions if the TA is in the include state. (b) Reward or inaction transitions if the TA is in the exclude state and literal is 0. (c) Penalty or inaction transitions if the TA is in the exclude state and literal is 1.

Now, following Fig. 4(f)(iv), if Type II feedback is provided and the clause output is 0, then the TAs of the current clause all receive an inaction regardless of their state and the values of their literals. This results in the Petri net segment shown in Fig. 12, where the appropriate transition is fired to go to the next clause, the next class, or the next datapoint.

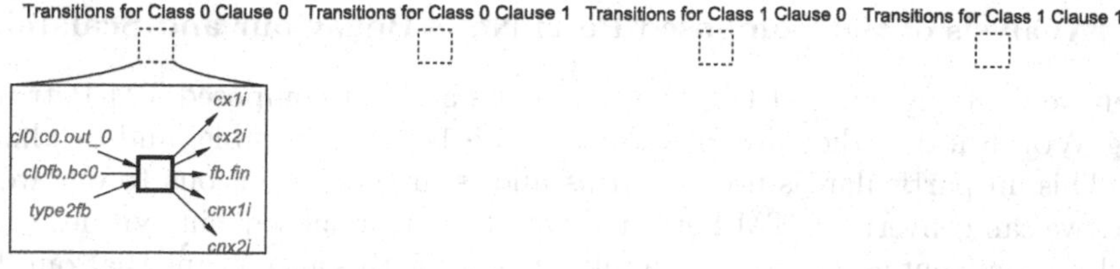

Fig. 12. Petri net segment for Type II Feedback when clause output is 0.

Again, following Fig. 4(f)(iv), if Type II feedback is provided and the clause output is 1, then any of the following happens for each literal:

- If the literal's TA is in the include state then the TA receives an inaction regardless of the literal's value, as shown in Fig. 13(a).
- If the TA is in the exclude state then:
  - The TA receives an inaction if the literal is 1, as shown in Fig. 13(b).
  - The TA receives a penalty if the literal is 0, as shown in Fig. 13(c).

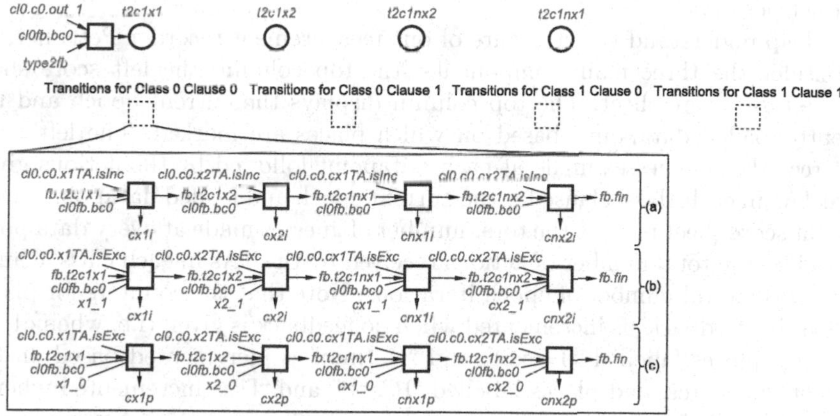

Fig. 13. Petri net segment for Type II Feedback when clause output is 1. (a) Inaction transitions if the TA is in the include state. (b) Inaction transitions if the TA is in the exclude state and literal is 1. (c) Penalty transitions if the TA is in the exclude state and literal is 0.

Now, by also completing the Petri net of the TM's feedback component, we can compose our two Petri nets together and create the necessary transitions to reset the TM for the reading the next datapoint, such that the tokens in the places corresponding to the features' values, literals' values, clause outputs, class sums and argmax output are emptied. Note that due to the large resulting size of the composed model, we decided to not include it to help preserve space.

3.4 Analysis of the Composed Petri Net's Behaviour and Scalability

When we verifying our TM Petri net segments and our composed TM Petri net using WORKCRAFT, they are all verified to be 1-safe, reversible, and deadlock-free. This, in particular, is useful as this allows further extensions to our work, where we can convert our TM Petri nets into STGs for subsequent synthesis into speed-independent asynchronous circuits. However, this also means that our TM Petri net is unrestricted meaning it can run indefinitely. This can be resolved by adding a place that corresponds to the TM's epochs, but this will also cause the TM Petri net to become unsafe. Thus, we decided to not include epochs in our blueprint designs of the TM Petri nets. Note that if one wishes to include epochs, they can add a place with k tokens, where $k \in \mathbb{N}$ is the user's desired number of epochs, and connect it to the first loading transition shown in Fig. 5.

To show that our TM Petri net is correctly learning the provided pattern and that their accuracy increases per epoch, we create another Petri net specifying a measurement recorder that can visually collect analytical data of the TM Petri net to measure its complexity and even some custom properties. This results in our measurement recorder Petri net shown in Fig. 14, where we collect analytical behaviour of our composed TM Petri net by recording the actions taken at each datapoint for each clause per class, before we calculate the accuracy of our TM Petri net per epoch.

To help understand the structure of our measurement recorder Petri net, let us consider the three main components: the top column, the left score sheet and the right score sheet. The top column displays the current epoch and the currently loaded datapoint, based on which places are marked. The left score sheet records the guesses made at each datapoint followed by the actions given to the TAs in each class' clause for the current epoch and loaded datapoint, while the right score sheet records the total number of guesses made at every data point followed by the total number of actions received by the TAs in each class' clause, based on the total number of epoch iterations. Note that the counters for places labelled "No feedback" is incremented when no feedback is given (i.e. when $c1 = 0$ or $c2 = 0$), places labelled "0 (exc)" and "(1 (inc)" is incremented based on the TA's current states, and places labelled "R", "P" and "I" is incremented when a transition labelled reward, penalty or inaction is fired within the TA respectively.

Also, to show that our TM Petri net scales linearly with respect to the number of components (i.e. the Petri net segments of TAs, clauses, and TM classes) that is required, we calculate the sizes of the blueprint Petri net segments (i.e. singular TM components) shown in Table 2 and the sizes of the used Petri net segments for our composed TM Petri net (i.e. n TM components) shown in Table 3, before comparing them. Note that in Table 2, the TA component contains six states (i.e. three include states and three exclude states), the clause component contains four TAs, and the TM class component contains two clauses. Additionally, the components named "Type I Clause 0", "Type I Clause 1", "Type II Clause 0" and "Type II Clause 1" are all part of the whole feedback module and also contain one clause. Moreover, we did not include the size of the inference's datapoint loader nor the size of the feedback's input reset mechanism, as they not only

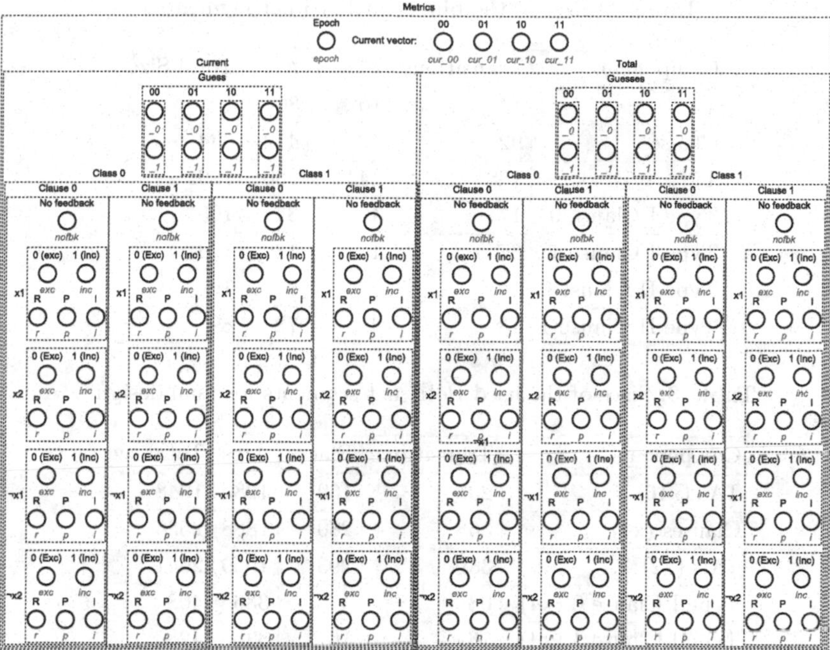

Fig. 14. Petri net of Measurement Recorder.

depend on a non-uniform number of features but also a non-uniform number of datapoints that do not follow a set structure, which may result in inaccurate size representations, e.g. we may have many features but very few datapoints or many datapoints but very few features. Note that the scalability of datasets is discussed as part of our future work in Sect. 5, where we consider other types of datasets like images.

When evaluating Tables 2 and 3, we can see that our composed TM Petri net scales well and grows relatively linear to the number of components that are required, e.g. the total size of a single TA is 103 while the total size of the sixteen TAs used in our TM Petri net is 1648 which is exactly a scale factor of 16. Also, in some instances, the actual size of some of the used components required less than the expected size, e.g. the total size of the "Type I Clause 1" feedback component for one clause is 173, whereas for four clauses this is 428.

Table 2. Sizes of the blueprint Petri net segments

Component	Transitions	Places	Arcs	Total Size
TA	13	10	80	103
Clause	92	70	472	634
TM	188	144	960	1292
Type I Clause 0	17	4	84	105
Type I Clause 1	37	16	120	173
Type II Clause 0	1	1	8	10
Type II Clause 1	13	4	72	89

Table 3. Sizes of the used Petri net segments for final design

Component	Transitions	Places	Arcs	Total size
TAs (x16)	208	160	1280	1648
Clauses (x4)	368	280	1888	2536
TMs (x2)	389	293	1959	2641
Type I Clause 0 (x4)	68	4	336	408
Type I Clause 1 (x4)	76	16	336	428
Type II Clause 0 (x4)	4	1	32	37
Type II Clause 1 (x4)	52	4	288	344

4 Benchmark

In our experiment, we used WORKCRAFT to simulate our composed TM Petri net and analysed its behaviour using our measurement recorder shown in Sect. 3.4. The size of our TM Petri net includes two classes that each contain two clauses, where each clause contains four TAs that corresponds to the four Boolean literals converted from the two Boolean features of every datapoint.

For our simulation, we completed 100 epochs and recorded the guess (i.e. the winning TM of the argmax function), the state of every TA, and the feedback provided to every TA at each datapoint per epoch. The former is used to determine the accuracy of our TM model by comparing it to the expected value, while the latter two are used to help understand the behaviour of our TM Petri net and the action taken at a specified datapoint and/or epoch. For each epoch, our TM Petri net required approximately 272 to 423 steps to complete, where the former step counter corresponds to when feedback is always not given and the latter step counter corresponds to when feedback is always given.

This led to the results shown in Fig. 15, where we calculated the average running accuracy for our TM Petri net. Note that we also include the average running accuracy of an XOR-based TM demo [17], where we use this TM as a reference to see how the behaviour of our TM Petri net correlates with the behaviour of the reference TM. To ensure fairness, we changed the parameters

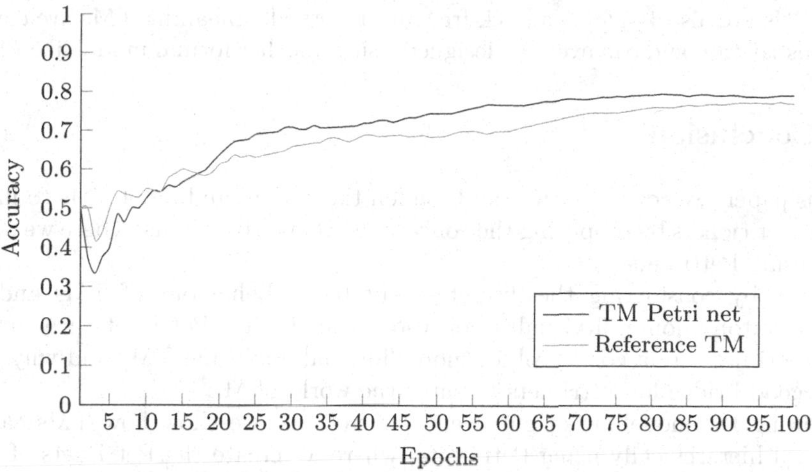

Fig. 15. Correlation between the Running Average Accuracy of the TM Petri net and the Running Average Accuracy of the Reference TM per Epoch.

of reference TM to match the TM Petri net's parameters in the WORKCRAFT simulation (e.g. the same initial states for the TAs). One exception is that the TM Petri net cannot match the probability checks of the reference TM due to the use of a random number generator, so we assume that the outcome of the TM Petri net's probability checks are what the TM would have decided at runtime.

In our graph, we included the accuracy of the TM ranging from 0% to 100% in normalised form (i.e. 0.0 to 1.0) on the left and the number of epochs of the TM (i.e. the number of completed iterations) on the bottom. The blue plot is the recorded average running accuracy of our TM Petri net, while the red plot is the recorded running accuracy of the reference TM. Note that the average running accuracy is calculated by summing the total accuracy and dividing it by the epoch (e.g. at epoch 20, our average running accuracy is $\frac{Acc_1+...+Acc_{20}}{20}$).

When analysing our experimental results, we can see that the behaviour of our TM Petri net matches the behaviour of the reference TM. In particular, we can see that the accuracy of both TMs decreases within the first 10 epochs, as they are required to fix the states of the TAs that were contradictory due to the randomly set initial states. But, after epoch 11, we can see how the accuracy for both TMs sharply increase due to more rewards from correct guesses, until the 70th epoch where the accuracy begins to stabilise around 75%.

An important observation from our results is that, rather than comparing the two TMs, we can see how our TM Petri net correctly imitates the reference TM, based on the correlating accuracy and even the behaviour. In fact, because our TM Petri net is 1-safe (with the removal of the measurement recorder), we can even convert our TM Petri net into another model to enable further extensions to the work (e.g. we can convert our TM Petri net into an STG for subsequent synthesis into an asynchronous circuit). Furthermore, we also show that TMs are also 1-safe, deadlock-free and reversible, meaning TMs are correct by construction and can even be designed using another formal model like FSMs.

5 Conclusion

In this paper, we cover how we can broaden the understanding of TMs to industrial practitioners by proposing the concept of event-driven TMs, where we model TMs using Petri nets.

Here, by considering the discrete-event based behaviour of TMs and the various automation and visualisation tools available for Petri nets, we surmise why Petri nets are a good tool for modelling and analysing TMs and why TMs are a good bridge for Petri nets to enter the world of ML.

During the modelling of our event-driven TMs, we show how TMs can be designed hierarchically using Petri nets, where we create the Petri nets of both the TM's inference and feedback components in stages, before composing them into the complete TM Petri net.

We then create a Petri net that specifies a measurement recorder, which we use to analyse the behaviour of our TM Petri net and record several metrics including the current epoch, the current loaded data point, the states of the TAs and the feedback given to each class' clause.

We also analyse the size of our TM Petri net, where we show that it scales well and grows linearly to the number of components (i.e. the blueprint Petri net segments) that are required.

In our experiment results, we show how the behaviour of our TM Petri net correlates with the behaviour of a reference TM, where the accuracy of the TMs decrease as it fixes the TAs' states before it sharply increases due to subsequent rewards from correct guesses. We also show that our TM Petri net is 1-safe, deadlock-free and reversible, meaning the TMs are correct by construction and can be easily designed using another formal model like FSMs.

Future Work

While further steps have been made towards the design of event-driven TMs, there are still several improvements that can be made to our approach.

Firstly, we consider the development of a new feature and/or plugin for WORKCRAFT, where the process of building a TM Petri net can be automated by importing a file that contains the TM's parameters (e.g. epochs, classes, clauses, TAs, and literals). This, in turn, simplifies the building process and even scales up the design of our TM Petri nets.

Secondly, we consider a dynamic approach in loading the TM's datapoints rather than the static approach used in our example to support other data types like images and to further scale up our TM Petri nets. Note that some initial investigations for this are underway as we analyse how a set of places and a set of complementary places can be used to represent a pixel of an image.

Thirdly, we consider the modelling of a random number generator at the level of Petri nets to support the probability functions used to determine if the TAs receive feedback or not, as WORKCRAFT's simulation does not support the use of random number generators, meaning the choices made to provide feedback had an equal chance. Note that some initial investigations for this are also underway as we explore how a set of transitions can be enabled in proportion to the value of s to mimic the probability functions $s1$ and $s2$, and how places can be used to represent individual digits that can be concatenated to generate some form of a randomised real number.

Finally, we consider how the parametric scaling of our TM Petri nets can be reached or even potentially optimised, by adopting some of the ideas from the Coloured Petri net formalism [10].

Acknowledgements. The authors would like to thank the anonymous reviewers for their feedback on this paper. This work was supported by EPSRC EP/X036006/1 as part of the Scalability Oriented Novel Network of Event Triggered Systems (SONNETS) project and by EPSRC EP/X039943/1 as part of the UKRI-RCN: Exploiting the dynamics of self-timed machine learning hardware (ESTEEM) project.

References

1. Workcraft (2006). https://workcraft.org/. Accessed 09 Jan 2024
2. Berge, G.T., Granmo, O.C., Tveit, T.O., Goodwin, M., Jiao, L., Matheussen, B.V.: Using the Tsetlin machine to learn human-interpretable rules for high-accuracy text categorization with medical applications. IEEE Access **7**, 115134–115146 (2019). https://doi.org/10.1109/ACCESS.2019.2935416
3. Bhattarai., B., Granmo., O., Jiao., L.: Measuring the novelty of natural language text using the conjunctive clauses of a Tsetlin machine text classifier. In: International Conference on Agents and Artificial Intelligence (ICAART), pp. 410–417. INSTICC, SciTePress (2021). https://doi.org/10.5220/0010382204100417
4. Blakely, C.D.: Tsetlin LOB: realtime regime learning and interpretable prediction in financial limit orderbooks using convolutional Tsetlin machines. In: International Symposium on the Tsetlin Machine (ISTM), pp. 13–20 (2022). https://doi.org/10.1109/ISTM54910.2022.00012
5. Blakely, C.D., Granmo, O.-C.: Closed-form expressions for global and local interpretation of Tsetlin machines. In: Fujita, H., Selamat, A., Lin, J.C.-W., Ali, M. (eds.) IEA/AIE 2021. LNCS (LNAI), vol. 12798, pp. 158–172. Springer, Cham (2021). https://doi.org/10.1007/978-3-030-79457-6_14
6. Chu, T.: Synthesis of self-timed VLSI circuits from graph-theoretic specifications. Technical report, Massachusetts Institute of Technology, Department of Electrical Engineering and Computer Science (1987)
7. Cortadella, J., Kishinevsky, M., Kondratyev, A., Lavagno, L., Yakovlev, A.: Petrify: a tool for manipulating concurrent specifications and synthesis of asynchronous controllers. IEICE Trans. Inf. Syst. **E80-D**, 315–325 (1997)
8. Gill, A.: Introduction to the Theory of Finite-State Machines. McGraw-Hill, New York (1962)
9. Granmo, O.C.: The Tsetlin machine - a game theoretic bandit driven approach to optimal pattern recognition with propositional logic. arXiv preprint arXiv:1804.01508 (2018). https://arxiv.org/abs/1804.01508

10. Jensen, K.: Coloured Petri Nets: Basic Concepts, Analysis Methods and Practical Use, vol. 2. Springer, Heidelberg (1995). https://doi.org/10.1007/978-3-662-03241-1
11. Khomenko, V., Koutny, M., Yakovlev, A.: Logic synthesis for asynchronous circuits based on Petri net unfoldings and incremental SAT. In: International Conference Application of Concurrency to System Design (ACSD), pp. 16–25 (2004)
12. Lan, T., Mao, G., Xia, F., Yu, S., Shafik, R., Yakovlev, A.: An asynchronous Tsetlin automaton architecture with integrated non-volatile memory. In: International Symposium on the Tsetlin Machine (ISTM), pp. 37–40 (2022). https://doi.org/10.1109/ISTM54910.2022.00015
13. Maheshwari, S., et al.: REDRESS: generating compressed models for edge inference using Tsetlin machines. IEEE Trans. Pattern Anal. Mach. Intell. **45**(9), 11152–11168 (2023). https://doi.org/10.1109/TPAMI.2023.3268415
14. Mao, G., Yakovlev, A., Xia, F., Lan, T., Yu, S., Shafik, R.: Automated synthesis of asynchronous Tsetlin machines on FPGA. In: IEEE International Conference on Electronics, Circuits and Systems (ICECS), pp. 1–4 (2022). https://doi.org/10.1109/ICECS202256217.2022.9970999
15. Mokhov, A., Rykunov, M., Sokolov, D., Yakovlev, A.: Towards reconfigurable processors for power-proportional computing. In: IEEE Faible Tension Faible Consommation (FTFC), pp. 1–4 (2013). https://doi.org/10.1109/FTFC.2013.6577770
16. Petri, C.A.: Kommunikation mit automaten (Communicating with automata). Ph.D. thesis (1962)
17. Rahman, T.: Tsetlin machine XOR demo (2021). https://github.com/tousifrahman/Tsetlin_Machine_XOR_demo. Accessed 16 Jan 2024
18. Rosenblum, L., Yakovlev, A.: Signal graphs: from self-timed to timed ones. In: International Workshop on Timed Petri Nets, pp. 199–206 (1985)
19. Tsetlin, M.L.: On behaviour of finite automata in random medium. Avtomat. i Telemekh **22**(10), 1345–1354 (1961)
20. Wheeldon, A., Yakovlev, A., Shafik, R.: Self-timed reinforcement learning using Tsetlin machine. In: International Symposium Asynchronous Circuits and Systems (ASYNC), pp. 40–47 (2021). https://doi.org/10.1109/ASYNC48570.2021.00014

Identifying Duplicates in Large Collections of Petri Nets and Nested-Unit Petri Nets

Pierre Bouvier[1,2] and Hubert Garavel[2(✉)]

[1] Kalray S.A, Montbonnot-Saint-Martin, France
[2] Univ. Grenoble Alpes, Inria, CNRS, Grenoble INP, LIG, 38000 Grenoble, France
{pierre.bouvier,hubert.garavel}@inria.fr

Abstract. We propose efficient techniques for detecting isomorphism between nets, i.e., for identifying, in large collections of (safe) Petri nets or Nested-Unit Petri Nets, all the nets that are identical modulo a permutation of places, a permutation of transitions, and/or a permutation of units. Our approach relies upon the successive application of diverse algorithms of increasing complexities: net signatures, net canonizations, and characterization of isomorphic nets in terms of isomorphic graphs. We implemented this approach in a complete tool chain that we successfully assessed on four collections, the largest of which comprises 241,000+ nets with many duplicates.

1 Introduction

The present work deals with large collections of Petri nets developed for non-regression testing or software competitions. Building and properly maintaining on the long run such collections, which gather hundreds or thousands of nets, requires a substantial amount of work. A common problem is the presence of *duplicates*, i.e., multiple occurrences of nets that are identical or very similar. Duplicates may be present for three reasons: (i) the collection consists of nets sent by different contributors; (ii) the collection is managed by several persons, who may insert the same net independently; (iii) duplicates may arise from transformations applied to existing models, e.g., conversion of colored Petri nets to P/T nets, removal of dead places or dead transitions [3], etc.

In practice, duplicates are undesirable for at least four reasons: (i) they waste disk space and backup storage, especially when nets are encoded in XML-based formats (such as the standard PNML format [11]), which are particularly verbose; (ii) they waste processor time in redundant calculations, which may be expensive due to state-space explosion issues; (iii) they may introduce biases in benchmarking experiments and software competitions by increasing the weight of certain nets unduly; (iv) their presence often raises time-consuming questions and debates between users and administrators of net collections.

The present article addresses this problem by proposing methods and tools to detect duplicates that may be present in existing net collections, and to prevent

L. M. Kristensen and J. M. van der Werf (Eds.): PETRI NETS 2024, LNCS 14628, pp. 379–401, 2024.
https://doi.org/10.1007/978-3-031-61433-0_18

duplicates from being created when new nets are inserted in collections under construction.

The classes of nets considered are one-safe P/T nets and NUPNs (Nested-Units Petri Nets) [7], an extension of Petri nets with the concept of *units*, which provide for modularity and hierarchy. Non-safe P/T nets are also partially supported in the sense that, for such nets, our approach produces over-approximated results, i.e., reports a superset of duplicates, possibly including false positives.

To formalize the problem, we consider that two nets are duplicates if there exist a bijective mapping between their places, their transitions and, in the case of NUPNs, their units. This mapping should preserve the arcs, the initial markings, and, in the case of NUPNs, the root units, the nesting of units, and the location of places in units. Such a definition extends the classical notion of *graph isomorphism* to Petri nets and NUPNs, keeping in mind that Petri nets are directed bipartite graphs, which NUPNs extend with a tree of units.

The detection of duplicates therefore amounts to the efficient partition of a set of nets according to such a *net isomorphism* relation. To the best of our knowledge, there is little prior work on this problem, probably because the construction of large net collections is a recent phenomenon.

The present article is organized as follows. Section 2 gives preliminary definitions. The next sections introduce the various approaches we developed for detecting duplicates: Sect. 3 exposes how the problem can be reduced to graph isomorphism; Sect. 4 presents the concept of net signatures; and Sect. 5 discusses the idea of net canonization. Section 6 presents the integration of all these approaches in a coherent tool chain. Section 7 gives experimental results obtained on four collections of Petri nets and NUPNs. Finally, Sect. 8 gives a few concluding remarks.

2 Definitions

2.1 Petri Nets and Nested-Unit Petri Nets

We briefly recall the usual definitions of Petri nets and refer the reader to classical surveys, e.g., [13], for a more detailed presentation of Petri nets.

Definition 1. *A (marked)* Petri Net *is a 4-tuple* (P, T, F, M_0) *where:*

1. *P is a finite, non-empty set; the elements of P are called* places.
2. *T is a finite set such that $P \cap T = \varnothing$; the elements of T are called* transitions.
3. *F is a subset of $(P \times T) \cup (T \times P)$; the elements of F are called* arcs.
4. *M_0 is a non-empty subset of P; M_0 is called the* initial marking.

Notice that the above definition only covers *ordinary* nets (i.e., it assumes all arc weights are equal to one). Also, it only considers *safe* nets (i.e., each place contains at most one token), which enables the initial marking to be defined as a subset of P, rather than a function $P \to \mathbb{N}$ as in the usual definition of P/T nets. We now recall the basic definition of a NUPN, referring the interested reader to [7] for a complete presentation of this model of computation.

Definition 2. *A (marked)* Nested-Unit Petri Net *(acronym: NUPN) is a 8-tuple* $(P, T, F, M_0, U, u_0, \sqsubseteq, \mathsf{unit})$ *where* (P, T, F, M_0) *is a Petri net, and where:*

5. U *is a finite, non-empty set such that* $U \cap T = U \cap P = \varnothing$*; the elements of* U *are called* units.
6. u_0 *is an element of* U*;* u_0 *is called the* root unit.
7. $\sqsubseteq$ *is a binary relation over* U *such that* $(U, \sqsupseteq)$ *is a tree with a single root* u_0*, where* $(\forall u_1, u_2 \in U)\ u_1 \sqsupseteq u_2 \stackrel{\text{def}}{=} u_2 \sqsubseteq u_1$*; intuitively*[1]*,* $u_1 \sqsubseteq u_2$ *expresses that unit* u_1 *is transitively nested in or equal to unit* u_2.
8. unit *is a function* $P \to U$ *such that* $(\forall u \in U \setminus \{u_0\})\ (\exists p \in P)\ \mathsf{unit}(p) = u$*; intuitively,* $\mathsf{unit}(p) = u$ *expresses that unit* u *directly contains place* p.

We now recall a few usual definitions for ordinary safe nets.

Definition 3. *Let* (P, T, F, M_0) *be a Petri Net.*

- *A* marking M *is defined as a set of places* $(M \subseteq P)$*. Each place belonging to a marking* M *is said to be* marked *or, also, to possess a* token.
- *The* pre-set *of a transition* t *is the set of places* ${}^\bullet t \stackrel{\text{def}}{=} \{p \in P \mid (p, t) \in F\}$.
- *The* post-set *of a transition* t *is the set of places* $t^\bullet \stackrel{\text{def}}{=} \{p \in P \mid (t, p) \in F\}$.
- *The* pre-set *of a place* p *is the set of transitions* ${}^\bullet p \stackrel{\text{def}}{=} \{t \in T \mid (t, p) \in F\}$.
- *The* post-set *of a place* p *is the set of transitions* $p^\bullet \stackrel{\text{def}}{=} \{t \in T \mid (p, t) \in F\}$.

Because NUPNs merely extend Petri nets by grouping places into units, Petri-net properties (including the standard firing rules for transitions) are preserved when NUPN information is added. Thus, all the concepts of Definition 3 for Petri nets also apply to NUPNs. The next definition provides useful notations used throughout this article.

Definition 4. *Let* $N = (P, T, F, M_0, U, u_0, \sqsubseteq, \mathsf{unit})$ *be a NUPN.*

- $u_1 \sqsubset u_2 \stackrel{\text{def}}{=} (u_1 \sqsubseteq u_2) \wedge (u_1 \neq u_2)$ *is the strict nesting partial order.*
- $\mathsf{disjoint}(u_1, u_2) \stackrel{\text{def}}{=} (u_1 \not\sqsubseteq u_2) \wedge (u_2 \not\sqsubseteq u_1)$ *characterizes pairs of units neither equal nor nested one in the other.*
- $\mathsf{places}(u) \stackrel{\text{def}}{=} \{p \in P \mid \mathsf{unit}(p) = u\}$ *gives all places directly contained in* u*; these are called the* local places *(or* proper places*) of* u.
- $\mathsf{places}^*(u) \stackrel{\text{def}}{=} \{p \in P \mid (\exists u' \in U)\ (u' \sqsubseteq u) \wedge (\mathsf{unit}(p) = u')\}$ *gives all places transitively contained in* u *or its sub-units.*
- $\mathsf{subunits}(u) \stackrel{\text{def}}{=} \{u' \in U \mid (u' \sqsubset u) \wedge (\nexists u'' \in U)\ (u' \sqsubset u'') \wedge (u'' \sqsubset u)\}$ *gives all units directly nested in* u.
- $\mathsf{subunits}^*(u) \stackrel{\text{def}}{=} \{u' \in U \mid (u' \sqsubset u)\}$ *gives all units transitively nested in* u.
- $\mathsf{leaf}(u) \stackrel{\text{def}}{=} (\mathsf{subunits}(u) = \varnothing)$ *characterizes the units having no nested sub-unit, i.e., the minimal elements of* $(U, \sqsubseteq)$.

[1] $\sqsubseteq$ is reflexive, antisymmetric, transitive, and u_0 is the greatest element of U for $\sqsubseteq$.

- $\mathsf{depth}(u)$ *is the length of the longest chain* $u \sqsubset ... \sqsubset u_0$ *of nested units from* u *to the root unit* u_0. *In particular,* $\mathsf{depth}(u_0) = 0$.
- $\mathsf{height}(u)$ *is the length, plus one, of the longest chain* $u_n \sqsubset ... \sqsubset u$ *of nested units from any leaf unit* u_n *to* u. *In particular,* $\mathsf{leaf}(u) \Leftrightarrow \mathsf{height}(u) = 1$.
- $\mathsf{width}(u)$ *is the number of leaf units contained in* $\{u\} \cup \mathsf{subunits}^*(u)$.
- *A* trivial *NUPN is such that* $\mathsf{width}(u_0)$ *equals the number of places* $\mathsf{card}(P)$, *meaning that the net carries no more NUPN information than a Petri net; in a trivial NUPN, each unit has a single local place, except the root unit* u_0, *which has either zero or one.*

Finally, a NUPN N is said to be *unit safe* [7] iff its underlying Petri net (P, T, F, M_0) is one-safe and, in any reachable marking M, all the places of M are contained in disjoint units.

2.2 Graph and Net Isomorphisms

We first recall the classical definitions of *graph* and *graph isomorphism.*

Definition 5. *A* vertex-colored directed graph *(or* colored graph *for short) is a 3-tuple* (V, E, c) *such that:* V *is a set of vertices,* $E \subseteq V \times V$ *is a set of (directed) edges, and* $c : V \to \mathbb{N}$ *is a function associating for each vertex a natural number representing a color. If the relation* E *is symmetric, then the graph is said to be* undirected.

Definition 6. *Two colored graphs* $G = (V, E, c)$ *and* $G' = (V', E', c')$ *are* isomorphic *iff there exists a bijection* $\pi_v : V \to V'$ *such that:*

- $(\forall v_1, v_2 \in V)\ (v_1, v_2) \in E \Leftrightarrow (\pi_v(v_1), \pi_v(v_2)) \in E'$.
- $(\forall v \in V)\ c(v) = c'(\pi_v(v))$.

We then define the concept of *net isomorphism* used throughout this article.

Definition 7. *Let* $N = (P, T, F, M_0, U, u_0, \sqsubseteq, \mathsf{unit})$ *and* $N' = (P', T', F', M'_0, U', u'_0, \sqsubseteq', \mathsf{unit}')$ *be two NUPNs.* N *and* N' *are said to be* isomorphic *iff there exist three bijections* $\pi_p : P \to P'$, $\pi_t : T \to T'$, *and* $\pi_u : U \to U'$ *such that:*

- $(\forall (p, t) \in P \times T)\ (p, t) \in F \Leftrightarrow (\pi_p(p), \pi_t(t)) \in F'$.
- $(\forall (t, p) \in T \times P)\ (t, p) \in F \Leftrightarrow (\pi_t(t), \pi_p(p)) \in F'$.
- $(\forall p \in P)\ p \in M_0 \Leftrightarrow \pi_p(p) \in M'_0$.
- $u'_0 = \pi_u(u_0)$.
- $(\forall u_1, u_2 \in U)\ u_1 \sqsubseteq u_2 \Leftrightarrow \pi_u(u_1) \sqsubseteq' \pi_u(u_2)$*—or, expressed in an equivalent way:* $(\forall u_1, u_2 \in U)\ u_1 \in \mathsf{subunits}(u_2) \Leftrightarrow \pi_u(u_1) \in \mathsf{subunits}'(\pi_u(u_2))$.
- $(\forall p \in P)\ \mathsf{unit}'(\pi_p(p)) = \pi_u(\mathsf{unit}(p))$.

Alternative definitions of net isomorphism can be found in the literature. The definition given in [10] takes into account places, place labels, transitions, transition labels, and arc weights, but not the initial marking. In [5,6] and [8], a less general definition of net isomorphism is given, in which only places are considered: two isomorphic Petri nets may have their places permuted but must have identical transitions. In [1,4], and [2], two nets are said to be isomorphic iff their marking graphs are isomorphic; notice that the left-to-right implication of this behavioural definition is also ensured by our purely structural Definition 7.

Finally, we recall the notion of disjoint union on sets and functions.

Definition 8. *Let S and S' be two disjoint sets. Let f and f' be two functions respectively defined on S and S'.*

- $S \uplus S'$ *denotes the* set union $S \cup S'$, *knowing that* $S \cap S' = \varnothing$.
- $f \uplus f'$ *denotes the* function union *of f and f', i.e., the function defined on $S \uplus S'$ such that $(f \uplus f')(x) = f(x)$ if $x \in S$ and $(f \uplus f')(x) = f'(x)$ if $x \in S'$.*

3 Net Isomorphism in Terms of Graph Isomorphism

Our first approach expresses net isomorphism in terms of graph isomorphism, a problem for which software tools are available [9].

3.1 Theoretical Aspects

Definition 9. *Let $N = (P, T, F, M_0, U, u_0, \sqsubseteq, \mathsf{unit})$ be a NUPN. We associate to N a colored directed graph $\mathcal{G}_N = (V, E, c)$ such that:*

- $V \stackrel{\text{def}}{=} P \uplus T \uplus U$.
- $E \stackrel{\text{def}}{=} F \uplus \{(p, u) \in P \times U \mid u = \mathsf{unit}(p)\} \uplus \{(u, u') \in U \times U \mid u \in \mathsf{subunits}(u')\}$.
- $(\forall v \in V)\ c(v) \stackrel{\text{def}}{=} 0$ *if* $v \in P \setminus M_0, 1$ *if* $v \in M_0, 2$ *if* $v \in T$, *or* 3 *if* $v \in U$.

The function $N \to \mathcal{G}_N$ is an injection, but not a surjection (as not every graph corresponds to a NUPN). The following definition computes the inverse function.

Definition 10. *Let $G = (V, E, c)$ be the colored directed graph associated to some NUPN via Definition 9. Let $\mathcal{N}_G \stackrel{\text{def}}{=} (P, T, F, M_0, U, u_0, \sqsubseteq, \mathsf{unit})$ be defined as follows:*

- $P \stackrel{\text{def}}{=} \{v \in V \mid c(v) \leq 1\}$.
- $T \stackrel{\text{def}}{=} \{v \in V \mid c(v) = 2\}$.
- $F \stackrel{\text{def}}{=} E \cap ((P \times T) \cup (T \times P))$.
- $M_0 \stackrel{\text{def}}{=} \{v \in V \mid c(v) = 1\}$.
- $U \stackrel{\text{def}}{=} \{v \in V \mid c(v) = 3\}$.
- u_0 *is the unique element such that* $(u_0 \in U) \wedge (E \cap (\{u_0\} \times U) = \varnothing)$.
- $\sqsubseteq$ *is the reflexive transitive closure of the relation* $E \cap (U \times U)$.
- unit *is the function* $p \mapsto u$ *such that* $(p, u) \in E \cap (P \times U)$.

It is easy to see that $\mathcal{N}_G$ is a NUPN, i.e., that: P, T, and U are pairwise disjoint; M_0 is not empty and contained in P; and $\sqsubseteq$ is a tree with a single root u_0.

Proposition 1. *Two NUPNs N and N' are isomorphic iff their corresponding graphs $\mathcal{G}_N$ and $\mathcal{G}_{N'}$ are isomorphic.*

Proof. Let $G \stackrel{\text{def}}{=} \mathcal{G}_N = (V, E, c)$ and let $G' \stackrel{\text{def}}{=} \mathcal{G}_{N'} = (V', E', c')$. By double implication. Direct: Let $N = (P, T, F, M_0, U, u_0, \sqsubseteq, \mathsf{unit})$ and let $N' = (P', T', F', M_0', U', u_0', \sqsubseteq', \mathsf{unit}')$. If N and N' are isomorphic, there exist three bijections π_p, π_t, and π_u satisfying the conditions of Definition 7. The function $\pi_v \stackrel{\text{def}}{=} \pi_p \uplus \pi_t \uplus \pi_u$ is a bijection from V to V' and satisfies both conditions of Definition 6: it preserves the edges (proven by disjunction of cases, depending whether each edge of E belongs to $P \times T$, $T \times P$, $U \times P$, or $U \times U$) and preserves the colors (also proven by disjunction of cases, depending whether each vertex of V belongs to $P \backslash M_0$, M_0, T, or U). Converse: Let $\mathcal{N}_G = (P, T, F, M_0, U, u_0, \sqsubseteq, \mathsf{unit})$ and let $\mathcal{N}_{G'} = (P', T', F', M_0', U', u_0', \sqsubseteq', \mathsf{unit}')$. If G and G' are isomorphic, there exists a bijection $\pi_v : V \to V'$ satisfying the two conditions of Definition 6. The second condition, the bijective nature of π_v, and the definitions of c and c' in Definition 9 imply that $\text{card}\,(P) = \text{card}\,(P')$, $\text{card}\,(T) = \text{card}\,(T')$, $\text{card}\,(M_0) = \text{card}\,(M_0)$, and $\text{card}\,(U) = \text{card}\,(U')$. Let $\pi_p \stackrel{\text{def}}{=} \pi_v|_P$, $\pi_t \stackrel{\text{def}}{=} \pi_v|_T$, and $\pi_u \stackrel{\text{def}}{=} \pi_v|_U$ be the restrictions of π_v to P, T, and U, respectively. Because π_v is a bijection, π_p is an injection from P to P', and even a bijection since $\text{card}\,(P) = \text{card}\,(P')$; similarly, π_t is a bijection from T to T' and π_u a bijection from U to U'. The six conditions of Definition 7 then follow from the combined assumptions of Definition 6 and Definition 9.

3.2 Practical Aspects

To assess on concrete examples the efficiency of the approach presented in Sect. 3.1, we selected the two reference tools dedicated to graph isomorphism, NAUTY and TRACES[2] [12] because of their high reputation of efficiency. These tools, which provide both an API and a command-line interface, can put a graph under canonical form or decide whether two graphs are isomorphic (i.e., iff their respective canonical graphs are identical).

As for benchmark, we selected the 1387 (non-colored) Petri Nets and NUPNs used for the 2022 edition of the Model Checking Contest, knowing that duplicates are present in this collection. Using a Python script implementing Definition 9, each net was translated to a graph in NAUTY/TRACES input format. We ran our experiments in parallel on the French Grid'5000 testbed[3], allocating to each model a dedicated server with 96 GB RAM and one hour of wallclock time.

The results were disappointing: NAUTY managed to put 310 graphs under canonical form (success rate: 22.4%) but failed on all other models, either due to lack of memory (on 8 graphs) or by hitting the one-hour timeout (on 1069 graphs). Furthermore, no duplicate was detected.

To improve these results, we did additional attempts in two directions: (i) devising alternative translations to the one of Definition 9, taking advantage

[2] https://pallini.di.uniroma1.it.

[3] https://www.grid5000.fr.

of the specificities of NAUTY to get as much performance as possible, and (ii) experimenting also with TRACES, which is more recent and slightly faster (by a few percents, as we observed) than NAUTY.

Rather than assigning to vertices four colors only (i.e., $\{0...3\}$ in Definition 9), one may increase the number of colors to better distinguish between the various vertices v_p associated to all places $p \in P$. For instance, one may choose $c(v_p) \stackrel{\text{def}}{=} \mathsf{depth}(\mathsf{unit}(p))$, together with $c(v) \stackrel{\text{def}}{=} \mathsf{height}(u_0)$ if $v \in T$ and $c(v) \stackrel{\text{def}}{=} \mathsf{height}(u_0) + 1$ if $v \in U$—keeping in mind that, for each $u \in U$, $\mathsf{depth}(u) < \mathsf{height}(u_0)$. With such colors, the information that a place p belongs to the initial marking M_0 can be expressed differently, e.g., by adding a looping arc (v_p, v_p) to E (noticing that E contains no arcs from P to P) or by adding a special vertex v_0 with a unique color and an arc (v_p, v_0).

Another idea is to reduce the number of vertices by no longer associating a vertex to each unit (i.e., $V \stackrel{\text{def}}{=} \{v_0\} \uplus P \uplus T$). In this approach, the root unit u_0, the function unit, and the relation $\sqsubseteq$ can be encoded by extra arcs, e.g., by adding an arc (v_0, v_p) for each place $p \in \mathsf{places}(u_0)$, and by adding an arc $(v_p, v_{p'})$ for each pair of places p and p' such that $\mathsf{unit}(p) \in \mathsf{subunits}(\mathsf{unit}(p'))$.

Because TRACES does not support directed graphs, we adapt the translation of Definition 9 by associating two unique vertices v_p and v'_p to each place $p \in P$, assigning distinctive colors to v_p and v'_p (e.g., $c(v_p) \stackrel{\text{def}}{=} 2 \times \mathsf{depth}(\mathsf{unit}(p))$ and $c(v'_p) \stackrel{\text{def}}{=} c(v_p) + 1$), and adding an edge $\{v_p, v'_p\}$ to express that both vertices are related to the same place. As before, a unique vertex v_t is associated to each transition $t \in T$. Then, each arc $(t, p) \in F$ is represented by an edge $\{v_t, v_p\}$ and each arc $(p, t) \in F$ is represented by an edge $\{v'_p, v_t\}$.

We implemented these ideas in five different translations, which we assessed on the aforementioned benchmark (2022 edition of the Model Checking Contest). In the most effective approach, NAUTY managed to put 498 graphs under canonical form (success rate: 35.9%) but failed due to lack of memory (on 15 graphs) or by hitting the one-hour timeout (on 874 graphs). Again, no duplicate was detected.

Thus, even if net isomorphism can theoretically be expressed in terms of graph isomorphism, this does not seem to be a practical solution. We now present alternative approaches specifically tailored for Petri nets and NUPNs.

4 Net Signatures

Our second approach is based on the idea of *net signature*, which borrows from the concepts of hash and checksum functions.

Definition 11. *A* net signature *(or* signature *for short) is a function* sig *defined on Petri nets or NUPNs, such that, for any two nets N and N', if N and N' are isomorphic, then* $\mathsf{sig}(N) = \mathsf{sig}(N')$.

In practice, one uses the contraposition of this implication: two nets having different signatures are not isomorphic. The reverse implication is not required:

two nets having the same signature are not necessarily isomorphic (there is a risk of collision between their signatures).

Proposition 2. *If* sig *is a signature, then for any net N and any permutation π of places, transitions, and/or units,* $\mathsf{sig}(\pi(N)) = \mathsf{sig}(N)$.

To discriminate as many nets as possible, a signature should be extensive enough to contain all information that is invariant by permutations, but it should also be fast to compute. We now introduce a few definitions upon which an effective signature can be built.

4.1 Multiset Hashing

Multisets are an extension of sets and may contain several instances of each element. We note multisets $\{\!|\ ...\ |\!\}$ to distinguish them from (normal) sets, noted $\{...\}$. Given a NUPN $N = (P, T, F, M_0, U, u_0, \sqsubseteq, \mathsf{unit})$, two examples of multisets are $\{\!|\ \mathsf{card}\,({}^\bullet t) \mid t \in T\ |\!\}$ and $\{\!|\ \mathsf{height}\,(u) \mid u \in U\ |\!\}$. To check the equality of such multisets, the lengths of which can be fixed or variable, we adopt a hash-based approach that converts each multiset into a (fixed-size) *digest*, such that the equality of two multisets implies the equality of the two corresponding digests. Thus, the chosen hash function is not assumed to be perfect (hash collisions may exist among digests). Yet, it is desirable that this function returns a result independent from the order of elements (multisets are not lists). A simple solution would be to sort the elements of a multiset and concatenate them to form a bit string on which some standard (cryptographic or not) hash function would be applied. However, this approach is slow (due to sorting, at least) and produces hash results that are not meaningful to humans. We thus adopt an alternative approach based on the following hash function, which does not require sorting and returns a tuple, many fields of which can be easily checked by inspection.

Definition 12. *Let a digest be a 5-tuple of natural numbers.*

- *Let* $\mathbb{D} \stackrel{\text{def}}{=} \mathbb{N}^5$ *denote the set of digests.*
- *For* $d \in \mathbb{D}$, *let* $d.\mathsf{card}$, $d.\mathsf{min}$, $d.\mathsf{max}$, $d.\mathsf{sum}$, *and* $d.\mathsf{prod}$ *denote, respectively, each of the five components of* d.
- *Let* $\mathcal{H}$: (**multiset of** $\mathbb{N}$) $\rightarrow \mathbb{D}$ *be the hash function defined as follows:* $\mathcal{H}(\varnothing) \stackrel{\text{def}}{=} (0, 0, 0, 0, 1)$ *and, for any natural* $n \geq 1$, $\mathcal{H}(\{\!|\ x_1, ..., x_n\ |\!\}) \stackrel{\text{def}}{=} (n, min(x_1, ..., x_n), max(x_1, ..., x_n), x_1 + ... + x_n, (2x_1 + r) \times ... \times (2x_n + r)/2)$, *where* r *is the constant* $2, 654, 435, 769$.
- *Let* $\mathcal{M}$: (**multiset of** $\mathbb{D}$) $\rightarrow \mathbb{D}$ *be the "hash-merge" function defined as follows:* $\mathcal{M}(\varnothing) \stackrel{\text{def}}{=} (0, 0, 0, 0, 1)$ *and, for any* $n \geq 1$, $\mathcal{M}(\{\!|\ d_1, ..., d_n\ |\!\}) \stackrel{\text{def}}{=} (d_1.\mathsf{card} + ... + d_n.\mathsf{card}, min(d_1.\mathsf{min}, ..., d_n.\mathsf{min}), max(d_1.\mathsf{max}, ..., d_n.\mathsf{max}), d_1.\mathsf{sum} + ... + d_n.\mathsf{sum}, (2 \times d_1.\mathsf{prod} + 1) \times ... \times (2 \times d_n.\mathsf{prod} + 1)/2)$.

Function $\mathcal{H}$ handles multisets of natural numbers, whereas function $\mathcal{M}$, at a higher level ("hash of hashes"), handles multisets of digests. Both functions can be computed by induction on the size n of their input multisets: there is no need for preliminary sorting, as all the operations involved in $\mathcal{H}$ and $\mathcal{M}$ are commutative and associative. In practice, the fields of $\mathbb{D}$ are implemented using machine integers, so that all arithmetical calculations are done modulo, e.g., 2^{32} or 2^{64}. All components of $\mathbb{D}$, but prod, are readable by humans and express meaningful properties of the corresponding multiset. Instead, prod uses a form of multiplicative hashing[4] that seeks to enhance dispersion for large multisets. Notice that, all factors of prod being odd, their product never becomes zero, even under modular arithmetic; the final division of this product by two eliminates the least significant bit, which is always equal to one.

4.2 Signature Function

We can propose a particular sig function defined on NUPNs; this function supports ordinary, safe Petri nets as a particular case (i.e., trivial NUPNs).

Definition 13. *Let* $N = (P, T, F, M_0, U, u_0, \sqsubseteq, \mathsf{unit})$ *be a NUPN. We define* $\mathsf{sig}(N)$ *to be a fixed-size tuple, each component of which is a natural number or a digest computed from* N*. The components are divided into three parts, respectively based on the places, transitions, and units of* N*. These parts, noted* $(h_1, ...)$*,* $(k_1, ...)$*, and* $(l_1, ...)$*, are defined below (see Definition 18, 19, and 20).*

4.3 Attributes for Places and Transitions

The definition of our signature function relies on various attributes computed for each place and transition of the net. These attributes contain information that helps differentiating the various places (resp. transitions). At first sight, all places are seemingly alike (except those of the initial marking), but they can be distinguished using local information (e.g., the number of arcs and transitions connected to them) as well as global information (e.g., their distance to other remarkable places of the net: initial places, sink places, etc.).

Definition 14. *Let* $N = (P, T, F, M_0)$ *be a Petri net. A set of places* $p_0, ..., p_n$ *and a set of transitions* $t_1, ..., t_n$ *are said to be a* chain of length n from p_0 to p_n *iff* $(\forall i \in \{1, ..., n\})\ (p_{i-1}, t_i) \in F \wedge (t_i, p_i) \in F$*. Given two places* p *and* p'*, the* distance from p to p' *is defined as the length of the shortest chain from* p *to* p'*; if no such chain exists, this distance is equal to* $\mathrm{card}\,(P) + 1$.

Definition 15. *To each place* p*, one associates three attributes:*

- $\mathsf{distance1}(p)$ *is defined as the minimal distance from* p *to any place of the initial marking* M_0.
- $\mathsf{distance2}(p)$ *is defined as the minimal distance from any place of the initial marking to* p.

[4] https://stackoverflow.com/questions/1536393/good-hash-function-for-permutations.

– $\mathsf{distance3}(p)$ *is defined as the minimal distance from p to any sink place (i.e., any place p' such that $p'^{\bullet} = \varnothing$).*

Definition 16. *To each transition t, one associates three Boolean attributes and six attributes of type $\mathbb{D}$:*

– $\mathsf{decreasing}(t) \stackrel{\text{def}}{=} (\mathrm{card}\,({}^{\bullet}t) > \mathrm{card}\,(t^{\bullet}))$
– $\mathsf{conservative}(t) \stackrel{\text{def}}{=} (\mathrm{card}\,({}^{\bullet}t) = \mathrm{card}\,(t^{\bullet}))$
– $\mathsf{increasing}(t) \stackrel{\text{def}}{=} (\mathrm{card}\,({}^{\bullet}t) < \mathrm{card}\,(t^{\bullet}))$
– *for* $i \in \{1,2,3\}$, $\mathsf{input\_distance}\,i(t) \stackrel{\text{def}}{=} \mathcal{H}(\{\!|\, \mathsf{distance}\,i(p) \mid p \in {}^{\bullet}t \,|\!\})$
– *for* $i \in \{1,2,3\}$, $\mathsf{output\_distance}\,i(t) \stackrel{\text{def}}{=} \mathcal{H}(\{\!|\, \mathsf{distance}\,i(p) \mid p \in t^{\bullet} \,|\!\})$

Definition 17. *To each place p, one associates seven natural-number attributes and sixteen attributes of type $\mathbb{D}$:*

– $\mathsf{nb\_loops}(p) \stackrel{\text{def}}{=} \mathrm{card}\,({}^{\bullet}p \cap p^{\bullet})$
– $\mathsf{nb\_decreasing\_input\_transitions}(p) \stackrel{\text{def}}{=} \mathrm{card}\,(\{t \in {}^{\bullet}p \mid \mathsf{decreasing}(t)\})$
– $\mathsf{nb\_conservative\_input\_transitions}(p) \stackrel{\text{def}}{=} \mathrm{card}\,(\{t \in {}^{\bullet}p \mid \mathsf{conservative}(t)\})$
– $\mathsf{nb\_increasing\_input\_transitions}(p) \stackrel{\text{def}}{=} \mathrm{card}\,(\{t \in {}^{\bullet}p \mid \mathsf{increasing}(t)\})$
– $\mathsf{nb\_decreasing\_output\_transitions}(p) \stackrel{\text{def}}{=} \mathrm{card}\,(\{t \in p^{\bullet} \mid \mathsf{decreasing}(t)\})$
– $\mathsf{nb\_conservative\_output\_transitions}(p) \stackrel{\text{def}}{=} \mathrm{card}\,(\{t \in p^{\bullet} \mid \mathsf{conservative}(t)\})$
– $\mathsf{nb\_increasing\_output\_transitions}(p) \stackrel{\text{def}}{=} \mathrm{card}\,(\{t \in p^{\bullet} \mid \mathsf{increasing}(t)\})$
– $\mathsf{pred\_nb\_input\_places}(p) \stackrel{\text{def}}{=} \mathcal{H}(\{\!|\, \mathrm{card}\,({}^{\bullet}t) \mid t \in {}^{\bullet}p \,|\!\})$
– $\mathsf{pred\_nb\_output\_places}(p) \stackrel{\text{def}}{=} \mathcal{H}(\{\!|\, \mathrm{card}\,(t^{\bullet}) \mid t \in {}^{\bullet}p \,|\!\})$
– $\mathsf{succ\_nb\_input\_places}(p) \stackrel{\text{def}}{=} \mathcal{H}(\{\!|\, \mathrm{card}\,({}^{\bullet}t) \mid t \in p^{\bullet} \,|\!\})$
– $\mathsf{succ\_nb\_output\_places}(p) \stackrel{\text{def}}{=} \mathcal{H}(\{\!|\, \mathrm{card}\,(t^{\bullet}) \mid t \in p^{\bullet} \,|\!\})$
– *for* $i \in \{1,2,3\}$, $\mathsf{pred\_input\_distance}\,i(p) \stackrel{\text{def}}{=} \mathcal{M}(\{\!|\, \mathsf{input\_distance}\,i(t) \mid t \in {}^{\bullet}p \,|\!\})$ *and* $\mathsf{pred\_output\_distance}\,i(p) \stackrel{\text{def}}{=} \mathcal{M}(\{\!|\, \mathsf{output\_distance}\,i(t) \mid t \in {}^{\bullet}p \,|\!\})$
– *for* $i \in \{1,2,3\}$, $\mathsf{succ\_input\_distance}\,i(p) \stackrel{\text{def}}{=} \mathcal{M}(\{\!|\, \mathsf{input\_distance}\,i(t) \mid t \in p^{\bullet} \,|\!\})$ *and* $\mathsf{succ\_output\_distance}\,i(p) \stackrel{\text{def}}{=} \mathcal{M}(\{\!|\, \mathsf{output\_distance}\,i(t) \mid t \in p^{\bullet} \,|\!\})$

4.4 Signature Part Based on Places

The first part of our signature function is defined as follows.

Definition 18. *Let $N = (P, T, F, M_0)$ be a Petri net. The* place-based part *of the* $\mathsf{sig}(N)$ *function of Definition 13 is a tuple $(h_1, ..., h_{16})$ of natural numbers or values of type $\mathbb{D}$. The components of this tuple are the following:*

– $h_1 \stackrel{\text{def}}{=} \mathrm{card}\,(P)$, *i.e., the number of places.*
– $h_2 \stackrel{\text{def}}{=} \mathcal{H}(\{\!|\, \mathsf{distance1}(p) \mid p \in P \,|\!\})$.
– $h_3 \stackrel{\text{def}}{=} \mathcal{H}(\{\!|\, \mathsf{distance2}(p) \mid p \in P \,|\!\})$.

- $h_4 \stackrel{\text{def}}{=} \mathcal{H}(\{\!|\ \mathsf{distance3}(p) \mid p \in P \,|\!\})$.
- $h_5 \stackrel{\text{def}}{=} \mathcal{H}(\{\!|\ \mathsf{nb\_loops}(p) \mid p \in P \,|\!\})$.
- $h_6 \stackrel{\text{def}}{=} \mathcal{H}(\{\!|\ \mathsf{nb\_decreasing\_input\_transitions}(p) \mid p \in P \,|\!\})$.
- $h_7 \stackrel{\text{def}}{=} \mathcal{H}(\{\!|\ \mathsf{nb\_conservative\_input\_transitions}(p) \mid p \in P \,|\!\})$.
- $h_8 \stackrel{\text{def}}{=} \mathcal{H}(\{\!|\ \mathsf{nb\_increasing\_input\_transitions}(p) \mid p \in P \,|\!\})$.
- $h_9 \stackrel{\text{def}}{=} \mathcal{H}(\{\!|\ \mathsf{nb\_decreasing\_output\_transitions}(p) \mid p \in P \,|\!\})$.
- $h_{10} \stackrel{\text{def}}{=} \mathcal{H}(\{\!|\ \mathsf{nb\_conservative\_output\_transitions}(p) \mid p \in P \,|\!\})$.
- $h_{11} \stackrel{\text{def}}{=} \mathcal{H}(\{\!|\ \mathsf{nb\_increasing\_output\_transitions}(p) \mid p \in P \,|\!\})$.
- $h_{12} \stackrel{\text{def}}{=} \mathcal{H}(\{\!|\ \mathsf{pair}(\mathrm{card}\,({}^\bullet p), \mathrm{card}\,(p^\bullet)) \mid p \in P \,|\!\})$, *where* $\mathsf{pair} : \mathbb{N} \times \mathbb{N} \to \mathbb{N}$ *is a pairing function that maps two natural numbers to a single one.*
- $h_{13} \stackrel{\text{def}}{=} \mathcal{M}(\{\!|\ \mathsf{pred\_nb\_input\_places}(p) \mid p \in P \,|\!\})$.
- $h_{14} \stackrel{\text{def}}{=} \mathcal{M}(\{\!|\ \mathsf{pred\_nb\_output\_places}(p) \mid p \in P \,|\!\})$.
- $h_{15} \stackrel{\text{def}}{=} \mathcal{M}(\{\!|\ \mathsf{succ\_nb\_input\_places}(p) \mid p \in P \,|\!\})$.
- $h_{16} \stackrel{\text{def}}{=} \mathcal{M}(\{\!|\ \mathsf{succ\_nb\_output\_places}(p) \mid p \in P \,|\!\})$.

The following components are excluded from the place-based part of the signature: $h_2.\mathsf{card}$, $h_3.\mathsf{card}$, ..., $h_{12}.\mathsf{card}$ *(because they are all equal to* h_1*);* $h_{14}.\mathsf{card}$ *(which is equal to* $h_{13}.\mathsf{card}$*);* $h_{16}.\mathsf{card}$ *(which is equal to* $h_{15}.\mathsf{card}$*); and* $h_{12}.\mathsf{sum}$ *(which a linear combination of* $h_6.\mathsf{sum}$, ..., $h_{11}.\mathsf{sum}$*).*

4.5 Signature Part Based on Transitions

The second part of our signature function is defined as follows.

Definition 19. *Let* $N = (P, T, F, M_0)$ *be a Petri net. The* transition-based part *of the* $\mathsf{sig}(N)$ *function of Definition 13 is a tuple* $(k_1, ..., k_3)$ *of natural numbers or values of type* $\mathbb{D}$*. The components of this tuple are the following:*

- $k_1 \stackrel{\text{def}}{=} \mathrm{card}\,(T)$, *i.e., the number of transitions.*
- $k_2 \stackrel{\text{def}}{=} \mathcal{H}(\{\!|\ \mathrm{card}\,({}^\bullet t) \mid t \in T \,|\!\})$.
- $k_3 \stackrel{\text{def}}{=} \mathcal{H}(\{\!|\ \mathrm{card}\,(t^\bullet) \mid t \in T \,|\!\})$.

The following components are excluded from the transition-based part of the signature: $k_2.\mathsf{card}$ *and* $k_3.\mathsf{card}$ *(because they are equal to* k_1*);* $k_2.\mathsf{sum}$ *(which is equal to* $h_9.\mathsf{sum} + h_{10}.\mathsf{sum} + h_{11}.\mathsf{sum}$*); and* $k_3.\mathsf{sum}$ *(which is equal to* $h_6.\mathsf{sum} + h_7.\mathsf{sum} + h_8.\mathsf{sum}$*).*

4.6 Signature Part Based on Units

The third part of our signature function is defined as follows.

Definition 20. *Let* $N = (P, T, F, M_0, U, u_0, \sqsubseteq, \mathsf{unit})$ *be a NUPN. The* unit-based part *of the* $\mathsf{sig}(N)$ *function of Definition 13 is a tuple* $(l_1, ..., l_{13})$ *of natural numbers or values of type* $\mathbb{D}$*. The components of this tuple are the following:*

- $l_1 \stackrel{\text{def}}{=} \text{card}(U)$, *i.e., the number of units.*
- $l_2 \stackrel{\text{def}}{=} \mathcal{H}(\{\!\!\{ \text{card}(\mathsf{subunits}(u)) \mid u \in U \}\!\!\})$.
- $l_3 \stackrel{\text{def}}{=} \mathcal{H}(\{\!\!\{ \text{card}(\mathsf{subunits}^*(u)) \mid u \in U \}\!\!\})$.
- $l_4 \stackrel{\text{def}}{=} \mathcal{H}(\{\!\!\{ \text{card}(\mathsf{places}(u)) \mid u \in U \}\!\!\})$.
- $l_5 \stackrel{\text{def}}{=} \mathcal{H}(\{\!\!\{ \text{card}(\mathsf{places}^*(u)) \mid u \in U \}\!\!\})$.
- $l_6 \stackrel{\text{def}}{=} \mathcal{H}(\{\!\!\{ \text{card}(\mathsf{places}(u) \cap M_0) \mid u \in U \}\!\!\})$.
- $l_7 \stackrel{\text{def}}{=} \mathcal{H}(\{\!\!\{ \text{card}(\mathsf{places}^*(u) \cap M_0) \mid u \in U \}\!\!\})$.
- $l_8 \stackrel{\text{def}}{=} \mathcal{H}(\{\!\!\{ \mathsf{depth}(u) \mid u \in U \}\!\!\})$.
- $l_9 \stackrel{\text{def}}{=} \mathcal{H}(\{\!\!\{ \mathsf{height}(u) \mid u \in U \}\!\!\})$.
- $l_{10} \stackrel{\text{def}}{=} \mathcal{H}(\{\!\!\{ \mathsf{width}(u) \mid u \in U \}\!\!\})$,
- $l_{11} \stackrel{\text{def}}{=} \mathcal{H}(\{\!\!\{ \mathsf{in}(u) \mid u \in U \}\!\!\})$, *where* $\mathsf{in}(u) \stackrel{\text{def}}{=} \sum_{t \in T} \text{card}({}^\bullet t \cap \mathsf{places}(u))$.
- $l_{12} \stackrel{\text{def}}{=} \mathcal{H}(\{\!\!\{ \mathsf{out}(u) \mid u \in U \}\!\!\})$, *where* $\mathsf{out}(u) \stackrel{\text{def}}{=} \sum_{t \in T} \text{card}(t^\bullet \cap \mathsf{places}(u))$.
- $l_{13} \stackrel{\text{def}}{=} \mathcal{H}(\{\!\!\{ \mathsf{mix}(\text{card}(\mathsf{subunits}(u)), \text{card}(\mathsf{subunits}^*(u)), \text{card}(\mathsf{places}(u)), \text{card}(\mathsf{places}^*(u)), \text{card}(\mathsf{places}(u) \cap M_0), \text{card}(\mathsf{places}^*(u) \cap M_0), \mathsf{depth}(u), \mathsf{height}(u), \mathsf{width}(u), \mathsf{in}(u), \mathsf{out}(u) \}\!\!\}$, *where* $\mathsf{mix} : \mathbb{N}^{11} \rightarrow \mathbb{N}$ *is a generalized pairing function.*

The following components are excluded from the unit-based part of the signature because their presence would be redundant: $l_2.\mathsf{card} = l_1$, $l_2.\mathsf{min} = 0$, $l_2.\mathsf{sum} = l_1 - 1$, $l_3.\mathsf{card} = l_1$, $l_3.\mathsf{min} = 0$, $l_4.\mathsf{card} = l_1$, $l_4.\mathsf{sum} = \text{card}(P)$, $l_5.\mathsf{card} = l_1$, $l_6.\mathsf{card} = l_1$, $l_6.\mathsf{min} \leq 1$ *and* $l_6.\mathsf{max} \leq 1$ *if N is unit safe,* $l_7.\mathsf{card} = l_1$, $l_7.\mathsf{min} \leq 1$ *if N is unit safe,* $l_7.\mathsf{max} = \text{card}(M_0)$, $l_8.\mathsf{card} = l_1$, $l_8.\mathsf{min} = 0$, $l_9.\mathsf{card} = l_1$, $l_9.\mathsf{min} = 1$, $l_9.\mathsf{max} = l_8.\mathsf{max} + 1$, $l_{10}.\mathsf{card} = l_1$, $l_{10}.\mathsf{min} = 1$, $l_{11}.\mathsf{card} = l_1$, $l_{12}.\mathsf{card} = l_1$, *and* $l_{13}.\mathsf{card} = l_1$.

5 Net Canonization

Our third approach relies on an idea derived from the concept of normal form.

Definition 21. *A* net canonization *(or* canonization *for short) is a function* can *defined on Petri nets or NUPNs, such that, for any two nets N and N', if* $\mathsf{can}(N) = \mathsf{can}(N')$, *then N and N' are isomorphic.*

The reverse implication is not required: two isomorphic nets do not have necessarily the same image by canonization. In the sequel, we propose a particular can function defined as the composition of three successive permutations of units, places, and transitions. Units are permuted first, because in a non-trivial NUPN, there are less units than places (in a trivial NUPN, $\text{card}(U) \leq \text{card}(P) + 1$); transitions are permuted last, because there are usually more transitions than places in a net.

Definition 22. *Let* $N = (P, T, F, M_0, U, u_0, \sqsubseteq, \mathsf{unit})$ *be a NUPN. We assume in this section that P (resp. T, U) is the natural range* $\{1, ..., \text{card}(P)\}$ *(resp.* $\{1, ..., \text{card}(T)\}$, $\{1, ..., \text{card}(U)\}$*) and that each place p (resp. each transition t, each unit u) is represented by a unique number noted $\#p$ (resp. $\#t$, $\#u$).*

- *Let* $\pi_{u[N]} : U \to U$ *denote a permutation of the units of* N*; the definition we chose for* $\pi_{u[N]}$ *is given below in Sect. 5.1.*
- *Let* $\pi_{p[N]} : P \to P$ *denote a permutation of the places of* N*; the definition we chose for* $\pi_{p[N]}$ *is given below in Sect. 5.2.*
- *Let* $\pi_{t[N]} : T \to T$ *denote a permutation of the transitions of* N*; the definition we chose for* $\pi_{t[N]}$ *is given below in Sect. 5.3.*
- *Let* N_1 *be the NUPN obtained by permuting the units of* N *with* $\pi_{u[N]}$*.*
- *Let* N_2 *be the NUPN obtained by permuting the places of* N_1 *with* $\pi_{p[N_1]}$*.*
- *Let* N_3 *be the NUPN obtained by permuting the transitions of* N_2 *with* $\pi_{t[N_2]}$*.*

Finally, we define can *to be the function that maps* N *to* N_3*.*

5.1 Unit Sorting

The unit-permutation function $\pi_{u[N]}$ mentioned in Definition 22 is defined as follows.

Definition 23. *Let* $N = (P, T, F, M_0, U, u_0, \sqsubseteq, \mathsf{unit})$ *be a NUPN. For each unit* u*, one builds a tuple* $m(u) \stackrel{\text{def}}{=} (m_1(u), ..., m_{35}(u))$ *of natural numbers or values of type* $\mathbb{D}$*. The components of this tuple are the following:*

- $m_1(u) \stackrel{\text{def}}{=} \mathsf{depth}\,(u)$.
- $m_2(u) \stackrel{\text{def}}{=} \mathsf{card}\,(\mathsf{subunits}\,(u))$.
- $m_3(u) \stackrel{\text{def}}{=} \mathsf{card}\,(\mathsf{places}\,(u))$.
- $m_4(u) \stackrel{\text{def}}{=} \mathsf{card}\,(\mathsf{places}\,(u) \cap M_0)$.
- $m_5(u) \stackrel{\text{def}}{=} \mathsf{card}\,(\mathsf{subunits}^*(u))$.
- $m_6(u) \stackrel{\text{def}}{=} \mathsf{card}\,(\mathsf{places}^*(u))$.
- $m_7(u) \stackrel{\text{def}}{=} \mathsf{card}\,(\mathsf{places}^*(u) \cap M_0)$.
- $m_8(u) \stackrel{\text{def}}{=} \mathsf{height}\,(u)$.
- $m_9(u) \stackrel{\text{def}}{=} \mathsf{width}\,(u)$.
- $m_{10}(u) \stackrel{\text{def}}{=} \mathcal{H}(\{\!|\, \mathsf{distance1}(p) \mid p \in \mathsf{places}\,(u) \,|\!\})$.
- $m_{11}(u) \stackrel{\text{def}}{=} \mathcal{H}(\{\!|\, \mathsf{distance2}(p) \mid p \in \mathsf{places}\,(u) \,|\!\})$.
- $m_{12}(u) \stackrel{\text{def}}{=} \mathcal{H}(\{\!|\, \mathsf{distance3}(p) \mid p \in \mathsf{places}\,(u) \,|\!\})$.
- $m_{13}(u) \stackrel{\text{def}}{=} \mathcal{H}(\{\!|\, \mathsf{nb\_loops}(p) \mid p \in \mathsf{places}\,(u) \,|\!\})$.
- $m_{14}(u) \stackrel{\text{def}}{=} \mathcal{H}(\{\!|\, \mathsf{nb\_decreasing\_input\_transitions}(p) \mid p \in \mathsf{places}\,(u) \,|\!\})$.
- $m_{15}(u) \stackrel{\text{def}}{=} \mathcal{H}(\{\!|\, \mathsf{nb\_conservative\_input\_transitions}(p) \mid p \in \mathsf{places}\,(u) \,|\!\})$.
- $m_{16}(u) \stackrel{\text{def}}{=} \mathcal{H}(\{\!|\, \mathsf{nb\_increasing\_input\_transitions}(p) \mid p \in \mathsf{places}\,(u) \,|\!\})$.
- $m_{17}(u) \stackrel{\text{def}}{=} \mathcal{H}(\{\!|\, \mathsf{nb\_decreasing\_output\_transitions}(p) \mid p \in \mathsf{places}\,(u) \,|\!\})$.
- $m_{18}(u) \stackrel{\text{def}}{=} \mathcal{H}(\{\!|\, \mathsf{nb\_conservative\_output\_transitions}(p) \mid p \in \mathsf{places}\,(u) \,|\!\})$.
- $m_{19}(u) \stackrel{\text{def}}{=} \mathcal{H}(\{\!|\, \mathsf{nb\_increasing\_output\_transitions}(p) \mid p \in \mathsf{places}\,(u) \,|\!\})$.
- $m_{20}(u) \stackrel{\text{def}}{=} \mathcal{M}(\{\!|\, \mathsf{pred\_nb\_input\_places}(p) \mid p \in \mathsf{places}\,(u) \,|\!\})$.

- $m_{21}(u) \stackrel{\text{def}}{=} \mathcal{M}(\{\!|\, \mathsf{pred\_nb\_output\_places}(p) \mid p \in \mathsf{places}\,(u) \,|\!\})$.
- $m_{22}(u) \stackrel{\text{def}}{=} \mathcal{M}(\{\!|\, \mathsf{succ\_nb\_input\_places}(p) \mid p \in \mathsf{places}\,(u) \,|\!\})$.
- $m_{23}(u) \stackrel{\text{def}}{=} \mathcal{M}(\{\!|\, \mathsf{succ\_nb\_output\_places}(p) \mid p \in \mathsf{places}\,(u) \,|\!\})$.
- $m_{24}(u) \stackrel{\text{def}}{=} \mathcal{M}(\{\!|\, \mathsf{pred\_input\_distance1}(p) \mid p \in \mathsf{places}\,(u) \,|\!\})$.
- $m_{25}(u) \stackrel{\text{def}}{=} \mathcal{M}(\{\!|\, \mathsf{pred\_output\_distance1}(p) \mid p \in \mathsf{places}\,(u) \,|\!\})$.
- $m_{26}(u) \stackrel{\text{def}}{=} \mathcal{M}(\{\!|\, \mathsf{succ\_input\_distance1}(p) \mid p \in \mathsf{places}\,(u) \,|\!\})$.
- $m_{27}(u) \stackrel{\text{def}}{=} \mathcal{M}(\{\!|\, \mathsf{succ\_output\_distance1}(p) \mid p \in \mathsf{places}\,(u) \,|\!\})$.
- $m_{28}(u) \stackrel{\text{def}}{=} \mathcal{M}(\{\!|\, \mathsf{pred\_input\_distance2}(p) \mid p \in \mathsf{places}\,(u) \,|\!\})$.
- $m_{29}(u) \stackrel{\text{def}}{=} \mathcal{M}(\{\!|\, \mathsf{pred\_output\_distance2}(p) \mid p \in \mathsf{places}\,(u) \,|\!\})$.
- $m_{30}(u) \stackrel{\text{def}}{=} \mathcal{M}(\{\!|\, \mathsf{succ\_input\_distance2}(p) \mid p \in \mathsf{places}\,(u) \,|\!\})$.
- $m_{31}(u) \stackrel{\text{def}}{=} \mathcal{M}(\{\!|\, \mathsf{succ\_output\_distance2}(p) \mid p \in \mathsf{places}\,(u) \,|\!\})$.
- $m_{32}(u) \stackrel{\text{def}}{=} \mathcal{M}(\{\!|\, \mathsf{pred\_input\_distance3}(p) \mid p \in \mathsf{places}\,(u) \,|\!\})$.
- $m_{33}(u) \stackrel{\text{def}}{=} \mathcal{M}(\{\!|\, \mathsf{pred\_output\_distance3}(p) \mid p \in \mathsf{places}\,(u) \,|\!\})$.
- $m_{34}(u) \stackrel{\text{def}}{=} \mathcal{M}(\{\!|\, \mathsf{succ\_input\_distance3}(p) \mid p \in \mathsf{places}\,(u) \,|\!\})$.
- $m_{35}(u) \stackrel{\text{def}}{=} \mathcal{M}(\{\!|\, \mathsf{succ\_output\_distance3}(p) \mid p \in \mathsf{places}\,(u) \,|\!\})$.

The following components are excluded from $m(u)$*:* $m_{10}(u).\mathsf{card}$, $m_{11}(u).\mathsf{card}$, ..., $m_{18}(u).\mathsf{card}$ *(because they are all equal to* $m_3(u)$*);* $m_{21}(u).\mathsf{card}$ *(which is equal to* $m_{20}(u).\mathsf{card}$*);* $m_{23}(u).\mathsf{card}$ *(which is equal to* $m_{22}(u).\mathsf{card}$*);* $m_{24}(u).\mathsf{card}$, $m_{28}(u).\mathsf{card}$, *and* $m_{32}(u).\mathsf{card}$ *(which are all equal to* $m_{20}(u).\mathsf{sum}$*);* $m_{25}(u).\mathsf{card}$, $m_{29}(u).\mathsf{card}$, *and* $m_{33}(u).\mathsf{card}$ *(which are all equal to* $m_{21}(u).\mathsf{sum}$*);* $m_{26}(u).\mathsf{card}$, $m_{30}(u).\mathsf{card}$, *and* $m_{34}(u).\mathsf{card}$ *(which are all equal to* $m_{22}(u).\mathsf{sum}$*); and* $m_{27}(u).\mathsf{card}$, $m_{31}(u).\mathsf{card}$, *and* $m_{35}(u).\mathsf{card}$ *(which are all equal to* $m_{23}(u).\mathsf{sum}$*).*

Definition 24. *Let* $N = (P, T, F, M_0, U, u_0, \sqsubseteq, \mathsf{unit})$ *be a NUPN. The function* $\pi_{u[N]}$ *mentioned in Definition 22 is defined to be any permutation* $\pi : U \to U$ *such that* $(\forall u, u' \in U)\ \#u \leq \#u' \Rightarrow m(\pi(u)) \preceq m(\pi(u'))$*, where* $\preceq$ *is the lexicographic order over tuples; thus,* π *sorts all units* u *by increasing values of* $m(u)$*.*

In practice, one can obtain a unique permutation π by extending the tuple $m(u)$ with an extra component $m_{36}(u) \stackrel{\text{def}}{=} \#u$. Doing so guarantees that π is a stable sort, i.e., does not permute indistinguishable units needlessly.

5.2 Place Sorting

The place-permutation function $\pi_{p[N]}$ mentioned in Definition 22 is defined as follows.

Definition 25. *Let* $N = (P, T, F, M_0, U, u_0, \sqsubseteq, \mathsf{unit})$ *be a NUPN. For each place* u*, one builds a tuple* $n(p) \stackrel{\text{def}}{=} (n_1(p), ..., n_{27}(p))$ *of natural numbers or values of type* $\mathbb{D}$*. The components of this tuple are the following:*

- $n_1(p) \stackrel{\text{def}}{=} \#(\mathsf{unit}\,(p))$.
- $n_2(p) \stackrel{\text{def}}{=} \mathsf{distance1}(p)$.
- $n_3(p) \stackrel{\text{def}}{=} \mathsf{distance2}(p)$.
- $n_4(p) \stackrel{\text{def}}{=} \mathsf{distance3}(p)$.
- $n_5(p) \stackrel{\text{def}}{=} \mathsf{nb\_loops}(p)$.
- $n_6(p) \stackrel{\text{def}}{=} \mathsf{nb\_decreasing\_input\_transitions}(p)$.
- $n_7(p) \stackrel{\text{def}}{=} \mathsf{nb\_conservative\_input\_transitions}(p)$.
- $n_8(p) \stackrel{\text{def}}{=} \mathsf{nb\_increasing\_input\_transitions}(p)$.
- $n_9(p) \stackrel{\text{def}}{=} \mathsf{nb\_decreasing\_output\_transitions}(p)$.
- $n_{10}(p) \stackrel{\text{def}}{=} \mathsf{nb\_conservative\_output\_transitions}(p)$.
- $n_{11}(p) \stackrel{\text{def}}{=} \mathsf{nb\_increasing\_output\_transitions}(p)$.
- $n_{12}(p) \stackrel{\text{def}}{=} \mathsf{pred\_nb\_input\_places}(p)$.
- $n_{13}(p) \stackrel{\text{def}}{=} \mathsf{pred\_nb\_input\_places}(p)$.
- $n_{14}(p) \stackrel{\text{def}}{=} \mathsf{succ\_nb\_input\_places}(p)$.
- $n_{15}(p) \stackrel{\text{def}}{=} \mathsf{succ\_nb\_input\_places}(p)$.
- $n_{16}(p) \stackrel{\text{def}}{=} \mathsf{pred\_input\_distance1}(p)$.
- $n_{17}(p) \stackrel{\text{def}}{=} \mathsf{pred\_output\_distance1}(p)$.
- $n_{18}(p) \stackrel{\text{def}}{=} \mathsf{succ\_input\_distance1}(p)$.
- $n_{19}(p) \stackrel{\text{def}}{=} \mathsf{succ\_output\_distance1}(p)$.
- $n_{20}(p) \stackrel{\text{def}}{=} \mathsf{pred\_input\_distance2}(p)$.
- $n_{21}(p) \stackrel{\text{def}}{=} \mathsf{pred\_output\_distance2}(p)$.
- $n_{22}(p) \stackrel{\text{def}}{=} \mathsf{succ\_input\_distance2}(p)$.
- $n_{23}(p) \stackrel{\text{def}}{=} \mathsf{succ\_output\_distance2}(p)$.
- $n_{24}(p) \stackrel{\text{def}}{=} \mathsf{pred\_input\_distance3}(p)$.
- $n_{25}(p) \stackrel{\text{def}}{=} \mathsf{pred\_output\_distance3}(p)$.
- $n_{26}(p) \stackrel{\text{def}}{=} \mathsf{succ\_input\_distance3}(p)$.
- $n_{27}(p) \stackrel{\text{def}}{=} \mathsf{succ\_output\_distance3}(p)$.

The following components are excluded from $n(p)$: $n_{12}(p)$.`card` *and $n_{13}(p)$.*`card` *(because they are equal to $n_6(p) + n_7(p) + n_8(p)$); $n_{14}(p)$.*`card` *and $n_{15}(p)$.*`card` *(because they are equal to $n_9(p)+n_{10}(p)+n_{11}(p)$); $n_{16}(p)$.*`card`*, $n_{20}(p)$.*`card`*, and $n_{24}(p)$.*`card` *(which are all equal to $n_{12}(p)$.*`sum`*); $n_{17}(p)$.*`card`*, $n_{21}(p)$.*`card`*, and $n_{25}(p)$.*`card` *(which are all equal to $n_{13}(p)$.*`sum`*); $n_{18}(p)$.*`card`*, $n_{22}(p)$.*`card`*, and $n_{26}(p)$.*`card` *(which are all equal to $n_{14}(p)$.*`sum`*); and $n_{19}(p)$.*`card`*, $n_{23}(p)$.*`card`*, and $n_{27}(p)$.*`card` *(which are all equal to $n_{15}(p)$.*`sum`*).*

Definition 26. *Let $N = (P, T, F, M_0, U, u_0, \sqsubseteq, \mathsf{unit}\,)$ be a NUPN. The function $\pi_{p[N]}$ mentioned in Definition 22 is defined to be any permutation $\pi : P \rightarrow P$ such that $(\forall p, p' \in P)\ \#p \leq \#p' \Rightarrow n(\pi(p)) \preceq n(\pi(p'))$, where $\preceq$ is the lexicographic order over tuples; thus, π sorts all places p by increasing values of $n(p)$.*

In practice, an extra component $n_{28}(p) \stackrel{\text{def}}{=} \#p$ can be added to tuple $n(p)$ to obtain a unique permutation π that is also a stable sort.

5.3 Transition Sorting

The transition-permutation function $\pi_{t[N]}$ of Definition 22 is defined as follows.

Definition 27. *Let* $N = (P, T, F, M_0, U, u_0, \sqsubseteq, \mathsf{unit})$ *be a NUPN. For each transition* t*, one builds an tuple* $o(t) \stackrel{\text{def}}{=} (o_1(t), o_2(t))$ *of natural numbers or values of type* $\mathbb{D}$*. The components of this tuple are the following:*

- $o_1(t) \stackrel{\text{def}}{=} \mathcal{H}(\{\!|\, \#p \mid p \in {}^{\bullet}t \,|\!\})$*, noticing that* $o_1(t).\mathtt{card} = \text{card}\,({}^{\bullet}t)$*.*
- $o_2(t) \stackrel{\text{def}}{=} \mathcal{H}(\{\!|\, \#p \mid p \in t^{\bullet} \,|\!\})$*, noticing that* $o_2(t).\mathtt{card} = \text{card}\,(t^{\bullet})$*.*

Definition 28. *Let* $N = (P, T, F, M_0, U, u_0, \sqsubseteq, \mathsf{unit})$ *be a NUPN. The function* $\pi_{t[N]}$ *mentioned in Definition 22 is defined to be any permutation* $\pi : T \to T$ *such that* $(\forall t, t' \in T)\ \#t \leq \#t' \Rightarrow o(\pi(t)) \preceq o(\pi(t'))$*, where* $\preceq$ *is the lexicographic order over tuples; thus,* π *sorts all transitions* t *by increasing values of* $o(t)$*.*

In practice, an extra component $o_3(t) \stackrel{\text{def}}{=} \#t$ can be added to tuple $o(t)$ to obtain a unique permutation π that is also a stable sort.

5.4 Unique Sorting

As mentioned above, the reverse implication of Definition 21 is not guaranteed: the canonization function can, applied to two isomorphic nets, may return different results. This is especially the case when the nets contain fragments that are locally symmetric (e.g., circular rings, complete subgraphs, etc.), for which several permutations exist. In other cases, however, the reverse implication may hold.

Proposition 3. *Let* N *and* N' *be two NUPNs. If each of the three permutations* $\pi_{u[N]}$*,* $\pi_{p[N_1]}$*, and* $\pi_{t[N_2]}$ *mentioned in Definition 22 to compute* $\mathsf{can}(N)$ *is unique i.e., if lexicographic order* $\preceq$ *over the tuples* m*,* n*, and* o *(see Definition 24, 26, and 28) defines three total order relations, then* N *and* N' *are isomorphic iff* $\mathsf{can}(N) = \mathsf{can}(N')$*.*

Proof. Let $N = (P, T, F, M_0, U, u_0, \sqsubseteq, \mathsf{unit})$ and $N' = (P', T', F', M_0', U', u_0', \sqsubseteq', \mathsf{unit}')$ be two isomorphic NUPNs. Let (π_p, π_t, π_u) be the three bijections of Definition 7 relating N and N'. First: for each $u \in U$ and $i \in \{1, ..., 35\}$, one can prove that $m_i(u) = m_i'(\pi_u(u))$ by combining Definition 7 and Definition 23; if $\pi_u[N]$ is unique, then $\pi_u[N']$ is unique too, and the two NUPNs N_1 and N_1' obtained from N and N' by applying $\pi_u[N]$ and $\pi_u[N']$, respectively, are isomorphic and related by the three bijections (π_p, π_t, id), where id is the identity function on $\mathbb{N}$. Second: for each $p \in P$ and $i \in \{1, ..., 27\}$, one can prove that $n_i(p) = n_i'(\pi_p(p))$ by combining Definition 7 and Definition 25; similarly, the two NUPNs N_2 and N_2' obtained from N_1 and N_1' by applying $\pi_p[N_1]$ and $\pi_p[N_1']$, respectively, are isomorphic and related by the three bijections (id, π_t, id). Third: for each $t \in T$ and $i \in \{1, 2\}$, one can prove that $o_i(t) = o_i'(\pi_t(t))$ by combining Definition 7 and Definition 27; similarly, the two NUPNs N_3 and N_3' obtained from N_2 and N_2' by applying $\pi_t[N_2]$ and $\pi_t[N_2']$, respectively, are related by the three bijections (id, id, id). Therefore, $\mathsf{can}(N) = \mathsf{can}(N')$.

6 Implementation

We implemented these ideas in a software tool chain that combines: (i) tools specifically developed for the purpose of the present article; (ii) tools already developed at INRIA Grenoble, which we extended for the same purpose; and (iii) third-party tools that we reused without modification.

Our tool chain takes as input a collection of nets and determines which ones are isomorphic. It accepts Petri nets and NUPNs given in the standard PNML format [11] or in the "`.nupn`" format[5] (conversion from PNML to "`.nupn`" format can be done using the PNML2NUPN translator[6]). Our tool chain implements all the approaches presented in Sects. 3–5, ordered by increasing complexities; the tool chain stops as soon has all the duplicates in a collection have been found. All steps of the tool chain have been carefully validated using miscellaneous techniques that cannot be presented here by lack of space.

6.1 File Deduplication

The first and simplest way to search for duplicates in a collection of nets is to search for identical files. Among the many tools for this purpose, we selected FDUPES[7] which is a fast, reliable Unix command-line tool; the alternative (seemingly faster) tool JDUPES[8] is also a good option.

Obviously, this approach is very limited. For instance, inserting an extra space in a PNML file may prevent two isomorphic nets from being detected this way. There is thus a clear trade-off between the flexibility of a net format and the ability to detect duplicates using mere file comparison. In this respect, the "`.nupn`" format is preferable to PNML because it is more stringent: places, transitions, and units are named using natural numbers instead of alphanumeric identifiers; lexical tokens must be separated using exactly one space; blank lines are forbidden, as well as trailing spaces before end of lines, etc. For this reason, our toolchain employs the "`.nupn`" format rather than PNML. In practice, translation from PNML to "`.nupn`", followed by an invocation of FDUPES, is often sufficient to detect duplicates that do not involve permutations.

6.2 Pre-canonization

Even if the "`.nupn`" format is more stringent than PNML, it still offers a degree of flexibility that allows a given net to be expressed under different forms, even in absence of any permutation of places, transitions, or units. To address this problem, the NUPN_INFO tool[9] has been extended with a "`-precanonical-nupn`" option that takes as input a net in "`.nupn`" format and produces as output the

[5] https://cadp.inria.fr/man/nupn.html.
[6] http://pnml.lip6.fr/pnml2nupn.
[7] https://github.com/adrianlopezroche/fdupes.
[8] https://github.com/jbruchon/jdupes.
[9] https://cadp.inria.fr/man/nupn_info.html.

same net in which: (i) all places (resp. transitions and units) are renumbered starting from zero; (ii) all lists of places (resp. transitions and units) are sorted by increasing numbers; (iii) all labels of places (resp. transitions and units) are deleted; and (iv) all pragmas are removed. After putting all nets under such pre-canonical form, FDUPES is invoked to detect duplicate files.

6.3 Signatures

We extended the CÆSAR.BDD tool[10] with a "`-signature`" option that computes the signature (as defined in Sect. 4) of a net given in "`.nupn`" format. Written in C, the computation is fast (0.12 s per net on average) and always succeeds. A shell script identifies classes of nets with the same signatures.

To check the correctness of signatures, we developed a Python script that, given a NUPN $N = (P, T, F, M_0, U, u_0, \sqsubseteq, \mathsf{unit})$, generates random permutations $\pi_p : P \to P$, $\pi_t : T \to T$, and $\pi_u : U \to U$. Since the "`.nupn`" format requires that places of the same unit have contiguous numbers in a range, each generated function π_p only permutes places within their units, i.e., for each $p \in P$, $\mathsf{unit}(\pi_p(p)) = \mathsf{unit}(p)$, without loss of generality. The NUPN_INFO tool (with options "`-place-permute`", "`-transition-permute`", and "`-unit-permute`") is then invoked on N to produce as output a permuted NUPN; when applying π_p, π_t and π_u to permute P, T, and U, NUPN_INFO updates F, M_0, u_0, $\sqsubseteq$, and unit to enforce the constraints of Definition 7. Finally, we validated Proposition 2 by checking, on tenths of thousands of NUPNs and tenths of millions of random permutations, that the signatures of the original and permuted nets are identical.

6.4 Canonization

We further extended the CÆSAR.BDD tool with three new options ("`-unit-order`", "`-place-order`", and "`-transition-order`") that compute, for a net given in "`.nupn`" format, all the tuples $m(u)$, $n(p)$, and $o(t)$ defined in Sect. 5. CÆSAR.BDD then invokes the Unix "`sort`" command to sort these tuples lexicographically and performs, for the "`-unit-order`" option only, further calculations that may help distinguishing units having the same $m(u)$ value. An Awk script is then invoked to transform these results into permutations of units, places, or transitions, and report whether such permutations are unique or not. The input net and the three permutations are then given to NUPN_INFO, which produces as output a canonized net. A new option "`-canonical-nupn`" that automates all these steps, including the three invocations to CÆSAR.BDD, was added to NUPN_INFO. Written in C and Awk, canonization is generally fast (8 s per net on average) but took, in the two worst cases, 8 min and 90 min on two nets of the Model Checking Contest having the largest number of places (143,908 and 537,708 places respectively). Finally, FDUPES is invoked to detect file duplicates in the set of canonized nets.

[10] https://cadp.inria.fr/man/caesar.bdd.html.

To increase confidence in our implementation of canonization, we checked on each net that canonization is idempotent, meaning that two successive invocations of NUPN_INFO with its "`-canonical-nupn`" option produce the same output as one single invocation. Also, for tenths of thousands of NUPNs $N = (P, T, F, M_0, U, u_0, \sqsubseteq, \mathsf{unit})$, we generated tenths of millions of random permutations π_p, π_t, and π_u, and verified that: (i) for each $i \in \{1, ..., 35\}$, for each $u \in U$, $m_i(u) = m'_i(\pi_u(u))$, where m' is the tuple of Definition 23 computed on the net obtained from N by applying π_u, π_p, and π_t; (ii) for each $i \in \{1, ..., 27\}$, for each $p \in P$, $n_i(p) = n'_i(\pi_p(p))$, where n' is the tuple of Definition 25 computed on the net obtained from N by applying π_p and π_t; and (iii) for each $i \in \{1, 2\}$, for each $t \in T$, $o_i(t) = o'_i(\pi_t(t))$, where n' is the tuple of Definition 27 computed on the net obtained from N by applying π_t; the nets are permuted using NUPN_INFO and the tuples m, m', n, n', o, and o' are computed using CÆSAR.BDD.

6.5 Graph Isomorphism

As mentioned in Sect. 3.2, we selected the TRACES software, which is deemed to be a reference tool for graph isomorphism. We developed a Python script that converts a net in "`.nupn`" format to a colored graph (see Sect. 3), and then invokes TRACES to put this graph under canonical form. To process a collection, our script is first invoked on each net of the collection; FDUPES is then used to detect duplicate files among the canonized graphs. This detection may be incomplete since TRACES sometimes aborts or times out on large graphs.

We validated our implementation as follows: when two nets N and N' have been found isomorphic via their associated graphs $\mathcal{G}_N$ and $\mathcal{G}_{N'}$, TRACES produces two bijections π and π' that map the vertices of $\mathcal{G}_N$ and $\mathcal{G}_{N'}$ to the vertices of their respective canonical graphs (which are identical). Let $\pi_v \stackrel{\text{def}}{=} \pi'^{-1} \circ \pi$; from π_v, we compute three bijections π_p, π_t, and π_u as explained in the proof of Proposition 1, easily adapted to our optimized translation mentioned in Sect. 3.2; we finally check that $(\pi_t \circ \pi_p \circ \pi_u)(N) = N'$. We also cross-checked the results of the net-canonization approach with those of the graph-isomorphism approach by validating Definition 21, i.e., if two NUPNs N and N' satisfy $\mathsf{can}(N) = \mathsf{can}(N')$, then their associated graphs $\mathcal{G}_N$ and $\mathcal{G}_{N'}$ should be found isomorphic by TRACES (when this tool can handle them).

6.6 Tool Chain

The five approaches of Sects. 6.1 to 6.5 are successively applied, in this order. Each approach only considers the problems not solved by prior approaches, and one stops as soon as all problems have been solved. Figure 1 depicts the application of our tool chain to a collection of 10 nets named from 'a' to 'j'. Some approaches (identical files, pre-canonization, canonization, and graph isomorphism) detect certain nets that are isomorphic: we represent this information using solid boxes that gather isomorphic nets. Other approaches (signatures,

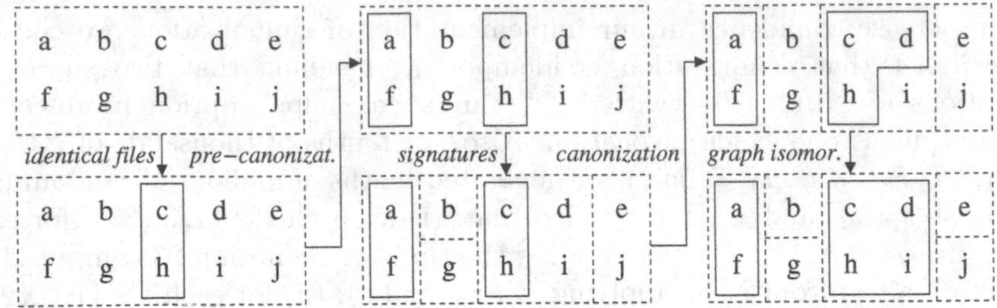

Fig. 1. Pipelined application of our 5 approaches to a collection of 10 nets

canonization when the assumptions of Proposition 3 hold, and graph isomorphism) detect certain nets that are not isomorphic: we represent this information by partitioning the collection into dashed boxes (using the partition-refinement idea), such that all nets belonging to distinct dashed boxes are pairwise non-isomorphic.

7 Experiments

We assessed our tool chain on four collections of nets listed in Table 1 below:

- *Collection 1*: 244 Petri nets (without NUPN structure) made available by the University of Zielona Góra[11];
- *Collection 2*: 1387 Petri nets obtained by taking all the (non-colored) models used for the 2022 edition of the Model Checking Contest[12]; 44% of these nets have initial places with more than one tokens and/or arcs with multiplicity greater than one, whereas 50% of these nets are non-trivial unit-safe NUPNs;
- *Collection 3*: 16,200 unit-safe NUPNs from diverse origins, containing few duplicates, gathered at INRIA Grenoble to be used in scientific experiments;
- *Collection 4*: 241,657 unit-safe NUPNs (135 GB of disk space) produced at INRIA Grenoble by removing, using FDUPES, all identical files from a larger set of 840,838 NUPNs that was obtained after extending collection 3 with additional NUPNs and applying numerous permutations to all these nets; therefore, collection 4 contains many duplicates (i.e., isomorphic NUPNs).

As mentioned above, our experiments were performed on the Grid'5000 testbed, each server having an Intel Xeon Gold 5220 (2.2 GHz) processor, 96 GB RAM, and running Linux Debian 11 with a shared NFS filesystem. To reduce the variability in results, each server was executing only one experiment at a time.

The results of applying our tool chain to these four collections are displayed in Table 2 below. After each step of the tool chain, we give three figures: *dupl.* is the percentage of nets that can be removed, since they were found isomorphic to

[11] http://www.hippo.iie.uz.zgora.pl (retrieved on January 23, 2023).
[12] http://mcc.lip6.fr.

Table 1. Numerical statistics about the four net collections used in experiments

	collection 1		collection 2		collection 3		collection 4	
	avg.	max.	avg.	max.	avg.	max.	avg.	max.
#places	15.4	200	2,801.5	537,708	345.8	131,216	740.8	131,216
#trans	11.8	51	10,798	1,070,836	7,998.1	16,967,720	15,645	16,967,720
#arcs	34.2	400	83,384	25,615,632	71,217.9	146,528,584	113,102.9	146,528,584
#units	—	—	1,970	537,709	123.4	78,644	270.4	78,644
height	—	—	15.4	2,891	4.3	2,891	6.3	2,891
width	—	—	1,959.1	537,708	117.6	78,643	259.9	78,643

other nets that will be kept in the collection; *uniq.* is the percentage of nets found to be unique in the collection after removing all duplicates; *unkn.* is the remaining percentage of nets whose status is not yet determined[13]. Notice that the values of *dupl.* and *uniq.* in Table 2 increase from top to bottom, as each line builds upon the cumulated successes reported in upper lines; thus, the contribution of each approach can be obtained as the difference between the percentage given on the corresponding line and the percentage given on the previous line.

The main finding is that our tool chain was conclusive for 99%–100% of each collection. More detailed remarks can be made:

Table 2. Results obtained by our tool chain on the four net collections

	collection 1			collection 2			collection 3			collection 4		
	dupl. (%)	uniq. (%)	unkn (%)	dupl. (%)	uniq. (%)	unkn. (%)	dupl. (%)	uniq. (%)	unkn. (%)	dupl. (%)	uniq. (%)	unkn. (%)
identical files	4.10	0.00	95.90	0.00	0.00	100.0	0.00	0.00	100.0	0.00	0.00	100.0
pre-canonizat	4.10	0.00	95.90	—	—	—	0.17	0.00	99.83	22.35	0.00	77.65
signatures	4.10	86.88	9.02	0.00	98.56	1.44	0.17	92.87	6.96	22.35	0.12	77.53
canonization	5.74	91.39	2.87	0.58	98.84	0.58	2.26	94.87	2.87	79.44	4.74	15.82
graph isomor.	6.97	93.03	0.00	0.58	99.42	0.00	2.79	97.20	0.01	90.05	9.01	0.94

- The simple application of FDUPES detected 10 duplicate files in collection 1.
- Pre-canonization detected 54,018 duplicates in collection 4; pre-canonization was not applied to collection 2 in order to preserve those "`.nupn`"-format pragmas giving information about multiple arcs and multiple initial tokens.
- Signatures massively identified unique nets in collections 1–3, but had no impact on collection 4, in which each net has at least one duplicate.
- Canonization was effective, both in identifying duplicate and unique nets.

---

13 Our success statistics could be slightly improved by considering that, among each set of $n > 0$ nets with undetermined status, there is at least one unique net.

- Use of graph isomorphism (with a one-hour timeout) clarified the case of most nets whose status remained unknown after canonization.
- Interestingly, our tool chain detected 17 duplicates in collection 1 and eight duplicates in collection 2; for the latter, one duplicate is certain (the corresponding nets are one-safe) and seven are uncertain, but very likely.

8 Conclusion

Starting from the concrete problem of finding duplicate models in large collections of Petri nets or NUPNs, we devised three approaches for the detection of isomorphic nets: reduction to graph isomorphism, net signatures (an over-approximation), and net canonization (an under-approximation).

These approaches, which draw on the careful examination of thousands of concrete benchmarks, have been fully implemented in an efficient tool chain, the successive steps of which are ordered by increasing complexity. To process large collections of nets, the calculations can easily be distributed on computer clusters or grids, as most steps deal with individual nets. Only the detection of identical files is not easy to parallelize, but did not cause bottlenecks, since the tool we selected is fast enough.

We assessed our tool chain on four collections ranging from 244 to 241,657 nets and containing either few or many duplicates. We observed a success rate of 99%–100% in the detection of isomorphic nets.

The present work could be pursued in, at least, two directions: (i) one could try shortening the component lists used in signatures and canonization to retain only those components that are most effective in practice; (ii) one could extend the proposed approaches to support wider classes of nets, such as non-safe Petri nets (these are currently handled using over-approximations) and colored nets.

Acknowledgements. Our experiments have been performed using the French Grid'5000 testbed.

References

1. Badouel, E., Darondeau, P.: Theory of regions. In: Reisig, W., Rozenberg, G. (eds.) ACPN 1996. LNCS, vol. 1491, pp. 529–586. Springer, Heidelberg (1998). https://doi.org/10.1007/3-540-65306-6_22
2. Best, E., Devillers, R.R.: Synthesis and reengineering of persistent systems. Acta Inf. **52**, 35–60 (2015)
3. Bouvier, P., Garavel, H.: Efficient algorithms for three reachability problems in safe Petri nets. In: Buchs, D., Carmona, J. (eds.) PETRI NETS 2021. LNCS, vol. 12734, pp. 339–359. Springer, Cham (2021). https://doi.org/10.1007/978-3-030-76983-3_17
4. Desel, J., Reisig, W.: The synthesis problem of Petri nets. Acta Inf. **33**, 297–315 (1996)
5. Devillers, R.R.: Articulations and products of transition systems and their applications to Petri net synthesis. CoRR **abs/2111.00202** (2021). https://arxiv.org/abs/2111.00202

6. Devillers, R., Schlachter, U.: Factorisation of Petri net solvable transition systems. In: Khomenko, V., Roux, O.H. (eds.) PETRI NETS 2018. LNCS, vol. 10877, pp. 82–98. Springer, Cham (2018). https://doi.org/10.1007/978-3-319-91268-4_5
7. Garavel, H.: Nested-unit Petri nets. J. Log. Algebr. Methods Program. **104**, 60–85 (2019)
8. Gorrieri, R.: Team equivalences for finite-state machines with silent moves. Inf. Comput. **275**, 104603 (2020)
9. Grohe, M., Schweitzer, P.: The graph isomorphism problem. Commun. ACM **63**(11), 128–134 (2020). https://doi.org/10.1145/3372123
10. Hesketh, M., Koutny, M.: An axiomatisation of duplication equivalence in the Petri box calculus. In: Desel, J., Silva, M. (eds.) ICATPN 1998. LNCS, vol. 1420, pp. 165–184. Springer, Heidelberg (1998). https://doi.org/10.1007/3-540-69108-1_10
11. ISO/IEC: High-level Petri Nets – Part 2: Transfer Format. International Standard 15909-2:2011, International Organization for Standardization – Information Technology – Systems and Software Engineering, Geneva (2011)
12. McKay, B.D., Piperno, A.: Practical graph isomorphism, II. J. Symb. Comput. **60**, 94–112 (2014)
13. Murata, T.: Petri nets: analysis and applications. Proc. IEEE **77**(4), 541–580 (1989)

Remote Debugger: A Tool to Remotely Monitor and Operate IOPT-Nets Controllers

Fernando Pereira[1,4,5], João-Paulo Barros[3,4,5], Filipe Moutinho[2,4,5], Anikó Costa[2,4,5], Rogério Campos-Rebelo[1,3,4], and Luis Gomes[2,4,5](✉)

[1] Instituto Superior de Engenharia de Lisboa, Lisbon, Portugal
fjp@deea.isel.ipl.pt

[2] NOVA School of Sciences and Technology, Caparica, Portugal
{fcm,akc,lugo}@fct.unl.pt

[3] Polytechnic Institute of Beja, Beja, Portugal

[4] Centre of Technology and Systems-UNINOVA, Caparica, Portugal
{jpb,rcr}@uninova.pt

[5] Intelligent Systems Associate LAboratory (LASI), Guimarães, Portugal

Abstract. This paper describes an interactive tool to remotely debug, control, and monitor controllers designed using IOPT Petri nets. The controllers run on physical hardware devices using code automatically generated from a Petri net model. A web-based user interface allows remote controllers to analyze, test, and debug, present a graphical view of the original Petri net model, and mark an evolution in quasi-real-time. Debugging capabilities include execution pausing, step-by-step execution and continuous execution, and the definition of break-points associated with transition firing. Users may remotely force values on input signals and inspect the values of output signals to test conditions that would otherwise require physical interaction with the hardware systems. The tool records the evolution of all signals and state variables. These are presented as waveforms, and the respective data can be exported, allowing posterior analysis with external tools of systems running at high-speed execution rates, The tool is already integrated into the IOPT-Tools framework, runs in a standard web browser, and does not require additional software installation. The tool employs a lightweight protocol based on HTTP to communicate with the code running on the embedded controllers. The protocol was designed to minimize CPU, memory, and bandwidth resources. Thus, the controller code generated automatically by the IOPT-Tools framework includes a minimal HTTP server for communication with the debugger tool running on a browser. The paper presents an example of an application employing a simple car-park controller model.

Keywords: Test · Debug · Remote Operation · Design Automation · Code Generation · Petri nets

This work was partially financed by Portuguese Agency FCT – Fundação para a Ciência e Tecnologia, in the framework of project UIDB/00066/2020.

L. M. Kristensen and J. M. van der Werf (Eds.): PETRI NETS 2024, LNCS 14628, pp. 402–412, 2024.
https://doi.org/10.1007/978-3-031-61433-0_19

1 Introduction

Graphical specification languages play an important role in systems specification and modeling. Additionally, the so-called formal languages also have the advantage of precise semantics, which, besides fostering additional reflection of the system requirements and behaviors, also allow automatic executable code generation. Creating models can help modelers better understand the system's behavior and identify potential issues before they arise. Code generation allows the use of the models as high-level programming languages. Another benefit of visual specification languages is that they can help developers communicate more effectively with other team members and stakeholders. Visual models are often easier to understand than written descriptions, so they can help share ideas and collaborate on complex projects.

The visual representation of Petri nets is especially appealing as it allows a readable representation of resources, conditions, concurrency, and synchronization. However, to create controllers, it is also essential to be able to generate executable code from those specifications. This can save time and reduce the risk of errors, which is especially important in developing controllers, where even small mistakes can have significant consequences.

The IOPT-Tools framework [1] is a set of freely available tools that allow the specification of discrete event controllers using a non-autonomous class of Petri nets, named IOPT nets, for *Input-Output Place-Transition Petri nets* [2]. Those tools include an editor, state-space generator, code generator, and simulator. Additionally, after deployment of the generated code (for the controller), testing the controller in real-time can be helpful. To that end, a tool named "remote debugger" was added to the IOPT-Tools framework. This new tool, here presented, communicates with a local micro HTTP server added to the controller-generated code to force values in the input signals, replacing the values read from the input pins and observing the values in the output pins. This way, it is now possible to test the code already running in real-time, allowing exploration of different scenarios and even stress tests regarding random or malicious inputs.

The structure of the paper is as follows. Section 2 presents the related work. Section 3 briefly presents the IOPT Petri nets class and Sect. 4 the IOPT-Tools framework where the presented debugger was integrated. The remote debugger tool is presented in Sect. 5, and Sect. 6 illustrates its application to a car-park controller. Finally, Sect. 7 concludes.

2 Related Work

The IOPT-Tools framework provides a unique set of free tools to create models for discrete event controllers from which code can be generated and deployed to multiple low-cost controllers (e.g., Arduino) and single-board computers. The explicit modeling of input and output signals, which are assigned to physical pins, provide an intrinsic and automatic back annotation and, consequentially, a detailed visualization of all controller states, signals, and execution steps. More specifically, the Petri net model, from which the runtime code was generated, visually represents the controller behavior at runtime: the model is also a code visualizer. The model also provides an interface to force input signal changes, allowing different scenarios to be tested.

Remote debuggers are a relatively common tool, especially when applied to low-resource machines like PLC and embedded computers. An example is the work in [3] that presents building a GDB-based PLC debugger with an embedded CPU using a specific PLC protocol instead of GDB's remote serial protocol (RSP). Other works also rely on GDB to debug remote machines (e.g., [4,5]). Yet, these debuggers, even when a GUI is made available, are text-based and do not visually represent the controller behavior, as there was no initial model from which the running code was generated. In [6], an Arduino debugger also relies on GDB, and the debug functionality is based on a program library that is added to the user application. As mentioned, in the IOPT-Tools remote debugger, an HTTP server is added to the controller code.

Sometimes, Petri nets are used to visualize distributed systems without the running code having been generated from the Petri net model used for visualization (e.g., [7,8]). The debugger presented here is specific to this class of Petri nets. It allows real-time visualization for markings, signals, and events in the same model that generated the code, hence completing the model-driven development with the possibility of visually debugging the generated code using the original model.

3 IOPT Nets and Their Deployment

The IOPT Petri net class [2] was created to support embedded system controller design. To that end, IOPT nets add some non-autonomous elements to place/-transition Petri nets (e.g., [9]): in addition to places, transitions, and arcs, IOPT nets also include the concepts of input and output signals (which can be Boolean or multi-valued signals) and events, that enable the communication of a Petri net model with the environment, and allow the use of priorities in transitions as one way to resolve conflicts and avoid non-determinism.

For the development of embedded controllers, explicit modeling of the dependencies on the environment and execution determinism assumes paramount importance. The firing rule, as in other non-autonomous nets (e.g., [10]) adds the concept of "ready" for a transition to fire and a maximal step semantics (in order to assure a deterministic execution at the physical controller level). Therefore, for a transition to fire, it must be not only `enabled`, by the marking in its input places as in place/transition nets, but also `ready`. A transition is `ready` if and only if it meets two conditions: (1) its associated guards, which are functions of the input signal values, are true; (2) its associated input events, which are functions of changes in the input signal values, are also true. Then, according to the maximal step semantics, all enabled and ready transitions will fire in each step. Also, contributing to a-priori conflicts arbiters, test-arcs can be used to check if a place is marked, but do not remove tokens.

When a controller developed using an IOPT model is deployed on an embedded hardware platform, the signals and events are associated with analog or digital input and output pins to read data from external switches and sensors or to control LEDs, actuators, motors, or user interface widgets. Then, these `signals` and `events` interact with the Petri net nodes in a bidirectional way: input `signals` and input `events` are used to inhibit the firing of transitions; expressions may be associated with places to assign values to `output signals` depending on the net marking; `output signals` can also be changed by `output events` when a transition fires.

4 IOPT-Tools

The IOPT-Tools framework is a cloud-based integrated development environment that supports all embedded system controller development steps. It contains a set of tools that enable model edition, simulation, model-checking, automatic code generation for several hardware targets, and remote debugging of the resulting controllers. All tools offer a web interface, are available online at http://gres.uninova.pt, and do not require any software installation.

The tools include an extensible graphical editor for model design and edition (see Fig. 1). In addition to the standard Petri net edition functionality, it is possible to associate events and assign physical input and output hardware pins to each external signal. It is also possible to execute external plug-ins, for instance, to split models into several distributed sub-systems, as well as applying a net-addition operation, allowing the composition of models.

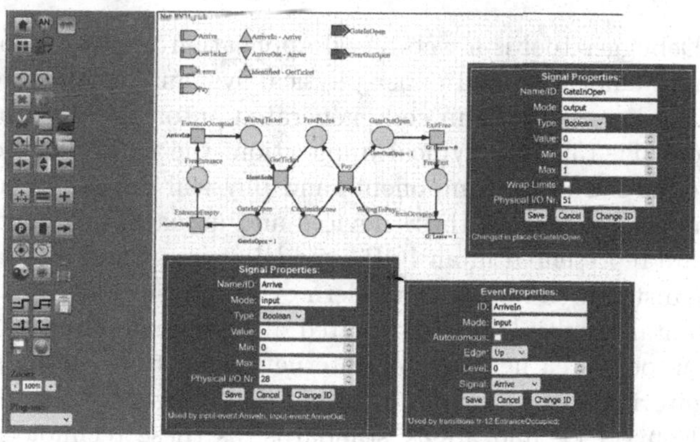

Fig. 1. IOPT-Tools Editor.

A Simulator tool enables the execution of IOPT models directly on the web browser. The user interacts with the models by changing the values of input and output signals and events and observing the system state evolution according to these changes: transitions fire, producing changes in place marking, resulting in new values associated with output signals and events. The simulator may execute the models step-by-step, continuously at predefined execution speeds that may be automatically stopped by assigning breakpoints to specific transitions.

In addition to the simulator, a model-checking sub-system based on a state-space generator is available. This tool automatically detects deadlock conditions and calculates the maximal bound for each place, which is useful for automatic code generation tools. The resulting state-space graphs can be visualized graphically, but when the number of states grows to many thousands or millions, it is possible to define queries to automate property checking. For instance, it is possible to automatically search for states that correspond to problematic fault conditions, check the reachability of certain desired states, etc.

When a model has been successfully simulated and model-checked, a set of automatic code generators produces code to run on several types of hardware platforms and implement the model behavior. At this point, there are automatic generators that produce software code in the C/C++, Matlab, and PLC Instruction List languages. Another automatic code generator produces VHDL hardware descriptions to run on FPGA and ASIC chips.

Finally, the Remote Debugger tool presented in this paper allows the remote debugging and monitoring of controllers deployed on embedded boards running the code produced automatically by the C code generator. This code has been tested on several embedded boards, including Arduino, Raspberry Pi 1-4, Red-Pitaya, Intel Edison, Coral-dev-board, and other boards based on the Linux operating system.

5 The Remote Debugger Tool

The Remote Debugger tool is a web-based application used to monitor, debug, and troubleshoot embedded and cyber-physical system controllers deployed on remote hardware boards that run code generated automatically from an IOPT Petri net model [11–13]. In a typical application, the controller board is connected to external hardware components and physical devices, containing sensors to read information from these devices and actuators to control physical systems that are represented in an IOPT model as input and output signals.

Due to the distributed nature of modern cyber-physical systems, a complete system may contain multiple boards located at different remote locations, connected through local area networks or Internet connections, supporting debugging and monitoring from a single location. The chosen communication protocol is based on HTTP and JSON standards, as these technologies are well-supported on all web browsers, and most firewalls do not usually filter the HTTP protocol. The "C" code produced automatically by the IOPT-Tools framework includes a minimalist HTTP server and infrastructure to support remote debugging.

When the Remote Debugger tool connects to an embedded board, it opens two communication channels. This first connection sends commands to the board, including debug instructions like step-by-step execution, undo execution steps, define breakpoints, force the value of input and output signals, system resets, and read the system state. The second channel monitors the system evolution in quasi-real-time, reporting any changes in the system status and I/O signals and events.

After connecting and authenticating into a remote board, the debugger tool immediately presents the system status, displaying the corresponding IOPT Petri net model, including the instantaneous values of place marking and input and output signals, along with information about which transitions are ready to fire. Figure 2a) presents the Remote Debugger tool running on a web browser. The remote debugger offers a user interface almost identical to the IOPT Simulator tool, except that the models run on physical devices instead of being just simulated on the browser.

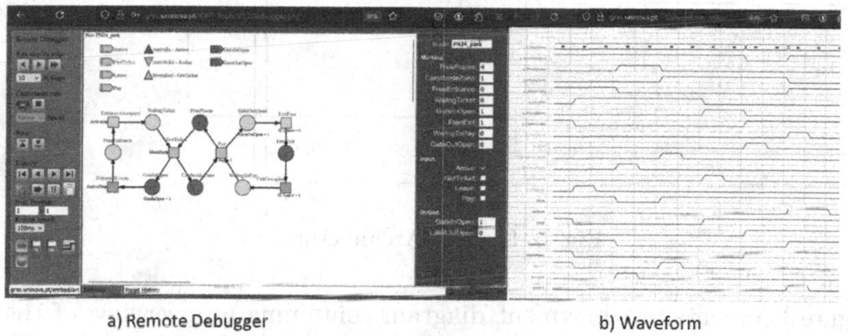

Fig. 2. Remote Debugger Simulator.

In the same way as the Simulator tool, the toolbox contains buttons to pause execution, step-by-step execution, and perform continuous execution. The evolution of the system state is reflected in the model's graphical representation, employing different colors to depict marked places, enabled signals and events, and even the transitions that are enabled and ready to fire on the next execution step. A form on the right presents the same information using numeric values.

As the controller boards typically run at high-speed execution rates, thousands of steps per second, it is impossible to monitor all state evolution details visually. However, the second communication channel transmits all information to the debugger, which stores it internally. This information may be visualized as graphical waveforms (Fig. 2b)) or exported as CSV files for later processing on spreadsheets or other analysis software. In the same way, when the system evolves at high-speed execution rates, it may not be possible to detect faulty situations visually. To help solve this problem, the user may assign breakpoints to specific transitions that pause execution when they fire. The user may also navigate through the saved execution history, moving back and forth to replay «interesting» execution sequences slowly.

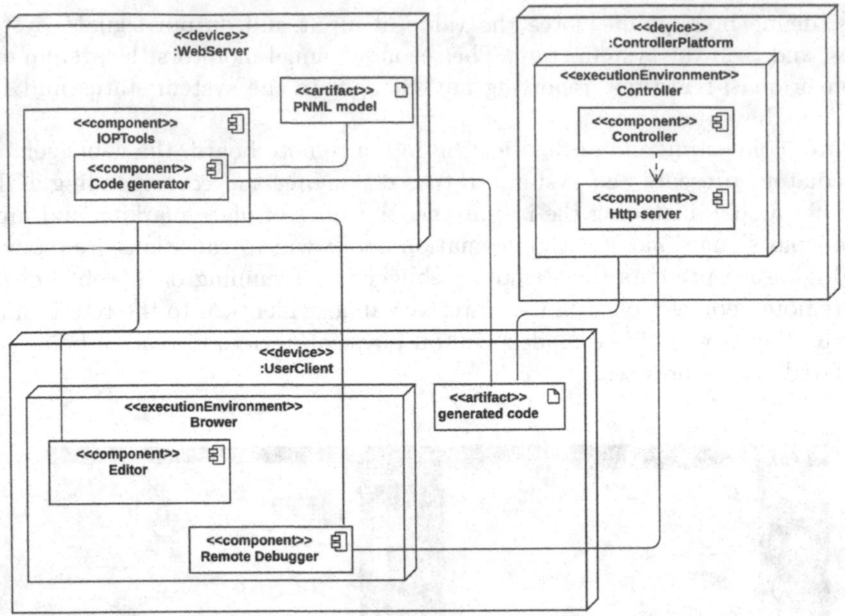

Fig. 3. System Architecture.

Figure 3 presents a deployment diagram containing an overview of the proposed system architecture, including a server running the IOPT-Tools framework, the developer's personal workstation running a web browser, and an embedded board where the controller is deployed. The IOPT-Tools web server manages all back-end operations, including storing the web applications and IOPT model files, performing automatic code generation, and running the model-checking subsystem. Other tools, such as the model editor, model simulator, and remote debugger, are executed directly on the user's web browser using Javascript and AJAX technologies. Finally, the C code produced by the automatic code generator is deployed on an embedded board connected to physical hardware devices (the controlled system). All three computer systems, the server, user workstation, and embedded board, must be connected through a local network or the Internet.

The C code produced automatically is divided into two parts: code that implements the IOPT model semantics and a small web server that communicates with the remote debugger and may also be used to support the creation of remote web user interfaces. To support remote debugging, the semantics execution code also contains functionality to pause execution, run a specific number of steps, check transition breakpoints, and force input and output values.

As the C code produced by the automatic code generator may be deployed into very diverse embedded single-board computers, the code is usually downloaded from the server to the developer's workstation to be compiled using the target board-specific tool-chain and produce binary files to run on the boards.

However, the embedded-board development tools may also be installed on the server to produce the binary files or executed directly on the embedded board. For example, an "Arduino-cli" tool may be used to compile C code and upload binary files directly from the server to Arduino boards. Furthermore, target boards based on the Linux operating system, such as Raspberry Pi SBCs, can locally compile and run their own code.

The Remote Debugger tool requires the cooperation of three systems running simultaneously: the user must log into the server, open the desired model, and run the debugger tools. This step is necessary as the debugger application and the IOPT model file are stored on the server. Next, the Remote Debugger application running on the user's web browser connects to the embedded board by specifying a URL containing the board address and a password. As soon as the Remote Debugger has retrieved the model information from the server, the debugging session proceeds with the user's web browser interaction and the code running on the embedded board.

6 Using the Remote Debugger Tool

This section summarizes the deployment of controller code with support for the remote debugger and an application example where IOPT-Tools and the remote debugger were used to develop a remote laboratory.

6.1 Automatic Generation of Controllers Supporting Remote Debugging

After model edition, simulation, and verification, the code can be automatically generated just by pressing the "C Code" button. However, before generating the code, it is required to ensure that the model input and output signals will be connected to the board pins, and to achieve it, it is required to specify a "Physical I/O Nr" for each signal. The automatically generated code, which supports diverse boards and remote debugging, is downloaded in a ZIP file containing multiple files. To implement not only the controller (left part of Fig. 4) but also an HTTP server that supports remote debugging (right part of Fig. 4), the following steps must be carried out. For Arduino boards:

- To implement the HTTP server used by the remote debugger, edit file net_types.h and uncomment the definition "#define HTTP_SERVER";
- The default IP address is 192.168.1.177 and the default password is "1234";
- To define the Arduino Ethernet IP address, edit "net_server.h" and change the ARDUINO_IP_ADDR macro. It is important to note that commas must separate the numbers (,) instead of dots (.);
- To define the password, change the "PASSWORD" definition in the file "net_server.h";
- The default port addresses are 80 and 81 (definitions "SYNCPORT" and "FEEDPORT" in "net_server.h" file). "SYNCPORT" is used to send comments to the board, whereas "FEEDPORT" is used to monitor the board;

- The file "net_main.c" must be renamed to "<PROJ>.ino", where <PROJ> is the Arduino project folder name.

For embedded boards based on the Linux operating system (including Raspberry Pi boards):

- Linux-based boards use the files "http_server.h" and "http_server.c" instead of "net_server.h" and "net_server.cpp";
- To implement the HTTP server used by the remote debugging, uncomment the -DHTTP_SERVER line on the "Makefile", or if the "Makefile" is not used, add this definition to the file "net_types.h";
- The default port address is 8000. To change the port, edit "http_server.h" and change the "DEF_PORT" definition.
- To define the password, change the "PASSWORD" definition in the file "http_server.h";
- A "Makefile" is supplied to simplify the project compilation (in the project directory, run the "make" command);
- After compilation, the code can be executed, implementing a controller that supports remote debugging.

As an example, a Raspberry Pi running the controller of a car parking lot can be remotely monitored and controlled through the Remote Debugger tool of IOPT-Tools using the model "INDIN07_park1in1out" available in the area "atpn2024" (password "atpn2024") and following further instructions available at http://gres.uninova.pt/boards.html.

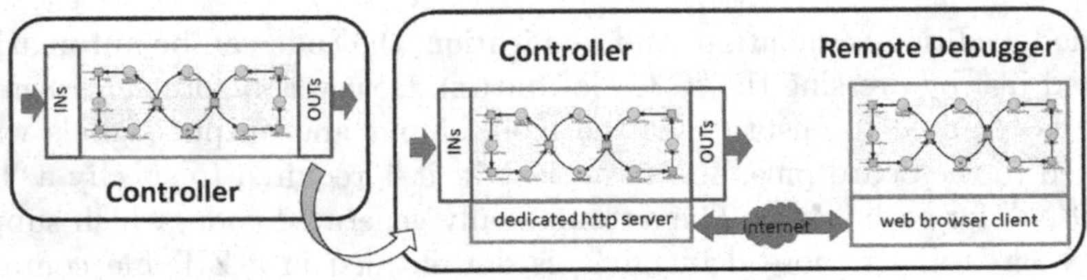

Fig. 4. From isolated operation to remote monitoring and operation of controllers.

6.2 Remote Monitoring and Operation: A Remote Laboratory Using Arduino Boards

This section presents an example of an application where the IOPT-Tools with the remote debugger were used to develop a remote laboratory. This laboratory has two Arduino boards, as Fig. 5 illustrates. Both Arduino boards have HTTP servers connected to two instances of the remote debugger web clients running on web browsers), meaning both can be remotely monitored and controlled.

In the presented scenario (Fig. 5), Arduino2 is running the controller of a car parking lot, whereas Arduino1 will be used to emulate the behavior of the car parking lot under control. Arduino1 outputs, which its remote user will force, are connected to Arduino2 inputs to emulate the car parking lot sensors' changed state. Arduino2 outputs, which are used to actuate on the car parking lot, are connected to Arduino1 inputs to be monitored by its remote user. It is important to note that it is possible to remotely monitor and control the controller of the car parking lot just using one Arduino (Arduino2); however, with this solution, it is possible to do it without interfering with the execution of the controller of the car parking lot.

This remote laboratory can be remotely monitored and controlled through the Remote Debugger tool of IOPT-Tools using the model "PN24_ParkSensors" for Arduino1 and the model "PN24_park" for Arduino2, both available in the area "atpn2024" (password "atpn2024") and following further instructions available at http://gres.uninova.pt/boards.html.

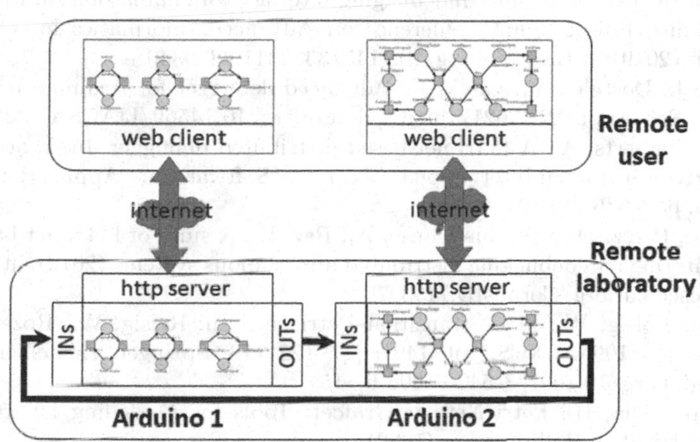

Fig. 5. Remote Debugger for the support of remote laboratories.

7 Conclusions

The freely available web-based IOPT-Tools already allow the creation, verification, and testing of Petri net models for discrete event controllers. From those models, generating code ready to run on multiple platforms, namely single-board computers and low-resource controllers, is also possible. The presented tool adds an important testing capability to IOPT-Tools as the same model can now be used to test and visualize the generated code execution in those platforms. The tool explicitly allows direct real-time actuation and the corresponding visual mapping between the model and physical controller states, thus providing an additional important form of testing.

References

1. Pereira, F., Moutinho, F., Costa, A., Barros, J.P., Campos-Rebelo, R., Gomes, L.: IOPT-tools - from executable models to automatic code generation for embedded controllers development. In: Bernardinello, L., Petrucci, L. (eds.) PETRI NETS 2022. LNCS, vol. 13288, pp. 127–138. Springer, Cham (2022). https://doi.org/10.1007/978-3-031-06653-5_7
2. Gomes, L., Barros, J.P.: Refining IOPT Petri nets class for embedded system controller modeling. In: IECON 2018 - 44th Annual Conference of the IEEE Industrial Electronics Society, pp. 4720–4725 (2018)
3. Yang, W., Yang, W., Lee, J., Lee, J.: PLC remote debugger development using GDB. In: International SoC Design Conference (2009)
4. Ji, J., Woo, G., Park, H., Park, J.: Design and implementation of retargetable software debugger based on GDB. In: 2008 Third International Conference on Convergence and Hybrid Information Technology, vol. 1, pp. 737–740 (2008)
5. Vaish, N., Khosla, C.: Uniform debugging interface for simulators. In: Proceedings of the Third International Conference on Advanced Informatics for Computing Research (2019). https://doi.org/10.1145/3339311.3339340
6. Dolinay, J., Dostálek, P., Vašek, V.: Advanced debugger for Arduino. Int. J. Adv. Comput. Sci. Appl. **12** (2021). http://dx.doi.org/10.14569/IJACSA.2021.0120204
7. Liu, A., Engberts, A.: A Petri net-based distributed debugger. In: Proceedings of the Fourteenth Annual International Computer Software and Applications Conference, pp. 639–646 (1990)
8. López, J., Pérez, D., Santana-Alonso, A., Paz, E.: A suite of Petri net based tools for monitoring and debugging distributed autonomous systems (2012). https://api.semanticscholar.org/CorpusID:7193973
9. Desel, J., Reisig, W.: Place/transition Petri nets. In: Reisig, W., Rozenberg, G. (eds.) ACPN 1996. LNCS, vol. 1491, pp. 122–173. Springer, Heidelberg (1998). https://doi.org/10.1007/3-540-65306-6_15
10. David, R., Alla, H.: Petri Nets & Grafcet: Tools for Modelling Discrete Event Systems. Prentice Hall, London (1992)
11. Pereira, F., Gomes, L.: Minimalist architecture to generate embedded system web user interfaces. In: Camarinha-Matos, L.M., Tomic, S., Graça, P. (eds.) DoCEIS 2013. IAICT, vol. 394, pp. 239–249. Springer, Heidelberg (2013). https://doi.org/10.1007/978-3-642-37291-9_26
12. Pereira, F., Melo, A., Gomes, L.: Remote operation of embedded controllers designed using IOPT Petri-nets. In: 2015 IEEE 13th International Conference on Industrial Informatics (INDIN), Cambridge, UK, pp. 572–579 (2015). https://doi.org/10.1109/INDIN.2015.7281797
13. Pereira, F., Gomes, L.: A JSON/HTTP communication protocol to support the development of distributed cyber-physical systems. In: 2018 IEEE 16th International Conference on Industrial Informatics (INDIN), Porto, Portugal, pp. 23–30 (2018). https://doi.org/10.1109/INDIN.2018.8472084

Using Petri Nets for Digital Twins Modeling and Deployment: A Power Wheelchair System Case Study

Carolina Lagartinho-Oliveira(✉), Filipe Moutinho, and Luís Gomes

NOVA School of Science and Technology, Center of Technology and Systems (UNINOVA-CTS) and Associated Lab of Intelligent Systems (LASI), NOVA University Lisbon, 2829-516 Caparica, Portugal
ci.oliveira@campus.fct.unl.pt, {fcm,lugo}@fct.unl.pt

Abstract. Ideally, safety-critical systems should be designed to avoid or be resilient in handling failures that may occur during their lifetime. For dependability purposes, IEC 62551 provides guidance on using the Petri net formalism for modeling and analysis of systems. Another concept that has been considered to ensure the reliability of systems and contribute to their overall safety is the digital twin (DT). A DT is a virtual counterpart that is seamlessly linked to a physical asset, both relying on data exchange for mirroring each other. DT has been used for the tracking, management, maintenance, and optimization of different systems. In some implementations, the DT emphasizes only the geometric models and their animation. To fully benefit from their usage, considering associated behavioral models is of paramount importance to allow full validation of the system. This paper proposes the application of Input-Output Place-Transition Petri Nets (IOPT-nets) to model and deploy both the physical and the virtual entities of the DT, contributing to a comprehensive use of Petri nets in the development of systems. The case study presented concerns the development of digital twins for power wheelchair systems using the IOPT-Tools framework to specify, validate, and implement it.

Keywords: Digital Twin · Petri Net · Real-Time Information · Reliability Analysis · Remote Monitoring and Control

1 Introduction

Modern systems are becoming more complex, requiring appropriate techniques to verify and validate their dependability, especially with regard to safety-critical systems (SCSs) [1]. A SCS is a system where non-desired properties can cause serious harm to people or the environment. In other words, a system is considered

This work was financed by the Portuguese Agency FCT (Fundação para a Ciência e Tecnologia), in the framework of project UIDB/00066/2020, and under the PhD scholarship 2020.08462.BD.

L. M. Kristensen and J. M. van der Werf (Eds.): PETRI NETS 2024, LNCS 14628, pp. 413–431, 2024.
https://doi.org/10.1007/978-3-031-61433-0_20

safety-critical based on the results of its failures[1] [2] which often stem from human actions.

The SCS covered in this paper pertains to a power wheelchair system. This type of system is particularly relevant, as it promotes the mobility of individuals with physical disabilities and involves the interaction of different actors: users, caregivers, therapists, technicians, etc. In addition to the complexity of developing these systems, challenges arise when using and maintaining them. In particular, many users have the misconception that wheelchairs do not require maintenance; and maintenance typically occurs with users seeking assistance only after a problem arises.

In order to minimize the risks and possible negative outcomes associated with power wheelchairs (PWCs) and SCSs in general, it is important to tackle faults during both the systems development phase and their operation. In this context, it is desirable to ensure the overall reliability[2] of the systems, as it can also contribute to improving their functional safety[3]. For PWCs, this allows them to be properly maintained and helps extend their lifespan, ensuring that they remain safe for the user.

IEC 61508 [3] is an international standard that facilitates the development of safety-related systems (also critical), and refers to the use of Petri nets (PNs) to model relevant aspects of system behavior, and to assess and possibly improve safety and operational requirements through analysis and re-design. There is also IEC 62551 [4] that focuses on dependability analysis techniques using exclusively Petri nets.

The application of these standards can also be supported by advances in digital technologies, strongly triggered by the Industry 4.0 (I4.0). I4.0 delves deep into connectivity, real-time data acquisition and processing, and digitization, allowing the connection of physical objects, people, and the internet.

In this realm, a concept that has been gaining prominence is the digital twin (DT) [5], which has brought a new approach to face various challenges [6] by creating always connected physical and virtual twins. A DT is a virtual counterpart seamlessly linked to a real physical asset or prototype (e.g. product, service, or machine), both sharing the same properties, characteristics, and behavior by means of data and information [7] (see Fig. 1). The benefit of creating and manipulating virtual copies lies in the ease of anticipating challenges, detecting problems, and increasing efficiency of the systems.

With respect to digital twin implementation, this is a complex task, with several approaches and models available to realize all its capabilities [8]. Geometric models are commonly used for their visual appeal, but their practicality is limited without corresponding models that define the behavior and rules of the DT.

---

[1] A failure is the consequence of an error caused by a fault.

[2] Reliability is the ability of a system to perform the function for which it was designed.

[3] Safety is the system's ability to behave safely in the presence of unacceptable failures.

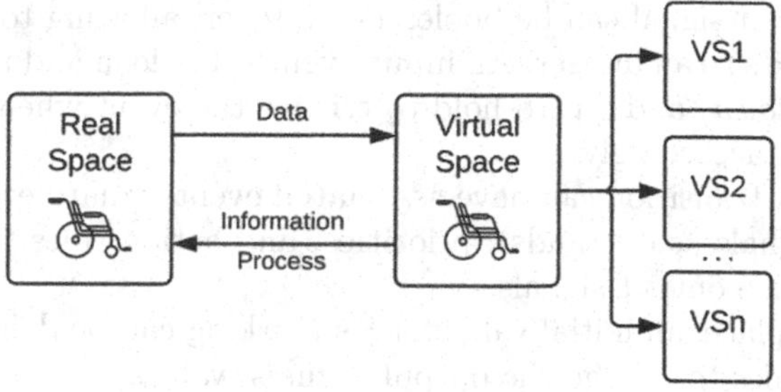

Fig. 1. DT conceptual model with VS1-VSn representing multiple virtual spaces associated with a real space [7].

In view of this, this paper proposes the creation of digital twins using Petri nets. Specifically, the class of non-autonomous Input-Output Place-Transition Petri net (IOPT-net) [9] is used to model and deploy both the physical and virtual entities of a DT. The objective is to fully leverage IOPT-nets in the development of systems through the digital twin concept: while Petri nets offer a formal structure for modeling and analysis, the digital twin enables real-time simulation and control. In this sense, one can benefit from the Petri nets body of knowledge, namely using model checking fully automatic techniques in aspects related with verification of properties of the system, namely through the construction of the associated reachability tree [10,11].

The case study presented then concerns the development of a digital twin for a power wheelchair. The main focus is on monitoring and controlling the behavior of the motors used to drive and accommodate different seating positions, as is found in many wheelchairs. The Input-Output Place-Transition Tools (IOPT-Tools) framework [12] was used to support the DT design through IOPT-net models, as well as its validation, implementation, and remote operation.

The paper has the following structure: Sect. 2 introduces basic concepts related to IOPT-nets; Sect. 3 delves into the digital twin concept and how to use IOPT-nets to enable it; Sect. 4 shows the case study and presents some discussion around the proposal; and finally, in Sect. 5, conclusions and future work are presented.

2 Preliminaries

Formal definitions of the IOPT-nets class can be found in [9]. The Input-Output Place-Transition net is a class of low-level Petri net that was extended from the well-known Place-Transition net (P/T-net) [13] to allow the modeling of systems capable of interacting with the environment. The IOPT-net is, therefore, a non-autonomous class in which the behavioral models of the systems can be conditioned through input and output signals or events. The following characteristics of a IOPT-net stand out:

- *Inputs*: An input signal can be boolean or integer, allowing to set the default value, and limits to an integer. An input event is boolean and requires an edge value (*up* or *down*) and a threshold to trigger the event when the associated signal changes accordingly.
- *Transitions*: A transition can have associated events, guard expressions (with places and signals as operands), priorities, and action rules that can change the values of the output signals.
- *Places*: For a place, an initial value for its marking can be defined. It can also have action rules to change the output signals' values.
- *Outputs*: Outputs have the characteristics of the inputs and can have defined limits for signals.
- *Arcs*: Arcs are weighted and can be normal arcs or test (read) arcs.

Considering this, at each execution step, all enabled transitions can be fired if their input events and guards are verified; but when multiple transitions compete for the same tokens, priorities and test arcs can prevent some of them from firing. Upon firing, the output signals' values may change due to the occurrence of output events or actions defined at transitions or places. The state of the net is then given by the marking of all places and the values of the output signals.

Figure 2 shows a simple IOPT-net model with 2 places (P1 and P2), 2 transitions (T1 and T2), 1 input signal (IS), 1 input event (IEu), 2 output events (OEu and OEd) and 2 output signals (OS and OS2). In this model, T1 can fire if IEu occurs with IS going from 0 to 1. If T1 fires, P1 becomes unmarked, P2 becomes marked, and both OS and OS2 become equal to 1; OS due to the occurrence of OEu, and OS2 due to the action rule "OS2 = 1" in P2. So, T2 can fire if the guard "G: IS = 0" is true. If T2 fires, OEd will occur making OS equal to 0; and since P2 is unmarked, OS2 returns to 0. The net then returns to its initial state.

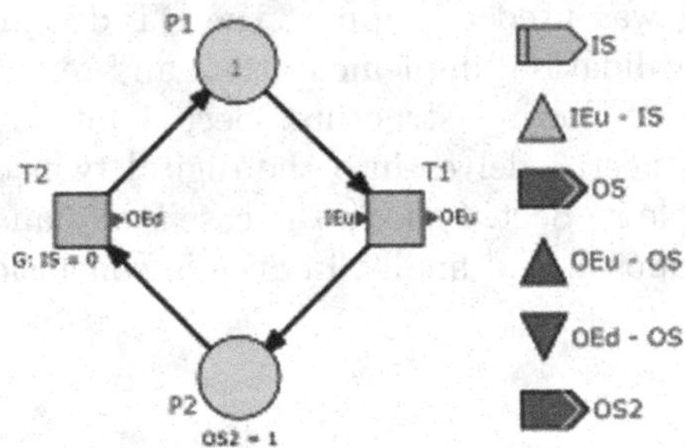

Fig. 2. A simple IOPT-net model.

The IOPT-Tools is a free, cloud-based tool framework accessible online at http://gres.uninova.pt/IOPT-Tools/, that supports the creation of system controllers specified with the IOPT-net formalism. This framework supports all stages of system development, including design, verification, validation, implementation, and even allows for remote debugging of deployed systems. The provided tools encompass an interactive graphical editor of IOPT-net models, a simulator, a state-space generator with a query engine for searching properties, an automatic code generator for software or hardware platforms, and a remote debugger [14–17].

3 IOPT-Nets Enabling Digital Twins

The concept of digital twin is sometimes mistaken for digital model (DM) or digital shadow (DS); but the distinction between the three becomes evident when it comes to the data flow between the physical entity and the virtual counterpart [18] (Fig. 3). While DM lacks data interaction between both entities, and DS presents an automatic one-way data flow from the physical entity to the virtual counterpart; in the DT there is an automatic and bidirectional data flow, to and from the physical entity. Here, changes that occur in both the physical entity (PE) and the virtual entity (VE) directly lead to changes in the other [19].

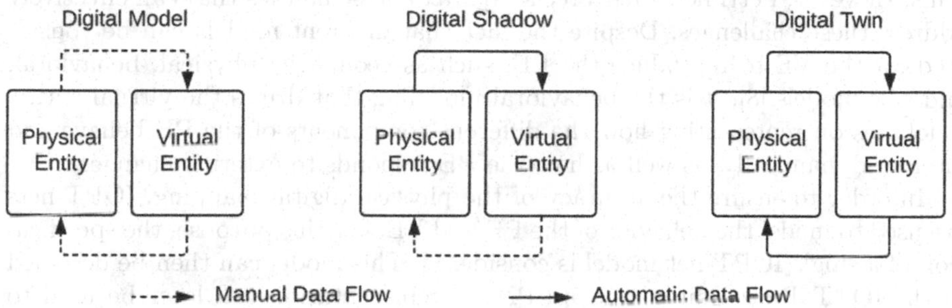

Fig. 3. Data flow in Digital Model, Digital Shadow, and Digital Twin [18].

Therefore, DM is suitable when the intention is to have static representations of physical objects; DS is used for monitoring and tracking their conditions; but DT takes it a step further. VE not only receives real-time data from a physical object, but can also send commands back to influence its behavior.

According to [20], the digital twin can support the representation of a physical entity through a virtual entity if it reproduces the relevant attributes of the PE in the context, ensures the entanglement of entities, and maps the PE values to the VE in time. Another relevant proposal presented in [21] highlights the physical entity and the virtual entity interacting with services; and services are said to be a vital element of the digital twin. On the one hand, the creation of the DT involves the use of several services and, on the other hand, its operation revolves around various platform services to facilitate simulation, verification, monitoring, optimization, and more.

Consequently, the development of a digital twin focuses on physical-digital mapping, data flow and connectivity, and provision of end-user services. Below its is explained how IOPT-nets enable digital twin and support each of these aspects, together with IOPT-Tools.

3.1 Modeling of Physical and Virtual Entities

Two lifecycles for the DT are described in [22]: one in which the PE and the VE are conceived at the same time, and designers can model them with the same characteristics, properties, functions, and models; and another where an unknown PE already exists, but the VE has to be implemented. This proposal focuses on the first scenario.

If VE fails to accurately mirror its physical counterpart, it will not accurately represent the true state of the physical system. Consequently, this can result in inaccurate measurements being fed into the VE, causing it to deviate from the intended representation of the PE. This is common in complex and dynamic systems. However, Petri nets offer precise syntax and semantics that can effectively address these challenges. Despite the fact that different models can be considered for the VE to reproduce the PE, such as geometric, physical, behavioral, and rule models [8], it is the behavioral modeling that drives the virtual entity. It focuses on representing how the different components of the PE behave and interact dynamically, as well as how the PE responds to external changes.

In order to ensure the accuracy of the physical-digital mapping, IOPT-nets are used to model the behavior of the PE and VE. For this purpose, the specification of a single IOPT-net model is considered. This model can then be designed in the IOPT-Tools Editor, and its PNML representation can later be used to support the deployment of both the PE and VE (see Fig. 4).

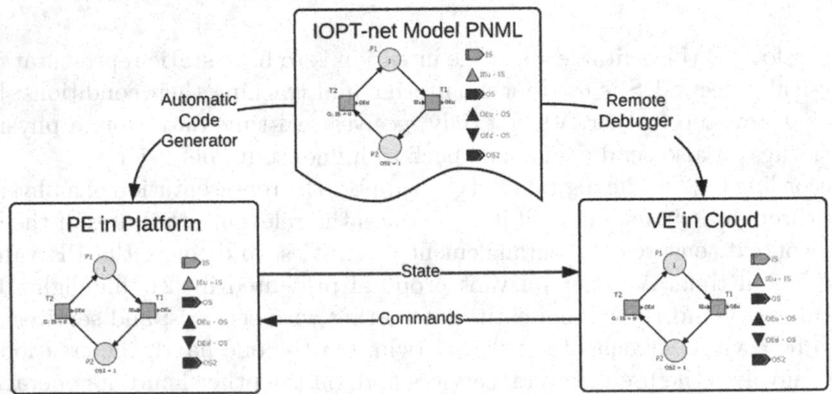

Fig. 4. One IOPT-net model can cover both PE and VE specifications.

This approach is advantageous because it can guarantee consistency throughout the DT lifecycle; allows entities to evolve and align seamlessly due to the precise syntax and semantics of Petri nets. For the PE, the IOPT-net model can effectively represent it and its physical interactions, meeting functional and safety requirements. For the VE, the model supports dependability issues, being used to monitor and analyze the state of the PE once deployed and predict its behavior, empowering users to make informed decisions.

After supporting the modeling of physical and virtual entities, the IOPT-Tools framework can also facilitate their implementation. In the real space, PE can be deployed on different platforms using the IOPT-Tools automatic code generators [16] that transforms the PNML file from the IOPT-net model into software code for microcontrollers or hardware descriptions for FPGAs. In the virtual space, VE and its connection to PE are deployed using the IOPT-tools remote debugger [17], fed by the same PNML.

3.2 Cognition and Control of the Physical World

The one-to-one connection between the real space and the virtual space in the DT is essentially established to allow the VE to monitor and/or control the PE. From the perspective of the VE, particularly in the context of safety-critical systems, the objective is not just to observe and anticipate PE behavior; it also intended that it can issue commands to the real space when necessary, enhancing the remote intervention capacity of the VE over PE.

As a result, it is crucial to cognize and control the elements that relate to the PE in real space, namely sensors and actuators, respectively. The sensors continuously produce input signals to the PE, conveying commands to the model or readings of real-world conditions. Actuators, in turn, convert the model's output signals into tangible actions or changes in the environment.

The IOPT-nets support this particular feature of DT through their inputs and outputs. Each signal can be assigned a GPIO number for deployment purposes, with integer range signals representing analog values, and boolean signals representing digital values. Upon implementation, the automatic code generator [16] associates the inputs and outputs of the IOPT-net model with the platform pins, allowing the PE to interact with the environment. This way, the PE is able to read sensors, command the status of mechanical actuators or read button/switch values, and control displays/LEDs to communicate with the users.

When the VE intends to exert control over the PE, such as affecting an actuator on the asset, this can be done by the VE forcing the values of the inputs and outputs. This involves overriding the actual values read or written by the GPIO pins. When the VE issues a new input, the PE evolves toward the desired state or behavior. This underscores another important aspect, which is ensuring connectivity and data flow between the PE and VE.

3.3 Connectivity and Data Flow

The continuous synchronization between the PE and VE is then driven by the real-time or near-real-time data flow, guaranteed by the connection between real and virtual spaces.

When an IOPT-net-based system is deployed, IOPT-Tools provides a remote debugger [17] that facilitates interaction with these systems from another location. This proves particularly useful in the case of models running on devices that are not locally accessible; and it also enables and introduces remote operation capabilities to the DT deployed with IOPT-nets, providing the VE with a way to mirror and interact with the PE in execution.

To elaborate, an HTTP server is defined in the automatically generated code for the PE, which establishes the connection with the remote debugger where the VE is located (see Fig. 5). The remote debugger initiates two HTTP connections with the platform: a data channel to receive notifications about changes in the PE state, whenever they occur; and a parallel control channel to send commands from the VE. These connections allow the transmission of JSON objects to support monitoring and control.

By capturing all changes in the PE, including changes in input and output signals, changes in net marking, autonomous events, and fired transitions, the VE graphically mirrors this information in the IOPT-net model, continuously presenting the evolution of the PE. This can be used to inspect the PE at any time and identify when eventual problems occur. On the VE side, the remote debugger also includes a tracing mechanism with step-by-step execution and breakpoint definition, along with the ability to remotely force the value of input and output signals in the PE.

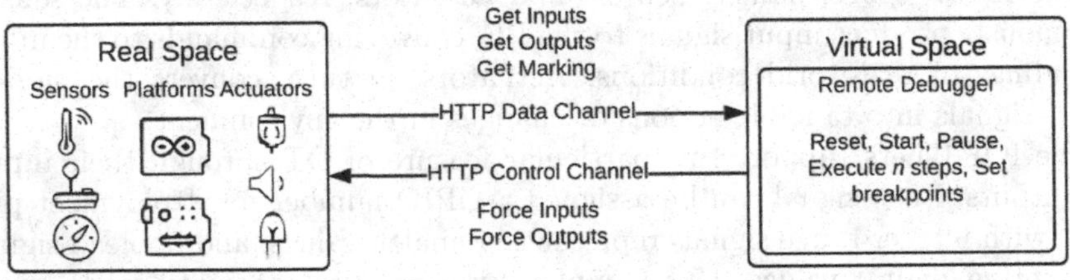

Fig. 5. Connectivity and data flow between the real and virtual spaces in an IOPT-net-based DT.

3.4 Provision of End-User Services

The IOPT-Tools have been mentioned in the different phases of the development of the digital twin (see Fig. 4 and Fig. 5). The editor, the automatic code generator [16] and the remote debugger [17] can be seen as services and an integral part of the DT.

As a matter of fact, by using the same PNML representation for PE and VE, the designed IOPT-net model can power multiple IOPT tools for DT end-user services. Other services (tools) can therefore be considered: the simulator [14] and the state-space generator [15] (see Fig. 6).

Specifically, the simulator [14] can be used to execute and debug the PE/VE model, relying on the token-game and the semantics of the IOPT-net class (see Sect. 2). It allows to manipulate the values of the input signals at each execution step and observe the evolution of the net, along with the resulting markings and outputs. This is similar to what happens with the remote debugger, but the analysis is done offline against the PE connection.

On the other hand, during each PE and VE connection, the remote debugger enables a history recording mechanism, saving the evolution of the net. This recorded history can be replayed later using the simulator, allowing for a more in-depth examination of PE conditions.

Combined with the state-space generator [15], these tools serve the dual purpose of: verifying and validating the desirable requirements for PE/VE; and check if the model can reach undesired states that may represent dangerous situations, to resolve them before implementation. This approach also provides a means for the VE to offer feedback on the dependability of the PE, and contribute to the refinement of the entities, supporting their redesign.

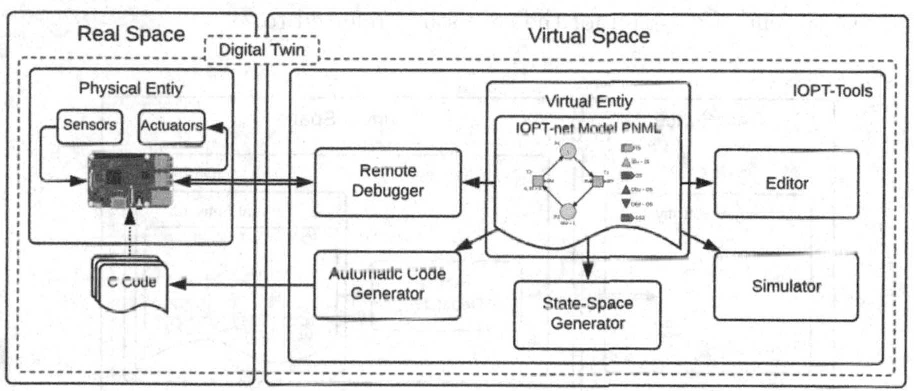

Fig. 6. Framework for the modeling and deployment of digital twins using IOPT-nets.

Ultimately, an IOPT-net-based digital twin can support:

- A precise way for a VE to represent and mirror a PE throughout its lifecycle;
- Real-time monitoring and analysis of a PE to support its design, maintenance, control, etc.;
- Preventive maintenance based on VE simulations and model-checking that can predict situations in which a PE is likely to fail;
- Informed decision-making about the operation and maintenance of a PE, potentially reducing downtime and costs;
- Remote access for human interaction with a PE, allowing interaction from remote locations via VE;
- Intelligent control via VE over PE, providing instructions and control signals when necessary.

4 Digital Twin of a Power Wheelchair: A Case Study and Discussion

The case study presented in the following is within the scope of research that has been carried out to apply the digital twin to power wheelchairs [23–25]. The case study can significantly contribute to the dissemination of best practices in the field, benefiting stakeholders across the sector. By modeling the intricate interactions within power wheelchairs, Petri nets offer a structured approach to evaluating system performance under various circumstances, including normal operation, failure conditions, and emergency situations. The use of DT in this context provides for a continuous remote connection between power wheelchairs and their digital counterparts ideally implemented in service centers of companies in the sector. The aim would be to support regular inspections, prompt repairs, and adjustments of wheelchair parameters.

4.1 Case-Study

Figure 7 presents a diagram for the case study referred to.

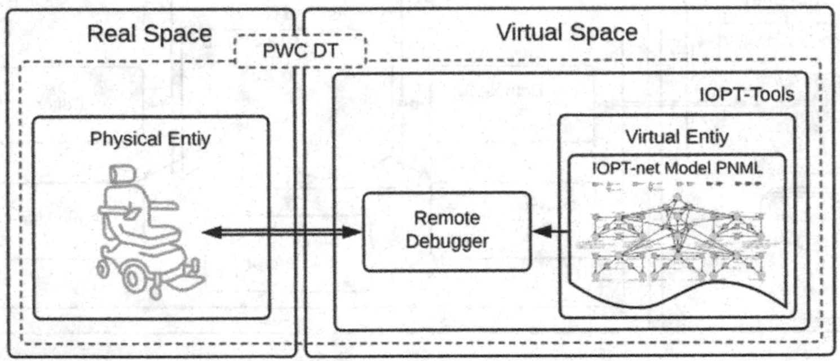

Fig. 7. A digital twin of a power wheelchair supported by IOPT-Tools.

The physical entity here focuses on the main elements that allow the wheelchair to move or change seating positions, and that significantly impact the wheelchair's overall functionality, safety, and user experience; specifically:

- Analog joystick;
- Right and left drive motors;
- Actuator for seat elevation;
- Actuator for backrest recline;
- Actuator for footrest elevation;
- Mode change button;
- Seat position change button.

Figure 8 presents the interface for cognition/control in the PWC DT, that is, the inputs and outputs of the model specifying PE and VE. From top to bottom: joystick signals "x" and "y"; "mode" signal and events to switch between driving and position change; "position" signal and events for change between positions; and 5 output signals for each motor and actuator, "r" (right), "l" (left), "b" (backrest), "s" (seat), and "f" (foot).

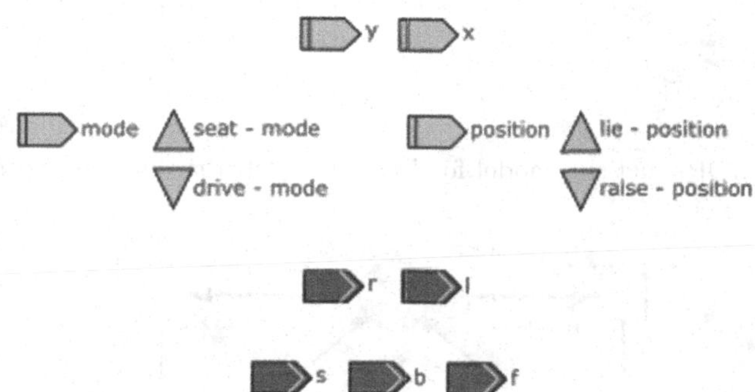

Fig. 8. Inputs and outputs of the IOPT-net model that implements the PWC DT.

Based on these signals and events, the model in Fig. 9 was designed to control the behavior of a drive motor, in this case, the left one (for the right motor, it is similar). For each drive motor, there are 3 possible actions: either the motor stops, moves forward, or moves backward. To control this, the net's initial marking is for motor to start from a standstill, and then different guards were defined to reflect the user's driving intent.

Here, the evolution of the net and the firing of transitions then depend on the values of the "x" and "y". In case of motion, there are 4 action rules at the places to set the value of the output signal for motor speed control. The action rules are the same for each place, as they only depend on the values of "x" and "y"; they are based on work presented in [26]. If the intention is to stop the motor, the net returns to its initial state, and the motor assumes its default value, which is 0.

A similar model is shown in Fig. 10. In this case, it models an actuator to change the backrest position (similar for controlling the seat and footrest). The difference in relation to the previous model is the fact that this model is for limit switch motors. The guards are simpler, as are the action rules, which assign the value 1 or −1 to the output signal. This is because regardless of the joystick throw, the speed of these actuators is not variable for safety reasons. The net does not present initial marking, as it is actually part of model from Fig. 12, whose evolution can lead to the marking of places in Fig. 10.

The last sub model being presented (see Fig. 11) controls the way the wheelchair operates: in driving mode or seating control mode. For this reason, the place "$X_$stop" of each motor and actuator is shown.

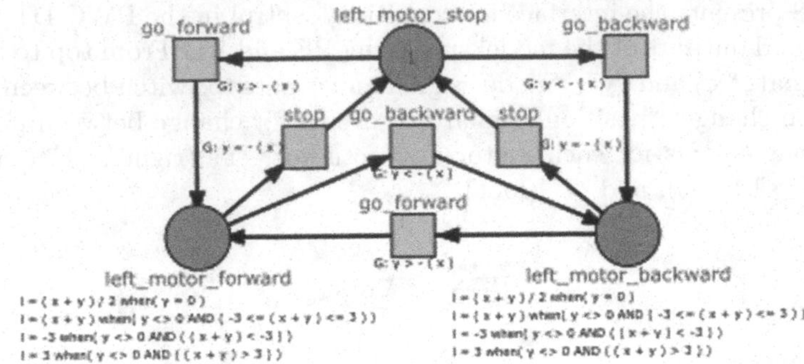

Fig. 9. IOPT-net sub model for PWC DT: left drive motor control.

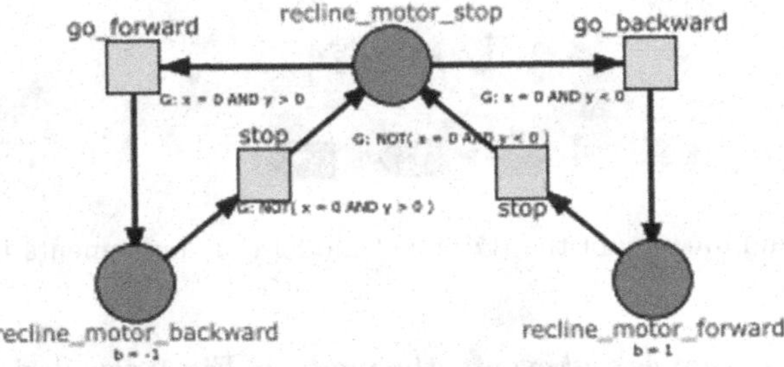

Fig. 10. IOPT-net sub model for PWC DT: power seat actuator control.

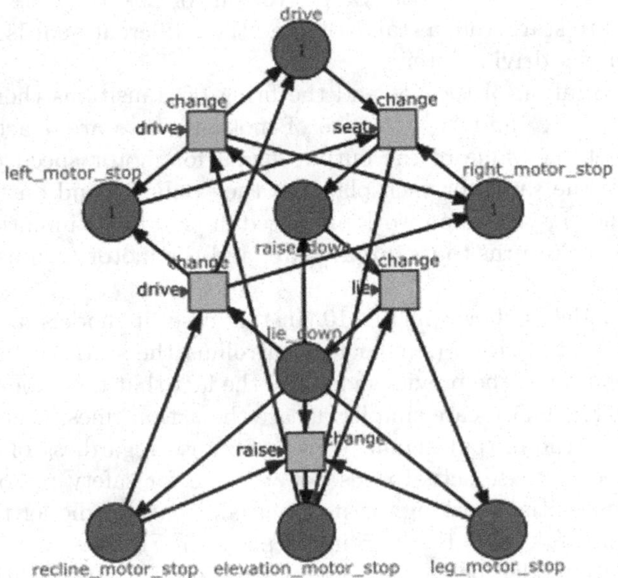

Fig. 11. IOPT-net sub model for PWC DT: operating mode control.

If both the right and left motors are stopped, the user can adjust the seat elevation by selecting "mode" ("seat" occurs); then, the user can also adjust the lying position, by clicking "position" ("lie" occurs) to act on the actuators to recline the backrest and to elevate the footrest. To drive again, the user needs to stop the actuators and deselect "mode" ("drive" occurs). This control enables the joystick to serve different purposes, while ensuring users do not change their seat position while driving as this would be unsafe.

Thus, the main model is presented in Fig. 12, and consists of 2 models in Fig. 9, 3 models in Fig. 10, and the model in Fig. 11.

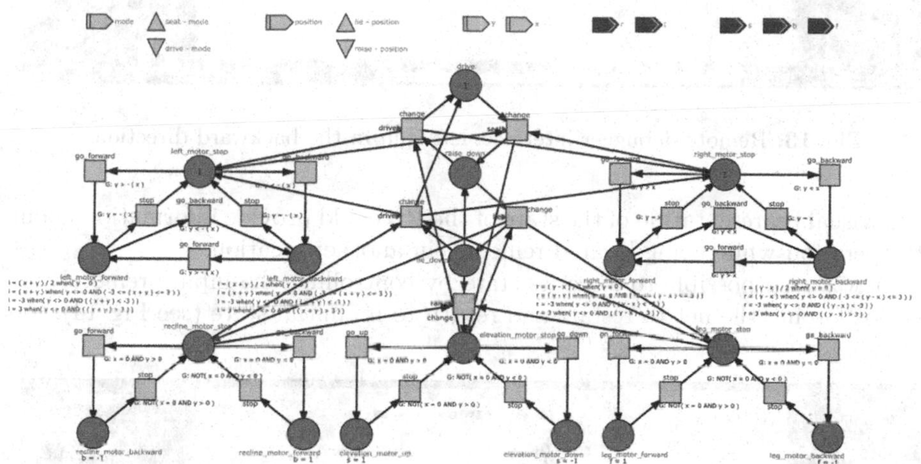

Fig. 12. IOPT-net model that implements the PWC DT.

To ensure the functioning of the model of Fig. 12 and the implementation of the DT, two phases of validation were undertaken. First, using the IOPT-Tools simulator to verify and validate the model's evolution by manipulating the inputs. In this phase, the state-space generator was also used to confirm that the model was deadlock-free. Future interest lies in exploring additional behavioural properties such as safety, boundedness, liveness, and reversibility.

Further validation involved a physical experiment using the automatic code generation tool to produce the C code of the model. The code was used to deploy the PE, built and compiled using a Raspberry Pi 3 Model B V1.2 board. The board also established a secure communication with the remote debugger through an HTTPS connection with user authentication.

Figure 13 shows the VE in the remote debugger, showing the state of the PE running on the platform with the IP address 192.168.1.254:9000. The state of the PE in this case showed that the wheelchair was in drive mode. The movement was related to a left spin in the backward direction, as the right motor was stopped and the left motor was moving backwards, producing "l" equal to −3.

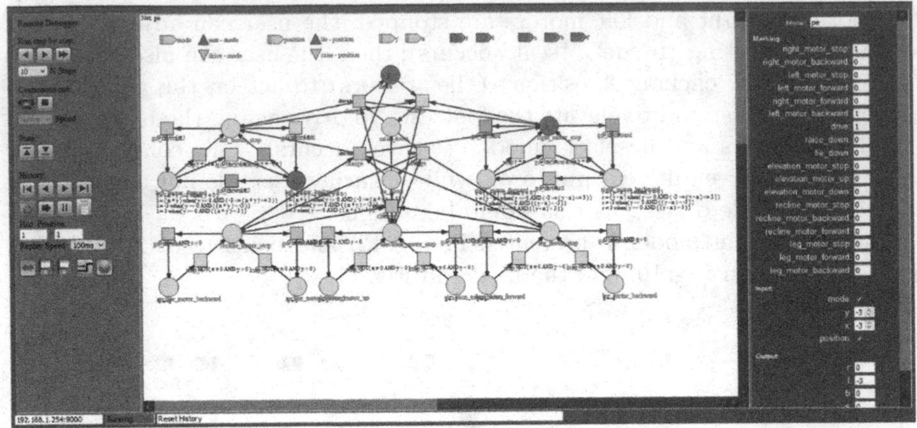

Fig. 13. Remote debugger interface: left spin in the backward direction.

This visual representation of the state of the PE could provide information about its operation while being located remotely in another location.

It was also possible to force its state by controlling the inputs remotely, as well as forcing the net to restart and return to its initial state (see Fig. 14).

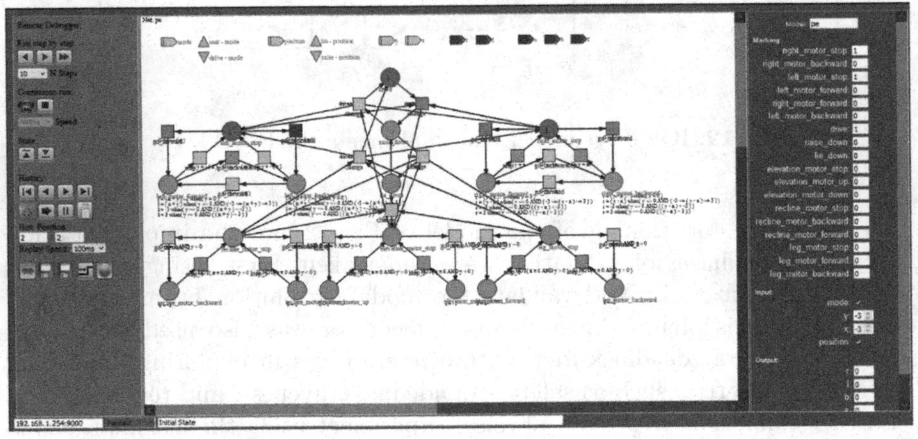

Fig. 14. Remote debugger interface: forcing the initial state.

After this, it was possible to accomplish the deployment of the PWC DT and validate the proposal.

4.2 Discussion

There are users who have used the same power wheelchair for longer than the recommended 5 years; and chronic use of wheelchairs leads to wear and tear, affecting their overall performance. To ensure the longevity and reliability of power wheelchairs, a systematic and comprehensive approach to maintenance is required, with both preventive and corrective measures.

In this way, there have been efforts to remotely connect the latest models of power wheelchairs with service centers of some companies in the sector. They use specialized software applications to collect and analyze data about wheelchair usage and performance. This data provides valuable insights into the wheelchair's condition and can be used to improve remote diagnostics [27–30]. However, they have limited control over the wheelchairs themselves. In many cases, they trust clients and caregivers to follow technicians' instructions on how to proceed.

The proposed solution and case study aim to bridge this gap by leveraging the concept of a digital twin for power wheelchairs. From enhancing customization to optimizing maintenance, a digital twin can revolutionize wheelchair design and management. It can simultaneously monitor and provide feedback on the wheelchair's condition and the environment in which it operates, as well as enabling technicians to take action when necessary. In that sense, they can also empower users with tailored solutions that promote their full participation in society, ultimately contributing to sustainability.

The case study presented focuses on a single wheelchair, but in a real-world scenarios, companies may seek to extend the application of DT to the various wheelchairs they manage. The architecture illustrated in Fig. 15 showcases how service providers can use DT for real-time connectivity with multiple power wheelchairs.

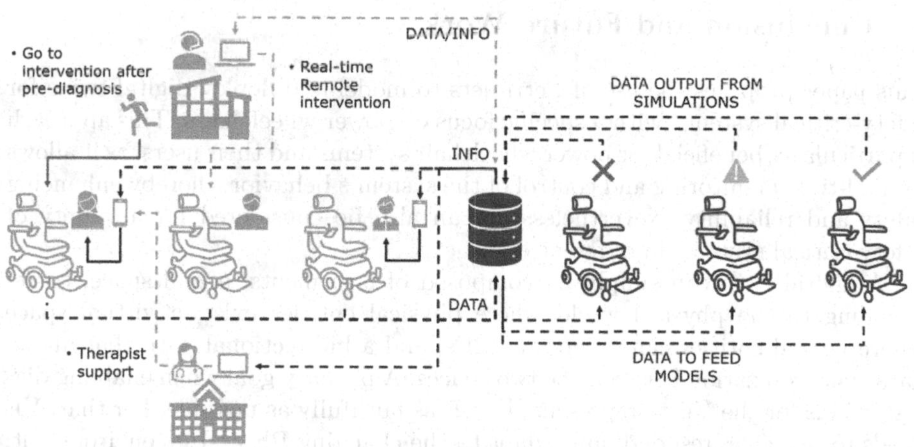

Fig. 15. Scenario of multiple power wheelchair digital twins.

Specific applications in this context could be:

- Real-time tracking of power wheelchair locations;
- Data collection on each user's driving style and use;

- Tips on safe and efficient power wheelchair driving;
- Access wheelchair data and information on how to optimize its performance;
- Sending instructions or updates directly to the power wheelchair;
- Establish a maintenance schedule and perform pre-diagnostics;
- Generation of cost simulations based on required maintenance;
- Based on the information from used wheelchairs, configure new or replacement wheelchairs;
- Automatic reports on power wheelchair usage and maintenance.

For that purpose, and based on the work presented, IOPT-nets and IOPT-Tools will continue to be taken into consideration. They have proven to be adequate for modeling and deploying digital twins. However, additional complexity may arise when dealing with even more intricate situations. An example might be the case of a wheelchair with more power seat functions; or even the scenario of Fig. 15. This cases would lead to the exponential growth of net elements and pose challenges in managing safety requirements.

As such, an extended IOPT-net class with high-level characteristics may be considered in the future. This may have the potential to reduce modeling complexity and support data processing. For example, as with colors, it is possible to distinguish between different processes, even if their sub nets have been folded into a single sub net. In this view, it would be possible to aggregate the activity of 3 or more actuators in the case of one wheelchair; or 3 or more wheelchairs in the view of their company.

5 Conclusion and Future Work

This paper proposes the use of Petri nets to model and deploy digital twins for safety-critical systems, with a specific focus on power wheelchairs. This approach is particularly beneficial for power wheelchair systems and their users, as it allows for real-time monitoring and control of the system's behavior, thereby enhancing safety and reliability. Nevertheless, it can also be considered for a variety of safety-critical systems in different areas.

A digital twin can simply be composed of 3 elements: a real space, corresponding to the physical world where physical entities exist; a virtual space where virtual entities can mirror the PEs; and a bidirectional path that allows data synchronization between the two spaces. A primary goal when enabling digital twin is for the VE to represent the PE as faithfully as possible. For that, VE needs to perceive, respond, and adapt to the changing PE and its environment, to promote insights in the evolution and performance of the PE and control the physical world.

The IOPT-nets were used to enhance the advantages of DT applied to model PE/VE behavior and interaction with the environment. In particular, in addition to dealing with concurrency, conflicts, synchronization, and resource sharing, their non-autonomous characteristic based on inputs and outputs could give support to the different stages in the development of a DT: modeling of PE and

VE, cognition and control of the physical world, connectivity and data-flow, and the provision of end-user services.

In fact IOPT-nets benefit from automated tools that support all phases of development, which is an advantage compared to other formalisms and state-oriented model languages. The IOPT-Tools framework was used for the design, verification, validation, implementation, and remote debugging of PE/VE. Using the simulator and the state-space generator it was possible to debug and validate the model, where most errors were detected in the early design stages. The automatic code generator allowed for PE implementation and ensured code free of errors. And the remote debugger enabled the PWC DT with PE continuous monitoring and control.

The IOPT-nets and IOPT-Tools framework have been shown to be effective in the formal modeling and analysis of DT, and real-time simulation and control of PE. However, there are possible limitations to this approach when dealing with more complex systems. To address these, an extended IOPT-net class with high-level characteristics will be considered for future work.

In addition to behavioral modeling, future work intends to use IOPT-nets also in the modeling of VE rule models to make DT able to reason, judge, evaluate. And a more complex case study will be undertaken, focusing on specific aspects such as obstacle navigation, wheelchair activity tracking, accessible route planning, emergency assistance, user health logs, wheelchair maintenance records, and training and education. This will provide valuable insights into the practical applications and benefits of the proposed Petri net and Digital Twin integration for dependability proposes.

Acknowledges. The authors of this manuscript would like to thank the reviewers for the thoughtful comments and efforts they have put into improving it.

References

1. Singh, L.K., Rajput, H.: Dependability analysis of safety critical real-time systems by using Petri nets. IEEE Trans. Control Syst. Technol. **26**(2), 415–426 (2018). https://doi.org/10.1109/TCST.2017.2669147
2. Rausand, M.: Reliability of Safety-Critical Systems. Wiley, New York (2014). https://doi.org/10.1002/9781118776353
3. IEC 61508:2010 CMV. https://webstore.iec.ch/publication/22273. Accessed 16 Jan 2024
4. IEC 62551:2012. https://webstore.iec.ch/publication/7191. Accessed 16 Jan 2024
5. Tao, F., Zhang, H., Liu, A., Nee, A.Y.C.: Digital twin in industry: state-of-the-art. IEEE Trans. Industr. Inf. **15**(4), 2405–2415 (2019). https://doi.org/10.1109/TII.2018.2873186
6. Fuller, A., Fan, Z., Day, C., Barlow, C.: Digital twin: enabling technologies, challenges and open research. IEEE Access **8**, 108952–108971 (2020). https://doi.org/10.1109/ACCESS.2020.2998358
7. Grieves, M.W.: Product lifecycle management: the new paradigm for enterprises. Int. J. Prod. Dev. **2**(1/2), 71–84 (2005). https://doi.org/10.1504/IJPD.2005.006669

8. Tao, F., Zhang, M., Nee, A.Y.C.: Five-dimension digital twin modeling and its key technologies. Digit. Twin Driven Smart Manuf. 63–81 (2019). https://doi.org/10.1016/B978-0-12-817630-6.00003-5
9. Gomes, L., Barros, J.P.: Refining IOPT Petri nets class for embedded system controller modeling. In: Proceedings of the IECON 2018 - 44th Annual Conference of the IEEE Industrial Electronics Society, pp. 4720–4725. IEEE (2018). https://doi.org/10.1109/IECON.2018.8592921
10. Girault, C., Valk, R.: Petri Nets for Systems Engineering - A Guide to Modeling, Verification, and Applications. Springer, Heidelberg (2002). https://doi.org/10.1007/978-3-662-05324-9
11. Wolf, K.: Petri net model checking with LoLA 2. In: Khomenko, V., Roux, O.H. (eds.) PETRI NETS 2018. LNCS, vol. 10877, pp. 351–362. Springer, Cham (2018). https://doi.org/10.1007/978-3-319-91268-4_18
12. Pereira, F., Moutinho, F., Costa, A., Barros, J.P., Campos-Rebelo, R., Gomes, L.: IOPT-tools - from executable models to automatic code generation for embedded controllers development. In: Bernardinello, L., Petrucci, L. (eds.) PETRI NETS 2022. LNCS, vol. 13288, pp. 127–138. Springer, Cham (2022). https://doi.org/10.1007/978-3-031-06653-5_7
13. Desel, J., Reisig, W.: Place/transition Petri nets. In: Reisig, W., Rozenberg, G. (eds.) ACPN 1996. LNCS, vol. 1491, pp. 122–173. Springer, Heidelberg (1998). https://doi.org/10.1007/3-540-65306-6_15
14. Pereira, F., Gomes, L.: Cloud based IOPT Petri net simulator to test and debug embedded system controllers. In: Camarinha-Matos, L.M., Baldissera, T.A., Di Orio, G., Marques, F. (eds.) DoCEIS 2015. IAICT, vol. 450, pp. 165–175. Springer, Cham (2015). https://doi.org/10.1007/978-3-319-16766-4_18
15. Pereira, F., Moutinho, F., Gomes, L., Ribeiro, J., Campos-Rebelo, R.: An IOPT-net state-space generator tool. In: Proceedings of the INDIN 2011 - 9th IEEE International Conference on Industrial Informatics, pp. 383–389. IEEE (2011). https://doi.org/10.1109/INDIN.2011.6034907
16. Pereira, F., Moutinho, F., Gomes, L.: A syntax-independent code generation tool for IOPT-Petri net. In: Proceedings of the PN4TT 2023 - Algorithms Theories for the Analysis of Event Data and Petri Nets for Twin Transition. CEUR-WS (2023). https://ceur-ws.org/Vol-3424/paper6.pdf
17. Pereira, F., Melo, A., Gomes, L.: Remote operation of embedded controllers designed using IOPT Petri-nets. In: Proceedings of the INDIN 2015 - 13th IEEE International Conference on Industrial Informatics, pp. 572–579. IEEE (2015). https://doi.org/10.1109/INDIN.2015.7281797
18. Kritzinger, W., Karner, M., Traar, G., Henjes, J., Sihn, W.: Digital twin in manufacturing: a categorical literature review and classification. IFAC-PapersOnLine **51**(11), 1016–1022 (2018). https://doi.org/10.1016/j.ifacol.2018.08.474
19. Singh, M., Fuenmayor, E., Hinchy, E.P., Qiao, Y., Murray, N., Devine, D.: Digital twin: origin to future. Appl. Syst. Innov. **4**(2), 36 (2021). https://doi.org/10.3390/asi4020036
20. Minerva, R., Lee, G.M., Crespi, N.: Digital twin in the IoT context: a survey on technical features, scenarios, and architectural models. Proc. IEEE **108**(10), 1785–1824 (2020). https://doi.org/10.1109/JPROC.2020.2998530
21. Qi, Q., et al.: Enabling technologies and tools for digital twin. J. Manuf. Syst. **58**(B), 3–21 (2021). https://doi.org/10.1016/j.jmsy.2019.10.001
22. Barricelli, B.R., Casiraghi, E., Fogli, D.: A survey on digital twin: definitions, characteristics, applications, and design implications. IEEE Access **7**, 167653–167671 (2019). https://doi.org/10.1109/ACCESS.2019.2953499

23. Lagartinho-Oliveira, C., Moutinho, F., Gomes, L.: Digital twin in the provision of power wheelchairs context: support for technical phases and conceptual model. Computers **11**(11), 166–180 (2022). https://doi.org/10.3390/computers11110166
24. Alves, A., Lagartinho-Oliveira, C., Moutinho, F., Gomes, L.: ROS-based digital twin for power wheelchair. In: Proceedings of the ONCON 2022 - 1st Industrial Electronics Society Annual On-Line Conference. IEEE (2022). https://doi.org/10.1109/ONCON56984.2022.10127002
25. Lagartinho-Oliveira, C., Moutinho, F., Gomes, L.: Support operation and maintenance of power wheelchairs with digital twins: the IoT and cloud-based data exchange. In: Camarinha-Matos, L.M., Ferrada, F. (eds.) DoCEIS 2023. IFIPAICT, vol. 678, pp. 191–202. Springer, Cham (2023). https://doi.org/10.1007/978-3-031-36007-7_14
26. Faria, B.M., Ferreira, L., Reis, L.P., Lau, N., Petry, M., Soares, J.C.: Manual control for driving an intelligent wheelchair: a comparative study of joystick mapping methods. In: Proceedings of the IROS 2012 - Workshop on Progress, Challenges and Future Perspectives in Navigation and Manipulation Assistance for Robotic Wheelchairs (2012). https://paginas.fe.up.pt/~niadr/PUBLICATIONS/LIACC_publications_2011_12/pdf/OC59_Manual_Control_Driving_IW_Comparative_Study_Joystick_Mapping_Methods.pdf
27. Mylinx Resources Hub. https://www.dynamiccontrols.com/resource-hub/mylinx-resources-hub. Accessed 29 Jan 2024
28. MyPermobil App. https://permobilwebcdn.azureedge.net/media/v5vgqmbp/mypermobil_brochure_uk_200525_web.pdf. Accessed 29 Jan 2024
29. Fleet Management. https://permobilwebcdn.azureedge.net/media/tyen1e5w/fleet-management-brochure.pdf. Accessed 29 Jan 2024
30. Interactive Assist. https://www.quantumrehab.com/quantum-electronics/interactive-assist.asp. Accessed 29 Jan 2024

CosyVerif: The Path to Formalisms Cohabitation

Étienne André[1], Jaime Arias[1(✉)], Benoît Barbot[2], Francis Hulin-Hubard[3], Fabrice Kordon[3], Van-François Le[1], and Laure Petrucci[1]

[1] LIPN, CNRS UMR 7030, Université Sorbonne Paris Nord, Villetaneuse, France
arias@lipn.univ-paris13.fr

[2] Univ Paris Est Creteil, LACL, 94010 Creteil, France

[3] Sorbonne Université, CNRS, LIP6, Paris, France

Abstract. More and more model checking approaches rely nowadays on several inputs, potentially expressed in different formalisms. Tools implementing these usually include only the expected formalisms. Thus, such tools are ad-hoc and lack extensibility and interoperability features, especially when new formalisms are needed.

The challenge is then to design a generic and easy way for several formalisms to cohabit in such verification software. Creation, exchange, and interoperability between formalisms would be facilitated, thus saving numerous development efforts.

The originality of the CosyVerif platform lies in its capability to easily and rapidly gather diverse formalisms within a same framework, and to provide extension facilities to integrate new ones.

1 Introduction

The CosyVerif platform (http://www.cosyverif.org) is a meta-tool, in the form of a web platform, which allows to put to work and synchronise several tools for the analysis of formal models, such as Petri nets or automata. It is a distributed open-source environment, which features capabilities to easily and rapidly gather diverse formalisms in a same framework, and to provide extension facilities to integrate new ones.

Requirements. A wide palette of tools for model checking concurrent systems exists nowadays. Each of these implements its own algorithms from models expressed in a dedicated format. Therefore such tools are often ad-hoc and lack interoperability features.

Several Petri net tools successfully rely on PNML (*Petri Net Markup Language*) [17,19], which provides a normalised format that allows for using multiple tools on the same models. This is key to the MCC (*Model Checking Contest*) [22] for participating tools. However, PNML only handles variants of Petri nets, while models to be used together may be expressed in several different formalisms, e.g. a Petri net and a property expressed as an automaton. The current command-line

L. M. Kristensen and J. M. van der Werf (Eds.): PETRI NETS 2024, LNCS 14628, pp. 432–444, 2024.
https://doi.org/10.1007/978-3-031-61433-0_21

verification tools often expect several files as input, not necessarily of the same nature. Both tools experimented with in Sect. 4 are in this category.

A challenge is then to design a generic and easy way for formalisms to cohabit within a single verification platform. Creation, exchange, and interoperability between formalisms would then be greatly facilitated, thus saving extensive development efforts. Moreover, using a shared Graphical User Interface (GUI) to manipulate all formalisms lowers not only the development effort but also the learning curve for the modeller, who only has a single GUI to master.

At this stage, this interoperability between formalisms is achieved at the syntactic level, provided that the semantics is consistent among the tools that share it. A direct advantage is to establish a hierarchy between formalisms that reuse other ones; this is of particular interest for Petri nets and their variants, as well as networks of automata.

Cohabitation of models can take the form of a set of modules to be handled together. Several well-known model checking software enable modules. GreatSPN [2] and Cosmos [11] require a Petri net and an automaton modelling a temporal logic formula. In IMITATOR the different modules are not standalone, but do collaborate by sharing e.g. synchronised transitions.

CPNTools [1] allows for analysing Hierarchical Coloured Petri Nets with several modules (named *pages*). UPPAAL [4] is dedicated to networks of timed automata. These latter tools have their own ad-hoc GUI and implementation; they do not provide easy extensibility.

Several verification tools deal with generic formalisms. IOPT-Tools [26] generates various output codes (C, VHDL, Simulink, ...); this multi-formalism approach does not concern various inputs, but addresses code generation. ePNK [21] allows for Petri net tools plugins, in different flavours, thanks to the PNML transfer format [18,20]. However, it does not support formalisms other than Petri nets. ITS-Tools [23,27] also handles numerous inputs from various tools, thanks to an internal generic representation (Instantiable Transition Systems) suitable for model-checking, but does not provide a GUI.

Outline. Section 2 presents the main features provided by the CosyVerif platform, in particular the use of heterogeneous modules. Then, Sect. 3 details the tool architecture and its flexibility for welcoming new formalisms or analysis tools. Two tools serve in Sect. 4 as examples which exhibit different characteristics. Finally, Sect. 5 summarises and provides directions for future work.

2 Desired Features Provided by CosyVerif

An open and extensible verification platform should provide the user with facilities to model his/her project and analyse it. It should furthermore provide a verification tool with easy means for new formalism dialect and tool integration. We now discuss those that are already supported by CosyVerif.

Flexibility for the User with Zero-Install. The web-based GUI of CosyVerif advantages are twofold. First, drawing models only requires a web browser to connect to the platform. Our laboratory provides a free-access server (https://cosyverif.lipn.univ-paris13.fr) for trials. Second, it is multi-platform, so it can be used with usual operating systems, on a computer, a laptop, a tablet or even a smartphone. Currently, CosyVerif is distributed as a Docker image for easy deployment either locally on a machine or on a server.

Different Formalisms as Modules. As detailed further in Sect. 3, a compositional model can be designed, by gathering a collection of models within the same project. These models can be expressed in similar or different formalisms. The collection bases on already existing formalisms that can be used for standalone models or as part of a project.

Handling Compositionality. For the moment, compositionality is kept simple, mainly by using inclusion of formalisms and models. A new formalism allows for describing the links between the different modules in the project. It can be seen as an additional formalism or language meant for this specific purpose. For example, networks of Parametric Timed Automata, as used by IMITATOR (see Sect. 4), is a collection of PTAs together with a set of synchronised transitions, which pairs labels of transitions in the different modules.

Creation of New Formalisms. Users can define their new formalisms in the FML (*Formalism Markup Language*) format, which was introduced in [7]. A FML description typically bases on an existing one for similar but simpler models. For example the FML for Symmetric Petri Nets uses the one for Place/Transition nets. Models themselves are stored in GrML (*Graph Markup Language*) [7], an XML-like language to describe models according to a FML formalism definition. The designer of a new formalism also has the possibility to parametrise the associated graphics. For example, the current version of CosyVerif supports ADTrees (*Attack-Defense Trees*) [24] in which the different types of nodes look similar to electronic gates (see such a model in Fig. 6). The reader can find some FML files created by the community in [3].

Interface with Command-Line (CLI) Tools. CLI tools can be embedded as plugins in CosyVerif's computation engine. Thus, verification can be launched using either the standalone CLI version or via the platform's GUI. Moreover, export to some tools ad-hoc input language is provided (e.g. Roméo [25], IMITATOR [6]).

Asynchronous Computation. Users can launch a verification and continue working on the platform, without waiting for the result (which may come after a lengthy period of time). When the computation is completed, the user can access the results of these finished jobs.

Distinct Editing and Computation Engines. As will be shown in Sect. 3, the platform implements a client/server paradigm. Thus the editing and computation are separate and can run on different machines.

3 Architecture

CosyVerif is constituted of the following two separated components that communicate which each other using `http` requests.

- **CosyDraw:** it is a web-based graphical user interface written in Javascript for the design of models (e.g. Petri nets, automata, ADTrees). At any time, users can send their models to Alligator for their analysis.
- **Alligator:** it is an application that provides a wide range of tools as web services (e.g. `prod`, IMITATOR, Cosmos, adt2amas). The number of tools can be easily extended thanks to the fact that Alligator handles them as plugins. The server containing Alligator also provides functionalities such as user management and file storage (e.g. models, formalisms).
- **3rd-party tools:** they are connected to Alligator, thanks to dedicated drivers (which usually require limited development effort). This is similar to a MDE approach as implemented in Eclipse/EMF, but with much less dependencies to the metamodels; conversion is completely handled by the driver (the tool may thus keep its own internal representation).

Figure 1 shows the architecture of CosyVerif. Each component can be installed locally and used offline, or can connect to an external host where one or both of them have been deployed. For the sake of facilitating the installation, each component is also distributed as a Docker image.

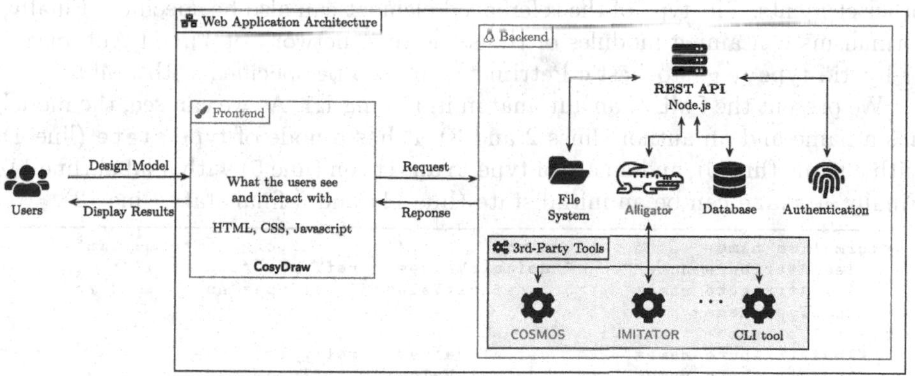

Fig. 1. Architecture of CosyVerif

3.1 Specification of a Formalism

A formalism is a structured document (XML) that has to comply with the *Formalism Markup Language* (FML). The elements (i.e. XML tags) supported by FML are given in Fig. 2. It allows for specifying any type of node (e.g. places and transitions of a Petri net) and arcs to connect them (e.g. inhibitor arcs,

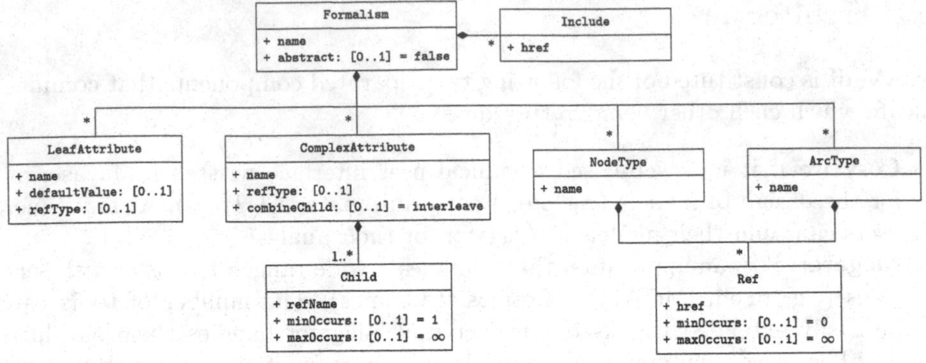

Fig. 2. FML schema

read arcs). It also allows for specifying the different types of attributes related to the model (e.g. name of the model) and to the different types of elements (e.g. number of tokens in a place).

A formalism can be defined as *abstract* to be later extended by another formalism. An abstract model cannot be used by the user to create a model. FML allows to include the definition of another formalism (abstract or not) thus permitting the definition of complex formalisms in a compositional and easy way.

FML allows also to define the types of hierarchies permitted by a hierarchical formalism by specifying which types of nodes and arcs can contain references to other elements. The type of the referenced element can also be specified. Finally, formalisms containing modules of the same (e.g. network of Timed Automata) or hybrid type (e.g. stochastic Petri net) can also be specified with FML.

We present the FML of an automaton in Listing 1.1. As we can see, the model has a name and an author (lines 2 and 3), it has a node of type `state` (line 4) with a name (line 6), and an arc of type `transition` (line 5) with a label (line 7). Finally, a `state` can be an initial state (line 11) and a final state (line 12).

```
1  <formalism name="Automaton" xmlns="http://cosyverif.org/ns/formalism">
2    <leafAttribute name="name" defaultValue="" refType="Automaton"/>
3    <leafAttribute name="author" defaultValue="" refType="Automaton" />
4    <nodeType name="state"/>
5    <arcType name="transition"/>
6    <leafAttribute name="name" defaultValue="" refType="state"/>
7    <leafAttribute name="label" defaultValue="" refType="transition"/>
8    <leafAttribute name="initialState" />
9    <leafAttribute name="finalState" />
10   <complexAttribute name="type" refType="state">
11     <child refName="initialState" minOccurs="0" maxOccurs="1"/>
12     <child refName="finalState" minOccurs="0" maxOccurs="1"/>
13   </complexAttribute>
14 </formalism>
```

Listing 1.1. FML for an automaton

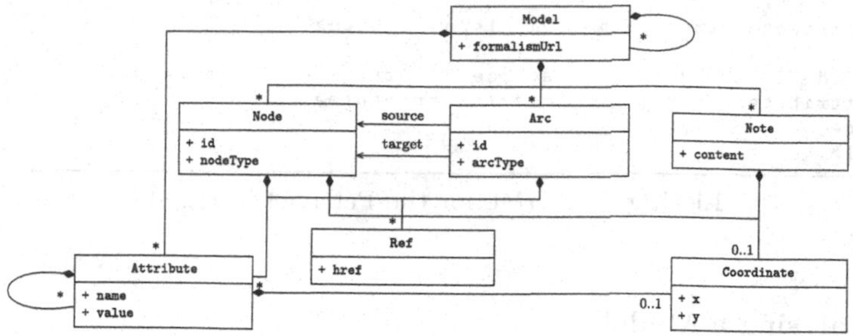

Fig. 3. GrML schema

3.2 Creating a Model

Once a formalism is defined (i.e. an FML), it is possible to create instances of it (i.e. models). A model is a structured document (XML) written using the *Graph Markup Language* (GrML) that is shown in Fig. 3. As we can see, a model contains nodes and arcs. The arcs make it possible to connect the nodes. Attributes are attached to the model as well as to all elements of the model. These attributes define information about the model (e.g. title, authors) or the elements (e.g. name, arc valuation, number of tokens).

The schema allows to manage hierarchical models and composition can be done with elementary references. The `formalismUrl`, `nodeType`, and `arcType` attributes allow the tool to independently determine the structure of the format that is used as well as the different types of elements that constitute the model.

In Listing 1.2, we present the GrML of the Petri net model shown in Fig. 4. As we can see, it is composed of two places (lines 7, and 11) where p_1 has one token (line 8) and p_2 has zero tokens (line 12). It also has one transition (line 15), and two arcs with valuation 1 (lines 20 and 23). Finally, it contains the name, the version, and the authors of the model (lines 3 to 5).

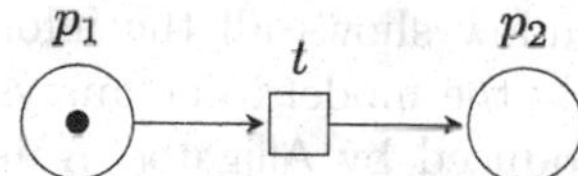

Fig. 4. Example of a Petri net

```
1  <model formalismUrl="http://formalisms.cosyverif.org/pt-net.fml"
2    xmlns="http://cosyverif.org/ns/model">
3    <attribute name="name">Example</attribute>
4    <attribute name="version">0.1</attribute>
5    <attribute name="author">CosyTeam</attribute>
6
7    <node id="1" nodeType="place" x="220" y="328">
8      <attribute name="marking">1</attribute>
9      <attribute name="name">p1</attribute>
10   </node>
11   <node id="2" nodeType="place" x="408" y="324">
12     <attribute name="marking">0</attribute>
13     <attribute name="name">p2</attribute>
14   </node>
15   <node id="3" nodeType="transition" x="330" y="323">
16     <attribute name="name">t</attribute>
17   </node>
18
19   <arc id="4" arcType="arc" source="1" target="3" order="0">
```

```
      <attribute name="valuation">1</attribute>
   </arc>
   <arc id="5" arcType="arc" source="3" target="2" order="0">
      <attribute name="valuation">1</attribute>
   </arc>
</model>
```

Listing 1.2. GrML for the Petri net in Fig. 4

3.3 Analysing a Model

To show the generality of CosyDraw, we have defined the FML for the Attack-Defense Tree (ADTree) formalism [24]. Figure 6 shows the model of the case study *treasure hunters* presented in [9]. We can observe on the left-hand side of Fig. 6 that CosyDraw dynamically loads the different gates (i.e. type of nodes) as well as their visual appearance, both specified in the FML.

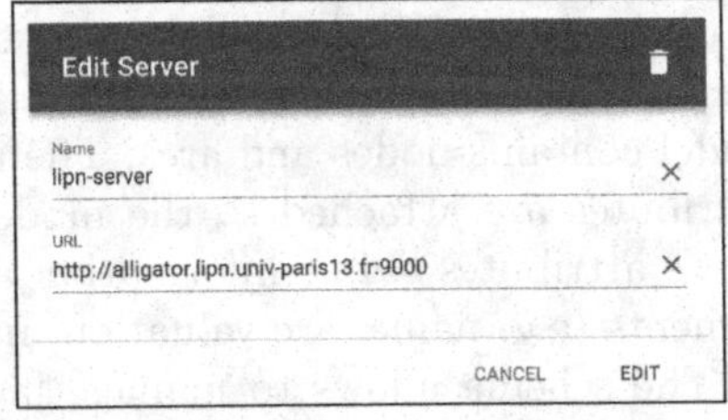

Fig. 5. Selecting an instance of Alligator

In order to analyse this kind of models, we added the tool adt2amas [10] as a service of the Alligator instance hosted by the Laboratoire d'Informatique de Paris Nord (see Fig. 5). Figure 7a shows that the queried instance of Alligator provides three services for ADTree models. For the sake of simplicity, we show in Fig. 7b only the service *EAMAS translator* that encodes an ADTree model in a multi-agent system. This window shows all the information and the inputs of the service (in this case only the model to be analysed). In section *output*, CosyDraw collects the results produced by Alligator. Running a service is a non-blocking process called *job*. This means that users can continue modifying their models and return later to this window to check the results. For instance, in Fig. 7b two files were generated and can be downloaded: an IMITATOR file and a LaTeX file.

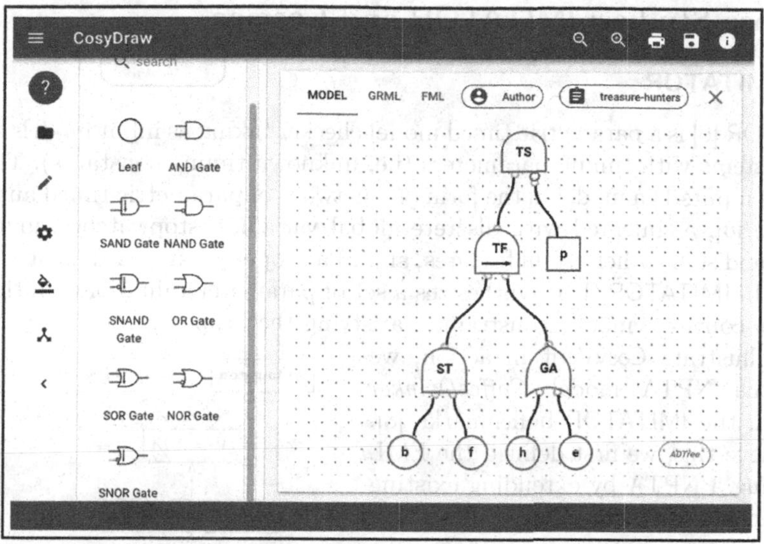

Fig. 6. Case study *treasure hunters* [9]

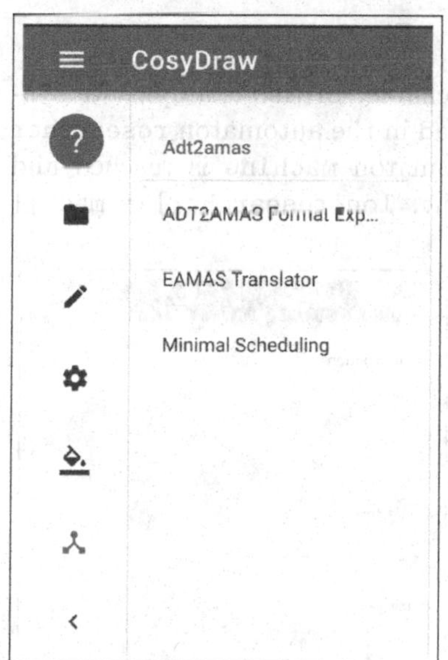

(a) Services offered by Alligator for the ADTree formalism

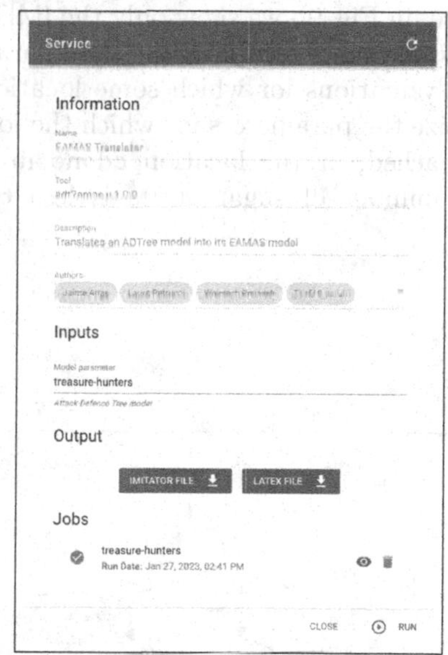

(b) Output of the *EAMAS translator* service

Fig. 7. Analysing ADTree models in CosyVerif

4 Case Studies: IMITATOR and Cosmos

4.1 IMITATOR

IMITATOR [6] is a parametric timed model checker taking as input models of real-time systems with timing parameters (i.e. unknown timing constants). The tool takes as inputs 1) a model in the form of a network of parametric timed automata (NPTA) [5], augmented with discrete global variables, stopwatches, multi-rate clocks, and some other useful features, and 2) a property expressed using a subset of TCTL. IMITATOR then synthesizes a set of parameter valuations (in the form of a non-convex symbolic constraint) satisfying the property.

To illustrate CosyVerif in action, we draw the NPTA model *CoffeeDrinker* found in the IMITATOR benchmarks [8]. To achieve this, we first defined the FML specifying a NPTA by extending existing formalisms (see Fig. 8). As we can observe at the bottom of the paper of Fig. 9a, the model is composed of two automata: `machine` and `researcher`. For lack of space, in Fig. 9b we show only the IMITATOR service to compute the set of parameter valuations for which some location is reachable. In this example, we synthesize the parameters for which the location `mad` in the automaton `researcher` is reached, or the location `cdone` in the automaton `machine` is reached and the number of sugar cubes is less than 3 (i.e. `loc[researcher] = mad ||`

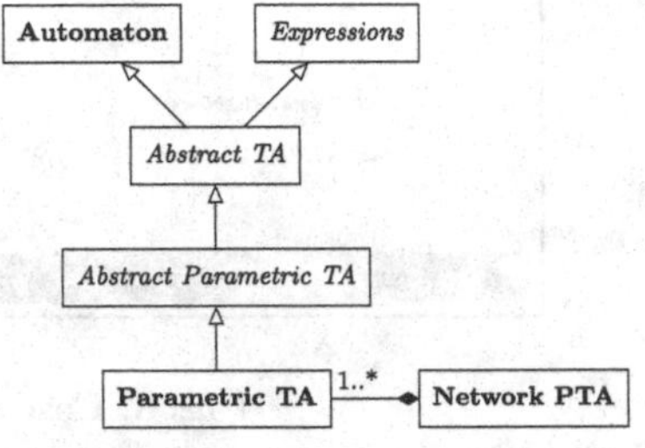

Fig. 8. FML for an NPTA

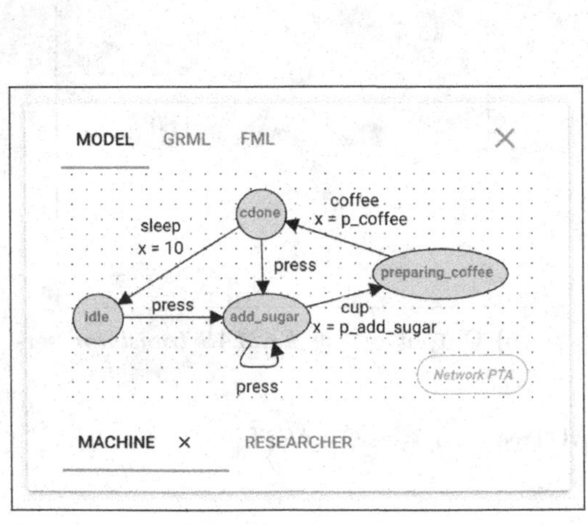

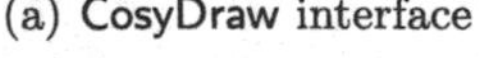
(a) CosyDraw interface

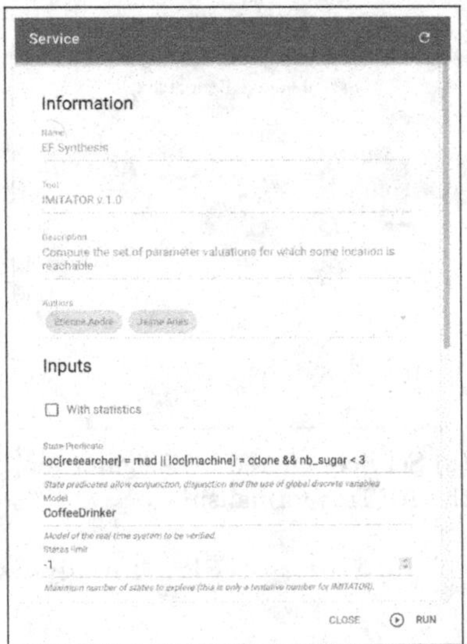

(b) Alligator service

Fig. 9. *CoffeeDrinker* NPTA model drawn on CosyVerif

`loc[machine] = cdone && nb_sugar < 3`). Checking the output file generated by the IMITATOR service, we got 8 convex sets of solutions. One of them is `p_button > 0 & p_coffee > 0 & p_add_sugar >= 15`.

4.2 Cosmos

Cosmos [12] is a statistical model checker for stochastic Petri nets against a specification given as a linear hybrid automaton. The automaton is used to specify both qualitative and quantitative measures over traces of the net. Internally, Cosmos computes a synchronised simulation of both the Petri net and the automaton. Cosmos takes as input the Petri net, the automaton, the synchronisation information between the two, and the property (i.e. a HASL logic formula [12]).

Taking advantage of CosyVerif's flexibility, we wrote the FML for Cosmos that is composed of two modules of different natures: the FML of *stochastic Petri nets* (SPN) and the FML of *linear hybrid automata* (LHA). Figure 10a shows the SPN modelling the shared memory system presented in [13]. Figure 10a shows the Petri net model, and Fig. 10b shows the automaton, both of them needed by Cosmos.

We ran the *Statistical Model Checker* provided by the Cosmos service with `width=0.01` as parameter. With these inputs, Cosmos performs simulations until the confidence interval for the HASL formula has a width smaller than 0.01.

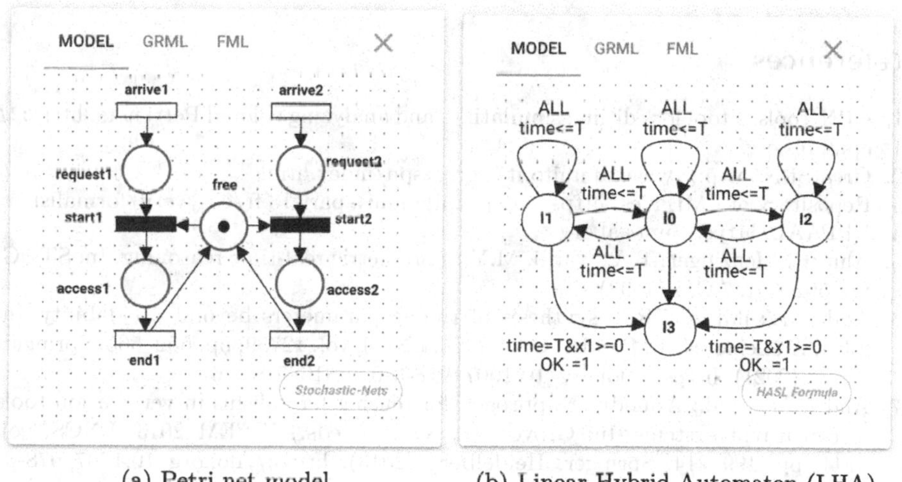

(a) Petri net model (b) Linear Hybrid Automaton (LHA)

Fig. 10. Stochastic Petri net model of a shared memory system in CosyVerif

5 Conclusions

We presented here the main features of CosyVerif that enables cohabitation between different types of formalisms. This is achieved through a modular structure embedding several models, within a same project. CosyVerif also provides extensibility at different levels: formalisms, GUI, verification tools integration. It supports graph-based formalisms, but could as well embed some languages within a single global attribute.

However, for the moment, the modularity is achieved at one level only. Its extension to a fully hierarchical modular structure is under development. Structural verification of formalisms constraints remains limited. More sophisticated semantical rules, such as consistency checks, could be added.

CosyVerif is used in summer schools and Master level courses. Thus, to improve the portability of the Petri net models designed on the platform, an export to PNML must be added from an older version of CosyVerif.

The working plans for property verification are extensive. First, CosyVerif could propose some level of compatibility with lower level models by discarding some attributes (e.g. from Symmetric Petri Nets to place/transition nets). Second, there is a lack of generic formalism for expressing properties and sharing results between tools. This is both a theoretical and practical challenge, which is tackled in [14–16] for software verification. Third, a nice addition would be—when possible—to present a schematic view of results on the initial model in the GUI by e.g. adding colours or labels to highlight solutions.

References

1. CPN Tools: a tool for editing, simulating, and analyzing colored Petri nets. https://cpntools.org
2. GreatSPN. http://www.di.unito.it/~greatspn/index.html
3. Repository of FML files. https://depot.lipn.univ-paris13.fr/cosyverif/formalisms
4. UPPAAL. https://uppaal.org
5. Alur, R., Henzinger, T.A., Vardi, M.Y.: Parametric real-time reasoning. In: STOC, pp. 592–601. ACM (1993)
6. André, É.: IMITATOR 3: synthesis of timing parameters beyond decidability. In: Silva, A., Leino, K.R.M. (eds.) CAV 2021. LNCS, vol. 12759, pp. 552–565. Springer, Cham (2021). https://doi.org/10.1007/978-3-030-81685-8_26
7. André, É., et al.: A modular approach for reusing formalisms in verification tools of concurrent systems. In: Groves, L., Sun, J. (eds.) ICFEM 2013. LNCS, vol. 8144, pp. 199–214. Springer, Heidelberg (2013). https://doi.org/10.1007/978-3-642-41202-8_14
8. André, É., Marinho, D., van de Pol, J.: A benchmarks library for extended parametric timed automata. In: Loulergue, F., Wotawa, F. (eds.) TAP 2021. LNCS, vol. 12740, pp. 39–50. Springer, Cham (2021). https://doi.org/10.1007/978-3-030-79379-1_3
9. Arias, J., Budde, C.E., Penczek, W., Petrucci, L., Sidoruk, T., Stoelinga, M.: Hackers vs. security: attack-defence trees as asynchronous multi-agent systems. In: Lin, S.-W., Hou, Z., Mahony, B. (eds.) ICFEM 2020. LNCS, vol. 12531, pp. 3–19. Springer, Cham (2020). https://doi.org/10.1007/978-3-030-63406-3_1

10. Arias, J., Penczek, W., Petrucci, L., Sidoruk, T.: ADT2AMAS: managing agents in attack-defence scenarios. In: AAMAS, pp. 1749–1751. ACM (2021)
11. Ballarini, P., Barbot, B.: Cosmos: evolution of a statistical model checking platform. SIGMETRICS Perform. Eval. Rev. **49**(4), 65–69 (2022)
12. Ballarini, P., Barbot, B., Duflot, M., Haddad, S., Pekergin, N.: HASL: a new approach for performance evaluation and model checking from concepts to experimentation. Perform. Eval. **90**, 53–77 (2015)
13. Ballarini, P., Djafri, H., Duflot, M., Haddad, S., Pekergin, N.: COSMOS: a statistical model checker for the hybrid automata stochastic logic. In: QEST, pp. 143–144. IEEE Computer Society (2011)
14. Beyer, D.: Cooperative verification: towards reliable safety-critical systems (invited talk). In: FTSCS, pp. 1–2. ACM (2022)
15. Beyer, D., Kanav, S.: CoVeriTeam: on-demand composition of cooperative verification systems. In: TACAS 2022. LNCS, vol. 13243, pp. 561–579. Springer, Cham (2022). https://doi.org/10.1007/978-3-030-99524-9_31
16. Beyer, D., Kanav, S., Richter, C.: Construction of verifier combinations based on off-the-shelf verifiers. In: FASE 2022. LNCS, vol. 13241, pp. 49–70. Springer, Cham (2022). https://doi.org/10.1007/978-3-030-99429-7_3
17. Hillah, L., Kindler, E., Kordon, F., Petrucci, L., Treves, N.T.: A primer on the Petri net markup language and ISO/IEC 15909-2. Petri Net Newsl. **76**, 9–28 (2009). (Originally Presented at the 10th International Workshop on Practical Use of Coloured Petri Nets and the CPN Tools - CPN 2009)
18. Hillah, L., Kordon, F., Petrucci, L., Trèves, N.: PN standardisation: a survey. In: Najm, E., Pradat-Peyre, J.-F., Donzeau-Gouge, V.V. (eds.) FORTE 2006. LNCS, vol. 4229, pp. 307–322. Springer, Heidelberg (2006). https://doi.org/10.1007/11888116_23
19. Hillah, L., Kordon, F., Lakos, C., Petrucci, L.: Extending PNML scope: a framework to combine Petri nets types. Trans. Petri Nets Other Model. Concurr. **6**, 46–70 (2012)
20. Hillah, L.M., Kordon, F., Petrucci, L., Trèves, N.: PNML framework: an extendable reference implementation of the petri net markup language. In: Lilius, J., Penczek, W. (eds.) PETRI NETS 2010. LNCS, vol. 6128, pp. 318–327. Springer, Heidelberg (2010). https://doi.org/10.1007/978-3-642-13675-7_20
21. Kindler, E.: ePNK applications and annotations: a simulator for YAWL nets. In: Khomenko, V., Roux, O.H. (eds.) PETRI NETS 2018. LNCS, vol. 10877, pp. 339–350. Springer, Cham (2018). https://doi.org/10.1007/978-3-319-91268-4_17
22. Kordon, F., Hillah, L., Hulin-Hubard, F., Jezequel, L., Paviot-Adet, E.: Study of the efficiency of model checking techniques using results of the MCC from 2015 to 2019. Int. J. Softw. Tools Technol. Transf. **23**(6), 931–952 (2021)
23. Kordon, F., Leuschel, M., van de Pol, J., Thierry-Mieg, Y.: Software architecture of modern model checkers. In: Steffen, B., Woeginger, G. (eds.) Computing and Software Science. LNCS, vol. 10000, pp. 393–419. Springer, Cham (2019). https://doi.org/10.1007/978-3-319-91908-9_20
24. Kordy, B., Mauw, S., Radomirovic, S., Schweitzer, P.: Attack-defense trees. J. Log. Comput. **24**(1), 55–87 (2014)
25. Lime, D., Roux, O.H., Seidner, C., Traonouez, L.-M.: Romeo: a parametric model-checker for petri nets with stopwatches. In: Kowalewski, S., Philippou, A. (eds.) TACAS 2009. LNCS, vol. 5505, pp. 54–57. Springer, Heidelberg (2009). https://doi.org/10.1007/978-3-642-00768-2_6

26. Pereira, F., Moutinho, F., Costa, A., Barros, J., Campos-Rebelo, R., Gomes, L.: IOPT-tools - from executable models to automatic code generation for embedded controllers development. In: Bernardinello, L., Petrucci, L. (eds.) PETRI NETS 2022. LNCS, vol. 13288, pp. 127–138. Springer, Cham (2022)
27. Thierry-Mieg, Y.: Symbolic model-checking using ITS-tools. In: Baier, C., Tinelli, C. (eds.) TACAS 2015. LNCS, vol. 9035, pp. 231–237. Springer, Heidelberg (2015). https://doi.org/10.1007/978-3-662-46681-0_20

Author Index

L. M. Kristensen and J. M. van der Werf (Eds.): PETRI NETS 2024, LNCS 14628, pp. 445–446, 2024.
https://doi.org/10.1007/978-3-031-61433-0

Printed in the United States
by Baker & Taylor Publisher Services

Printed in the United States
by Baker & Taylor Publisher Services